VOLUME ONE HUNDRED AND TWENTY FIVE

INTERNATIONAL REVIEW OF NEUROBIOLOGY

Endocannabinoids

INTERNATIONAL REVIEW OF NEUROBIOLOGY

VOLUME 125

SERIES EDITORS

R. ADRON HARRIS

Waggoner Center for Alcohol and Drug Addiction Research
The University of Texas at Austin
Austin, Texas, USA

PETER JENNER

Division of Pharmacology and Therapeutics
GKT School of Biomedical Sciences
King's College, London, UK

VOLUME ONE HUNDRED AND TWENTY FIVE

INTERNATIONAL REVIEW OF NEUROBIOLOGY

Endocannabinoids

Edited by

LOREN PARSONS

MATTHEW HILL

ELSEVIER

AMSTERDAM • BOSTON • HEIDELBERG • LONDON
NEW YORK • OXFORD • PARIS • SAN DIEGO
SAN FRANCISCO • SINGAPORE • SYDNEY • TOKYO
Academic Press is an imprint of Elsevier

Academic Press is an imprint of Elsevier
225 Wyman Street, Waltham, MA 02451, USA
525 B Street, Suite 1800, San Diego, CA 92101-4495, USA
The Boulevard, Langford Lane, Kidlington, Oxford OX5 1GB, UK
125 London Wall, London, EC2Y 5AS, UK

First edition 2015

ISBN: 978-0-12-801278-9
ISSN: 0074-7742

For information on all Academic Press publications
visit our website at http://store.elsevier.com/

CONTENTS

Contributors *ix*
Preface *xi*

1. The Endocannabinoid Signaling System in the CNS: A Primer **1**

Cecilia J. Hillard

1. Introduction 3
2. The Endocannabinoids 4
3. Endocannabinoid Receptors 21
4. CB1R and Retrograde Regulation of Synaptic Activity 25
5. Summary 28

Acknowledgments 28
References 28

2. Evidence for a Role of Adolescent Endocannabinoid Signaling in Regulating HPA Axis Stress Responsivity and Emotional Behavior Development **49**

Tiffany T.-Y. Lee and Boris B. Gorzalka

1. Adolescence and Pubertal Maturation 52
2. The Neurobiology of Stress 53
3. Adolescent HPA Axis Development 55
4. The Developmental Influence of Gonadal Hormones on the HPA Axis 58
5. The Endocannabinoid System 60
6. Endocannabinoid System Regulation of the HPA Axis in Adulthood 61
7. Ontogeny of the Endocannabinoid System 64
8. Adolescent HPA Axis Function: A Regulatory Role for the Endocannabinoid System? 68
9. Concluding Remarks 74

Acknowledgments 75
References 75

3. The Endocannabinoid System and Its Role in Regulating the Intrinsic Neural Circuitry of the Gastrointestinal Tract **85**

Samantha M. Trautmann and Keith A. Sharkey

1. Introduction 87
2. The Enteric Nervous System and the Brain–Gut Axis 88

3. The Endocannabinoid System 91
4. Endocannabinoid Signaling Mechanisms 98
5. Other Receptors, Agonists, and Antagonists 100
6. Endocannabinoid Tone 105
7. Cannabinoid Receptor Metaplasticity in the ENS 107
8. Cannabinoid Receptor Properties 107
9. Enteric Glia 110
10. Neurotransmission in the ENS and GI Tract 110
11. Brain–Gut Axis 114
12. The Endocannabinoid System of the ENS in GI Inflammation 115
13. Summary and Future Directions 116
Disclosures 116
References 117

4. Endocannabinoid Mechanisms Influencing Nausea 127

Martin A. Sticht, Erin M. Rock, Cheryl L. Limebeer, and Linda A. Parker

1. Endocannabinoids in Acute Nausea 132
2. Endocannabinoids in Anticipatory Nausea 145
3. Conclusions 152
References 152

5. Endocannabinoid Regulation of Neuroendocrine Systems 163

Jeffrey G. Tasker, Chun Chen, Marc O. Fisher, Xin Fu, Jennifer R. Rainville, and Grant L. Weiss

1. The Endocannabinoids and Their Receptors 164
2. Cannabinoid Regulation of the Neuroendocrine Stress Response 166
3. Endocannabinoid Regulation of Energy Homeostasis 172
4. Endocannabinoids and Fluid Homeostasis 177
5. Endocannabinoid Regulation of Reproduction 183
References 189

6. The Role of the Brain's Endocannabinoid System in Pain and Its Modulation by Stress 203

Louise Corcoran, Michelle Roche, and David P. Finn

1. Introduction 205
2. The Endocannabinoid System 207
3. The Endocannabinoid System in the Brain Regulates Pain 209
4. The Modulation of Pain by Stress: Role for the Brain's Endocannabinoid System 211

5. The Role of the Endocannabinoid System in the Rostral Ventromedial Medulla in Pain, Stress-Induced Analgesia, and Stress-Induced Hyperalgesia 213
6. The Role of the Endocannabinoid System in the Periaquaductal Gray in Pain, Stress-Induced Analgesia, and Stress-Induced Hyperalgesia 228
7. The Role of the Endocannabinoid System in the Amygdala in Pain, Stress-Induced Analgesia, and Stress-Induced Hyperalgesia 232
8. The Role of the Endocannabinoid System in the Prefrontal Cortex in Pain, Stress-Induced Analgesia, and Stress-Induced Hyperalgesia 235
9. Less Characterized Supraspinal Endocannabinoid Mechanisms in Pain Modulation 238
10. Concluding Remarks 241
Acknowledgments 242
References 242

7. Endocannabinoid Signaling in Motivation, Reward, and Addiction: Influences on Mesocorticolimbic Dopamine Function 257

Claudia Sagheddu, Anna Lisa Muntoni, Marco Pistis, and Miriam Melis

1. Introduction 258
2. The Neurobiology of Reward 260
3. The Influence of Endocannabinoid Signaling on Reward 263
4. The Role of the Endocannabinoid System in Motivation 271
5. Drug Addiction and the Endocannabinoid System: A Reciprocal Modulation of Synaptic Plasticity 277
6. Concluding Remarks 283
Acknowledgments 285
References 285

Index *303*
Contents of Recent Volumes *309*

CONTRIBUTORS

Chun Chen
Department of Cell and Molecular Biology, Tulane University, New Orleans, Louisiana, USA

Louise Corcoran
Pharmacology and Therapeutics, School of Medicine, and Galway Neuroscience Centre and Centre for Pain Research, NCBES, National University of Ireland, Galway, Ireland

David P. Finn
Pharmacology and Therapeutics, School of Medicine, and Galway Neuroscience Centre and Centre for Pain Research, NCBES, National University of Ireland, Galway, Ireland

Marc O. Fisher
Neuroscience Program, Tulane University, New Orleans, Louisiana, USA

Xin Fu
Neuroscience Program, Tulane University, New Orleans, Louisiana, USA

Boris B. Gorzalka
Department of Psychology, University of British Columbia, Vancouver, British Columbia, Canada

Cecilia J. Hillard
Neuroscience Research Center, and Department of Pharmacology, Medical College of Wisconsin, Milwaukee, Wisconsin, USA

Tiffany T.-Y. Lee
Department of Psychology, University of British Columbia, Vancouver, British Columbia, Canada

Cheryl L. Limebeer
Department of Psychology, Collaborative Neuroscience Graduate Program, University of Guelph, Guelph, Ontario, Canada

Miriam Melis
Division of Neuroscience and Clinical Pharmacology, Department of Biomedical Sciences, Cittadella Universitaria di Monserrato, University of Cagliari, Monserrato, (CA), Italy

Anna Lisa Muntoni
Neuroscience Institute, Section of Cagliari, National Research Council of Italy, Monserrato, (CA), Italy

Linda A. Parker
Department of Psychology, Collaborative Neuroscience Graduate Program, University of Guelph, Guelph, Ontario, Canada

Marco Pistis
Division of Neuroscience and Clinical Pharmacology, Department of Biomedical Sciences, Cittadella Universitaria di Monserrato, University of Cagliari, and Neuroscience Institute, Section of Cagliari, National Research Council of Italy, Monserrato, (CA), Italy

Jennifer R. Rainville
Department of Cell and Molecular Biology, Tulane University, New Orleans, Louisiana, USA

Michelle Roche
Galway Neuroscience Centre and Centre for Pain Research, NCBES, and Physiology, School of Medicine, National University of Ireland, Galway, Ireland

Erin M. Rock
Department of Psychology, Collaborative Neuroscience Graduate Program, University of Guelph, Guelph, Ontario, Canada

Claudia Sagheddu
Division of Neuroscience and Clinical Pharmacology, Department of Biomedical Sciences, Cittadella Universitaria di Monserrato, University of Cagliari, Monserrato, (CA), Italy

Keith A. Sharkey
Hotchkiss Brain Institute; Snyder Institute for Chronic Diseases, and Department of Physiology and Pharmacology, Cumming School of Medicine, University of Calgary, Calgary, Alberta, Canada

Martin A. Sticht
Department of Psychology, Collaborative Neuroscience Graduate Program, University of Guelph, Guelph, Ontario, and Department of Physiology and Pharmacology, Hotchkiss Brain Institute, University of Calgary, Calgary, Alberta, Canada

Jeffrey G. Tasker
Department of Cell and Molecular Biology, and Neuroscience Program, Tulane University, New Orleans, Louisiana, USA

Samantha M. Trautmann
Hotchkiss Brain Institute; Snyder Institute for Chronic Diseases, and Department of Physiology and Pharmacology, Cumming School of Medicine, University of Calgary, Calgary, Alberta, Canada

Grant L. Weiss
Department of Cell and Molecular Biology, Tulane University, New Orleans, Louisiana, USA

PREFACE

The endocannabinoid (eCB) system is a relative infant in the world of neuroscience, with the identification of some of its receptors, endogenous ligands, and associated metabolic pathways occurring within only the past 25 years. Similar to the endogenous opioid system, the discovery of the eCB system was predicated on the determination of the molecular target of a recreational drug. The discovery that the psychoactive constituent of cannabis, delta-9-tetrahydrocannabinol, exerts its effects through the activation of cannabinoid type 1 receptors was pivotal for the scientific exploration of this system. In further similarity to our understanding of endogenous opioid function, many of the processes found to be regulated by eCB signaling correspond quite well with the known physiological and psychological effects produced by exogenous cannabinoid drugs. For example, eCB signaling is known to be critically involved in an array of processes including memory, pain, inflammation, nausea, feeding, cognition, emotion, and reward. However, major distinctions are evident between the normal influence and function of eCB signaling and the less physiologically discriminate effects produced by broad disruption of cannabinoid signaling by exogenous cannabinoid agents. The growing knowledge of these distinctions has not only increased our understanding of human physiology but has also highlighted the eCB system as a viable therapeutic target for a variety of pathologies and disorders.

This volume of the *International Review of Neurobiology* focuses on the current understanding of eCB signaling, processing, and influence in a host of physiological processes within the brain and body. In the first chapter, Dr. Hillard presents a very clear and concise primer on the localization and function of the eCB system, along with comment on gaps in our knowledge and ongoing controversies in the field. Next, Drs. Lee and Gorzalka explain what we have learned in the last decade regarding the prominent eCB influence in sculpting the development of the adolescent brain and how disruptions in eCB signaling can induce lasting perturbations in this process. The subsequent chapters from Drs. Sharkey and Parker examine the importance of eCB signaling in the regulation of the neurogastrointestinal axis and the modulation of nausea, respectively. Dr. Tasker and colleagues then detail the intricate relationship between eCB signaling and neuroendocrine processes, including the neuroendocrine mechanisms influencing stress, energy balance, reproduction, and growth. Dr. Finn and colleagues continue this theme through an exposition on the eCB influence on pain processing and its modulation by stress and emotional state. Finally, Dr. Melis and colleagues provide discussion of how eCB signaling regulates motivation and reward through influences on the mesolimbic dopamine system and how eCB signaling modulates the effects of abused drugs on this system in the etiology of addiction. Collectively, these chapters provide both a broad overview and detailed insight into the multifaceted influence of eCB signaling in neurobiological processes governing the function of brain and body.

LOREN PARSONS
MATTHEW HILL

CHAPTER ONE

The Endocannabinoid Signaling System in the CNS: A Primer

Cecilia J. Hillard[1]
Neuroscience Research Center, and Department of Pharmacology, Medical College of Wisconsin, Milwaukee, Wisconsin, USA
[1]Corresponding author: e-mail address: chillard@mcw.edu

Contents

1. Introduction 3
2. The Endocannabinoids 4
   2.1 Definitions 4
   2.2 Mechanisms of AEA Biosynthesis 4
   2.3 Mechanisms of AEA Hydrolysis 9
   2.4 Mechanisms of 2-AG Biosynthesis 12
   2.5 Mechanisms of 2-AG Catabolism 15
   2.6 Other Inactivation Mechanisms for AEA and 2-AG 19
3. Endocannabinoid Receptors 21
   3.1 Introduction 21
   3.2 CB1 Cannabinoid Receptors 22
4. CB1R and Retrograde Regulation of Synaptic Activity 25
   4.1 The Basic Paradigms 25
5. Summary 28
Acknowledgments 28
References 28

Abstract

The purpose of this chapter is to provide an introduction to the mechanisms for the regulation of endocannabinoid signaling through CB1 cannabinoid receptors in the central nervous system. The processes involved in the synthesis and degradation of the two most well-studied endocannabinoids, 2-arachidonoylglycerol and *N*-arachidonylethanolamine are outlined along with information regarding the regulation of the proteins involved. Signaling mechanisms and pharmacology of the CB1 cannabinoid receptor are outlined, as is the paradigm of endocannabinoid/CB1 receptor regulation of neurotransmitter release. The reader is encouraged to appreciate the importance of the endocannabinoid/CB1 receptor signaling system in the regulation of synaptic activity in the brain.

International Review of Neurobiology, Volume 125
ISSN 0074-7742
http://dx.doi.org/10.1016/bs.irn.2015.10.001

ABBREVIATIONS

2-AG 2-arachidonoylglycerol
AA arachidonic acid
Abhd alpha–beta hydrolase domain protein
AEA *N*-arachidonylethanolamine
CB1R cannabinoid receptor subtype 1
CBD cannabidiol
CCK cholecystokinin
DAG diacylglycerol
DAGL diacylglycerol lipase
DSE depolarization-induced suppression of excitation
DSI depolarization-induced suppression of inhibition
EA ethanolamine
ECS endocannabinoid system
FAAH fatty acid amide hydrolase
GDE1 glycerophosphodiesterase 1
GP-AEA glycerophospho-*N*-arachidonylethanolamine
GPCR G protein-coupled receptor
GP-NAE glycerophospho-*N*-acylethanolamine
HFS high-frequency stimulation
IP_3 inositol triphosphate
LTD long-term depression
MAGL monoacylglycerol lipase
mGluR metabotropic glutamate receptor
NAAA *N*-acylethanolamine-hydrolyzing acid amidase
NAE *N*-acylethanolamine
NAPE *N*-acyl phosphatidylethanolamine
NAPE-PLD *N*-acyl phosphatidylethanolamine-specific phospholipase D
NAT *N*-acyl transferase
PA phosphatidic acid
PAP phosphatidic acid phosphatase
PC phosphatidylcholine
PE phosphatidylethanolamine
PIP_2 phosphatidylinositol 4,5-bisphosphate
PKA protein kinase A
PKC protein kinase C
PLA/AT phospholipase A/acyltransferase
PLC phospholipase C
PLD phospholipase D
PPAR peroxisome proliferation-activating receptor
PTPN22 protein tyrosine phosphatase, nonreceptor type 22
SHIP-1 Src homology 2-containing inositol phosphatase-1
THC Δ^9-tetrahydrocannabinol
THL tetrahydrolipstatin
VDCC voltage-dependent calcium channel

1. INTRODUCTION

Cannabis sativa has been used by humans for thousands of years as a medicinal agent and for its euphoric and relaxing properties. The active principal of *C. sativa* was isolated and identified as Δ^9-tetrahydrocannabinol (THC; Gaoni & Mechoulam, 1964), a landmark discovery that leads to the ultimate uncovering of the endocannabinoid system (ECS) which modulates virtually every brain region and thereby contributes to nearly every function of the CNS.

Using levo-nantradol, a structural analog of THC (Milne, Koe, & Johnson, 1979), Howlett and colleagues provided evidence that THC had biochemical effects consistent with activation of a G protein-coupled receptor (GPCR) (Howlett, 1984, 1985; Howlett & Fleming, 1984; Howlett, Qualy, & Khachatrian, 1986). Another THC analog, $[^3H]$ CP55940, was used to demonstrate that a high-affinity receptor for THC and structural analogs was present at high density throughout the brain (Devane, Dysarz, Johnson, Melvin, & Howlett, 1988). In a clever set of studies utilizing autoradiography in cerebellar neuron mutant mice, cannabinoid receptor was found to be enriched in axon terminals, which hinted that their activation would modulate neurotransmitter release (Herkenham, Groen, Lynn, De Costa, & Richfield, 1991). Molecular cloning of the cannabinoid receptor followed shortly after, officially introducing the CB1 cannabinoid receptor (*CB1R*) gene and its protein product to the scientific world (Matsuda, Lolait, Brownstein, Young, & Bonner, 1990). Subsequent studies have added considerably to our understanding of the ECS; however, none of these would have been possible without these and other seminal observations made between 1965 and 1990.

As the other chapters in this issue describe, there is strong evidence that dysregulation of the ECS contributes to many human maladies, including pain, psychiatric disorders, neurodegenerative diseases, and inflammation. Thus, therapies that alter the ECS could have usefulness as treatments for diseases and disorders that can significantly reduce quality of life. The purpose of this chapter is to introduce the ECS and review what is currently known about its regulation and role in synaptic function.

2. THE ENDOCANNABINOIDS

2.1 Definitions

The definition of endocannabinoid used in this chapter is "an endogenous molecule that activates CB1R signaling." This definition is somewhat arbitrary, since receptors in addition to the CB1R are activated by phytocannabinoids, and the endocannabinoids can bind to other receptors that are not activated by phytocannabinoids.

The first endocannabinoid identified is the very low-abundance brain lipid, *N*-arachidonylethanolamine (also called anandamide; AEA) (Devane et al., 1992). AEA is one of a family of *N*-acylethanolamines (NAEs), first identified by Udenfriend (Bachur, Masek, Melmon, & Udenfriend, 1965; Bachur & Udenfriend, 1966; Colodzin, Bachur, Weissbach, & Udenfriend, 1963) and studied in depth by Schmid and colleagues (Schmid et al., 1995; Schmid, Schmid, & Natarajan, 1990). Shortly after the identification of AEA, two laboratories simultaneously and independently reported that a high abundance 2-monoacylglycerol, 2-arachidonoylglycerol (2-AG), also bound and activated CB1R (Mechoulam et al., 1995; Sugiura et al., 1995). Other endogenous lipids that can bind CB1R include *O*-arachidonylethanolamine (virodhamine), which is a weak, partial agonist (Porter et al., 2002), and noladin ether, the ether of arachidonic acid (AA) and glycerol (Hanus et al., 2001). There is far less known about the roles of these lipids in the regulation of the ECS than is known about AEA and 2-AG.

In vitro assays demonstrated that the peptide, hemopressin, a nonapeptide derived from the α chain of hemoglobin, binds to the CB1R with high affinity and signals as an inverse agonist (Heimann et al., 2007). More recent studies suggest that hemopressin is not likely the primary signaling molecule but is a cleavage product of RVD-hemopressin (Bomar & Galande, 2013). RVD-hemopressin and several other hemopressin peptides have activity as negative allosteric modulators of the CB1R (Bomar & Galande, 2013).

The focus of this chapter will be on the lipid endocannabinoids, AEA and 2-AG, given the large amount of data supporting their role in brain function and as endocannabinoids.

2.2 Mechanisms of AEA Biosynthesis

2.2.1 Precursor Synthesis

Available evidence indicates that the primary mechanisms for AEA synthesis (and NAE synthesis in general) involve hydrolysis of a minor phospholipid class, *N*-acyl phosphatidylethanolamines (NAPEs; Fig. 1). NAPEs can be synthesized by an *N*-acyl transferase (NAT) that catalyzes transfer and formation of an amide bond between the fatty acyl moiety at the *sn*-1 position of a donor phospholipid and the ethanolamine of phosphatidylethanolamine (PE; Cadas, di Tomaso, & Piomelli, 1997; Schmid, Schmid, & Natarajan, 1996; Sugiura et al., 1996) or plasmenylethanolamine (Hansen, Moesgaard, Hansen, & Petersen, 2000). NAT activity is dependent on high micromolar concentrations of calcium and is present in membranes harvested from brain (Cadas et al., 1997; Cadas, Gaillet, Beltramo, Venance, & Piomelli, 1996; Schmid et al., 1990).

The phospholipase A/acyltransferase (PLA/AT) family of enzymes also carries out the acyl-transfer reaction needed to form the NAPEs (Golczak et al., 2012; Shinohara et al., 2011; Uyama et al., 2013). One member of this family, PLA/AT-1, is expressed in brain of human, mouse, and rat, and catalyzes the same reaction as NAT, but is calcium independent. Cellular overexpression of PLA/AT-1 results in significantly increased concentrations of NAPE, while its silencing in endogenously expressing cells leads to reduced NAPE (Uyama et al., 2013).

Since acyl groups in the *sn*-1 position are retained in the NAEs, and there is very little AA in this position, these mechanisms would be expected to produce very little *N*-arachidonylPE, the precursor of AEA. Indeed, the concentrations of *N*-arachidonylPE and AEA are very low compared to the saturated substrate and product pairs, such as *N*-palmitoylPE and palmitoylethanolamide (PEA; Ueda, Tsuboi, & Uyama, 2010).

2.2.2 NAPE Conversion to NAE: NAPE-PLD

Multiple pathways have been described for the conversion of NAPE into NAEs (Fig. 1). The first is a phosphodiesterase of the phospholipase D (PLD) family, *N*-acyl phosphatidylethanolamine-specific phospholipase D (NAPE-PLD), that converts NAPEs into phosphatidic acid (PA) and NAE (Schmid et al., 1990). NAPE-PLD does not exhibit selectivity for the *N*-acyl moiety, suggesting that it regulates formation of the entire family of NAEs (Wang et al., 2006). The majority of available evidence demonstrates a lack of calcium dependence by NAPE-PLD (Rahman et al., 2014), but little else is known about the mechanisms that regulate its activity. There are a few studies investigating NAPE-PLD mRNA expression. The transcription factor Sp1 regulates basal NAPE-PLD in macrophages, while endotoxin decreases its expression, supporting a link between NAE synthesis and inflammatory state (Zhu et al., 2011). NAPE-PLD expression in the

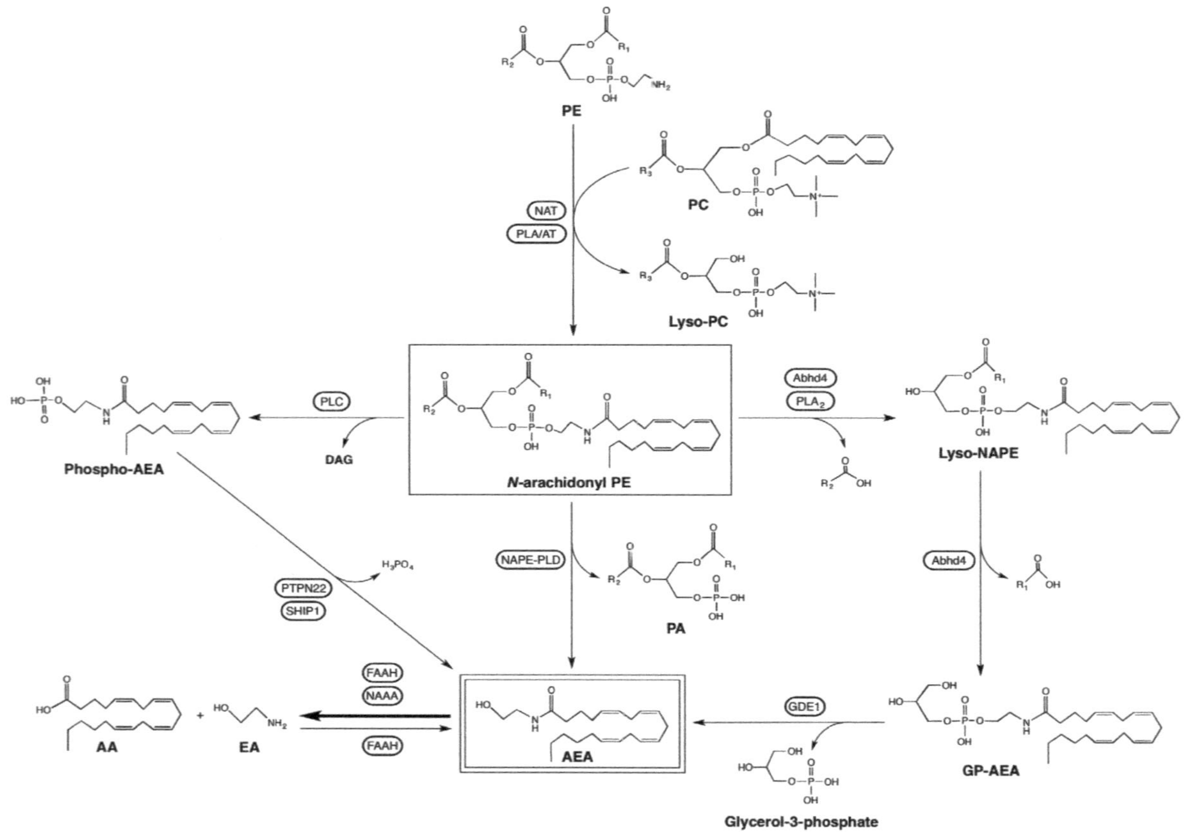

Figure 1 See legend on next page.

human endometrium changes across the menstrual cycle, suggesting hormonal regulation of expression (Scotchie, Savaris, Martin, & Young, 2015).

Genetic deletion of *NAPE-PLD* in mice (NAPE-PLD$^{-/-}$) results in a significant increase in the concentrations of the NAPEs, evidence that NAPE-PLD-mediated hydrolysis plays an important role in the regulation of NAPE concentrations (Leung, Saghatelian, Simon, & Cravatt, 2006; Tsuboi et al., 2011). Concentrations of saturated NAEs are significantly reduced, but not completely absent, in brains from NAPE-PLD$^{-/-}$ mice (Leung et al., 2006); and brain homogenates of NAPE-PLD$^{-/-}$ mice incubated with *N*-arachidonylPE exhibit only 25% of the conversion to AEA of wild-type mice (Liu et al., 2008). However, brain concentrations of AEA and other polyunsaturated NAEs in NAPE-PLD$^{-/-}$ mice are not significantly different from wild-type mice, which suggests that other pathways are important in the biosynthesis of AEA and that these pathways could be upregulated in the NAPE-PLD$^{-/-}$ mice.

NAPE-PLD is associated with intracellular membranes (Okamoto, Morishita, Tsuboi, Tonai, & Ueda, 2004). In the ventral pallidum, NAPE-PLD is in presynaptic terminals that are opposed to other axon terminals that express the CB1R, suggesting that AEA could contribute to axo-axonal regulation of CB1R signaling (Pickel, Shobin, Lane, & Mackie, 2012). Studies in the hippocampus also demonstrate presynaptic distribution of NAPE-PLD in CB1R-negative, glutamatergic terminals (Nyilas et al., 2008). NAPE-PLD is found in both axons and dendrites in the hypothalamus (Reguero et al., 2014).

2.2.3 NAPE Conversion to NAE: Multienzyme Pathways

Two alternative pathways have been discovered that can convert NAPE to NAE (Fig. 1). NAPE can be deacylated to lysoNAPE via phospholipase A_2

Figure 1—Cont'd AEA biosynthesis and degradation pathways. The pathways described in the text are shown here specifically for the endocannabinoid, AEA. Abbreviations: AA, arachidonic acid; Abhd4, alpha–beta hydrolase domain protein 4; AEA, *N*-arachidonylethanolamine; EA, ethanolamine; FAAH, fatty acid amide hydrolase; GDE1, glycerophosphodiesterase 1; GP-AEA, glycerophospho-*N*-arachidonylethanolamine; NAAA, *N*-acylethanolamine-hydrolyzing acid amidase; NAPE-PLD, *N*-acyl phosphatidylethanolamine-specific phospholipase D; NAT, *N*-acyl transferase; PC, phosphatidylcholine; PE, phosphatidylethanolamine; PLA/AT, phospholipase A/acyltransferase; PLC, phospholipase C; PTPN22, protein tyrosine phosphatase, non-receptor type 22; SHIP-1, Src homology 2-containing inositol phosphatase-1. *Scheme is a modification of Fig. 2 from Rahman, Tsuboi, Uyama, and Ueda (2014).*

family members (Sun et al., 2004) and by alpha–beta hydrolase domain-containing protein 4 (Abhd4; Simon & Cravatt, 2006). Abhd4 is expressed in mouse brain and can also deacylate lysoNAPE, resulting in the formation of glycerophospho-*N*-acylethanolamine (GP-NAE; Fig. 1).

GP-NAE is a substrate for glycerophosphodiester phosphodiesterase 1 (GDE1) which catalyzes its hydrolysis to glycerol-3-phosphate and NAE (Fig. 1; Simon & Cravatt, 2008). GDE1 is widely distributed in mammalian tissues, including brain and spinal cord. GDE1$^{-/-}$ mice have been used to examine the role of the Abhd4–GDE1 pathway in the synthesis of NAEs *in vivo* (Simon & Cravatt, 2010). Brains from GDE1$^{-/-}$ mice exhibit no detectable conversion of GP-NAE to NAE; however, brain concentrations of the NAEs, including AEA, are not different from wild-type mice indicating that GDE1 is not essential for NAE synthesis *in vivo*. GDE1$^{-/-}$/NAPE-PLD$^{-/-}$ double knockout mice demonstrate no conversion of NAPE to NAE in brain homogenates; however, cultured neurons from these mice can convert NAPE to NAE (Simon & Cravatt, 2010). These data indicate that there are mechanisms other than NAPE-PLD and Abhd4–GDE1 that can convert NAPE to NAE in an intact cell that are not operative when brain tissue is homogenized.

An additional, multistep pathway that generates NAE from NAPE which involves phospholipase C (PLC) has also been described (Fig. 1). This pathway, elucidated in macrophages, involves PLC-mediated conversion of NAPE to diacylglycerol (DAG) and phospho-NAE, which is subsequently dephosphorylated by several phosphatases, including protein tyrosine phosphatase, nonreceptor type 22 (PTPN22) and Src homology 2-containing inositol phosphatase-1 (SHIP-1; Liu et al., 2008, 2006). The PLC/phosphatase pathway was shown to be essential for endotoxin-induced synthesis of AEA in macrophages (Liu et al., 2006).

Inhibitor studies suggest both the Abhd4/GDE-1 and PLC/phosphatase pathways contribute to AEA synthesis in NAPE-PLD$^{-/-}$ mice (Liu et al., 2008). Interestingly, the kinetics of these pathways differ; the PLC/phosphatase pathway is active during the first 1–10 min after the addition of NAPE, and the Abhd4/GDE-1 pathway contributes to AEA accumulation only at later times of NAPE incubation (Liu et al., 2008). Phospho-AEA can be detected in brain and its concentration is significantly increased in the presence of vanadate, which provides further support for the possibility that the PLC/phosphatase pathway is involved in brain NAE formation (Liu et al., 2008).

2.2.4 AEA Synthesis from AA

A synthetic pathway for AEA has been identified in mammalian tissues that does not involve NAPE as a precursor (Fig. 1). In this pathway, the NAE-hydrolyzing enzyme, fatty acid amide hydrolase (FAAH), acts in "reverse" as an NAE synthase to directly couple AA and ethanolamine to form AEA (Arreaza et al., 1997; Katayama et al., 1997; Kurahashi, Ueda, Suzuki, Suzuki, & Yamamoto, 1997). This mechanism generates AEA during severe hepatic damage (Mukhopadhyay et al., 2011) and in *postmortem* brain (Patel et al., 2005), conditions that are characterized by large concentrations of ethanolamine, which is necessary to drive the synthase function of the enzyme.

2.2.5 Summary

The synthesis of AEA occurs by a number of possible routes, but it is not clear whether any of the known mechanisms contribute to stimulation of AEA production for the purpose of activating ECS. The answer to this question has not been provided by genetic deletion of possible enzymes, suggesting that the multiple mechanisms can compensate for one another. The development of effective and selective inhibitors of the various synthetic pathways is needed to delineate these processes.

2.3 Mechanisms of AEA Hydrolysis

2.3.1 Fatty Acid Amide Hydrolase

Early studies identified an enzymatic activity in liver microsomes that hydrolyzed NAEs to free fatty acid and ethanolamine (Schmid, Zuzarte-Augustin, & Schmid, 1985). The activity identified was likely due to the enzyme identified molecularly by Cravatt and colleagues and given the name FAAH (Cravatt et al., 1996). In addition to saturated and unsaturated, long- and short-chain NAEs, FAAH also hydrolyzes oleamide (a primary amine) and *N*-acyltaurines, which are very significantly increased brain tissue from FAAH$^{-/-}$ mice (Saghatelian & Cravatt, 2005). FAAH can also function as an esterase and hydrolyze 2-AG *in vitro* (Patricelli & Cravatt, 1999), although genetic deletion of FAAH in mice does not affect brain 2-AG contents (Patel et al., 2005), suggesting that FAAH does not play a prominent role in 2-AG hydrolysis.

FAAH is an integral membrane protein, present primarily on endoplasmic reticulum (Hillard, Wilkison, Edgemond, & Campbell, 1995) and mitochondria (Gulyas et al., 2004). FAAH is localized primarily in large, outflow neurons and is ubiquitously expressed throughout the brain (Tsou et al.,

1998). FAAH is active over a wide range of pH values and its activity is unaffected by either the addition or removal of divalent cations (Hillard et al., 1995; Schmid et al., 1985).

FAAH is constitutively active and several studies suggest that its activity can be regulated by posttranslational processes. For example, follicle-stimulating hormone treatment increases FAAH activity through a mechanism that requires increased protein kinase A (PKA) activity, which appears to phosphorylate an accessory protein, not FAAH itself (Grimaldi, Rossi, Catanzaro, & Maccarrone, 2009). Activation of receptors for corticotropin-releasing factor receptor 1 (CRF1) by corticotropin-releasing hormone (CRH) *in vivo* results in increased FAAH activity *ex vivo* through a mechanism consistent with a posttranslational modification (Gray et al., 2015). Given that CRF1 receptors can also couple to activation of PKA (Pollandt et al., 2006), it is possible that similar mechanisms are involved in the actions of both follicle-stimulating hormone and CRH.

The *FAAH* promoter has been analyzed; its mRNA is transcribed from multiple transcription start sites that lack a TATA box element (Puffenbarger, Kapulina, Howell, & Deutsch, 2001). The promoter region contains several estrogen receptor-binding elements (Grimaldi et al., 2012; Waleh, Cravatt, Apte-Deshpande, Terao, & Kilduff, 2002); and estrogen (Grimaldi et al., 2012) and the xenoestrogen, bisphenol A (Vermeer, Gregory, Winter, McCarson, & Berman, 2014) increase *FAAH* gene transcription. Progesterone also increases FAAH expression in T lymphocytes through the transcription factor, Ikaros (Maccarrone, Bari, Di Rienzo, Finazzi-Agro, & Rossi, 2003). Glucocorticoid receptor-binding sites are present in the *FAAH* promoter, and reporter studies demonstrate that glucocorticoid receptors regulate FAAH expression in a negative manner (Waleh et al., 2002). However, glucocorticoid administration in the drinking water results in increased FAAH activity without any effect on mRNA expression in brain (Bowles et al., 2012), suggesting that this process may not occur *in vivo*.

In vitro treatment with endotoxin reduces *FAAH* expression at a transcriptional level in human peripheral lymphocytes (Maccarrone et al., 2001) and in peripheral blood monocytes of mice (Wolfson et al., 2013). Recent data indicate that treatment of lymphocytes with IL6 results in increased FAAH expression and that a cAMP-response element in the *FAAH* promoter is likely involved in the regulatory mechanism (Gasperi et al., 2014). Leptin receptors activate the *FAAH* promoter in T lymphocytes through STAT3; perhaps also acting via a cAMP-responsive element (Maccarrone, Di Rienzo, Finazzi-Agro, & Rossi, 2003; Maccarrone, Gasperi, Fezza, Finazzi-Agro, & Rossi, 2004).

A single-nucleotide polymorphism has been identified in the coding region of the human *FAAH* gene that results in a missense mutation changing a conserved proline residue to threonine (Sipe, Chiang, Gerber, Beutler, & Cravatt, 2002). The amino acid change does not alter FAAH kinetics, but results in reduced protein amounts, likely as a result of protein instability (Chiang, Gerber, Sipe, & Cravatt, 2004). Recently, a transgenic mouse line was established in which a nucleotide substitution homologous to the rare allele in humans was introduced (Dincheva et al., 2015). The transgenic mice exhibit reduced FAAH protein amounts and higher brain AEA contents.

$FAAH^{-/-}$ mice exhibit 10-fold increases in basal concentrations of AEA and other NAEs in brain tissue (Cravatt et al., 2001), suggesting that FAAH activity is essential for the regulation of NAE concentrations in brain. $FAAH^{-/-}$ mice have been used in a large number of behavioral paradigms, and changes are generally ascribed to increased AEA tone. However, the other NAEs are increased to an even greater extent than AEA by FAAH deletion and these lipids have non-CB1R targets, most significantly, the peroxisome proliferation-activating receptors (PPARs; Panlilio, Justinova, & Goldberg, 2013). In addition, the *N*-acyl taurine family of lipids is very significantly increased in $FAAH^{-/-}$ mice (Saghatelian & Cravatt, 2005) can affect TRPV channel function (Saghatelian, McKinney, Bandell, Patapoutian, & Cravatt, 2006).

There are several widely used pharmacological inhibitors of FAAH that are very effective *in vivo*, including URB597 (Tarzia et al., 2003), PF-3845 (Ahn et al., 2009), and JNJ5003 (Hill et al., 2013). One FAAH inhibitor (PF-04457845) is currently in phase 2 clinical trials for the treatment of posttraumatic stress disorder (http://www.pfizer.com/sites/default/files/product-pipeline/February_27_2015_Pipeline_Update_Final2.pdf).

2.3.2 NAE-Hydrolyzing Acid Amidase: A Peripheral AEA Hydrolase

A second amidohydrolase, *N*-acylethanolamine-hydrolyzing acid amidase (NAAA), has been identified in peripheral tissues (Ueda, Yamanaka, & Yamamoto, 2001). NAAA is present in lysosomes, active at acidic pH, and prefers PEA as a substrate over AEA. Several inhibitors of NAAA have been developed and shown to increase endogenous concentrations of PEA (Rahman et al., 2014). There are no effects of NAAA inhibitors on AEA concentrations and no overlap or substitution of NAAA for FAAH-mediated hydrolysis of AEA in brain (Rahman et al., 2014).

2.4 Mechanisms of 2-AG Biosynthesis

2.4.1 Diacylglycerol Lipase

2-AG is present in brain tissue and microdialysates at nanomolar concentrations (Buczynski & Parsons, 2010). A well-supported mechanism for the synthesis of 2-AG involves hydrolysis of the ester bond at the *sn*-1 position of DAG by diacylglycerol lipase (DAGL; Fig. 2). Two isotypes of DAGL are expressed in mammals, DAGLα and DAGLβ, that share a transmembrane domain of four loops, coupled to a catalytic domain and regulatory loop (Bisogno et al., 2003; Reisenberg, Singh, Williams, & Doherty, 2012). DAGLα has an additional, large C-terminal tail. Both isotypes are expressed in brain; however, DAGLα is present in high density at perisynaptic regions of dendrites in many brain regions (Katona et al., 2006; Matyas et al., 2008; Uchigashima et al., 2007; Yoshida et al., 2006), which is consistent with a prominent role for this isoform in ECS-mediated regulation of synaptic plasticity (described below).

Mass spectrometry studies demonstrate that both isoforms of DAGL are phosphorylated within the catalytic and regulatory domains; and identify many phosphor residues in the C-terminal tail of DAGLα (Reisenberg et al., 2012). There is some evidence that both PKA and protein kinase C (PKC) can stimulate DAGL activity (Malcher-Lopes et al., 2006; Rosenberger, Farooqui, & Horrocks, 2007; Vellani et al., 2008). DAGLα is phosphorylated on two serines in the C-terminal region by calcium/calmodulin-dependent kinase II; this phosphorylation inhibits DAGLα activity and thereby provides a mechanism by which calcium can produce a feedback inhibition of 2-AG synthesis (Shonesy et al., 2013).

Proteomic analyses demonstrate that both DAGL isoforms can be palmitoylated (Kang et al., 2008; Martin & Cravatt, 2009; Yang, Di Vizio, Kirchner, Steen, & Freeman, 2010). Although both DAGLs are membrane intrinsic proteins, palmitoylation could contribute to membrane localization, particularly to specific membrane regions in which other palmitoylated proteins are localized. However, no studies that specifically explore the role of palmitoylation in localization or activity of DAGL have been published.

The catalytic activities of DAGLα and DAGLβ do not differ (Bisogno et al., 2003), so the additional C-terminal domain of DAGLα is apparently not important for enzymatic activity. In addition to being a target for phosphorylation, the DAGLα C-terminus contains a consensus motif for binding to Homer proteins (Jung et al., 2007), which are a family of adaptor proteins

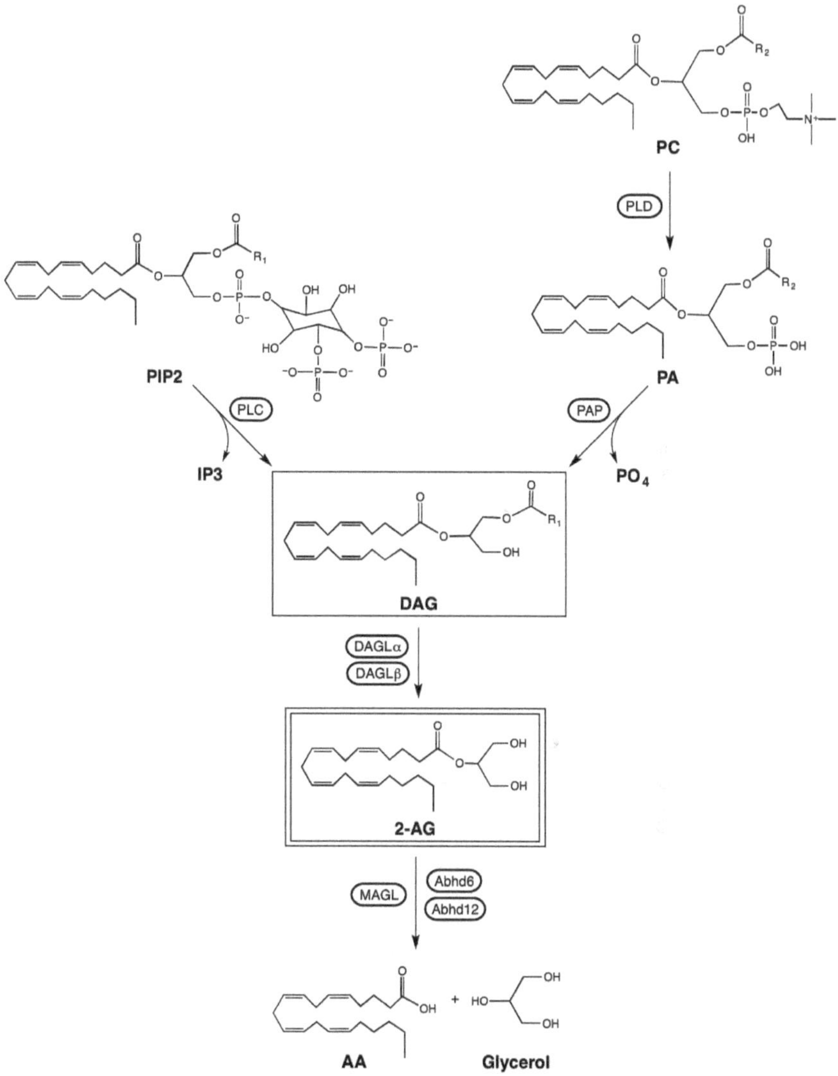

Figure 2 Biosynthesis and catabolism of 2-AG. Abbreviations: AA, arachidonic acid; abhd6, alpha–beta hydrolase domain-containing protein 6; abhd12, alpha–beta hydrolase domain-containing protein 12; 2-AG, 2-arachidonoylglycerol; DAG, diacylglycerol; DAGL, diacylglycerol lipase; IP_3, inositol triphosphate; MAGL, monoacylglycerol lipase; PA, phosphatidic acid; PAP, phosphatidic acid phosphatase; PC, phosphatidylcholine; PIP_2, phosphatidylinositol 4,5-bisphosphate; PLC, phospholipase C; PLD, phospholipase D.

that can tether synaptic proteins, such as metabotropic glutamate receptors (mGluRs) to the postsynaptic density (Gao, Tronson, & Radulovic, 2013). Although interaction between DAGLα and Homer proteins was found to be important for DAGLα association with the plasma membrane in a cell line, this interaction was not required for 2-AG synthesis (Jung et al., 2007). On the other hand, several studies have demonstrated that Homer is required for ECS-mediated changes in synaptic transmission (Fourgeaud et al., 2004; Roloff, Anderson, Martemyanov, & Thayer, 2010; Won, Puhl, & Ikeda, 2009), suggesting that DAGLα location within the cell is regulated by Homer proteins and that this process is required for efficient activation of 2-AG-mediated signaling.

Several DAGL inhibitors have been identified and used to implicate 2-AG synthesis in cellular and physiological function. Tetrahydrolipstatin (THL; orlistat) is a high-affinity inhibitor of DAGL at concentrations that are without effect on monoacylglycerol lipase (MAGL) or NAPE-PLD (Bisogno et al., 2006; Lee, Kraemer, & Severson, 1995) and is a potent inhibitor of 2-AG *in vitro* (Hashimotodani, Ohno-Shosaku, Maejima, Fukami, & Kano, 2008; Won et al., 2009). However, THL inhibits a broad range of gastric and pancreatic lipases and is irreversible (Guerciolini, 1997), which limits its use *in vivo*. RHC 80267 is also a DAGL inhibitor, but it interferes with the CB1R-mediated effects of exogenously administered 2-AG, suggesting direct effects on CB1R signaling (Hashimotodani et al., 2008). OMDM-188, a THL analog (Ortar et al., 2008), is also a selective and potent inhibitor of DAGL *in vitro* (Hashimotodani et al., 2013; Min et al., 2010). LEI105 is a reversible inhibitor of both DAGLα and DAGLβ, is without effect on the known 2-AG catabolic enzymes or FAAH, and reduces 2-AG but not AEA concentrations in Neuro2A cells (Baggelaar et al., 2015). If OMDM-188 and LEI105 have *in vivo* efficacy and selectivity, they will be important additions to the ECS pharmacological toolbox.

2.4.2 Mechanisms of DAG Synthesis

The DAG substrate for DAGL can be generated in multiple ways (Fig. 2). The PLC family of enzymes, particularly PLCβ and PLCγ isoforms, act on phosphatidylinositol 4,5-bisphosphate (PIP_2) to produce DAG and inositol triphosphate (IP_3). This pathway links the regulation of 2-AG synthesis to GPCRs, since the Gαq family of G proteins activates PLCβ. The group I mGluRs are Gαq coupled and their activation in brain slices has been shown to increase 2-AG concentrations (Jung et al., 2005) through a mechanism that requires DAGLα (Jung et al., 2007). Other Gαq-activating

GPCRs that have been shown to elevate 2-AG include muscarinic (Straiker & Mackie, 2007; Uchigashima et al., 2007), orexin 1 (Ho et al., 2011), angiotensin (Turu et al., 2007), alpha-1 adrenergic (Turu et al., 2009), vasopressin (Turu et al., 2009), bradykinin (Turu et al., 2009), and neurotensin (Kortleven, Bruneau, & Trudeau, 2012) receptors. Studies in the cerebellum demonstrate a close spatial relationship between PLCβ1, mGluRs, and DAGLα in dendritic spines (Fukaya et al., 2008), supporting the synthetic pathway outlined and specifically placing these components together in the perisynaptic region.

Like PLCβ, PLCγ isoforms also act on PIP_2 and generate DAG, but are downstream of tyrosine kinase linked, growth factor receptors, suggesting that 2-AG synthesis can also be regulated by that receptor class. In support of this notion, brain-derived neurotrophic factor induces 2-AG synthesis through a trkB receptor/PLCγ-requiring mechanism (Zhao & Levine, 2014). Thus, 2-AG synthesis can potentially be mediated by a wide variety of receptor types, including GPCRs and growth factor receptors. The proximity of DAGL determines whether 2-AG will be produced from DAG when PLC-activating receptors are engaged.

There is evidence in N18TG2 cells that 2-AG synthesis can be evoked by calcium without a requirement for PLC (Bisogno, Melck, De Petrocellis, & Di Marzo, 1999). A combination of inhibitor studies and analysis of intermediates indicates that calcium evokes production of DAG via a two-enzyme pathway in this cellular model: PLD resulting in production of PA, followed by removal of the phosphate group of PA via PA phosphatase (Fig. 2).

2.5 Mechanisms of 2-AG Catabolism

2.5.1 Monoacylglycerol Lipase

2-AG is catabolized by hydrolysis of the ester bond between the AA backbone and glycerol through the actions of several enzymes (Fig. 2). MAGL is responsible for more than 85% of 2-AG hydrolysis by brain homogenates (Blankman, Simon, & Cravatt, 2007) and is considered the dominant mechanism for the inactivation of 2-AG in its role as CB1R agonist in neurons (Murataeva, Straiker, & Mackie, 2014). MAGL is a serine hydrolase that hydrolyzes both 1(3)- and 2-monoacylglycerols with little ability to hydrolyze triacylglycerol or DAG (Tornqvist & Belfrage, 1976). MAGL is ubiquitously distributed throughout the body, including brain. Interestingly, mouse and rat brain MAGL (but not that of other tissues) appear as a doublet on Western blot (one at the predicted molecular weight of 33 and a second that migrates at approximately 35 kDa; Dinh et al., 2002; Karlsson et al., 2001). The reason for this is not known; it is possible that the proteins have the same amino acid sequence but differ in posttranslational modification. Alternatively, previous studies have shown that multiple mRNA species are present for mouse MAGL that could give rise to proteins of varying amino acid length (Karlsson et al., 2001). How this would occur in a tissue-specific manner is not clear.

In situ hybridization has been used to assess the distribution of MAGL mRNA expression in the brain, showing good agreement of brain regions expressing the CB1R and MAGL (Dinh et al., 2002). MAGL is found in presynaptic terminals (Gulyas et al., 2004; Horvath et al., 2014; Suarez et al., 2008), and enzymatic activity is enriched in synaptosomal preparations (Farooqui & Horrocks, 1997; Vyvoda & Rowe, 1973). A recent study provides evidence that MAGL is expressed in astrocytes as well as neurons; and astrocytic MAGL contributes to overall regulation of brain 2-AG content (Viader et al., 2015).

Although MAGL does not contain canonical transmembrane domains, subcellular fractionation studies find MAGL enzymatic activity in particulate (i.e., membrane) as well as cytosolic fractions (Bisogno et al., 1997; Di Marzo et al., 1999; Goparaju, Ueda, Taniguchi, & Yamamoto, 1999; Sakurada & Noma, 1981). Although there are no detectable differences in catalytic activity between cytosolic- and membrane-associated MAGL (Goparaju et al., 1999), recent studies using nanodisk models of phospholipid bilayers demonstrate that interactions of MAGL with the bilayer hold MAGL in a conformation that facilitates substrate access to the catalytic region (Nasr et al., 2013). This, together with the likelihood that 2-AG will partition to membranes rather than cytosol, provides evidence that MAGL associated with the plasma membrane is more important for termination of 2-AG action at the CB1R than cytosolic MAGL. There are several studies demonstrating compartment-selective changes in MAGL protein or activity. Cytosolic MAGL activity was found to be reduced by 50% in adipocytes from fasted rats, while particulate MAGL activity was not changed (Sakurada & Noma, 1981). Importantly for its function in regulating brain ECS, chronic stress selectively reduced MAGL protein detected in membrane but not cytosolic fractions of the basolateral amygdala (Sumislawski, Ramikie, & Patel, 2011).

Several studies demonstrate decreased *MAGL* mRNA expression in tissues that are inflamed (Engeli et al., 2014; Lappas, 2014; Mai et al., 2015). Sustained elevation of neuregulin-1 increases *MAGL* expression in hippocampal slices, likely through activation of the ErbB4 receptor (Du, Kwon, & Kim, 2013). The transcription factor and tumor suppressor Prdm5 is a repressor of *MAGL* expression and its loss acts synergistically with WNT pathway activation to increase *MAGL* expression and to increase adenoma formation (Galli et al., 2014). Genetic deletion of *PLCβ1* results in reduced *MAGL* expression, suggesting that its expression is regulated by 2-AG concentrations (Filis, Kind, & Spears, 2013). MAGL protein is degraded in developing cholinergic neurons through a process that involves nerve growth factor upregulation of the E3 ubiquitin ligase, BRCA1 (Keimpema et al., 2013).

Although there are no studies to date examining the role of posttranslational modifications in the regulation of MAGL activity, one study showing rapid changes in MAGL activity during ischemia is consistent with phosphorylation or other short-term regulatory mechanisms (Strosznajder, Singh, & Horrocks, 1984).

Several inhibitors of MAGL have been developed. URB602 is a relatively weak inhibitor of recombinant MAGL *in vitro* (King et al., 2007). URB602 increases 2-AG concentrations when injected into the periaqueductal gray (Hohmann et al., 2005) and significantly increases depolarization-induced increases in 2-AG in microdialysates of rat nucleus accumbens (Wiskerke et al., 2012). JZL184 is a more potent inhibitor of mouse MAGL (Long, Li, et al., 2009) but is not as an effective inhibitor of rat MAGL (Long, Nomura, & Cravatt, 2009; Pan et al., 2009). Microdialysis studies in nucleus accumbens of rat and mice confirm that JZL184 is effective at elevation of 2-AG in mice but not rats (Wiskerke et al., 2012). Although these findings could suggest that JZL184 is not a good indirect agonist of CB1R signaling in rat, it has been shown to increase CB1R-mediated behavioral effects in rats without affecting tissue 2-AG concentrations (e.g., Woodhams et al., 2012). Since both URB602 and JZL184 show evidence of being efficacious *in vivo* in spite of relatively poor inhibition of MAGL *in vitro*, it is likely that even modest reductions of MAGL activity can significantly enhance 2-AG actions. There are several recent reports of other MAGL inhibitors that have not been used as widely as JZL184 and URB602 (Ignatowska-Jankowska et al., 2014; Kapanda et al., 2012; Tuccinardi et al., 2014).

Although acute inhibition of MAGL activity potentiates CB1R signaling (Pan et al., 2009), genetic deletion (Chanda et al., 2010), and chronic pharmacological inhibition (Schlosburg et al., 2010) of MAGL both result in functional reductions in CB1R signaling rather than activation.

Interestingly, the MAGL$^{-/-}$ mice have increased tonic CB1R activity (Pan et al., 2011) but exhibit region-specific desensitization of CB1R agonist-induced activation of G proteins (Navia-Paldanius et al., 2015). Given that both genetic deletion and chronic inhibition of MAGL produce very large increases in brain 2-AG contents, these data are consistent with agonist (i.e., 2-AG)-induced desensitization of CB1R signaling. The profound effects of reduced MAGL activity on 2-AG suggest that 2-AG homeostasis is regulated more by its degradation than synthesis. In support of this hypothesis, recent data demonstrate that 2-AG is generated continuously at hippocampal GABA synapses and that MAGL activity is critical for opposing this steady stream of 2-AG to maintain the CB1R in a low-ligand state (Lee et al., 2015).

2.5.2 Other Enzymes that Hydrolyze 2-AG in the Brain

Abhd6 and Abhd12 were identified as potential 2-AG hydrolases using a functional proteomic approach (Blankman et al., 2007). Subsequent studies have demonstrated that Abhd6 is expressed by neurons in postsynaptic compartments and its activity can regulate 2-AG-mediated activation of CB1R (Marrs et al., 2010). Given the postsynaptic distribution, Abhd6 has been proposed to regulate 2-AG concentrations at the site of synthesis (Savinainen, Saario, & Laitinen, 2012), which complements the role of MAGL to regulate 2-AG concentrations in the axon terminal. This function of Abhd6 could play a very important role in basal CB1R activity by controlling the amount of 2-AG that survives to exit the postsynaptic terminal.

Studies in brain homogenates suggest that Abhd12 accounts for approximately 9% of brain 2-AG hydrolysis (Blankman et al., 2007). *Abhd12* is mutated in the human neurodegenerative disorder, PHARC (Blankman, Long, Trauger, Siuzdak, & Cravatt, 2013). There is little information regarding the role of Abhd12 in brain 2-AG homeostasis; however, its mRNA is enriched in microglia, suggesting it may be important in the termination of 2-AG-mediated CB2 receptor activation (Fiskerstrand et al., 2010).

2.5.3 Contribution of 2-AG to AA Concentrations

Early studies in platelets demonstrated that free AA was increased by the metabolism of DAG and 2-AG (Bell, Kennerly, Stanford, & Majerus, 1979; Prescott & Majerus, 1983). MAGL$^{-/-}$ mice exhibit significantly lower brain tissue concentrations of AA in addition to elevated 2-AG (Schlosburg et al., 2010). This finding, together with data that acute inhibition of MAGL abolishes endotoxin-induced increases in brain AA and prostaglandin E_2 (Nomura et al., 2011), strongly indicates that 2-AG is a biologically significant precursor for AA and that MAGL is a critical enzyme in the provision of free AA for further metabolism. A recent study using cell-specific MAGL deletion suggests that astrocyte MAGL is mainly responsible for converting 2-AG to inflammatory prostaglandins in brain (Viader et al., 2015). Similar roles for 2-AG and MAGL have been demonstrated in hepatic injury (Cao et al., 2013) and for the synthesis of vasodilatory arachidonates in coronary arteries (Gauthier et al., 2005). Studies carried out using vascular tissue suggest that the AEA/FAAH pair can serve a parallel function in the synthesis of vasoactive AA metabolites (Pratt, Hillard, Edgemond, & Campbell, 1998).

These and other studies demonstrate integration and synergism between the endocannabinoid and AA signaling systems. This conclusion is further supported by the findings discussed immediately below that AEA and 2-AG can also serve as substrates for enzymes that metabolize AA.

2.6 Other Inactivation Mechanisms for AEA and 2-AG

2.6.1 Uptake, Accumulation, and Sequestration

The addition of labeled AEA to the outside of cells results in its cellular association in a manner that is consistent with accumulation and in some cases, intracellular sequestration (Hillard & Jarrahian, 2003). However, details of the mechanisms involved in this "uptake" process remain unclear. FAAH-mediated catabolism of AEA can maintain the concentration gradient and thereby enhance AEA uptake into the cells by either passive or facilitated diffusion, and many inhibitors of AEA uptake also inhibit FAAH-mediated catabolism of AEA (Deutsch et al., 2001; Glaser et al., 2003; Kaczocha, Hermann, Glaser, Bojesen, & Deutsch, 2006). However, AEA is also accumulated by cells that express very low or no FAAH (Hillard, Edgemond, Jarrahian, & Campbell, 1997; Hillard & Jarrahian, 2005; Nicolussi, Chicca, et al., 2014), suggesting that other processes can contribute to accumulation. Indeed, other intracellular proteins have been identified that bind AEA and could thereby serve as sequestration sites. These include fatty acid-binding proteins (Kaczocha, Glaser, & Deutsch, 2009), heat-shock proteins (Oddi et al., 2009), and sterol carrier protein 2 (Liedhegner, Vogt, Sem, Cunningham, & Hillard, 2014). In addition, AEA associates with membrane lipid rafts (McFarland et al., 2004; McFarland, Terebova, & Barker, 2006), which could also serve as a sequestration site following uptake. Importantly, it is likely that AEA uptake occurs via different processes in different cells (Hillard & Jarrahian, 2005).

There are several experiments that indirectly support a protein transporter that can translocate AEA across the plasma membrane in a bidirectional manner (Chicca, Marazzi, Nicolussi, & Gertsch, 2012; Hillard et al., 1997; Ligresti et al., 2004; Ronesi, Gerdeman, & Lovinger, 2004). These data suggest that an endocannabinoid transporter could participate in both inactivation (as a first step in a cellular sequestration process) and in the release of AEA. FLAT, a variant of FAAH that is without catalytic activity, has been suggested as a putative AEA transport protein (Fu et al., 2012). However, other studies dispute these findings (Leung, Elmes, Glaser, Deutsch, & Kaczocha, 2013). Although fewer experiments have been done, inhibitor studies suggest that 2-AG could also be subjected to similar regulation by uptake and/or sequestration (Bisogno et al., 2001; Nicolussi & Gertsch, 2015).

There are several inhibitors of the cellular accumulation of the endocannabinoids, although the lack of complete understanding of the processes involved has made it difficult to ascribe mechanisms to the inhibitors. Some inhibitors of AEA uptake, including AM404, VDM11, AM1172, and LY2183240, also inhibit or are substrates for FAAH (Alexander & Cravatt, 2006; Fowler, Tiger, Ligresti, Lopez-Rodriguez, & Di Marzo, 2004; Glaser et al., 2003; Vandevoorde & Fowler, 2005) and can compete with endocannabinoids for binding to intracellular proteins (Kaczocha, Vivieca, Sun, Glaser, & Deutsch, 2012; Liedhegner et al., 2014). Recent reports have identified the natural product, guineensine, as a potent inhibitor of AEA uptake in several cell types (Nicolussi, Viveros-Paredes, et al., 2014) and a series of *N*-alkylcarbamates as extremely potent inhibitors of both AEA uptake and FAAH activity (Nicolussi, Chicca, et al., 2014). Characteristics of the inhibitory effects of guineensine and the *N*-alkylcarbamates support the notion that FAAH activity contributes to AEA uptake, but it is not sufficient to explain all forms of uptake in all cell types. Development and study of inhibitors with novel structures will undoubtedly improve our understanding of the mechanisms involved in AEA and 2-AG uptake.

In vivo treatment with uptake inhibitors AM404 and UCM707 increases brain tissue concentrations of AEA and, to a lesser extent, 2-AG (de Lago et al., 2005; Di et al., 2005); however, AM404 only affects 2-AG, and UCM707 is ineffective at elevating either AEA or 2-AG in brain extracellular space, measured using microdialysis (Wiskerke et al., 2012). Since effective inhibitors of FAAH and MAGL have very significant effects on microdialysate AEA and 2-AG, respectively, these findings suggest that the processes of endocannabinoid uptake and accumulation may contribute only a small amount to their overall clearance. On the other hand, small changes in clearance, particularly of 2-AG, could result in a significant increase in CB1R activation and might even be a very useful approach because it should avoid inducing receptor desensitization.

2.6.2 Oxygenation of the Arachidonate Backbone

The acyl chain of AA can be modified by cyclooxygenases, lipoxygenases, and cytochrome P450s, resulting in the production of prostaglandins, leukotrienes, and epoxyeicosatrienoic acids, respectively. Subtypes of each of these enzyme classes can also utilize AEA and 2-AG as substrates, resulting in the formation of ethanolamide and glycerol ester analogs of the arachidonates (see Urquhart, Nicolaou, & Woodward, 2015 for an excellent recent review). The result of these processes is a large number of lipid mediators, some of which have been shown to have their own targets. On the other hand, metabolism of AEA and 2-AG along any of these pathways is likely to reduce affinity for CB1R, thus they are inactivation mechanisms for the CB1R signaling roles of these lipids.

In particular, 2-AG is an excellent substrate for COX-2, having K_m and k_{cat} values that are very similar to those of AA (Rouzer & Marnett, 2011). Based upon structure function studies showing that 2-AG and AA bind differentially to COX-2, substrate-specific inhibitors of COX-2 that selectively reduce 2-AG metabolism while preserving prostaglandin formation have been designed (Hermanson, Gamble-George, Marnett, & Patel, 2014). The *in vivo* effects of these inhibitors support the hypothesis that COX-2-mediated metabolism of 2-AG contributes in a significant manner to the regulation of CB1R signaling (Hermanson et al., 2013).

3. ENDOCANNABINOID RECEPTORS

3.1 Introduction

The focus of this review is on the CB1R. However, other receptors have been identified that can bind the endocannabinoids AEA and 2-AG, and it is highly likely that these receptors contribute to the biological effects of the lipids. Both AEA and 2-AG bind to CB2 cannabinoid receptors (CB2R; Gonsiorek et al., 2000), which are GPCRs (Munro, Thomas, & Abu-Shaar, 1993). Although 2-AG has the characteristics of a full CB2R agonist, AEA does not induce CB2R-mediated GDP/GTP exchange

(Gonsiorek et al., 2000; Hillard et al., 1999) and is likely a weak partial agonist of the CB2R. CB2R are expressed in circulating immune cells (Bouaboula et al., 1993), spleen (Galiegue et al., 1995), and tissue-resident macrophage populations, including microglial cells (Carlisle, Marciano-Cabral, Staab, Ludwick, & Cabral, 2002). CB2R are also expressed by some neuronal populations, although the expression levels are far lower than CB1R (Van Sickle et al., 2005; Zhang et al., 2014).

AEA is an agonist of the vanilloid type 1 receptor, also called TRPV1 (Kim et al., 2007; Ross et al., 2001; Saghatelian et al., 2006). TRPV1 is a nonselective cation channel expressed widely in the CNS. Intracellular AEA induces opening of the channel, and this function contributes to many of the non-CB1-mediated effects of AEA. AEA and other NAEs are agonists of PPARs, particularly PPARα (Bouaboula et al., 2005; Fu et al., 2003; Lo Verme et al., 2005). Recent data suggest that PPARα-mediated changes could contribute to the effects of FAAH inhibition (Panlilio et al., 2013).

3.2 CB1 Cannabinoid Receptors

CB1R are heterogeneously expressed throughout the CNS (Hu & Mackie, 2015). CB1R are present at extremely high density in the cingulate gyrus, frontal cortex, hippocampus, cerebellum, and the basal ganglia. Moderate receptor densities are found in the basal forebrain, amygdala, nucleus accumbens, periaqueductal gray, and hypothalamus; and low density is seen in the midbrain, pons, and medulla. Relatively, little receptor is found in primary motor cortex or thalamus. In the forebrain, CB1R mRNA is expressed at very high density in a restricted number of neurons (Marsicano & Lutz, 1999). These CB1R-expressing neurons project widely, resulting in a dense network of CB1R-positive processes. Double-labeling studies demonstrate that these highly expressing cells are GABAergic interneurons that also express the neuropeptide cholecystokinin (CCK; Katona et al., 1999). Other neurons express the CB1R at lower densities; these neurons are more heterogenous and consist of both non-CCK, GABAergic interneurons and glutamatergic neurons (Hu & Mackie, 2015).

The CB1R is also expressed by nonneuronal cells in the CNS, including astrocytes (Navarrete & Araque, 2008; Salio, Doly, Fischer, Franzoni, & Conrath, 2002), oligodendrocytes (Molina-Holgado et al., 2002); and by endothelial (Golech et al., 2004) and smooth muscle cells (Gebremedhin, Lange, Campbell, Hillard, & Harder, 1999) of the cerebral circulation.

3.2.1 CB1R Signaling

Activation of the CB1R results in inhibition of adenylyl cyclase activity in most tissues and cells via activation of Gαi-mediated signaling (Howlett, 1985; Howlett & Fleming, 1984). As is the case for most GPCRs that couple to Gαi, CB1R also engage Gαo-mediated signaling (Glass & Northup, 1999) which results in inhibition of the opening of voltage-dependent calcium channels (VDCCs) through the release of associated βγ subunits (Caulfield & Brown, 1992; Mackie & Hille, 1992). This mechanism is likely the primary process by which CB1R regulates short-term changes in neurotransmitter release (described below). Signaling through Gαi/o also results in activation of inward-rectifying potassium channels (Henry & Chavkin, 1995). There is some evidence that CB1R agonists can exhibit bias toward activation of specific G protein alpha subtypes (Turu & Hunyady, 2010), suggesting that they can selectively engage inhibition of calcium influx versus adenylyl cyclase, for example. If Gαi/o proteins are unavailable, CB1R will couple to Gαs and thereby enhance adenylyl cyclase activity (Glass & Felder, 1997).

CB1R activation is coupled to activation of p42/p44 and p38 mitogen-activated kinases and Jun N-terminal kinase through a variety of signaling mechanisms, including G proteins (Turu & Hunyady, 2010) and β-arrestin (Ahn, Mahmoud, & Kendall, 2012). CB1R activation has also been linked to the activation of PLC (Lograno & Romano, 2004) and Akt signaling in some cells (Gomez et al., 2011); through these pathways, CB1R activation can influence intracellular calcium concentrations, protein kinase activities, and other signaling cascades that regulate cell growth and differentiation.

There is evidence that the CB1R exhibits considerable constitutive activity, meaning that it can activate signaling cascades in the absence of agonist binding (Lee et al., 2015; Nie & Lewis, 2001; Savinainen, Saario, Niemi, Jarvinen, & Laitinen, 2003).

3.2.2 CB1R Pharmacology

The endocannabinoids 2-AG and AEA are arachidonates that bind to the CB1R with affinities in the mid- to high-nanomolar range. 2-AG acts as a full agonist at the CB1R, at least with respect to G protein activation, while AEA is a partial agonist (Hillard, 2000). Ethanolamides of several other long chain unsaturated fatty acids are also agonists of the CB1R, including the ethanolamide of docosahexaenoic acid (Hillard & Campbell, 1997). Two synthetic arachidonates, *N*-arachidonyl-2-chloroethylamide and arachidonylcyclopropylamide that have 500–1000 times greater affinity for the CB1R than CB2R, have been developed (Hillard et al., 1999) and are the only CB1R-selective ligands in wide use. Many other synthetic agonists for the CB1R have been synthesized and characterized. These include derivatives of THC, such as levo-nantradol, CP55940, and HU210, and the aminoalkylindole, WIN55212-2 (Pertwee, 2008).

The identification of a CB1R antagonist occurred much later than the synthesis of the agonists. The first antagonist discovered was SR141716A, later named rimonabant (Rinaldi-Carmona et al., 1994). Rimonabant and a structural analog, AM251, are both inverse agonists of the CB1R (Bouaboula et al., 1997; Savinainen et al., 2003) so will reduce both constitutive and endocannabinoid-activated CB1R signaling. Several neutral antagonists have been designed that can be used to differentiate these possible mechanisms (Kirilly, Gonda, & Bagdy, 2012).

Recent studies have identified multiple allosteric ligands that interact with the CB1R. ORG27569 was originally reported to function as a negative allosteric modulator of the CB1R (Price et al., 2005). Subsequent studies found that this compound is a biased allosteric modulator, inhibiting CB1R agonist activation of G protein signaling (Baillie et al., 2013) while increasing high-affinity orthosteric agonist-binding sites, β-arrestin 1 recruitment, and ERK activation (Ahn et al., 2012). ORG27569 attenuates CB1R-mediated signaling in hippocampal neurons in culture (Straiker, Mitjavila, Yin, Gibson, & Mackie, 2015), however, is not effective *in vivo*, perhaps because of a poor pharmacokinetic profile (Gamage et al., 2014).

PSNCBAM-1 is a negative allosteric modulator of CB1R-mediated activation of G protein signaling (Horswill et al., 2007). This compound exhibits *in vivo* efficacy to reduce body weight and feeding in a rat model, findings that are consistent with reduced CB1R activation. Further *in vitro* studies have demonstrated that PSNCBAM-1 reduces the effects of CP55940 to modulate synaptic CB1R signaling, but not the effects of WIN55212-2 (Wang, Horswill, Whalley, & Stephens, 2011). PSNCBAM-1 inhibits endocannabinoid-mediated activation of CB1R in cultured hippocampal neurons, suggesting that it negatively modulates 2-AG activation of CB1R signaling (Straiker et al., 2015).

Two endogenous compounds have been reported to function as CB1R negative allosteric modulators: a family of hemopressin-related peptides, particularly pepcan 12 (Bauer et al., 2012), which are widely expressed in brain (Hofer et al., 2015), and the steroid, pregnenolone (Vallee et al., 2014). While pepcan 12 is effective at reducing endogenous CB1R signaling in cultured hippocampal neurons, pregnenolone is not (Straiker et al., 2015). However, pregnenolone might be more effective at reducing the signaling of THC and therefore could reduce its adverse effects (Vallee et al., 2014).

The phytocannabinoid, cannabidiol (CBD), was reported to act as a negative allosteric modulator for CB1R-mediated β-arrestin 2 recruitment and ERK signaling (Laprairie, Bagher, Kelly, & Denovan-Wright, 2015). This is an interesting observation, given that CBD can dampen many of the effects of THC, including its anxiogenic effects.

Two positive allosteric modulators of the CB1R have been described. The first is the naturally occurring arachidonate, lipoxin A4 (Pamplona et al., 2012) and the second is a synthetic, ZCZ011 (Ignatowska-Jankowska et al., 2015). ZCZ011 potentiates binding of [^{3}H]CP55,940 to the CB1R and increases AEA-stimulated [^{35}S]GTPγS binding in mouse brain membranes and β-arrestin recruitment and ERK phosphorylation in CB1R-expressing cells. Lipoxin A4 also increases AEA signaling as well as its affinity for the CB1R.

4. CB1R AND RETROGRADE REGULATION OF SYNAPTIC ACTIVITY

4.1 The Basic Paradigms

4.1.1 Short-Term Depression of Synaptic Transmission

Three simultaneous reports in 2001 described endocannabinoid/CB1R signaling as a mediator of retrograde inhibition of neurotransmitter release (Kreitzer & Regehr, 2001; Ohno-Shosaku, Maejima, & Kano, 2001; Wilson & Nicoll, 2001). Many studies in multiple brain regions carried out since demonstrate that modulation of synaptic plasticity is a major mechanism by which CB1R signaling affects the brain (Freund, Katona, & Piomelli, 2003; Kano, 2014; Patel & Hillard, 2009). In this mechanism, endocannabinoids are mobilized from the postsynaptic neuron in response to triggers, such as depolarization and activation of NMDA receptors which increase calcium, and/or receptors that couple to an increase in PLC. These triggers result in the synthesis of 2-AG which diffuses from the postsynaptic neuron and binds to CB1R on the presynaptic terminal. The result of presynaptic CB1R activation is inhibition of neurotransmitter release; in the case of short-term modulation, the CB1R signaling mechanisms include inhibition of the opening of VDCCs and perhaps, also increased activation of potassium channels, resulting in hyperpolarization. In the case where depolarization of the postsynaptic neuron is the trigger and CB1R activation inhibits GABA release, the plasticity is called depolarization-induced suppression of inhibition (DSI). Evidence supporting the role of the CB1R in DSI include data that DSI is both mimicked and occluded by a CB1R agonist and that CB1R agonist and DSI increase the paired-pulse ratio and decrease spontaneous, calcium dependent, miniature IPSPs, both indicative of a reduction in presynaptic vesicular release probability.

Endocannabinoid/CB1R signaling also produces suppression of glutamate release by a parallel mechanism. Specifically, depolarization of CA1 pyramidal neurons results in transient suppression of excitatory postsynaptic potential amplitude that is dependent upon CB1R activation (Ohno-Shosaku et al., 2002). Depolarization-induced suppression of excitation (DSE) is also dependent upon presynaptic CB1R availability and is the result of activation of CB1R on glutamatergic terminals (Ohno-Shosaku et al., 2002; Ruehle et al., 2013).

DSI is not affected by botulinum toxin applied to the postsynaptic neuron, indicating that the retrograde messenger does not utilize a synaptic release mechanism. This finding is consistent with an "on demand" synthesis of the endocannabinoids (Freund et al., 2003). Further studies have demonstrated that acute inhibition of DAGL suppresses (Hashimotodani et al., 2013), while acute inhibition of MAGL prolongs (Pan et al., 2009) hippocampal DSI. DSI/DSE are completely absent in slices from hippocampus, cerebellum, striatum, and prefrontal cortex mice in which DAGLα has been genetically deleted (Gao et al., 2010; Tanimura et al., 2010; Yoshino et al., 2011). These findings, together with a lack of effect of FAAH inhibition on DSI (Pan et al., 2009), indicate that 2-AG is the endocannabinoid subserving this type of synaptic plasticity.

The primary triggers for the synthesis of endocannabinoid to produce short-term plasticity are increased postsynaptic calcium and activation of receptors that increase PLC activity (Kano, 2014). Receptor-driven, endocannabinoid-mediated plasticity was first demonstrated for the group I mGluRs (Maejima, Hashimoto, Yoshida, Aiba, & Kano, 2001), which couple to Gαq proteins and therefore increase PLCβ activity and generate DAG as described above. Subsequent studies demonstrated that many receptors that activate PLC can also evoke endocannabinoid-dependent short-term suppression of neurotransmitter release. Endocannabinoid signaling can be induced by strong Gαq-linked receptor activation alone, or by a combination of subthreshold Gαq activation with a simultaneous increase in postsynaptic calcium (Kim, Isokawa, Ledent, & Alger, 2002). It is not completely clear how increased postsynaptic calcium concentrations results in an increase in 2-AG; a two-enzyme pathway for DAG synthesis involving PLD and PAP is calcium sensitive and could serve this function (Bisogno et al., 1999). There is evidence that different DAGLα pools could be responsible for 2-AG generation form DAG synthesized in response to calcium versus PLC activation (Zhang, Wang, Bisogno, Di Marzo, & Alger, 2011).

4.1.2 Long-Term Depression of Transmission

The ECS has also been implicated in more persistent forms of synaptic plasticity. Stimulation protocols which induce LTP of excitatory synapses onto CA1 pyramidal neurons also produce long-term depression (LTD) of GABAergic inputs to the same neurons; a phenomenon called iLTD (Chevaleyre & Castillo, 2003). High-frequency stimulation (HFS) of inputs to CA1 pyramidal neurons depresses GABAergic transmission onto these neurons for almost 1 h, an effect that is presynaptic in nature, and requires activation of mGluR and CB1R (Chevaleyre & Castillo, 2003). Induction of iLTD is blocked by PLC and DAGL inhibition, supporting a role for 2-AG in the initiation of the plasticity (Edwards, Kim, & Alger, 2006). HFS of cortical afferents to the striatum induces LTD of glutamatergic transmission, an effect absent in $CB1R^{-/-}$ mice and mice treated with CB1R antagonist (Gerdeman, Ronesi, & Lovinger, 2002). Interestingly, CB1R antagonist application 10 min after HFS does not block LTD, suggesting that the ECS is required for the induction of LTD, but not its long-term maintenance (Ronesi et al., 2004). LTD at this synapse is dependent upon increases in postsynaptic intracellular calcium (Gerdeman et al., 2002).

The triggers for endocannabinoid-mediated LTD are the same as those for short-term synaptic suppression; i.e., increased postsynaptic calcium and increased PLC activity. What differentiates short- and long-term changes in neurotransmitter release are the presynaptic signaling mechanisms evoked by CB1R activation. For ECS-mediated LTD, inhibition of PKA is critical (Castillo, Younts, Chavez, & Hashimotodani, 2012), and synergistic changes in the presynaptic terminal are likely required, such as increased calcium concentrations or coactivation of other GPCRs (Kano, 2014).

There is evidence that AEA can also produce changes in synaptic activity at a more limited number of synapses and through TRPV1 signaling. For example, AEA-mediated postsynaptic TRPV1 activation produces LTD in the hippocampus (Chavez, Chiu, & Castillo, 2010), nucleus accumbens (Grueter, Brasnjo, & Malenka, 2010), and extended amygdala (Puente et al., 2011). These effects do not involve CB1R signaling.

5. SUMMARY

The ECS is a beautiful and fascinating neuromodulatory system that allows for moment-to-moment synapse-specific regulation of neurotransmission. The basic building blocks of the proteins involved in endocannabinoid synthesis and degradation together with the receptor targets are utilized throughout the brain, but in very diverse ways depending upon their relative locations and in relationship to other ongoing signaling events. The ECS modulates the functions of the major excitatory and inhibitory neurotransmitters, serving as a break on their release (described above). CB1R can also inhibit the release of biogenic amines (Haring, Guggenhuber, & Lutz, 2012) and neuropeptides (Hirasawa et al., 2004), thereby modulating the modulators! The ECS is likely the mechanism by which steroid hormones such as glucocorticoids (Di, Malcher-Lopes, Halmos, & Tasker, 2003) and estrogen (Huang & Woolley, 2012) alter synaptic plasticity. Recent evidence that a cycle of 2-AG → AA is tightly regulated by MAGL activity (discussed above) adds another layer of complexity to the role of 2-AG homeostasis to brain function. It is not surprising that increasing evidence indicates that dysregulation of ECS contributes to many forms of brain dysfunction, including psychopathology, developmental problems and plays a role in neurodegenerative diseases as well.

ACKNOWLEDGMENTS

I was supported during the writing of this review by NIH Grants DA038663, DA026996, and MH102838 and by the Research and Education Component of the Advancing a Healthier Wisconsin Endowment at the Medical College of Wisconsin. I am extremely grateful to Margaret Beatka for making the figures.

REFERENCES

Ahn, K., Johnson, D. S., Mileni, M., Beidler, D., Long, J. Z., McKinney, M. K., et al. (2009). Discovery and characterization of a highly selective FAAH inhibitor that reduces inflammatory pain. *Chemistry & Biology*, *16*(4), 411–420. http://dx.doi.org/10.1016/j.chembiol.2009.02.013.

Ahn, K. H., Mahmoud, M. M., & Kendall, D. A. (2012). Allosteric modulator ORG27569 induces CB1 cannabinoid receptor high affinity agonist binding state, receptor internalization, and Gi protein-independent ERK1/2 kinase activation. *The Journal of Biological Chemistry*, *287*(15), 12070–12082. http://dx.doi.org/10.1074/jbc.M111.316463.

Alexander, J. P., & Cravatt, B. F. (2006). The putative endocannabinoid transport blocker LY2183240 is a potent inhibitor of FAAH and several other brain serine hydrolases. *Journal of the American Chemical Society*, *128*(30), 9699–9704. http://dx.doi.org/10.1021/ja062999h.

Arreaza, G., Devane, W. A., Omeir, D. L., Sajnani, G., Kunz, J., Cravatt, B. F., et al. (1997). The cloned rat hydrolytic enzyme responsible for the breakdown of anandamide also catalyzes its formation via the condensation of arachidonic acid and ethanolamine. *Neuroscience Letters*, *234*, 59–62.

Bachur, N. R., Masek, K., Melmon, K. L., & Udenfriend, S. (1965). Fatty acid amides of ethanolamine in mammalian tissues. *The Journal of Biological Chemistry*, *240*, 1019–1024.

Bachur, N. R., & Udenfriend, S. (1966). Microsomal synthesis of fatty acid amides. *The Journal of Biological Chemistry*, *241*(6), 1308–1313.

Baggelaar, M. P., Chameau, P. J., Kantae, V., Hummel, J., Hsu, K. L., Janssen, F., et al. (2015). Highly selective, reversible inhibitor identified by comparative chemoproteomics modulates diacylglycerol lipase activity in neurons. *Journal of the American Chemical Society*, *137*(27), 8851–8857. http://dx.doi.org/10.1021/jacs.5b04883.

Baillie, G. L., Horswill, J. G., Anavi-Goffer, S., Reggio, P. H., Bolognini, D., Abood, M. E., et al. (2013). CB1 receptor allosteric modulators display both agonist and signaling pathway specificity. *Molecular Pharmacology*, *83*(2), 322–338. http://dx.doi.org/10.1124/mol.112.080879.

Bauer, M., Chicca, A., Tamborrini, M., Eisen, D., Lerner, R., Lutz, B., et al. (2012). Identification and quantification of a new family of peptide endocannabinoids (pepcans) showing negative allosteric modulation at CB1 receptors. *The Journal of Biological Chemistry*, *287*(44), 36944–36967. http://dx.doi.org/10.1074/jbc.M112.382481.

Bell, R. L., Kennerly, D. A., Stanford, N., & Majerus, P. W. (1979). Diglyceride lipase: A pathway for arachidonate release from human platelets. *Proceedings of the National Academy of Sciences of the United States of America*, *76*(7), 3238–3241.

Bisogno, T., Cascio, M. G., Saha, B., Mahadevan, A., Urbani, P., Minassi, A., et al. (2006). Development of the first potent and specific inhibitors of endocannabinoid biosynthesis. *Biochimica et Biophysica Acta*, *1761*(2), 205–212.

Bisogno, T., Howell, F., Williams, G., Minassi, A., Cascio, M. G., Ligresti, A., et al. (2003). Cloning of the first sn1-DAG lipases points to the spatial and temporal regulation of endocannabinoid signaling in the brain. *The Journal of Cell Biology*, *163*(3), 463–468.

Bisogno, T., Maccarrone, M., De Petrocellis, L., Jarrahian, A., Finazzi-Agro, A., Hillard, C., et al. (2001). The uptake by cells of 2-arachidonoylglycerol, an endogenous agonist of cannabinoid receptors. *European Journal of Biochemistry*, *268*(7), 1982–1989.

Bisogno, T., Melck, D., De Petrocellis, L., & Di Marzo, V. (1999). Phosphatidic acid as the biosynthetic precursor of the endocannabinoid 2-arachidonylglycerol in intact mouse neuroblastoma cells stimulated with ionomycin. *Journal of Neurochemistry*, *72*, 2113–2119.

Bisogno, T., Sepe, N., Melck, D., Maurelli, S., De Petrocellis, L., & Di Marzo, V. (1997). Biosynthesis, release and degradation of the novel endogenous cannabimimetic metabolite 2-arachidonoylglycerol in mouse neuroblastoma cells. *The Biochemical Journal*, *322*(Pt. 2), 671–677.

Blankman, J. L., Long, J. Z., Trauger, S. A., Siuzdak, G., & Cravatt, B. F. (2013). ABHD12 controls brain lysophosphatidylserine pathways that are deregulated in a murine model of the neurodegenerative disease PHARC. *Proceedings of the National Academy of Sciences of the United States of America*, *110*(4), 1500–1505. http://dx.doi.org/10.1073/pnas.1217121110.

Blankman, J. L., Simon, G. M., & Cravatt, B. F. (2007). A comprehensive profile of brain enzymes that hydrolyze the endocannabinoid 2-arachidonoylglycerol. *Chemistry & Biology*, *14*(12), 1347–1356. http://dx.doi.org/10.1016/j.chembiol.2007.11.006.

Bomar, M. G., & Galande, A. K. (2013). Modulation of the cannabinoid receptors by hemopressin peptides. *Life Sciences*, *92*(8–9), 520–524. http://dx.doi.org/10.1016/j.lfs.2012.07.028.

Bouaboula, M., Hilairet, S., Marchand, J., Fajas, L., Le Fur, G., & Casellas, P. (2005). Anandamide induced PPARgamma transcriptional activation and 3T3-L1 preadipocyte

differentiation. *European Journal of Pharmacology, 517*(3), 174–181. http://dx.doi.org/10.1016/j.ejphar.2005.05.032.
Bouaboula, M., Perrachon, S., Milligan, L., Canat, X., Rinaldi-Carmona, M., Portier, M., et al. (1997). A selective inverse agonist for central cannabinoid receptor inhibits mitogen-activated protein kinase activation stimulated by insulin or insulin-like growth factor 1. Evidence for a new model of receptor/ligand interactions. *The Journal of Biological Chemistry, 272*(35), 22330–22339.
Bouaboula, M., Rinaldi, M., Carayon, P., Carillon, C., Delpech, B., Shire, D., et al. (1993). Cannabinoid-receptor expression in human leukocytes. *European Journal of Biochemistry, 214*(1), 173–180.
Bowles, N. P., Hill, M. N., Bhagat, S. M., Karatsoreos, I. N., Hillard, C. J., & McEwen, B. S. (2012). Chronic, noninvasive glucocorticoid administration suppresses limbic endocannabinoid signaling in mice. *Neuroscience, 204*, 83–89. http://dx.doi.org/10.1016/j.neuroscience.2011.08.048.
Buczynski, M. W., & Parsons, L. H. (2010). Quantification of brain endocannabinoid levels: Methods, interpretations and pitfalls. *British Journal of Pharmacology, 160*(3), 423–442. http://dx.doi.org/10.1111/j.1476-5381.2010.00787.x. BPH787 [pii].
Cadas, H., di Tomaso, E., & Piomelli, D. (1997). Occurrence and biosynthesis of endogenous cannabinoid precursor, N-arachidonoyl phosphatidylethanolamine, in rat brain. *The Journal of Neuroscience, 17*(4), 1226–1242.
Cadas, H., Gaillet, S., Beltramo, M., Venance, L., & Piomelli, D. (1996). Biosynthesis of an endogenous cannabinoid precursor in neurons and its control by calcium and cAMP. *The Journal of Neuroscience, 16*(12), 3934–3942.
Cao, Z., Mulvihill, M. M., Mukhopadhyay, P., Xu, H., Erdelyi, K., Hao, E., et al. (2013). Monoacylglycerol lipase controls endocannabinoid and eicosanoid signaling and hepatic injury in mice. *Gastroenterology, 144*, 808–817. http://dx.doi.org/10.1053/j.gastro.2012.12.028.
Carlisle, S. J., Marciano-Cabral, F., Staab, A., Ludwick, C., & Cabral, G. A. (2002). Differential expression of the CB2 cannabinoid receptor by rodent macrophages and macrophage-like cells in relation to cell activation. *International Immunopharmacology, 2*(1), 69–82.
Castillo, P. E., Younts, T. J., Chavez, A. E., & Hashimotodani, Y. (2012). Endocannabinoid signaling and synaptic function. *Neuron, 76*(1), 70–81. http://dx.doi.org/10.1016/j.neuron.2012.09.020.
Caulfield, M. P., & Brown, D. A. (1992). Cannabinoid receptor agonists inhibit Ca current in NG108-15 neuroblastoma cells via a pertussis toxin-sensitive mechanism. *British Journal of Pharmacology, 106*, 231–232.
Chanda, P. K., Gao, Y., Mark, L., Btesh, J., Strassle, B. W., Lu, P., et al. (2010). Monoacylglycerol lipase activity is a critical modulator of the tone and integrity of the endocannabinoid system. *Molecular Pharmacology, 78*(6), 996–1003. http://dx.doi.org/10.1124/mol.110.068304. mol.110.068304 [pii].
Chavez, A. E., Chiu, C. Q., & Castillo, P. E. (2010). TRPV1 activation by endogenous anandamide triggers postsynaptic long-term depression in dentate gyrus. *Nature Neuroscience, 13*(12), 1511–1518. http://dx.doi.org/10.1038/nn.2684. nn.2684 [pii].
Chevaleyre, V., & Castillo, P. E. (2003). Heterosynaptic LTD of hippocampal GABAergic synapses: A novel role of endocannabinoids in regulating excitability. *Neuron, 38*(3), 461–472.
Chiang, K. P., Gerber, A. L., Sipe, J. C., & Cravatt, B. F. (2004). Reduced cellular expression and activity of the P129T mutant of human fatty acid amide hydrolase: Evidence for a link between defects in the endocannabinoid system and problem drug use. *Human Molecular Genetics, 13*(18), 2113–2119.

Chicca, A., Marazzi, J., Nicolussi, S., & Gertsch, J. (2012). Evidence for bidirectional endocannabinoid transport across cell membranes. *The Journal of Biological Chemistry, 287*(41), 34660–34682. http://dx.doi.org/10.1074/jbc.M112.373241.

Colodzin, M., Bachur, N. R., Weissbach, H., & Udenfriend, S. (1963). Enzymatic formation of fatty acid amides of ethanolamine by rat liver microsomes. *Biochemical and Biophysical Research Communications, 10*(2), 165–170.

Cravatt, B. F., Demarest, K., Patricelli, M. P., Bracey, M. H., Giang, D. K., Martin, B. R., et al. (2001). Supersensitivity to anandamide and enhanced endogenous cannabinoid signaling in mice lacking fatty acid amide hydrolase. *Proceedings of the National Academy of Sciences of the United States of America, 98*(16), 9371–9376.

Cravatt, B. F., Giang, D. K., Mayfield, S. P., Boger, D. L., Lerner, R. A., & Gilula, N. B. (1996). Molecular characterization of an enzyme that degrades neuromodulatory fatty-acid amides. *Nature, 384*(6604), 83–87.

de Lago, E., Petrosino, S., Valenti, M., Morera, E., Ortega-Gutierrez, S., Fernandez-Ruiz, J., et al. (2005). Effect of repeated systemic administration of selective inhibitors of endocannabinoid inactivation on rat brain endocannabinoid levels. *Biochemical Pharmacology, 70*(3), 446–452.

Deutsch, D. G., Glaser, S. T., Howell, J. M., Kunz, J. S., Puffenbarger, R. A., Hillard, C. J., et al. (2001). The cellular uptake of anandamide is coupled to its breakdown by fatty-acid amide hydrolase. *The Journal of Biological Chemistry, 276*(10), 6967–6973.

Devane, W. A., Dysarz, F. A., 3rd., Johnson, M. R., Melvin, L. S., & Howlett, A. C. (1988). Determination and characterization of a cannabinoid receptor in rat brain. *Molecular Pharmacology, 34*(5), 605–613.

Devane, W. A., Hanus, L., Breuer, A., Pertwee, R. G., Stevenson, L. A., Griffin, G., et al. (1992). Isolation and structure of a brain constituent that binds to the cannabinoid receptor. *Science, 258*(5090), 1946–1949.

Di Marzo, V., Bisogno, T., De Petrocellis, L., Melck, D., Orlando, P., Wagner, J. A., et al. (1999). Biosynthesis and inactivation of the endocannabinoid 2-arachidonoylglycerol in circulating and tumoral macrophages. *European Journal of Biochemistry, 264*(1), 258–267.

Di, S., Boudaba, C., Popescu, I. R., Weng, F. J., Harris, C., Marcheselli, V. L., et al. (2005). Activity-dependent release and actions of endocannabinoids in the rat hypothalamic supraoptic nucleus. *The Journal of Physiology, 569*(Pt. 3), 751–760.

Di, S., Malcher-Lopes, R., Halmos, K. C., & Tasker, J. G. (2003). Nongenomic glucocorticoid inhibition via endocannabinoid release in the hypothalamus: A fast feedback mechanism. *The Journal of Neuroscience, 23*(12), 4850–4857.

Dincheva, I., Drysdale, A. T., Hartley, C. A., Johnson, D. C., Jing, D., King, E. C., et al. (2015). FAAH genetic variation enhances fronto-amygdala function in mouse and human. *Nature Communications, 6*, 6395. http://dx.doi.org/10.1038/ncomms7395.

Dinh, T. P., Carpenter, D., Leslie, F. M., Freund, T. F., Katona, I., Sensi, S. L., et al. (2002). Brain monoglyceride lipase participating in endocannabinoid inactivation. *Proceedings of the National Academy of Sciences of the United States of America, 99*(16), 10819–10824.

Du, H., Kwon, I. K., & Kim, J. (2013). Neuregulin-1 impairs the long-term depression of hippocampal inhibitory synapses by facilitating the degradation of endocannabinoid 2-AG. *The Journal of Neuroscience, 33*(38), 15022–15031. http://dx.doi.org/10.1523/JNEUROSCI.5833-12.2013.

Edwards, D. A., Kim, J., & Alger, B. E. (2006). Multiple mechanisms of endocannabinoid response initiation in hippocampus. *Journal of Neurophysiology, 95*(1), 67–75.

Engeli, S., Lehmann, A. C., Kaminski, J., Haas, V., Janke, J., Zoerner, A. A., et al. (2014). Influence of dietary fat intake on the endocannabinoid system in lean and obese subjects. *Obesity (Silver Spring), 22*(5), E70–E76. http://dx.doi.org/10.1002/oby.20728.

Farooqui, A. A., & Horrocks, L. A. (1997). Nitric oxide synthase inhibitors do not attenuate diacylglycerol or monoacylglycerol lipase activities in synaptoneurosomes. *Neurochemical Research*, *22*(10), 1265–1269.

Filis, P., Kind, P. C., & Spears, N. (2013). Implantation failure in mice with a disruption in Phospholipase C beta 1 gene: Lack of embryonic attachment, aberrant steroid hormone signalling and defective endocannabinoid metabolism. *Molecular Human Reproduction*, *19*(5), 290–301. http://dx.doi.org/10.1093/molehr/gas067.

Fiskerstrand, T., H'Mida-Ben Brahim, D., Johansson, S., M'Zahem, A., Haukanes, B. I., Drouot, N., et al. (2010). Mutations in ABHD12 cause the neurodegenerative disease PHARC: An inborn error of endocannabinoid metabolism. *American Journal of Human Genetics*, *87*(3), 410–417. http://dx.doi.org/10.1016/j.ajhg.2010.08.002. S0002-9297 (10)00414-3 [pii].

Fourgeaud, L., Mato, S., Bouchet, D., Hemar, A., Worley, P. F., & Manzoni, O. J. (2004). A single in vivo exposure to cocaine abolishes endocannabinoid-mediated long-term depression in the nucleus accumbens. *The Journal of Neuroscience*, *24*(31), 6939–6945.

Fowler, C. J., Tiger, G., Ligresti, A., Lopez-Rodriguez, M. L., & Di Marzo, V. (2004). Selective inhibition of anandamide cellular uptake versus enzymatic hydrolysis—A difficult issue to handle. *European Journal of Pharmacology*, *492*(1), 1–11.

Freund, T. F., Katona, I., & Piomelli, D. (2003). Role of endogenous cannabinoids in synaptic signaling. *Physiological Reviews*, *83*(3), 1017–1066.

Fu, J., Bottegoni, G., Sasso, O., Bertorelli, R., Rocchia, W., Masetti, M., et al. (2012). A catalytically silent FAAH-1 variant drives anandamide transport in neurons. *Nature Neuroscience*, *15*(1), 64–69. http://dx.doi.org/10.1038/nn.2986.

Fu, J., Gaetani, S., Oveisi, F., Lo Verme, J., Serrano, A., Rodriguez De Fonseca, F., et al. (2003). Oleylethanolamide regulates feeding and body weight through activation of the nuclear receptor PPAR-alpha. *Nature*, *425*(6953), 90–93.

Fukaya, M., Uchigashima, M., Nomura, S., Hasegawa, Y., Kikuchi, H., & Watanabe, M. (2008). Predominant expression of phospholipase Cbeta1 in telencephalic principal neurons and cerebellar interneurons, and its close association with related signaling molecules in somatodendritic neuronal elements. *The European Journal of Neuroscience*, *28*(9), 1744–1759. http://dx.doi.org/10.1111/j.1460-9568.2008.06495.x. EJN6495 [pii].

Galiegue, S., Mary, S., Marchand, J., Dussossoy, D., Carriere, D., Carayon, P., et al. (1995). Expression of central and peripheral cannabinoid receptors in human immune tissues and leukocyte subpopulations. *European Journal of Biochemistry*, *232*(1), 54–61.

Galli, G. G., Multhaupt, H. A., Carrara, M., de Lichtenberg, K. H., Christensen, I. B., Linnemann, D., et al. (2014). Prdm5 suppresses Apc(Min)-driven intestinal adenomas and regulates monoacylglycerol lipase expression. *Oncogene*, *33*(25), 3342–3350. http://dx.doi.org/10.1038/onc.2013.283.

Gamage, T. F., Ignatowska-Jankowska, B. M., Wiley, J. L., Abdelrahman, M., Trembleau, L., Greig, I. R., et al. (2014). In-vivo pharmacological evaluation of the CB1-receptor allosteric modulator Org-27569. *Behavioural Pharmacology*, *25*(2), 182–185. http://dx.doi.org/10.1097/FBP.0000000000000027.

Gao, C., Tronson, N. C., & Radulovic, J. (2013). Modulation of behavior by scaffolding proteins of the post-synaptic density. *Neurobiology of Learning and Memory*, *105*, 3–12. http://dx.doi.org/10.1016/j.nlm.2013.04.014.

Gao, Y., Vasilyev, D. V., Goncalves, M. B., Howell, F. V., Hobbs, C., Reisenberg, M., et al. (2010). Loss of retrograde endocannabinoid signaling and reduced adult neurogenesis in diacylglycerol lipase knock-out mice. *The Journal of Neuroscience*, *30*(6), 2017–2024. http://dx.doi.org/10.1523/JNEUROSCI.5693-09.2010. 30/6/2017 [pii].

Gaoni, Y., & Mechoulam, R. (1964). Isolation, structure and partial synthesis of an active constituent of hashish. *Journal of the American Chemical Society*, *86*, 1646–1647.

Gasperi, V., Ceci, R., Tantimonaco, M., Talamonti, E., Battista, N., Parisi, A., et al. (2014). The fatty acid amide hydrolase in lymphocytes from sedentary and active subjects. *Medicine and Science in Sports and Exercise, 46*(1), 24–32. http://dx.doi.org/10.1249/MSS.0b013e3182a10ce6.

Gauthier, K. M., Baewer, D. V., Hittner, S., Hillard, C. J., Nithipatikom, K., Reddy, D. S., et al. (2005). Endothelium-derived 2-arachidonylglycerol: An intermediate in vasodilatory eicosanoid release in bovine coronary arteries. *American Journal of Physiology Heart and Circulatory Physiology, 288*(3), H1344–H1351.

Gebremedhin, D., Lange, A. R., Campbell, W. B., Hillard, C. J., & Harder, D. R. (1999). Cannabinoid CB1 receptor of cat cerebral arterial muscle functions to inhibit L-type Ca^{2+} channel current. *American Journal of Physiology Heart and Circulatory Physiology, 276*(6 Pt. 2), H2085–H2093.

Gerdeman, G. L., Ronesi, J., & Lovinger, D. M. (2002). Postsynaptic endocannabinoid release is critical to long-term depression in the striatum. *Nature Neuroscience, 5*(5), 446–451.

Glaser, S. T., Abumrad, N. A., Fatade, F., Kaczocha, M., Studholme, K. M., & Deutsch, D. G. (2003). Evidence against the presence of an anandamide transporter. *Proceedings of the National Academy of Sciences of the United States of America, 100*(7), 4269–4274.

Glass, M., & Felder, C. C. (1997). Concurrent stimulation of cannabinoid CB1 and dopamine D2 receptors augments cAMP accumulation in striatal neurons: Evidence for a Gs linkage to the CB1 receptor. *The Journal of Neuroscience, 17*, 5327–5333.

Glass, M., & Northup, J. K. (1999). Agonist selective regulation of G proteins by cannabinoid CB(1) and CB(2) receptors. *Molecular Pharmacology, 56*(6), 1362–1369.

Golczak, M., Kiser, P. D., Sears, A. E., Lodowski, D. T., Blaner, W. S., & Palczewski, K. (2012). Structural basis for the acyltransferase activity of lecithin:Retinol acyltransferase-like proteins. *The Journal of Biological Chemistry, 287*(28), 23790–23807. http://dx.doi.org/10.1074/jbc.M112.361550.

Golech, S. A., McCarron, R. M., Chen, Y., Bembry, J., Lenz, F., Mechoulam, R., et al. (2004). Human brain endothelium: Coexpression and function of vanilloid and endocannabinoid receptors. *Brain Research Molecular Brain Research, 132*(1), 87–92.

Gomez, O., Sanchez-Rodriguez, A., Le, M., Sanchez-Caro, C., Molina-Holgado, F., & Molina-Holgado, E. (2011). Cannabinoid receptor agonists modulate oligodendrocyte differentiation by activating PI3K/Akt and the mammalian target of rapamycin (mTOR) pathways. *British Journal of Pharmacology, 163*(7), 1520–1532. http://dx.doi.org/10.1111/j.1476-5381.2011.01414.x.

Gonsiorek, W., Lunn, C., Fan, X., Narula, S., Lundell, D., & Hipkin, R. W. (2000). Endocannabinoid 2-arachidonyl glycerol is a full agonist through human type 2 cannabinoid receptor: Antagonism by anandamide. *Molecular Pharmacology, 57*(5), 1045–1050.

Goparaju, S. K., Ueda, N., Taniguchi, K., & Yamamoto, S. (1999). Enzymes of porcine brain hydrolyzing 2-arachidonoylglycerol, an endogenous ligand of cannabinoid receptors. *Biochemical Pharmacology, 57*(4), 417–423.

Gray, J. M., Vecchiarelli, H. A., Morena, M., Lee, T. T., Hermanson, D. J., Kim, A. B., et al. (2015). Corticotropin-releasing hormone drives anandamide hydrolysis in the amygdala to promote anxiety. *The Journal of Neuroscience, 35*(9), 3879–3892. http://dx.doi.org/10.1523/JNEUROSCI.2737-14.2015.

Grimaldi, P., Pucci, M., Di Siena, S., Di Giacomo, D., Pirazzi, V., Geremia, R., et al. (2012). The faah gene is the first direct target of estrogen in the testis: Role of histone demethylase LSD1. *Cellular and Molecular Life Sciences, 69*(24), 4177–4190. http://dx.doi.org/10.1007/s00018-012-1074-6.

Grimaldi, P., Rossi, G., Catanzaro, G., & Maccarrone, M. (2009). Modulation of the endocannabinoid-degrading enzyme fatty acid amide hydrolase by follicle-stimulating hormone. *Vitamins and Hormones, 81*, 231–261. http://dx.doi.org/10.1016/S0083-6729(09)81010-8. S0083-6729(09)81010-8 [pii].

Grueter, B. A., Brasnjo, G., & Malenka, R. C. (2010). Postsynaptic TRPV1 triggers cell type-specific long-term depression in the nucleus accumbens. *Nature Neuroscience, 13*(12), 1519–1525. http://dx.doi.org/10.1038/nn.2685. nn.2685 [pii].

Guerciolini, R. (1997). Mode of action of orlistat. *International Journal of Obesity and Related Metabolic Disorders, 21*(Suppl. 3), S12–S23.

Gulyas, A. I., Cravatt, B. F., Bracey, M. H., Dinh, T. P., Piomelli, D., Boscia, F., et al. (2004). Segregation of two endocannabinoid-hydrolyzing enzymes into pre- and postsynaptic compartments in the rat hippocampus, cerebellum and amygdala. *The European Journal of Neuroscience, 20*(2), 441–458.

Hansen, H. S., Moesgaard, B., Hansen, H. H., & Petersen, G. (2000). N-Acylethanolamines and precursor phospholipids—Relation to cell injury. *Chemistry and Physics of Lipids, 108*(1–2), 135–150.

Hanus, L., Abu-Lafi, S., Fride, E., Breuer, A., Vogel, Z., Shalev, D. E., et al. (2001). 2-Arachidonyl glyceryl ether, an endogenous agonist of the cannabinoid CB1 receptor. *Proceedings of the National Academy of Sciences of the United States of America, 98*(7), 3662–3665.

Haring, M., Guggenhuber, S., & Lutz, B. (2012). Neuronal populations mediating the effects of endocannabinoids on stress and emotionality. *Neuroscience, 204*, 145–158. http://dx.doi.org/10.1016/j.neuroscience.2011.12.035.

Hashimotodani, Y., Ohno-Shosaku, T., Maejima, T., Fukami, K., & Kano, M. (2008). Pharmacological evidence for the involvement of diacylglycerol lipase in depolarization-induced endocanabinoid release. *Neuropharmacology, 54*, 58–67.

Hashimotodani, Y., Ohno-Shosaku, T., Tanimura, A., Kita, Y., Sano, Y., Shimizu, T., et al. (2013). Acute inhibition of diacylglycerol lipase blocks endocannabinoid-mediated retrograde signalling: Evidence for on-demand biosynthesis of 2-arachidonoylglycerol. *The Journal of Physiology, 591*(Pt. 19), 4765–4776. http://dx.doi.org/10.1113/jphysiol.2013.254474.

Heimann, A. S., Gomes, I., Dale, C. S., Pagano, R. L., Gupta, A., de Souza, L. L., et al. (2007). Hemopressin is an inverse agonist of CB1 cannabinoid receptors. *Proceedings of the National Academy of Sciences of the United States of America, 104*(51), 20588–20593. http://dx.doi.org/10.1073/pnas.0706980105.

Henry, D. J., & Chavkin, C. (1995). Activation of inwardly rectifying potassium channels (GIRK1) by co-expressed rat brain cannabinoid receptors in Xenopus oocytes. *Neuroscience Letters, 186*, 91–94.

Herkenham, M., Groen, B. G., Lynn, A. B., De Costa, B. R., & Richfield, E. K. (1991). Neuronal localization of cannabinoid receptors and second messengers in mutant mouse cerebellum. *Brain Research, 552*(2), 301–310.

Hermanson, D. J., Gamble-George, J. C., Marnett, L. J., & Patel, S. (2014). Substrate-selective COX-2 inhibition as a novel strategy for therapeutic endocannabinoid augmentation. *Trends in Pharmacological Sciences, 35*(7), 358–367. http://dx.doi.org/10.1016/j.tips.2014.04.006.

Hermanson, D. J., Hartley, N. D., Gamble-George, J., Brown, N., Shonesy, B. C., Kingsley, P. J., et al. (2013). Substrate-selective COX-2 inhibition decreases anxiety via endocannabinoid activation. *Nature Neuroscience, 16*(9), 1291–1298. http://dx.doi.org/10.1038/nn.3480.

Hill, M. N., Kumar, S. A., Filipski, S. B., Iverson, M., Stuhr, K. L., Keith, J. M., et al. (2013). Disruption of fatty acid amide hydrolase activity prevents the effects of chronic stress on anxiety and amygdalar microstructure. *Molecular Psychiatry, 18*, 1125–1135. http://dx.doi.org/10.1038/mp.2012.90.

Hillard, C. J. (2000). Biochemistry and pharmacology of the endocannabinoids arachidonylethanolamide and 2-arachidonylglycerol. *Prostaglandins & Other Lipid Mediators*, *61*(1–2), 3–18.

Hillard, C. J., & Campbell, W. B. (1997). Biochemistry and pharmacology of arachidonylethanolamide, a putative endogenous cannabinoid. *Journal of Lipid Research*, *38*(12), 2383–2398.

Hillard, C. J., Edgemond, W. S., Jarrahian, A., & Campbell, W. B. (1997). Accumulation of N-arachidonoylethanolamine (anandamide) into cerebellar granule cells occurs via facilitated diffusion. *Journal of Neurochemistry*, *69*(2), 631–638.

Hillard, C. J., & Jarrahian, A. (2003). Cellular accumulation of anandamide: Consensus and controversy. *British Journal of Pharmacology*, *140*(5), 802–808.

Hillard, C. J., & Jarrahian, A. (2005). Accumulation of anandamide: Evidence for cellular diversity. *Neuropharmacology*, *48*(8), 1072–1078.

Hillard, C. J., Manna, S., Greenberg, M. J., DiCamelli, R., Ross, R. A., Stevenson, L. A., et al. (1999). Synthesis and characterization of potent and selective agonists of the neuronal cannabinoid receptor (CB1). *The Journal of Pharmacology and Experimental Therapeutics*, *289*(3), 1427–1433.

Hillard, C. J., Wilkison, D. M., Edgemond, W. S., & Campbell, W. B. (1995). Characterization of the kinetics and distribution of N-arachidonylethanolamine (anandamide) hydrolysis by rat brain. *Biochimica et Biophysica Acta*, *1257*(3), 249–256.

Hirasawa, M., Schwab, Y., Natah, S., Hillard, C. J., Mackie, K., Sharkey, K. A., et al. (2004). Dendritically released transmitters cooperate via autocrine and retrograde actions to inhibit afferent excitation in rat brain. *The Journal of Physiology*, *559*(Pt. 2), 611–624.

Ho, Y. C., Lee, H. J., Tung, L. W., Liao, Y. Y., Fu, S. Y., Teng, S. F., et al. (2011). Activation of orexin 1 receptors in the periaqueductal gray of male rats leads to antinociception via retrograde endocannabinoid (2-arachidonoylglycerol)-induced disinhibition. *The Journal of Neuroscience*, *31*(41), 14600–14610. http://dx.doi.org/10.1523/JNEUROSCI.2671-11.2011.

Hofer, S. C., Ralvenius, W. T., Gachet, M. S., Fritschy, J. M., Zeilhofer, H. U., & Gertsch, J. (2015). Localization and production of peptide endocannabinoids in the rodent CNS and adrenal medulla. *Neuropharmacology*. http://dx.doi.org/10.1016/j.neuropharm.2015.03.021.

Hohmann, A. G., Suplita, R. L., Bolton, N. M., Neely, M. H., Fegley, D., Mangieri, R., et al. (2005). An endocannabinoid mechanism for stress-induced analgesia. *Nature*, *435*(7045), 1108–1112.

Horswill, J. G., Bali, U., Shaaban, S., Keily, J. F., Jeevaratnam, P., Babbs, A. J., et al. (2007). PSNCBAM-1, a novel allosteric antagonist at cannabinoid CB(1) receptors with hypophagic effects in rats. *British Journal of Pharmacology*, *152*(5), 805–814.

Horvath, E., Woodhams, S. G., Nyilas, R., Henstridge, C. M., Kano, M., Sakimura, K., et al. (2014). Heterogeneous presynaptic distribution of monoacylglycerol lipase, a multipotent regulator of nociceptive circuits in the mouse spinal cord. *The European Journal of Neuroscience*, *39*(3), 419–434. http://dx.doi.org/10.1111/ejn.12470.

Howlett, A. C. (1984). Inhibition of neuroblastoma adenylate cyclase by cannabinoid and nantradol compounds. *Life Sciences*, *35*, 1803–1810.

Howlett, A. C. (1985). Cannabinoid inhibition of adenylate cyclase. Biochemistry of the response in neuroblastoma cell membranes. *Molecular Pharmacology*, *27*, 429–436.

Howlett, A. C., & Fleming, R. M. (1984). Cannabinoid inhibition of adenylate cyclase. Pharmacology of the response of neuroblastoma cell membranes. *Molecular Pharmacology*, *26*, 532–538.

Howlett, A. C., Qualy, J. M., & Khachatrian, L. L. (1986). Involvement of Gi in the inhibition of adenylate cyclase by cannabimimetic drugs. *Molecular Pharmacology*, *29*, 307–313.

Hu, S. S., & Mackie, K. (2015). Distribution of the endocannabinoid system in the central nervous system. *Handbook of Experimental Pharmacology*, *231*, 59–93. http://dx.doi.org/10.1007/978-3-319-20825-1_3.

Huang, G. Z., & Woolley, C. S. (2012). Estradiol acutely suppresses inhibition in the hippocampus through a sex-specific endocannabinoid and mGluR-dependent mechanism. *Neuron*, *74*(5), 801–808. http://dx.doi.org/10.1016/j.neuron.2012.03.035.

Ignatowska-Jankowska, B. M., Baillie, G. L., Kinsey, S., Crowe, M., Ghosh, S., Owens, R. A., et al. (2015). A cannabinoid CB1 receptor-positive allosteric modulator reduces neuropathic pain in the mouse with no psychoactive effects. *Neuropsychopharmacology*. http://dx.doi.org/10.1038/npp.2015.148.

Ignatowska-Jankowska, B. M., Ghosh, S., Crowe, M. S., Kinsey, S. G., Niphakis, M. J., Abdullah, R. A., et al. (2014). In vivo characterization of the highly selective monoacylglycerol lipase inhibitor KML29: Antinociceptive activity without cannabimimetic side effects. *British Journal of Pharmacology*, *171*(6), 1392–1407. http://dx.doi.org/10.1111/bph.12298.

Jung, K. M., Astarita, G., Zhu, C., Wallace, M., Mackie, K., & Piomelli, D. (2007). A key role for diacylglycerol lipase-{alpha} in metabotropic glutamate receptor-dependent endocannabinoid mobilization. *Molecular Pharmacology*, *72*(3), 612–621.

Jung, K. M., Mangieri, R., Stapleton, C., Kim, J., Fegley, D., Wallace, M., et al. (2005). Stimulation of endocannabinoid formation in brain slice cultures through activation of group I metabotropic glutamate receptors. *Molecular Pharmacology*, *68*, 1196–1202.

Kaczocha, M., Glaser, S. T., & Deutsch, D. G. (2009). Identification of intracellular carriers for the endocannabinoid anandamide. *Proceedings of the National Academy of Sciences of the United States of America*, *106*(15), 6375–6380. http://dx.doi.org/10.1073/pnas.0901515106. 0901515106 [pii].

Kaczocha, M., Hermann, A., Glaser, S. T., Bojesen, I. N., & Deutsch, D. G. (2006). Anandamide uptake is consistent with rate-limited diffusion and is regulated by the degree of its hydrolysis by fatty acid amide hydrolase. *The Journal of Biological Chemistry*, *281*(14), 9066–9075.

Kaczocha, M., Vivieca, S., Sun, J., Glaser, S. T., & Deutsch, D. G. (2012). Fatty acid-binding proteins transport N-acylethanolamines to nuclear receptors and are targets of endocannabinoid transport inhibitors. *The Journal of Biological Chemistry*, *287*(5), 3415–3424. http://dx.doi.org/10.1074/jbc.M111.304907.

Kang, R., Wan, J., Arstikaitis, P., Takahashi, H., Huang, K., Bailey, A. O., et al. (2008). Neural palmitoyl-proteomics reveals dynamic synaptic palmitoylation. *Nature*, *456*(7224), 904–909. http://dx.doi.org/10.1038/nature07605.

Kano, M. (2014). Control of synaptic function by endocannabinoid-mediated retrograde signaling. *Proceedings of the Japan Academy Series B, Physical and Biological Sciences*, *90*(7), 235–250.

Kapanda, C. N., Masquelier, J., Labar, G., Muccioli, G. G., Poupaert, J. H., & Lambert, D. M. (2012). Synthesis and pharmacological evaluation of 2,4-dinitroaryldithiocarbamate derivatives as novel monoacylglycerol lipase inhibitors. *Journal of Medicinal Chemistry*, *55*(12), 5774–5783. http://dx.doi.org/10.1021/jm3006004.

Karlsson, M., Reue, K., Xia, Y. R., Lusis, A. J., Langin, D., Tornqvist, H., et al. (2001). Exon-intron organization and chromosomal localization of the mouse monoglyceride lipase gene. *Gene*, *272*(1–2), 11–18.

Katayama, K., Ueda, N., Kurahashi, Y., Suzuki, H., Yamamoto, S., & Kato, I. (1997). Distribution of anandamide amidohydrolase in rat tissues with special reference to small intestine. *Biochimica et Biophysica Acta*, *1347*(2–3), 212–218.

Katona, I., Sperlagh, B., Sik, A., Kafalvi, A., Vizi, E. S., Mackie, K., et al. (1999). Presynaptically located CB1 cannabinoid receptors regulate GABA release from axon terminals of specific hippocampal interneurons. *The Journal of Neuroscience*, *19*(11), 4544–4558.

Katona, I., Urban, G. M., Wallace, M., Ledent, C., Jung, K. M., Piomelli, D., et al. (2006). Molecular composition of the endocannabinoid system at glutamatergic synapses. *The Journal of Neuroscience*, *26*(21), 5628–5637.

Keimpema, E., Tortoriello, G., Alpar, A., Capsoni, S., Arisi, I., Calvigioni, D., et al. (2013). Nerve growth factor scales endocannabinoid signaling by regulating monoacylglycerol lipase turnover in developing cholinergic neurons. *Proceedings of the National Academy of Sciences of the United States of America*, *110*(5), 1935–1940. http://dx.doi.org/10.1073/pnas.1212563110.

Kim, S. R., Chung, Y. C., Chung, E. S., Park, K. W., Won, S. Y., Bok, E., et al. (2007). Roles of transient receptor potential vanilloid subtype 1 and cannabinoid type 1 receptors in the brain: Neuroprotection versus neurotoxicity. *Molecular Neurobiology*, *35*(3), 245–254.

Kim, J., Isokawa, M., Ledent, C., & Alger, B. E. (2002). Activation of muscarinic acetylcholine receptors enhances the release of endogenous cannabinoids in the hippocampus. *The Journal of Neuroscience*, *22*(23), 10182–10191.

King, A. R., Duranti, A., Tontini, A., Rivara, S., Rosengarth, A., Clapper, J. R., et al. (2007). URB602 inhibits monoacylglycerol lipase and selectively blocks 2-arachidonoylglycerol degradation in intact brain slices. *Chemistry & Biology*, *14*(12), 1357–1365. http://dx.doi.org/10.1016/j.chembiol.2007.10.017.

Kirilly, E., Gonda, X., & Bagdy, G. (2012). CB1 receptor antagonists: New discoveries leading to new perspectives. *Acta Physiologica*, *205*(1), 41–60. http://dx.doi.org/10.1111/j.1748-1716.2012.02402.x.

Kortleven, C., Bruneau, L. C., & Trudeau, L. E. (2012). Neurotensin inhibits glutamate-mediated synaptic inputs onto ventral tegmental area dopamine neurons through the release of the endocannabinoid 2-AG. *Neuropharmacology*, *63*(6), 983–991. http://dx.doi.org/10.1016/j.neuropharm.2012.07.037.

Kreitzer, A. C., & Regehr, W. G. (2001). Retrograde inhibition of presynaptic calcium influx by endogenous cannabinoids at excitatory synapses onto Purkinje cells. *Neuron*, *29*(3), 717–727.

Kurahashi, Y., Ueda, N., Suzuki, H., Suzuki, M., & Yamamoto, S. (1997). Reversible hydrolysis and synthesis of anandamide demonstrated by recombinant rat fatty-acid amide hydrolase. *Biochemical and Biophysical Research Communications*, *237*(3), 512–515.

Lappas, M. (2014). Effect of pre-existing maternal obesity, gestational diabetes and adipokines on the expression of genes involved in lipid metabolism in adipose tissue. *Metabolism*, *63*(2), 250–262. http://dx.doi.org/10.1016/j.metabol.2013.10.001.

Laprairie, R. B., Bagher, A. M., Kelly, M. E., & Denovan-Wright, E. M. (2015). Cannabidiol is a negative allosteric modulator of the type 1 cannabinoid receptor. *British Journal of Pharmacology*. http://dx.doi.org/10.1111/bph.13250.

Lee, M. W., Kraemer, F. B., & Severson, D. L. (1995). Characterization of a partially purified diacylglycerol lipase from bovine aorta. *Biochimica et Biophysica Acta*, *1254*(3), 311–318.

Lee, S. H., Ledri, M., Toth, B., Marchionni, I., Henstridge, C. M., Dudok, B., et al. (2015). Multiple forms of endocannabinoid and endovanilloid signaling regulate the tonic control of GABA release. *The Journal of Neuroscience*, *35*(27), 10039–10057. http://dx.doi.org/10.1523/JNEUROSCI.4112-14.2015.

Leung, K., Elmes, M. W., Glaser, S. T., Deutsch, D. G., & Kaczocha, M. (2013). Role of FAAH-like anandamide transporter in anandamide inactivation. *PLoS One*, *8*(11), e79355. http://dx.doi.org/10.1371/journal.pone.0079355.

Leung, D., Saghatelian, A., Simon, G. M., & Cravatt, B. F. (2006). Inactivation of N-acyl phosphatidylethanolamine phospholipase D reveals multiple mechanisms for the biosynthesis of endocannabinoids. *Biochemistry*, *45*(15), 4720–4726.

Liedhegner, E. S., Vogt, C. D., Sem, D. S., Cunningham, C. W., & Hillard, C. J. (2014). Sterol carrier protein-2: Binding protein for endocannabinoids. *Molecular Neurobiology*, *50*, 149–158. http://dx.doi.org/10.1007/s12035-014-8651-7.

Ligresti, A., Morera, E., Van Der Stelt, M., Monory, K., Lutz, B., Ortar, G., et al. (2004). Further evidence for the existence of a specific process for the membrane transport of anandamide. *The Biochemical Journal*, *380*(Pt. 1), 265–272.

Liu, J., Wang, L., Harvey-White, J., Huang, B. X., Kim, H. Y., Luquet, S., et al. (2008). Multiple pathways involved in the biosynthesis of anandamide. *Neuropharmacology*, *54*(1), 1–7. http://dx.doi.org/10.1016/j.neuropharm.2007.05.020. S0028-3908(07)00157-8 [pii].

Liu, J., Wang, L., Harvey-White, J., Osei-Hyiaman, D., Razdan, R., Gong, Q., et al. (2006). A biosynthetic pathway for anandamide. *Proceedings of the National Academy of Sciences of the United States of America*, *103*(36), 13345–13350.

Lo Verme, J., Fu, J., Astarita, G., La Rana, G., Russo, R., Calignano, A., et al. (2005). The nuclear receptor peroxisome proliferator-activated receptor-alpha mediates the anti-inflammatory actions of palmitoylethanolamide. *Molecular Pharmacology*, *67*(1), 15–19. http://dx.doi.org/10.1124/mol.104.006353. mol.104.006353 [pii].

Lograno, M. D., & Romano, M. R. (2004). Cannabinoid agonists induce contractile responses through Gi/o-dependent activation of phospholipase C in the bovine ciliary muscle. *European Journal of Pharmacology*, *494*(1), 55–62.

Long, J. Z., Li, W., Booker, L., Burston, J. J., Kinsey, S. G., Schlosburg, J. E., et al. (2009). Selective blockade of 2-arachidonoylglycerol hydrolysis produces cannabinoid behavioral effects. *Nature Chemical Biology*, *5*(1), 37–44. http://dx.doi.org/10.1038/nchembio.129. nchembio.129 [pii].

Long, J. Z., Nomura, D. K., & Cravatt, B. F. (2009). Characterization of monoacylglycerol lipase inhibition reveals differences in central and peripheral endocannabinoid metabolism. *Chemistry & Biology*, *16*(7), 744–753. http://dx.doi.org/10.1016/j.chembiol.2009.05.009. S1074-5521(09)00176-8 [pii].

Maccarrone, M., Bari, M., Di Rienzo, M., Finazzi-Agro, A., & Rossi, A. (2003). Progesterone activates fatty acid amide hydrolase (FAAH) promoter in human T lymphocytes through the transcription factor Ikaros. Evidence for a synergistic effect of leptin. *The Journal of Biological Chemistry*, *278*(35), 32726–32732.

Maccarrone, M., De Petrocellis, L., Bari, M., Fezza, F., Salvati, S., Di Marzo, V., et al. (2001). Lipopolysaccharide downregulates fatty acid amide hydrolase expression and increases anandamide levels in human peripheral lymphocytes. *Archives of Biochemistry and Biophysics*, *393*(2), 321–328.

Maccarrone, M., Di Rienzo, M., Finazzi-Agro, A., & Rossi, A. (2003). Leptin activates the anandamide hydrolase promoter in human T lymphocytes through STAT3. *The Journal of Biological Chemistry*, *278*(15), 13318–13324.

Maccarrone, M., Gasperi, V., Fezza, F., Finazzi-Agro, A., & Rossi, A. (2004). Differential regulation of fatty acid amide hydrolase promoter in human immune cells and neuronal cells by leptin and progesterone. *European Journal of Biochemistry*, *271*(23–24), 4666–4676. http://dx.doi.org/10.1111/j.1432-1033.2004.04427.x.

Mackie, K., & Hille, B. (1992). Cannabinoids inhibit N-type calcium channels in neuroblastoma-glioma cells. *Proceedings of the National Academy of Sciences of the United States of America*, *89*, 3825–3829.

Maejima, T., Hashimoto, K., Yoshida, T., Aiba, A., & Kano, M. (2001). Presynaptic inhibition caused by retrograde signal from metabotropic glutamate to cannabinoid receptors. *Neuron*, *31*(3), 463–475.

Mai, P., Yang, L., Tian, L., Wang, L., Jia, S., Zhang, Y., et al. (2015). Endocannabinoid system contributes to liver injury and inflammation by activation of bone marrow-derived monocytes/macrophages in a CB1-dependent manner. *Journal of Immunology*, *195*(7), 3390–3401. http://dx.doi.org/10.4049/jimmunol.1403205.

Malcher-Lopes, R., Di, S., Marcheselli, V. S., Weng, F. J., Stuart, C. T., Bazan, N. G., et al. (2006). Opposing crosstalk between leptin and glucocorticoids rapidly modulates synaptic excitation via endocannabinoid release. *The Journal of Neuroscience, 26*(24), 6643–6650.

Marrs, W. R., Blankman, J. L., Horne, E. A., Thomazeau, A., Lin, Y. H., Coy, J., et al. (2010). The serine hydrolase ABHD6 controls the accumulation and efficacy of 2-AG at cannabinoid receptors. *Nature Neuroscience, 13*(8), 951–957. http://dx.doi.org/10.1038/nn.2601. nn.2601 [pii].

Marsicano, G., & Lutz, B. (1999). Expression of the cannabinoid receptor CB1 in distinct neuronal subpopulations in the adult mouse forebrain. *The European Journal of Neuroscience, 11*(12), 4213–4225.

Martin, B. R., & Cravatt, B. F. (2009). Large-scale profiling of protein palmitoylation in mammalian cells. *Nature Methods, 6*(2), 135–138. http://dx.doi.org/10.1038/nmeth.1293.

Matsuda, L. A., Lolait, S. J., Brownstein, M. J., Young, A. C., & Bonner, T. I. (1990). Structure of a cannabinoid receptor and functional expression of the cloned cDNA. *Nature, 346*(6284), 561–564.

Matyas, F., Urban, G. M., Watanabe, M., Mackie, K., Zimmer, A., Freund, T. F., et al. (2008). Identification of the sites of 2-arachidonoylglycerol synthesis and action imply retrograde endocannabinoid signaling at both GABAergic and glutamatergic synapses in the ventral tegmental area. *Neuropharmacology, 54*(1), 95–107. http://dx.doi.org/10.1016/j.neuropharm.2007.05.028. S0028-3908(07)00170-0 [pii].

McFarland, M. J., Porter, A. C., Rakhshan, F. R., Rawat, D. S., Gibbs, R. A., & Barker, E. L. (2004). A role for caveolae/lipid rafts in the uptake and recycling of the endogenous cannabinoid anandamide. *The Journal of Biological Chemistry, 279*(40), 41991–41997.

McFarland, M. J., Terebova, E. A., & Barker, E. L. (2006). Detergent-resistant membrane microdomains in the disposition of the lipid signaling molecule anandamide. *The AAPS Journal, 8*(1), E95–E100.

Mechoulam, R., Ben-Shabat, S., Hanus, L., Ligumsky, M., Kaminski, N. E., Schatz, A. R., et al. (1995). Identification of an endogenous 2-monoglyceride, present in canine gut, that binds to cannabinoid receptors. *Biochemical Pharmacology, 50*(1), 83–90.

Milne, G. M., Koe, B. K., & Johnson, M. R. (1979). Stereospecific and potent analgetic activity for nantradol—A structurally novel, cannabinoid-related analgetic. *NIDA Research Monograph, 27*, 84–92.

Min, R., Testa-Silva, G., Heistek, T. S., Canto, C. B., Lodder, J. C., Bisogno, T., et al. (2010). Diacylglycerol lipase is not involved in depolarization-induced suppression of inhibition at unitary inhibitory connections in mouse hippocampus. *The Journal of Neuroscience, 30*(7), 2710–2715. http://dx.doi.org/10.1523/JNEUROSCI.BC-3622-09.2010. 30/7/2710 [pii].

Molina-Holgado, E., Vela, J. M., Arevalo-Martin, A., Almazan, G., Molina-Holgado, F., Borrell, J., et al. (2002). Cannabinoids promote oligodendrocyte progenitor survival: Involvement of cannabinoid receptors and phosphatidylinositol-3 kinase/Akt signaling. *The Journal of Neuroscience, 22*(22), 9742–9753.

Mukhopadhyay, B., Cinar, R., Yin, S., Liu, J., Tam, J., Godlewski, G., et al. (2011). Hyperactivation of anandamide synthesis and regulation of cell-cycle progression via cannabinoid type 1 (CB1) receptors in the regenerating liver. *Proceedings of the National Academy of Sciences of the United States of America, 108*(15), 6323–6328. http://dx.doi.org/10.1073/pnas.1017689108. 1017689108 [pii].

Munro, S., Thomas, K. L., & Abu-Shaar, M. (1993). Molecular characterization of a peripheral receptor for cannabinoids. *Nature, 365*, 61–65.

Murataeva, N., Straiker, A., & Mackie, K. (2014). Parsing the players: 2-Arachidonoylglycerol synthesis and degradation in the CNS. *British Journal of Pharmacology, 171*(6), 1379–1391. http://dx.doi.org/10.1111/bph.12411.

Nasr, M. L., Shi, X., Bowman, A. L., Johnson, M., Zvonok, N., Janero, D. R., et al. (2013). Membrane phospholipid bilayer as a determinant of monoacylglycerol lipase kinetic profile and conformational repertoire. *Protein Science*, *22*(6), 774–787. http://dx.doi.org/10.1002/pro.2257.

Navarrete, M., & Araque, A. (2008). Endocannabinoids mediate neuron-astrocyte communication. *Neuron*, *57*(6), 883–893. http://dx.doi.org/10.1016/j.neuron.2008.01.029. S0896-6273(08)00116-5 [pii].

Navia-Paldanius, D., Aaltonen, N., Lehtonen, M., Savinainen, J. R., Taschler, U., Radner, F. P., et al. (2015). Increased tonic cannabinoid CB1R activity and brain region-specific desensitization of CB1R Gi/o signaling axis in mice with global genetic knockout of monoacylglycerol lipase. *European Journal of Pharmaceutical Sciences*, *77*, 180–188. http://dx.doi.org/10.1016/j.ejps.2015.06.005.

Nicolussi, S., Chicca, A., Rau, M., Rihs, S., Soeberdt, M., Abels, C., et al. (2014). Correlating FAAH and anandamide cellular uptake inhibition using N-alkylcarbamate inhibitors: From ultrapotent to hyperpotent. *Biochemical Pharmacology*, *92*(4), 669–689. http://dx.doi.org/10.1016/j.bcp.2014.09.020.

Nicolussi, S., & Gertsch, J. (2015). Endocannabinoid transport revisited. *Vitamins and Hormones*, *98*, 441–485. http://dx.doi.org/10.1016/bs.vh.2014.12.011.

Nicolussi, S., Viveros-Paredes, J. M., Gachet, M. S., Rau, M., Flores-Soto, M. E., Blunder, M., et al. (2014). Guineensine is a novel inhibitor of endocannabinoid uptake showing cannabimimetic behavioral effects in BALB/c mice. *Pharmacological Research*, *80*, 52–65. http://dx.doi.org/10.1016/j.phrs.2013.12.010.

Nie, J., & Lewis, D. L. (2001). Structural domains of the CB1 cannabinoid receptor that contribute to constitutive activity and G-protein sequestration. *The Journal of Neuroscience*, *21*(22), 8758–8764.

Nomura, D. K., Morrison, B. E., Blankman, J. L., Long, J. Z., Kinsey, S. G., Marcondes, M. C., et al. (2011). Endocannabinoid hydrolysis generates brain prostaglandins that promote neuroinflammation. *Science*, *334*(6057), 809–813. http://dx.doi.org/10.1126/science.1209200.

Nyilas, R., Dudok, B., Urban, G. M., Mackie, K., Watanabe, M., Cravatt, B. F., et al. (2008). Enzymatic machinery for endocannabinoid biosynthesis associated with calcium stores in glutamatergic axon terminals. *The Journal of Neuroscience*, *28*(5), 1058–1063. http://dx.doi.org/10.1523/JNEUROSCI.5102-07.2008. 28/5/1058 [pii].

Oddi, S., Fezza, F., Pasquariello, N., D'Agostino, A., Catanzaro, G., De Simone, C., et al. (2009). Molecular identification of albumin and Hsp70 as cytosolic anandamide-binding proteins. *Chemistry & Biology*, *16*(6), 624–632. http://dx.doi.org/10.1016/j.chembiol.2009.05.004. S1074-5521(09)00151-3 [pii].

Ohno-Shosaku, T., Maejima, T., & Kano, M. (2001). Endogenous cannabinoids mediate retrograde signals from depolarized postsynaptic neurons to presynaptic terminals. *Neuron*, *29*(3), 729–738.

Ohno-Shosaku, T., Tsubokawa, H., Mizushima, I., Yoneda, N., Zimmer, A., & Kano, M. (2002). Presynaptic cannabinoid sensitivity is a major determinant of depolarization-induced retrograde suppression at hippocampal synapses. *The Journal of Neuroscience*, *22*(10), 3864–3872.

Okamoto, Y., Morishita, J., Tsuboi, K., Tonai, T., & Ueda, N. (2004). Molecular characterization of a phospholipase D generating anandamide and its congeners. *The Journal of Biological Chemistry*, *279*(7), 5298–5305.

Ortar, G., Bisogno, T., Ligresti, A., Morera, E., Nalli, M., & Di Marzo, V. (2008). Tetrahydrolipstatin analogues as modulators of endocannabinoid 2-arachidonoylglycerol metabolism. *Journal of Medicinal Chemistry*, *51*, 6970–6979. http://dx.doi.org/10.1021/jm800978m.

Pamplona, F. A., Ferreira, J., Menezes de Lima, O., Jr., Duarte, F. S., Bento, A. F., Forner, S., et al. (2012). Anti-inflammatory lipoxin A4 is an endogenous allosteric enhancer of CB1 cannabinoid receptor. *Proceedings of the National Academy of Sciences of the United States of America, 109*(51), 21134–21139. http://dx.doi.org/10.1073/pnas.1202906109.

Pan, B., Wang, W., Long, J. Z., Sun, D., Hillard, C. J., Cravatt, B. F., et al. (2009). Blockade of 2-arachidonoylglycerol hydrolysis by selective monoacylglycerol lipase inhibitor 4-nitrophenyl 4-(dibenzo[d][1,3]dioxol-5-yl(hydroxy)methyl)piperidine-1-carboxylate (JZL184) enhances retrograde endocannabinoid signaling. *The Journal of Pharmacology and Experimental Therapeutics, 331*(2), 591–597. http://dx.doi.org/10.1124/jpet.109.158162. jpet.109.158162 [pii].

Pan, B., Wang, W., Zhong, P., Blankman, J. L., Cravatt, B. F., & Liu, Q. S. (2011). Alterations of endocannabinoid signaling, synaptic plasticity, learning, and memory in monoacylglycerol lipase knock-out mice. *The Journal of Neuroscience, 31*(38), 13420–13430. http://dx.doi.org/10.1523/JNEUROSCI.2075-11.2011.

Panlilio, L. V., Justinova, Z., & Goldberg, S. R. (2013). Inhibition of FAAH and activation of PPAR: New approaches to the treatment of cognitive dysfunction and drug addiction. *Pharmacology & Therapeutics, 138*(1), 84–102. http://dx.doi.org/10.1016/j.pharmthera.2013.01.003.

Patel, S., Carrier, E. J., Ho, W. S., Rademacher, D. J., Cunningham, S., Reddy, D. S., et al. (2005). The postmortal accumulation of brain N-arachidonylethanolamine (anandamide) is dependent upon fatty acid amide hydrolase activity. *Journal of Lipid Research, 46*(2), 342–349.

Patel, S., & Hillard, C. J. (2009). Endocannabinoids as modulators of synaptic signaling. In P. H. Reggio (Ed.), *The cannabinoid receptors* (pp. 281–308). New York: Humana Press.

Patricelli, M. P., & Cravatt, B. F. (1999). Fatty acid amide hydrolase competitively degrades bioactive amides and esters through a nonconventional catalytic mechanism. *Biochemistry, 38*, 14125–14130.

Pertwee, R. G. (2008). Ligands that target cannabinoid receptors in the brain: From THC to anandamide and beyond. *Addiction Biology, 13*(2), 147–159. http://dx.doi.org/10.1111/j.1369-1600.2008.00108.x. ADB108 [pii].

Pickel, V. M., Shobin, E. T., Lane, D. A., & Mackie, K. (2012). Cannabinoid-1 receptors in the mouse ventral pallidum are targeted to axonal profiles expressing functionally opposed opioid peptides and contacting N-acylphosphatidylethanolamine-hydrolyzing phospholipase D terminals. *Neuroscience, 227C*, 10–21. http://dx.doi.org/10.1016/j.neuroscience.2012.07.050.

Pollandt, S., Liu, J., Orozco-Cabal, L., Grigoriadis, D. E., Vale, W. W., Gallagher, J. P., et al. (2006). Cocaine withdrawal enhances long-term potentiation induced by corticotropin-releasing factor at central amygdala glutamatergic synapses via CRF, NMDA receptors and PKA. *The European Journal of Neuroscience, 24*(6), 1733–1743. http://dx.doi.org/10.1111/j.1460-9568.2006.05049.x.

Porter, A. C., Sauer, J. M., Knierman, M. D., Becker, G. W., Berna, M. J., Bao, J., et al. (2002). Characterization of a novel endocannabinoid, virodhamine, with antagonist activity at the CB1 receptor. *The Journal of Pharmacology and Experimental Therapeutics, 301*(3), 1020–1024.

Pratt, P. F., Hillard, C. J., Edgemond, W. S., & Campbell, W. B. (1998). N-arachidonylethanolamide relaxation of bovine coronary artery is not mediated by CB1 cannabinoid receptor. *The American Journal of Physiology, 274*(1 Pt. 2), H375–H381.

Prescott, S. M., & Majerus, P. W. (1983). Characterization of 1,2-diacylglycerol hydrolysis in human platelets. Demonstration of an arachidonoyl-monoacylglycerol intermediate. *The Journal of Biological Chemistry, 258*(2), 764–769.

Price, M. R., Baillie, G. L., Thomas, A., Stevenson, L. A., Easson, M., Goodwin, R., et al. (2005). Allosteric modulation of the cannabinoid CB1 receptor. *Molecular Pharmacology*, *68*(5), 1484–1495.

Puente, N., Cui, Y., Lassalle, O., Lafourcade, M., Georges, F., Venance, L., et al. (2011). Polymodal activation of the endocannabinoid system in the extended amygdala. *Nature Neuroscience*, *14*(12), 1542–1547. http://dx.doi.org/10.1038/nn.2974.

Puffenbarger, R. A., Kapulina, O., Howell, J. M., & Deutsch, D. G. (2001). Characterization of the 5′-sequence of the mouse fatty acid amide hydrolase. *Neuroscience Letters*, *314*(1–2), 21–24.

Rahman, I. A., Tsuboi, K., Uyama, T., & Ueda, N. (2014). New players in the fatty acyl ethanolamide metabolism. *Pharmacological Research*, *86*, 1–10. http://dx.doi.org/10.1016/j.phrs.2014.04.001.

Reguero, L., Puente, N., Elezgarai, I., Ramos-Uriarte, A., Gerrikagoitia, I., Bueno-Lopez, J. L., et al. (2014). Subcellular localization of NAPE-PLD and DAGL-alpha in the ventromedial nucleus of the hypothalamus by a preembedding immunogold method. *Histochemistry and Cell Biology*, *141*(5), 543–550. http://dx.doi.org/10.1007/s00418-013-1174-x.

Reisenberg, M., Singh, P. K., Williams, G., & Doherty, P. (2012). The diacylglycerol lipases: Structure, regulation and roles in and beyond endocannabinoid signalling. *Philosophical Transactions of the Royal Society of London Series B, Biological Sciences*, *367*(1607), 3264–3275. http://dx.doi.org/10.1098/rstb.2011.0387.

Rinaldi-Carmona, M., Barth, F., Heaulme, M., Shire, D., Calandra, B., Congy, C., et al. (1994). SR141716A, a potent and selective antagonist of the brain cannabinoid receptor. *FEBS Letters*, *350*(2–3), 240–244.

Roloff, A. M., Anderson, G. R., Martemyanov, K. A., & Thayer, S. A. (2010). Homer 1a gates the induction mechanism for endocannabinoid-mediated synaptic plasticity. *The Journal of Neuroscience*, *30*(8), 3072–3081. http://dx.doi.org/10.1523/JNEUROSCI.4603-09.2010. 30/8/3072 [pii].

Ronesi, J., Gerdeman, G. L., & Lovinger, D. M. (2004). Disruption of endocannabinoid release and striatal long-term depression by postsynaptic blockade of endocannabinoid membrane transport. *The Journal of Neuroscience*, *24*(7), 1673–1679.

Rosenberger, T. A., Farooqui, A. A., & Horrocks, L. A. (2007). Bovine brain diacylglycerol lipase: Substrate specificity and activation by cyclic AMP-dependent protein kinase. *Lipids*, *42*(3), 187–195. http://dx.doi.org/10.1007/s11745-007-3019-7.

Ross, R. A., Gibson, T. M., Brockie, H. C., Leslie, M., Pashmi, G., Craib, S. J., et al. (2001). Structure-activity relationship for the endogenous cannabinoid, anandamide, and certain of its analogues at vanilloid receptors in transfected cells and vas deferens. *British Journal of Pharmacology*, *132*(3), 631–640.

Rouzer, C. A., & Marnett, L. J. (2011). Endocannabinoid oxygenation by cyclooxygenases, lipoxygenases, and cytochromes P450: Cross-talk between the eicosanoid and endocannabinoid signaling pathways. *Chemical Reviews*, *111*(10), 5899–5921. http://dx.doi.org/10.1021/cr2002799.

Ruehle, S., Remmers, F., Romo-Parra, H., Massa, F., Wickert, M., Wortge, S., et al. (2013). Cannabinoid CB1 receptor in dorsal telencephalic glutamatergic neurons: Distinctive sufficiency for hippocampus-dependent and amygdala-dependent synaptic and behavioral functions. *The Journal of Neuroscience*, *33*(25), 10264–10277. http://dx.doi.org/10.1523/JNEUROSCI.4171-12.2013.

Saghatelian, A., & Cravatt, B. F. (2005). Discovery metabolite profiling—Forging functional connections between the proteome and metabolome. *Life Sciences*, 77(14), 1759–1766.

Saghatelian, A., McKinney, M. K., Bandell, M., Patapoutian, A., & Cravatt, B. F. (2006). A FAAH-regulated class of N-acyl taurines that activates TRP ion channels. *Biochemistry*, *45*(30), 9007–9015. http://dx.doi.org/10.1021/bi0608008.

Sakurada, T., & Noma, A. (1981). Subcellular localization and some properties of monoacylglycerol lipase in rat adipocytes. *Journal of Biochemistry*, *90*(5), 1413–1419.

Salio, C., Doly, S., Fischer, J., Franzoni, M. F., & Conrath, M. (2002). Neuronal and astrocytic localization of the cannabinoid receptor-1 in the dorsal horn of the rat spinal cord. *Neuroscience Letters*, *329*(1), 13–16.

Savinainen, J. R., Saario, S. M., & Laitinen, J. T. (2012). The serine hydrolases MAGL, ABHD6 and ABHD12 as guardians of 2-arachidonoylglycerol signalling through cannabinoid receptors. *Acta Physiologica*, *204*(2), 267–276. http://dx.doi.org/10.1111/j.1748-1716.2011.02280.x.

Savinainen, J. R., Saario, S. M., Niemi, R., Jarvinen, T., & Laitinen, J. T. (2003). An optimized approach to study endocannabinoid signaling: Evidence against constitutive activity of rat brain adenosine A1 and cannabinoid CB1 receptors. *British Journal of Pharmacology*, *140*(8), 1451–1459.

Schlosburg, J. E., Blankman, J. L., Long, J. Z., Nomura, D. K., Pan, B., Kinsey, S. G., et al. (2010). Chronic monoacylglycerol lipase blockade causes functional antagonism of the endocannabinoid system. *Nature Neuroscience*, *13*(9), 1113–1119. http://dx.doi.org/10.1038/nn.2616. nn.2616 [pii].

Schmid, P. C., Krebsbach, R. J., Perry, S. R., Dettmer, T. M., Maasson, J. L., & Schmid, H. H. (1995). Occurrence and postmortem generation of anandamide and other long-chain N-acylethanolamines in mammalian brain. *FEBS Letters*, *375*(1–2), 117–120.

Schmid, H. H., Schmid, P. C., & Natarajan, V. (1990). N-acylated glycerophospholipids and their derivatives. *Progress in Lipid Research*, *29*(1), 1–43.

Schmid, H. H., Schmid, P. C., & Natarajan, V. (1996). The N-acylation-phosphodiesterase pathway and cell signalling. *Chemistry and Physics of Lipids*, *80*(1–2), 133–142.

Schmid, P. C., Zuzarte-Augustin, M. L., & Schmid, H. H. O. (1985). Properties of rat liver N-acylethanolamine amidohydrolase. *The Journal of Biological Chemistry*, *260*, 14145–14149.

Scotchie, J. G., Savaris, R. F., Martin, C. E., & Young, S. L. (2015). Endocannabinoid regulation in human endometrium across the menstrual cycle. *Reproductive Sciences*, *22*(1), 113–123. http://dx.doi.org/10.1177/1933719114533730.

Shinohara, N., Uyama, T., Jin, X. H., Tsuboi, K., Tonai, T., Houchi, H., et al. (2011). Enzymological analysis of the tumor suppressor A-C1 reveals a novel group of phospholipid-metabolizing enzymes. *Journal of Lipid Research*, *52*(11), 1927–1935. http://dx.doi.org/10.1194/jlr.M015081.

Shonesy, B. C., Wang, X., Rose, K. L., Ramikie, T. S., Cavener, V. S., Rentz, T., et al. (2013). CaMKII regulates diacylglycerol lipase-alpha and striatal endocannabinoid signaling. *Nature Neuroscience*, *16*(4), 456–463. http://dx.doi.org/10.1038/nn.3353.

Simon, G. M., & Cravatt, B. F. (2006). Endocannabinoid biosynthesis proceeding through glycerophospho-N-acyl ethanolamine and a role for {alpha}/beta-hydrolase 4 in this pathway. *The Journal of Biological Chemistry*, *281*(36), 26465–26472.

Simon, G. M., & Cravatt, B. F. (2008). Anandamide biosynthesis catalyzed by the phosphodiesterase GDE1 and detection of glycerophospho-N-acyl ethanolamine precursors in mouse brain. *The Journal of Biological Chemistry*, *283*, 9341–9349. http://dx.doi.org/10.1074/jbc.M707807200. M707807200 [pii].

Simon, G. M., & Cravatt, B. F. (2010). Characterization of mice lacking candidate N-acyl ethanolamine biosynthetic enzymes provides evidence for multiple pathways that contribute to endocannabinoid production in vivo. *Molecular BioSystems*, *6*(8), 1411–1418. http://dx.doi.org/10.1039/c000237b.

Sipe, J. C., Chiang, K., Gerber, A. L., Beutler, E., & Cravatt, B. F. (2002). A missense mutation in human fatty acid amide hydrolase associated with problem drug use. *Proceedings of the National Academy of Sciences of the United States of America*, *99*(12), 8394–8399.

Straiker, A., & Mackie, K. (2007). Metabotropic suppression of excitation in murine autaptic hippocampal neurons. *The Journal of Physiology*, *578*(Pt. 3), 773–785.
Straiker, A., Mitjavila, J., Yin, D., Gibson, A., & Mackie, K. (2015). Aiming for allosterism: Evaluation of allosteric modulators of CB1 in a neuronal model. *Pharmacological Research*, *99*, 370–376. http://dx.doi.org/10.1016/j.phrs.2015.07.017.
Strosznajder, J., Singh, H., & Horrocks, L. A. (1984). Monoacylglycerol lipase. Regulation and increased activity during hypoxia and ischemia. *Neurochemical Pathology*, *2*(2), 139–147.
Suarez, J., Bermudez-Silva, F. J., Mackie, K., Ledent, C., Zimmer, A., Cravatt, B. F., et al. (2008). Immunohistochemical description of the endogenous cannabinoid system in the rat cerebellum and functionally related nuclei. *The Journal of Comparative Neurology*, *509*(4), 400–421. http://dx.doi.org/10.1002/cne.21774.
Sugiura, T., Kondo, S., Sukagawa, A., Nakane, S., Shinoda, A., Itoh, K., et al. (1995). 2-Arachidonoylglycerol: A possible endogenous cannabinoid receptor ligand in brain. *Biochemical and Biophysical Research Communications*, *215*(1), 89–97.
Sugiura, T., Kondo, S., Sukagawa, A., Tonegawa, T., Nakane, S., Yamashita, A., et al. (1996). Transacylase-mediated and phosphodiesterase-mediated synthesis of N-arachidonoylethanolamine, an endogenous cannabinoid-receptor ligand, in rat brain microsomes. Comparison with synthesis from free arachidonic acid and ethanolamine. *European Journal of Biochemistry*, *240*(1), 53–62.
Sumislawski, J. J., Ramikie, T. S., & Patel, S. (2011). Reversible gating of endocannabinoid plasticity in the amygdala by chronic stress: A potential role for monoacylglycerol lipase inhibition in the prevention of stress-induced behavioral adaptation. *Neuropsychopharmacology*, *36*(13), 2750–2761. http://dx.doi.org/10.1038/npp.2011.166.
Sun, Y. X., Tsuboi, K., Okamoto, Y., Tonai, T., Murakami, M., Kudo, I., et al. (2004). Biosynthesis of anandamide and N-palmitoylethanolamine by sequential actions of phospholipase A^2 and lysophospholipase D. *The Biochemical Journal*, *380*(Pt. 3), 749–756.
Tanimura, A., Yamazaki, M., Hashimotodani, Y., Uchigashima, M., Kawata, S., Abe, M., et al. (2010). The endocannabinoid 2-arachidonoylglycerol produced by diacylglycerol lipase alpha mediates retrograde suppression of synaptic transmission. *Neuron*, *65*(3), 320–327. http://dx.doi.org/10.1016/j.neuron.2010.01.021. S0896-6273(10)00047-4 [pii].
Tarzia, G., Duranti, A., Tontini, A., Piersanti, G., Mor, M., Rivara, S., et al. (2003). Design, synthesis, and structure-activity relationships of alkylcarbamic acid aryl esters, a new class of fatty acid amide hydrolase inhibitors. *Journal of Medicinal Chemistry*, *46*(12), 2352–2360.
Tornqvist, H., & Belfrage, P. (1976). Purification and some properties of a monoacylglycerol-hydrolyzing enzyme of rat adipose tissue. *The Journal of Biological Chemistry*, *251*(3), 813–819.
Tsou, K., Nogueron, M. I., Muthian, S., Sanudo-Pena, M. C., Hillard, C. J., Deutsch, D. G., et al. (1998). Fatty acid amide hydrolase is located preferentially in large neurons in the rat central nervous system as revealed by immunohistochemistry. *Neuroscience Letters*, *254*(3), 137–140.
Tsuboi, K., Okamoto, Y., Ikematsu, N., Inoue, M., Shimizu, Y., Uyama, T., et al. (2011). Enzymatic formation of N-acylethanolamines from N-acylethanolamine plasmalogen through N-acylphosphatidylethanolamine-hydrolyzing phospholipase D-dependent and -independent pathways. *Biochimica et Biophysica Acta*, *1811*(10), 565–577. http://dx.doi.org/10.1016/j.bbalip.2011.07.009. S1388-1981(11)00134-X [pii].
Tuccinardi, T., Granchi, C., Rizzolio, F., Caligiuri, I., Battistello, V., Toffoli, G., et al. (2014). Identification and characterization of a new reversible MAGL inhibitor. *Bioorganic & Medicinal Chemistry*, *22*(13), 3285–3291. http://dx.doi.org/10.1016/j.bmc.2014.04.057.

Turu, G., & Hunyady, L. (2010). Signal transduction of the CB1 cannabinoid receptor. *Journal of Molecular Endocrinology*, *44*(2), 75–85. http://dx.doi.org/10.1677/JME-08-0190. JME-08-0190 [pii].

Turu, G., Simon, A., Gyombolai, P., Szidonya, L., Bagdy, G., Lenkei, Z., et al. (2007). The role of diacylglycerol lipase in constitutive and angiotensin AT1 receptor-stimulated cannabinoid CB1 receptor activity. *The Journal of Biological Chemistry*, *282*(11), 7753–7757.

Turu, G., Varnai, P., Gyombolai, P., Szidonya, L., Offertaler, L., Bagdy, G., et al. (2009). Paracrine transactivation of the CB1 cannabinoid receptor by AT1 angiotensin and other Gq/11 protein-coupled receptors. *The Journal of Biological Chemistry*, *284*, 16914–16921. http://dx.doi.org/10.1074/jbc.M109.003681. M109.003681 [pii].

Uchigashima, M., Narushima, M., Fukaya, M., Katona, I., Kano, M., & Watanabe, M. (2007). Subcellular arrangement of molecules for 2-arachidonoyl-glycerol-mediated retrograde signaling and its physiological contribution to synaptic modulation in the striatum. *The Journal of Neuroscience*, *27*(14), 3663–3676.

Ueda, N., Tsuboi, K., & Uyama, T. (2010). Enzymological studies on the biosynthesis of N-acylethanolamines. *Biochimica et Biophysica Acta*, *1801*(12), 1274–1285. http://dx.doi.org/10.1016/j.bbalip.2010.08.010. S1388-1981(10)00188-5 [pii].

Ueda, N., Yamanaka, K., & Yamamoto, S. (2001). Purification and characterization of an acid amidase selective for *N*-palmitoylethanolamine, a putative endogenous anti-inflammatory substance. *The Journal of Biological Chemistry*, *276*(38), 35552–35557.

Urquhart, P., Nicolaou, A., & Woodward, D. F. (2015). Endocannabinoids and their oxygenation by cyclo-oxygenases, lipoxygenases and other oxygenases. *Biochimica et Biophysica Acta*, *1851*(4), 366–376. http://dx.doi.org/10.1016/j.bbalip.2014.12.015.

Uyama, T., Inoue, M., Okamoto, Y., Shinohara, N., Tai, T., Tsuboi, K., et al. (2013). Involvement of phospholipase A/acyltransferase-1 in N-acylphosphatidylethanolamine generation. *Biochimica et Biophysica Acta*, *1831*(12), 1690–1701. http://dx.doi.org/10.1016/j.bbalip.2013.08.017.

Vallee, M., Vitiello, S., Bellocchio, L., Hebert-Chatelain, E., Monlezun, S., Martin-Garcia,-E., et al. (2014). Pregnenolone can protect the brain from cannabis intoxication. *Science*, *343*(6166), 94–98. http://dx.doi.org/10.1126/science.1243985.

Van Sickle, M. D., Duncan, M., Kingsley, P. J., Mouihate, A., Urbani, P., Mackie, K., et al. (2005). Identification and functional characterization of brainstem cannabinoid CB2 receptors. *Science*, *310*(5746), 329–332.

Vandevoorde, S., & Fowler, C. J. (2005). Inhibition of fatty acid amide hydrolase and monoacylglycerol lipase by the anandamide uptake inhibitor VDM11: Evidence that VDM11 acts as an FAAH substrate. *British Journal of Pharmacology*, *145*(7), 885–893.

Vellani, V., Petrosino, S., De Petrocellis, L., Valenti, M., Prandini, M., Magherini, P. C., et al. (2008). Functional lipidomics. Calcium-independent activation of endocannabinoid/endovanilloid lipid signalling in sensory neurons by protein kinases C and A and thrombin. *Neuropharmacology*, *55*, 1274–1279. http://dx.doi.org/10.1016/j.neuropharm.2008.01.010. S0028-3908(08)00034-8 [pii].

Vermeer, L. M., Gregory, E., Winter, M. K., McCarson, K. E., & Berman, N. E. (2014). Exposure to bisphenol A exacerbates migraine-like behaviors in a multibehavior model of rat migraine. *Toxicological Sciences*, *137*(2), 416–427. http://dx.doi.org/10.1093/toxsci/kft245.

Viader, A., Blankman, J. L., Zhong, P., Liu, X., Schlosburg, J. E., Joslyn, C. M., et al. (2015). Metabolic interplay between astrocytes and neurons regulates endocannabinoid action. *Cell Reports*, *12*(5), 798–808. http://dx.doi.org/10.1016/j.celrep.2015.06.075.

Vyvoda, O. S., & Rowe, C. E. (1973). Glyceride lipases in nerve endings of guinea-pig brain and their stimulation by noradrenaline, 5-hydroxytryptamine and adrenaline. *The Biochemical Journal*, *132*(2), 233–248.

Waleh, N. S., Cravatt, B. F., Apte-Deshpande, A., Terao, A., & Kilduff, T. S. (2002). Transcriptional regulation of the mouse fatty acid amide hydrolase gene. *Gene*, *291*(1–2), 203–210.

Wang, X., Horswill, J. G., Whalley, B. J., & Stephens, G. J. (2011). Effects of the allosteric antagonist 1-(4-chlorophenyl)-3-[3-(6-pyrrolidin-1-ylpyridin-2-yl)phenyl]urea (PSNCBAM-1) on CB1 receptor modulation in the cerebellum. *Molecular Pharmacology*, *79*(4), 758–767. http://dx.doi.org/10.1124/mol.110.068197. mol.110.068197 [pii].

Wang, J., Okamoto, Y., Morishita, J., Tsuboi, K., Miyatake, A., & Ueda, N. (2006). Functional analysis of the purified anandamide-generating phospholipase D as a member of the metallo-beta-lactamase family. *The Journal of Biological Chemistry*, *281*(18), 12325–12335. http://dx.doi.org/10.1074/jbc.M512359200.

Wilson, R. I., & Nicoll, R. A. (2001). Endogenous cannabinoids mediate retrograde signalling at hippocampal synapses. *Nature*, *410*(6828), 588–592.

Wiskerke, J., Irimia, C., Cravatt, B. F., De Vries, T. J., Schoffelmeer, A. N., Pattij, T., et al. (2012). Characterization of the effects of reuptake and hydrolysis inhibition on interstitial endocannabinoid levels in the brain: An in vivo microdialysis study. *ACS Chemical Neuroscience*, *3*(5), 407–417. http://dx.doi.org/10.1021/cn300036b.

Wolfson, M. L., Aisemberg, J., Salazar, A. I., Dominguez Rubio, A. P., Vercelli, C. A., & Franchi, A. M. (2013). Progesterone reverts LPS-reduced FAAH activity in murine peripheral blood mononuclear cells by a receptor-mediated fashion. *Molecular and Cellular Endocrinology*, *381*(1–2), 97–105. http://dx.doi.org/10.1016/j.mce.2013.07.020.

Won, Y. J., Puhl, H. L., 3rd., & Ikeda, S. R. (2009). Molecular reconstruction of mGluR5a-mediated endocannabinoid signaling cascade in single rat sympathetic neurons. *The Journal of Neuroscience*, *29*(43), 13603–13612. http://dx.doi.org/10.1523/JNEUROSCI.2244-09.2009. 29/43/13603 [pii].

Woodhams, S. G., Wong, A., Barrett, D. A., Bennett, A. J., Chapman, V., & Alexander, S. P. (2012). Spinal administration of the monoacylglycerol lipase inhibitor JZL184 produces robust inhibitory effects on nociceptive processing and the development of central sensitization in the rat. *British Journal of Pharmacology*, *167*(8), 1609–1619. http://dx.doi.org/10.1111/j.1476-5381.2012.02179.x.

Yang, W., Di Vizio, D., Kirchner, M., Steen, H., & Freeman, M. R. (2010). Proteome scale characterization of human S-acylated proteins in lipid raft-enriched and non-raft membranes. *Molecular & Cellular Proteomics: MCP*, *9*(1), 54–70. http://dx.doi.org/10.1074/mcp.M800448-MCP200.

Yoshida, T., Fukaya, M., Uchigashima, M., Miura, E., Kamiya, H., Kano, M., et al. (2006). Localization of diacylglycerol lipase-alpha around postsynaptic spine suggests close proximity between production site of an endocannabinoid, 2-arachidonoyl-glycerol, and presynaptic cannabinoid CB1 receptor. *The Journal of Neuroscience*, *26*(18), 4740–4751.

Yoshino, H., Miyamae, T., Hansen, G., Zambrowicz, B., Flynn, M., Pedicord, D., et al. (2011). Postsynaptic diacylglycerol lipase alpha mediates retrograde endocannabinoid suppression of inhibition in mouse prefrontal cortex. *The Journal of Physiology*, *589*(Pt. 20), 4857–4884. http://dx.doi.org/10.1113/jphysiol.2011.212225. jphysiol.2011.212225 [pii].

Zhang, H. Y., Gao, M., Liu, Q. R., Bi, G. H., Li, X., Yang, H. J., et al. (2014). Cannabinoid CB2 receptors modulate midbrain dopamine neuronal activity and dopamine-related behavior in mice. *Proceedings of the National Academy of Sciences of the United States of America*, *111*(46), E5007–E5015. http://dx.doi.org/10.1073/pnas.1413210111.

Zhang, L., Wang, M., Bisogno, T., Di Marzo, V., & Alger, B. E. (2011). Endocannabinoids generated by Ca^{2+} or by metabotropic glutamate receptors appear to arise from different pools of diacylglycerol lipase. *PLoS One*, *6*(1), e16305. http://dx.doi.org/10.1371/journal.pone.0016305.

Zhao, L., & Levine, E. S. (2014). BDNF-endocannabinoid interactions at neocortical inhibitory synapses require phospholipase C signaling. *Journal of Neurophysiology*, *111*(5), 1008–1015. http://dx.doi.org/10.1152/jn.00554.2013.

Zhu, C., Solorzano, C., Sahar, S., Realini, N., Fung, E., Sassone-Corsi, P., et al. (2011). Proinflammatory stimuli control N-acylphosphatidylethanolamine-specific phospholipase D expression in macrophages. *Molecular Pharmacology*, *79*(4), 786–792. http://dx.doi.org/10.1124/mol.110.070201.

CHAPTER TWO

Evidence for a Role of Adolescent Endocannabinoid Signaling in Regulating HPA Axis Stress Responsivity and Emotional Behavior Development

Tiffany T.-Y. Lee, Boris B. Gorzalka[1]
Department of Psychology, University of British Columbia, Vancouver, British Columbia, Canada
[1]Corresponding author: e-mail address: bgorzalka@psych.ubc.ca

Contents

1. Adolescence and Pubertal Maturation 52
2. The Neurobiology of Stress 53
3. Adolescent HPA Axis Development 55
4. The Developmental Influence of Gonadal Hormones on the HPA Axis 58
5. The Endocannabinoid System 60
6. Endocannabinoid System Regulation of the HPA Axis in Adulthood 61
7. Ontogeny of the Endocannabinoid System 64
8. Adolescent HPA Axis Function: A Regulatory Role for the Endocannabinoid System? 68
   8.1 Adolescent Stress Exposure Modulates Corticolimbic Endocannabinoid Signaling 68
   8.2 Adolescent Cannabinoid Exposure Modifies Adult Stress-Induced HPA Axis Activity 69
   8.3 Long-Term Consequences of Adolescent Endocannabinoid System Dysregulation on the Developing Brain and Behavior 71
9. Concluding Remarks 74
Acknowledgments 75
References 75

Abstract

Adolescence is a period characterized by many distinct physical, behavioral, and neural changes during the transition from child- to adulthood. In particular, adolescent neural changes often confer greater plasticity and flexibility, yet with this comes the potential for heightened vulnerability to external perturbations such as stress exposure or recreational drug use. There is substantial evidence to suggest that factors such as adolescent stress exposure have longer lasting and sometimes more deleterious effects on an organism than stress exposure during adulthood. Moreover, the adolescent

International Review of Neurobiology, Volume 125
ISSN 0074-7742
http://dx.doi.org/10.1016/bs.irn.2015.09.002

neuroendocrine response to stress exposure is different from that of adults, suggesting that further maturation of the adolescent hypothalamic–pituitary–adrenal (HPA) axis is required. The endocannabinoid (eCB) system is a potential candidate underlying these age-dependent differences given that it is an important regulator of the adult HPA axis and neuronal development. Therefore, this review will focus on (1) the functionality of the adolescent HPA axis, (2) eCB regulation of the adult HPA axis, (3) dynamic changes in eCB signaling during the adolescent period, (4) the effects of adolescent stress exposure on the eCB system, and (5) modulation of HPA axis activity and emotional behavior by adolescent cannabinoid treatment. Collectively, the emerging picture suggests that the eCB system mediates interactions between HPA axis stress responsivity, emotionality, and maturational stage. These findings may be particularly relevant to our understanding of the development of affective disorders and the risks of adolescent cannabis consumption on emotional health and stress responsivity.

Adolescence is often referred to as a "perfect storm," given the numerous physical, neural, and behavioral changes occurring simultaneously. Behaviorally, the adolescent is somewhat erratic with increased intensity of emotional states, oppositional attitudes, and increased risk-taking behavior coupled to a limited ability to engage in self-control in order to override these emotions and behaviors (Casey, Jones, & Somerville, 2011; Walker, Sabuwalla, & Huot, 2004). These behaviors eventually decline and stabilize with the onset of young adulthood (Walker et al., 2004). However, in some cases, behavioral or adjustment problems are precipitated during the onset of puberty and lead to psychiatric disorders that persist in adulthood (Kim-Cohen et al., 2003; Pine, Cohen, Gurley, Brook, & Ma, 1998; Walker et al., 2004). In fact, more than 75% of adults with a fear/anxiety-related disorder met diagnostic criteria as children or adolescents (Kim-Cohen et al., 2003). Similarly, estimates indicate that one in five adolescents have a mental illness that will persist into adulthood (Paus, Keshavan, & Giedd, 2008), and depression and anxiety disorders occur in as many as one in 10 adolescents (Costello, Egger, & Angold, 2005; Kessler et al., 2005). Together, these findings support a sobering reality that the prevalence of mental illness, particularly emotional, and anxiety disorders is specifically heightened in adolescence.

Moreover, certain aspects of adolescent development appear nonlinear and unique relative to younger children and adults. Underlying these distinctive adolescent behavioral characteristics, numerous maturational processes such as synaptic pruning, reduction of gray matter, increased myelination, and strengthening of connectivity between neural structures, occur throughout the brain in a time-dependent and region-specific manner

(Crews, He, & Hodge, 2007; Guerry & Hastings, 2011). For example, human imaging studies indicate that adolescent neurodevelopment involves maturation of motor and sensory cortices earlier than association areas that govern higher order cognition and behavior (Gogtay et al., 2004). It has been suggested that a transient developmental imbalance occurs between the structural and functional maturity of neural circuitry involved in emotional- and reward-based behavior and circuitry governing cognition and impulse control, giving rise to adolescent-specific characteristics such as heightened emotional reactivity and increased risk-taking behavior (Somerville, Jones, & Casey, 2010). Moreover, it has been hypothesized that exacerbation of this transient structural and functional imbalance by various factors, such as stress exposure, contributes to risk for the development of mental illness (Lee, Heimer, et al., 2014). However, the exact mechanisms of action and conditions by which this kind of exacerbation can occur remain to be determined.

The hypothalamic–pituitary–adrenal (HPA) axis provides the body's major neuroendocrine response to stress exposure, with activity rising to promote processes (e.g., glucose mobilization) that meet the energy demands associated with resolving the imminent perceived or real threat (e.g., running away from the threat). In the short term, the HPA axis stress response is adaptive; however, studies examining the impact of chronic stress exposure indicate that prolonged glucocorticoid exposure is associated with a number of detrimental consequences ranging from cardiac disease to depression (Chrousos, 2009). Moreover, dysfunction of the adult HPA axis has been associated with the same emotional disorders that are heightened in adolescence, such as depression and anxiety disorders (Miller, Chen, & Zhou, 2007). Numerous studies indicate that maturation of the HPA axis also occurs during childhood/early life and adolescence before reaching the full functionality observed in adulthood (see review, Eiland & Romeo, 2013). Moreover, processes devoted to pubertal maturation coincide with HPA axis maturation and sharp elevations in circulating gonadal hormones are believed to modulate subsequent neural and HPA axis development (e.g., Sisk & Zehr, 2005). Deviations from the normative developmental trajectory induced by disruptions such as stress exposure appear to elicit profound long-term deficits or impairments to adult HPA axis stress reactivity and emotional behavior, providing confirmation that early life and adolescence are windows of susceptibility to stress.

There are separate but compelling lines of evidence supporting the idea that the endocannabinoid (eCB) system regulates both neurodevelopmental processes and adult HPA axis stress responsivity and emotional behavior. Moreover, there is evidence reviewed in this chapter, suggesting that corticolimbic eCB signaling is an important mediator of interactions between HPA axis stress responsivity and maturational stage (e.g., adolescence). Therefore, this chapter will provide a review of (1) the functionality of the adolescent HPA axis, (2) eCB regulation of the adult HPA axis, (3) dynamic changes in eCB signaling during the adolescent period, (4) the effects of adolescent stress exposure on the eCB system, and (5) modulation of HPA axis activity and emotional behavior by adolescent cannabinoid treatment. Taken together, the emerging picture suggests that the eCB system can mediate interactions between HPA axis stress responsivity, emotionality, and maturational stage and thus could also provide a substrate by which stress or cannabinoid exposure can exert its often detrimental effects.

1. ADOLESCENCE AND PUBERTAL MATURATION

Puberty is a key developmental process that occurs during the adolescent period and includes behavioral and gonadal maturation that allows transition of an organism from a nonreproductive to a reproductive state (Sisk & Foster, 2004). One of the biological signatures of pubertal maturation is the dramatic rise in circulating gonadal hormones that are not only necessary for achieving reproductive maturation but also trigger secondary sex characteristics such as breast development or facial hair growth. The developing adolescent brain is also exquisitely sensitive to elevated levels of gonadal hormones, particularly in the remodeling of cortical and limbic circuits (Guerry & Hastings, 2011; Johnson, Blum, & Giedd, 2009; Juraska, Sisk, & DonCarlos, 2013; Rice & Barone, 2000) implicated in adult cognition and behavior. However, the pubertal rise in gonadal hormones and the subsequent timing of maturational neural processes does vary by individual, thus making the developmental trajectory of the adolescent brain unique to each individual (Sisk & Zehr, 2005). For example, age of pubertal onset varies among individuals and creates variability in the point at which gonadal hormones act on the brain to modulate the developmental trajectory of neural and behavioral maturation (Sisk & Zehr, 2005). Timing of pubertal onset is tied to an increased frequency of gonadotropin-releasing hormone (GnRH) pulse secretions generated by the median eminence of the hypothalamus. In order to support full gonadal and reproductive functioning, GnRH pulses must have relatively short interpulse intervals, but the exact mechanisms triggering increases in GnRH pulse secretions from a quiescent state to initiate pubertal maturation remain to be determined (see review Sisk & Zehr, 2005). The available literature suggests that these permissive signals vary by species and sex but predominantly relate to energy balance (Sisk & Zehr, 2005). Although the terms, "puberty" and "adolescence" have often been used interchangeably, they do have distinct definitions. "Puberty" refers specifically to developmental processes of the hypothalamic–pituitary–gonadal axis that ensure gonadal maturation, whereas "adolescence" may include pubertal processes as well as maturation of neural, cognitive, social, and emotional processes (Sisk & Zehr, 2005).

Defining an individual's maturational stage from infancy to adulthood remains somewhat difficult, particularly when the criteria accompanying each developmental category are not always agreed upon and individuals mature at different rates (McCormick, Mathews, Thomas, & Waters, 2010). Moreover, considering maturational stage in the context of species comparisons can further complicate one's understanding of developmental findings. Many developmental rodent studies adhere to specific age ranges outlined in excellent reviews by Spear (2000) and Eiland and Romeo (2013) as well as McCormick et al. (2010). In the rodent literature, "juvenile" refers to the period spanning from the age of weaning (generally postnatal day (PND) 21) to early adolescence or prepuberty (Eiland & Romeo, 2013). Rodent "adolescence" is conservatively regarded as PNDs 28–42, although this period has been reported by some authors to extend until PNDs 55–60 (Eiland & Romeo, 2013; Spear, 2000). Moreover, age of pubertal onset is often used as an indicator of adolescence (Spear, 2000). In the male rodent, pubertal onset generally occurs around PND 42 and can be defined by balano-preputial separation in which the prepuce exhibits complete retraction from the head of the penis (Lewis, Barnett, Freshwater, Hoberman, & Christian, 2002). As in human females, pubertal onset occurs earlier in female rodents with vaginal opening in the rat appearing around PND 30 (Lewis et al., 2002). Although age ranges will vary by individual and strain of rodent, the current work adheres to the terminology described in previous work (i.e., McCormick et al., 2010; Spear, 2000) for consistency.

2. THE NEUROBIOLOGY OF STRESS

Stress is considered a state of strain elicited by a real or perceived threat to homeostatic functioning. The HPA axis is the major neuroendocrine axis responsible for the maintenance of homeostatic functioning in the face of stress exposure (Ulrich-Lai & Herman, 2009) by promoting glucose mobilization from muscle and liver to enhance cardiac function, and inhibit growth, reproductive, and immune responses in an effort to divert all energy stores toward dealing with the threat at hand (Tasker & Herman, 2011; Ulrich-Lai & Herman, 2009). The typical adult neuroendocrine HPA axis response to acute stress exposure involves hypothalamic activation of corticotropin-releasing hormone (CRH) neurosecretory cells in the paraventricular nucleus (PVN), with CRH stimulating anterior pituitary gland secretion of adrenocorticotropin-releasing hormone (ACTH) into the circulatory system. Detection of ACTH in the blood stimulates the adrenal cortex to release glucocorticoids, primarily corticosterone (CORT) in rodents and cortisol in humans. Cessation of this neuroendocrine stress response occurs via negative feedback with glucocorticoids crossing the blood–brain barrier and binding to glucocorticoid receptors (GR) and mineralocorticoid receptors found in the PVN and extrahypothalamic sites such as the amygdala, PFC, and hippocampus.

The brain also plays an important governing role in the regulation of the HPA axis based in part on the stressor type (Herman et al., 2003). Corticolimbic structures of the forebrain assist in regulating psychological stress-induced HPA axis responses; the PFC, amygdala, and hippocampus, provide input to the hypothalamus mostly via the bed nucleus of the stria terminalis (McEwen, 1992, 2005; Romeo & McEwen, 2006). The medial PFC is subdivided into two regions with opposing contributions to the stress response: the prelimbic region contributes to inhibition and termination of HPA axis and autonomic responses to psychological stressors, while the infralimbic region of the medial PFC is associated with initiating autonomic and HPA axis responses (Figueiredo, Bruestle, Bodie, Dolgas, & Herman, 2003; Radley, Arias, & Sawchenko, 2006). In contrast, stimulation of the hippocampus has an inhibitory effect on HPA axis activity (e.g., decreased glucocorticoid secretion), and lesion studies have demonstrated that the hippocampus contributes to termination of the stress response (Ulrich-Lai & Herman, 2009). The amygdala is also functionally heterogeneous with numerous downstream targets that modulate the HPA axis and autonomic system. The central nucleus of the amygdala is activated by physiological stressors and the integration of autonomic components to psychological stressors, whereas the medial and basolateral amygdalar nuclei (BLA) are activated by psychological stressors and do not appear to have a role in regulating autonomic responses (Ulrich-Lai & Herman, 2009).

The brain plays an integral role in terminating the stress response by promoting feedback inhibition of the HPA axis and restoring homeostasis. Negative feedback regulation is critical to normative HPA axis functioning since it prevents hypothalamic and pituitary stress hormone depletion, allowing for the mounting of successive stress responses, and prevents prolonged exposure to high glucocorticoid levels (Sapolsky, Krey, & McEwen, 1984). Furthermore, under conditions of high-circulating glucocorticoid levels such as those of chronic stress exposure, vasopressin (AVP), which potentiates the stimulatory effects of CRH on ACTH secretion, also contributes to maintaining corticotroph responsiveness (Aguilera & Rabadan-Diehl, 2000). While it is believed that in the short term, activation of the neuroendocrine stress response is adaptive and beneficial, chronically elevated levels of glucocorticoid hormones, whether resulting from impaired negative feedback or chronic stress exposure, can lead to many long-term harmful consequences (Chrousos, 2009). Chronic stress exposure can lead to pathological disorders ranging from diabetes and cardiac failure to depression and anxiety disorders (McEwen, 2007). In addition to HPA axis negative feedback mechanisms that likely act to combat these detrimental conditions from developing, an adaptive form of HPA axis plasticity emerges in response to repeated exposure to the same stressor, termed the habituation response. In this case, the organism learns that the stressor is no longer a threat to survival with repeated exposure, resulting in a decrease in activation of neural structures in the corticolimbic circuit as well as an inhibition of glucocorticoid synthesis and release (Grissom & Bhatnagar, 2009; Jaferi & Bhatnagar, 2006).

3. ADOLESCENT HPA AXIS DEVELOPMENT

The rat neonate has been reported to experience a "stress hyporesponsive period," whereby the HPA axis appears to exhibit comparatively low-basal CORT levels (PNDs 4–14) and stress exposure results in little or no increase in those CORT levels (Schoenfeld, Leadhem, & Rabii, 1980). Previous work demonstrates that maternal sensory stimulation during nursing and grooming is responsible for maintaining these relatively low CORT levels in pups (Stanton & Levine, 1990; Sullivan & Holman, 2011). Following the stress hyporesponsive period, the juvenile/adolescent HPA axis maintains basal glucocorticoid levels comparable to those of adults and the ability to launch a neuroendocrine response to stress exposure (Pignatelli, Xiao, Gouveia, Ferreira, & Vinson, 2006;

Romeo, 2010b; Romeo et al., 2006; Romeo & McEwen, 2006). However, numerous studies demonstrate unique stress-induced HPA axis functioning profiles between prepubertal/adolescent and adult rodents. Acute restraint stress (Doremus-Fitzwater, Varlinskaya, & Spear, 2009; Romeo, Lee, Chhua, McPherson, & McEwen, 2004; Romeo, Lee, & McEwen, 2004), intermittent exposure to foot shock (Goldman, Winget, Hollingshead, & Levine, 1973), or ether vapors (Vázquez & Akil, 1993) elicit comparable basal and peak CORT levels in juvenile/adolescent and adult rats. However, roughly twice as much time is required for CORT and ACTH to recover back to basal levels relative to that of adults under the same stress exposure. This effect is observed in both sexes exposed to acute restraint, with females exhibiting higher basal and stress-induced increases in CORT levels relative to males (Doremus-Fitzwater et al., 2009; Romeo, 2010a, 2010b). In male rats, adult-like stress responses develop between PNDs 30–40 and 40–50 for CORT and ACTH, respectively (Foilb, Lui, & Romeo, 2011).

Foundational work by Vázquez and colleagues (see review Vázquez, 1998) indicate that chronic intermittent stress exposure in both the adult and juvenile rodent (PND 25) yields elevated CORT levels; however, the way in which this is achieved is vastly different. Whereas adults increase the secretion of ACTH and reduce negative feedback at the pituitary gland to stimulate elevations in CORT, the developing rat increases pro-opiomelanocortin (the precursor of ACTH) processing to stimulate only a nominal increase in ACTH; however, the combination of reduced responsiveness to negative feedback and a more sensitive adrenal cortex renders the prepubertal animal capable of responding to even small amounts of ACTH, thus increasing CORT levels. At the same time, this combination also results in longer exposure to circulating glucocorticoids, while the HPA axis, itself, is undergoing maturation (Vázquez, 1998). Using acute restraint stress, Romeo et al. (2014) complemented the work of Vázquez by demonstrating that adrenal expression of the ACTH receptor, the melanocortin 2 receptor, and the expression of melanocortin receptor accessory protein, which chaperones this receptor to the cell surface, is greater in prepubertal rats relative to adults. Moreover, exogenous ACTH administration results in higher CORT levels at lower doses of ACTH in prepubertal animals 60 min following injection, indicating that the protracted prepubertal stress response is at least partly due to greater adrenal sensitivity to ACTH and also suggests that prolonged exposure to ACTH, itself, leads to greater CORT responsivity (Romeo et al., 2014).

Further evidence of age-dependent stress reactivity has been observed using a repeated restraint stress paradigm (7 consecutive days of 30 min restraint) in which adult rats exhibit habituated (blunted) peak CORT levels, whereas adolescent rats exhibit high peak CORT levels during the final restraint session, yet display accelerated recovery to basal levels (Doremus-Fitzwater et al., 2009; Romeo, Lee, Chhua, et al., 2004; Romeo, Lee, & McEwen, 2004). These hormonal observations are associated with region-specific patterns of neuronal activation in corticolimbic circuitry and increased activation of CRH neurons in the PVN of prepubertal, but not adult rats (Romeo et al., 2006). However, somewhat surprisingly, these age-dependent effects are not dependent on differences in hippocampal glucocorticoid mRNA or protein expression (Dziedzic, Ho, Adabi, Foilb, & Romeo, 2014; Romeo et al., 2008). Using another stress model, chronic variable stress exposure during early-, mid-, and late-adolescence periods as well as adulthood revealed age-dependent effects on HPA axis function and emotionality. Whereas the rats exposed to chronic variable stress in late adolescence exhibited elevated basal CORT and oxytocin levels, only the adult chronic variable stress group displayed an increase in stress-coping behavior in the forced swim test (Jankord et al., 2011). Collectively, these data generally indicate that adolescent rodents produce a protracted HPA axis response to acute stress and a failure to exhibit HPA axis habituation with repeated homotypic stress, both situations resulting in greater glucocorticoid exposure relative to that of the adult rodent. Although, somewhat paradoxically, when an ecologically relevant resident-intruder model of social stress is employed for 7 consecutive days, basal- and stress-induced CORT levels in adolescent (PND 28) rats do not differ from those of adults on the first day; however, by the final day of this paradigm, adults continue to display comparable CORT levels to day 1, whereas adolescent groups exhibited habituated CORT levels (Bingham et al., 2011).

These observations that adolescent stress exposure results in extended exposure to glucocorticoids have led to the theory that adolescence represents a period of heightened vulnerability to the negative effects of stress exposure. Indeed, there is evidence that adolescent stress exposure experience has long term, negative effects on subsequent stress reactivity, and corticolimbic neurocircuitry. Rats housed in social isolation (PNDs 30–50) were shown to exhibit long-term sex- and region-dependent consequences. There was an increased stress response (i.e., elevated CORT and hypothalamic AVP mRNA) to acute and repeated restraint as well as increased hippocampal brain-derived neurotrophic factor mRNA in females exposed to adolescent social isolation, whereas males in the same condition exhibited a reduced CORT response to acute restraint, reduced orexin (an arousal and attention promoting neuropeptide) mRNA, and reduced anxiety-like behavior in the elevated plus maze (Weintraub, Singaravelu, & Bhatnagar, 2010). Adolescent chronic variable stress exposure also produces decreases in hippocampal volume, downregulates basal GR expression, elicits a protracted CORT response to a 15 min session in the open arm of an elevated plus maze, as well as deficits in spatial learning in the Morris water maze in adulthood (Isgor, Kabbaj, Akil, & Watson, 2004).

Together, the literature indicates that adolescent HPA axis stress responsivity functions distinctively from that of adults. Moreover, it is currently thought that these age-dependent differences in HPA axis functionality contribute to greater sensitivity to stress exposure, both in the immediate and long term, in adolescents compared to adults. However, potential central mechanisms subserving age-dependent differences in negative feedback remain to be determined.

4. THE DEVELOPMENTAL INFLUENCE OF GONADAL HORMONES ON THE HPA AXIS

While pubertal onset is considered a key developmental event signaling maturation, the HPA axis also undergoes maturation during early adolescence as demonstrated by adrenarche (i.e., increased production and secretion of adrenal steroids). This process precedes increasing pulses of GnRH that stimulate luteinizing and follicle-stimulating hormone release from the pituitary gland to contribute to increases in gonadal steroid production. The sharp and rapid elevations in gonadal hormones facilitate reproductive maturation, physical growth, and development of secondary sex characteristics in the adolescent. Rising gonadal hormone levels also facilitate organizational and activational effects on the developing adolescent brain to influence the HPA axis, likely contributing to the emergence of sex differences in HPA axis functionality.

From what is known, increasing cortisol reactivity to stressors occurs in parallel with pubertal onset and similar to sex differences in the rodent, human females typically begin gonadarche (i.e., the early stages of puberty) 1–2 years earlier than males (Marshall & Tanner, 1969, 1970). Interestingly, stress-induced increases in cortisol are known to decrease as males progress through puberty (Di Luigi et al., 2006). As such, differences in age of pubertal onset may also contribute to differing developmental trajectories of the HPA axis and ultimately facilitate sex differences in HPA stress responsivity. This is consistent with neuroanatomical maturational processes occurring in adolescents. Cortical refinement is initiated earlier in females, sex differences in brain volume emerge, and existing sex differences in limbic volumes during childhood/early life are further amplified during adolescent development (see review Ordaz & Luna, 2012).

While it is generally accepted that sex differences in HPA axis activity begin to emerge with pubertal onset, there is a considerable gap in our understanding of the role of gonadal hormones on adolescent HPA axis activity development. The existing literature suggests that pubertal maturation, which is associated with rising levels of androgens, estradiol, and other estrogens, sets in motion maturational processes that contribute to the development of adult HPA axis stress responsivity. In support of this, when exposed to acute restraint stress, gonadectomized prepubertal males treated with testosterone continue to exhibit a protracted CORT response to acute restraint stress exposure (Romeo, Lee, Chhua, et al., 2004), suggesting that the effects of testosterone on HPA axis activity are functional only after pubertal onset (Romeo, Lee, Chhua, et al., 2004). Similarly, prepubertal females treated with estradiol and exposed to acute restraint 5 days later display reduced ACTH and CORT responses to restraint; however, adult females that were ovariectomized pre- or post-puberty and then given the same estradiol treatment display increased CORT levels to the same stressor, suggesting that estradiol differentially affects female HPA axis stress responsivity based on pubertal status (Evuarherhe, Leggett, Waite, Kershaw, & Lightman, 2009). Moreover, ovariectomized prepubertal female rats produce a comparable protracted stress-induced CORT response as intact prepubertal female rats, indicating that the protracted nature of the prepubertal stress response is independent of the influence of ovarian hormones (Romeo, Lee, & McEwen, 2004). Pubertal maturation appears to be a key developmental process contributing to HPA axis development; however, given the complexity of activational and organizational influences of sex hormones, the numerous brain regions involved, and the ever-changing nature of sex differences across development, a greater understanding of the role of sex steroids and their underlying mechanisms in the development of the HPA axis is warranted.

5. THE ENDOCANNABINOID SYSTEM

Cannabis consumption has been documented over centuries of human civilization, commonly used for its therapeutic, mood-enhancing, and stress-relieving properties. The main psychoactive component of cannabis, Δ-9-tetrahydrocannabinol (THC), was first discovered and characterized in 1964 (Gaoni & Mechoulam, 1964) and subsequently gave rise to the discovery of the endogenous cannabinoid system. This, in turn, has sparked an entire field of research devoted to understanding a myriad of cognitive, behavioral, emotional, and physiological processes under the regulation of the eCB system. This system includes two inhibitory G protein-coupled receptors, the CB_1 and CB_2 receptors. CB_1 receptors (CB_1Rs) are widely expressed in the brain. CB_2 receptors have been detected in the central nervous system relatively recently (Onaivi et al., 2006; Van Sickle et al., 2005), although the majority of these reports localize their presence in microglia rather than neurons (e.g., Cabral, Raborn, Griffin, Dennis, & Marciano-Cabral, 2008). CB_1Rs are found presynaptically on a number of neuronal populations including glutamate, gamma-aminobutyric acid (GABA), and monoamines (Haring, Guggenhuber, & Lutz, 2012). Moreover, they are widely expressed throughout the brain, with high densities in the hippocampus and neocortex, moderate densities within the amygdala and a relatively lower distribution in the hypothalamus (Herkenham et al., 1991). The wide distribution of CB_1Rs across different neuronal subtypes and structures of the brain make them ideally situated to exert regulation over a variety of cognitive, behavioral, and physiological processes including HPA axis stress responsivity and emotionality.

The eCB system also possesses two major endogenous ligands, *N*-arachidonoylethanolamine (anandamide; AEA) and 2-arachidonoylglycerol (2-AG), which are synthesized "on demand" and act as retrograde messengers to regulate the release of other neurotransmitters (Freund, Katona, & Piomelli, 2003; see review Jutras-Aswad, DiNieri, Harkany, & Hurd, 2009) and contribute to both short- and long-term synaptic plasticity (Mackie, 2006). The synthesis of 2-AG is achieved by the conversion of phosphatidylinositol by phospholipase C into diacylglycerol, then converted to 2-AG by diacylglycerol lipase-α (DAGL-α; Hillard, 2000; Sugiura, Kobayashi, Oka, & Waku, 2002). The biosynthetic pathways of AEA are less clear, although it is known that synthesis begins with the conversion of phospholipid precursors to *N*-acyl phosphatidylethanolamine

(NAPE) via a calcium-dependent transacylase enzyme; however, there are three known distinct and independent pathways of AEA production from NAPE involving enzymes such as phospholipase A2, C, and D (Ahn, McKinney, & Cravatt, 2008). AEA and 2-AG are synthesized following postsynaptic membrane depolarization and are released into the synapse, traveling retrogradely to activate presynaptic CB_1Rs. Activation of these presynaptic CB_1Rs hyperpolarizes the membrane and reduces postsynaptic currents thereby reducing subsequent neurotransmitter release (Di Marzo, 2011). Both ligands are subject to rapid intracellular degradation primarily by hydrolytic enzymes, fatty acid amide hydrolase (FAAH) for AEA and monoacylglycerol lipase (MAGL) for 2-AG (Di Marzo, 2011).

6. ENDOCANNABINOID SYSTEM REGULATION OF THE HPA AXIS IN ADULTHOOD

The eCB system has been shown to regulate the HPA axis in the maintenance of both basal- and stress-induced responses (e.g., Finn, 2010; Hill, McLaughlin, et al., 2010; Hill, Patel, et al., 2010). eCB signaling exerts inhibition on HPA axis activity, contributing to the maintenance of low glucocorticoid levels during basal conditions and functioning to restrict HPA axis activity in situations of acute stress (Gorzalka & Hill, 2009). In support of this, genetic studies have revealed that CB_1R-deficient mice have increased basal levels of CRH mRNA in the PVN, as well as elevated ACTH and CORT secretion (Cota et al., 2007; Finn, 2010; Steiner et al., 2008). This is further supported by pharmacological studies demonstrating that administration of the CB_1R antagonist, Rimonabant (SR141716), results in elevated circulating CORT levels in rodents (Patel, Roelke, Rademacher, Cullinan, & Hillard, 2004; Steiner et al., 2008). *In vitro* evidence also indicates that eCB signaling negatively regulates the HPA axis given that glucocorticoid application stimulates synthesis and release of AEA and 2-AG in the PVN. This, in turn, activates CB_1Rs on presynaptic glutamatergic terminals to suppress excitatory inputs to postsynaptic CRH neurons in the PVN, thus contributing to termination of the stress-induced HPA axis response (Di, Malcher-Lopes, Halmos, & Tasker, 2003; Evanson, Tasker, Hill, Hillard, & Herman, 2010; Malcher-Lopes et al., 2006).

It is currently thought that based on their separate biosynthetic/catabolic pathways and pharmacodynamic properties, AEA contributes to a "tonic-like" mechanism, whereas 2-AG promotes a "burst-like" mechanism in

CB_1R activation (Ahn et al., 2008; Gorzalka, Hill, & Hillard, 2008; Hill, McLaughlin, et al., 2010). The differing roles of 2-AG and AEA are evident in the regulation of the HPA axis. On a functional level, acute stress decreases tonic AEA levels by increasing FAAH-mediated hydrolysis in the corticolimbic system (see review Hill & McEwen, 2009; Hill, McLaughlin, et al., 2010), supporting the "gatekeeper" hypothesis which posits that AEA contributes to maintaining basal levels of glucocorticoids in an organism, while removal of this AEA tone facilitates activation of the HPA axis and increases the frequency of emotional and anxiety-like behaviors (Gray et al., 2015; Hill & McEwen, 2009; Hill et al., 2009; Patel et al., 2004). In contrast, stress exposure generally elicits a delayed increase in corticolimbic 2-AG (Hill et al., 2011; Hill & Tasker, 2012; Wang et al., 2011).

There is compelling genetic, pharmacological, and electrophysiological evidence that eCBs mediate fast-feedback inhibition actions within the PVN to modulate the HPA axis (Hill & Tasker, 2012). Glucocorticoid application to PVN slices results in a rapid suppression of glutamate-mediated excitatory synaptic currents in CRH neurons, whereas this effect is completely abolished with coapplication of a CB_1R antagonist (Di et al., 2003; Malcher-Lopes et al., 2006; Wamsteeker, Kuzmiski, & Bains, 2010b). Furthermore, glucocorticoids increase AEA and 2-AG content in PVN slices *in vitro* (Malcher-Lopes et al., 2006) as well as in whole tissue punches of the hypothalamus *in vivo* (Hill, Karatsoreos, Hillard, & McEwen, 2010). Moreover, 30 min exposure to acute restraint stress increases hypothalamic 2-AG content (Evanson et al., 2010). Lastly, genetic deletion of CB_1Rs results in elevated ACTH and CORT responses to acute stress, indicating that the loss of these receptors diminishes negative feedback, thus yielding an increase in HPA axis activity in response to acute stress exposure (Barna, Zelena, Arszovszki, & Ledent, 2004; Haller, Bakos, Szirmay, Ledent, & Freund, 2002; Hill et al., 2011).

Beyond the PVN, previous work from our laboratory has demonstrated a clear role for eCB signaling in the PFC in the regulation of glucocorticoid-mediated feedback inhibition of HPA axis activity. Acute stress exposure elicits a glucocorticoid-dependent delayed increase in PFC 2-AG content through genomic actions as the increase in 2-AG is blocked by administration of a GR antagonist (RU-486; Hill et al., 2011). Furthermore, local administration of a CB_1R antagonist in the PFC results in a protracted CORT response, although the peak magnitude of the response is not affected (Hill et al., 2011). PFC CB_1Rs were found situated on GABAergic terminals that impinged on layer V pyramidal neurons and PFC slices exposed to CORT exhibited a reduction in GABA release, which was blocked with CB_1R antagonist treatment (Hill et al., 2011). Collectively, these results suggest that glucocorticoids increase PFC 2-AG content via genomic mechanisms, which reduce inhibitory GABA tone, allowing increased PFC outputs to contribute to termination of the stress response and negative feedback (Hill et al., 2011).

There are also multiple lines of evidence indicating that amygdalar eCB signaling (particularly in the BLA) is key in the regulation of basal- and stress-induced HPA axis activity, functioning as a "gatekeeper" over the HPA axis (Hill & Tasker, 2012). Acute stress consistently decreases AEA content of the BLA by a rapid increase in FAAH activity (Hill et al., 2009; Patel, Roelke, Rademacher, & Hillard, 2005; Rademacher et al., 2008). Furthermore, local infusion of a FAAH inhibitor specifically into the BLA attenuates, while infusion of a CB_1R agonist suppresses stress-induced activation of the HPA axis (Ganon-Elazar & Akirav, 2009; Hill et al., 2009). Thus, the emerging picture indicates that BLA AEA tone gates excitatory glutamatergic inputs to principal neurons and when stress exposure disrupts this tone, this increases principal neuron activity, activation of the HPA axis, and subsequent release of glucocorticoid hormones into circulation (Hill & Tasker, 2012).

Amygdalar eCB signaling has also been shown to be a critical regulator of HPA axis stress habituation (Hill, McLaughlin, et al., 2010). Repeated exposure to 9 consecutive days of 30 min sessions of restraint stress produces reductions in corticolimbic AEA, contributing to a basal hypersecretion of CORT (Hill, McLaughlin, et al., 2010), whereas a region-specific elevation of 2-AG in the amygdala facilitates habituation of the HPA axis (i.e., reduced circulating CORT levels; Hill, McLaughlin, et al., 2010; Patel, Kingsley, Mackie, Marnett, & Winder, 2009; Patel et al., 2005). These findings suggest that repeated exposure to the same stressor results in an enhanced capacity to elevate amygdalar 2-AG content which acts to dampen HPA axis activity (Hill, McLaughlin, et al., 2010; Patel & Hillard, 2008; Patel et al., 2005). Given the previously discussed evidence indicating that the BLA gates HPA axis activity by regulating excitatory inputs to principal neurons in this structure, it is likely that repeated stress exposure induces transient 2-AG increases to dampen these excitatory inputs, thus reducing amygdalar principal neuron activity and resulting in HPA axis habituation (Hill, McLaughlin, et al., 2010). This theory is corroborated by a study demonstrating that CORT inhibits glutamatergic inputs to the BLA via an eCB-dependent mechanism, but only in rodents with a previous history of stress exposure (Karst, Berger, Erdmann, Schütz, & Joëls, 2010). In summary, the body of research discussed in this section indicates that adult corticolimbic eCB signaling exerts tight regulation over HPA activity and negative feedback through temporal-, region-, and eCB ligand-dependent mechanisms.

7. ONTOGENY OF THE ENDOCANNABINOID SYSTEM

Based on *in vitro* and *in vivo* studies employing pharmacological manipulation of eCB signaling as well as human and rodent descriptive work characterizing eCB system activity from prepuberty to adulthood (see Table 1), activity of the eCB system appears reliant on developmental stage (Borcel et al., 2004). On a cellular level, eCB signaling plays a multifaceted role in structural and functional neurodevelopment (see review Harkany, Keimpema, Barabás, & Mulder, 2008; Maccarrone, Guzmán, Mackie, Doherty, & Harkany, 2014), regulating proliferation of neural progenitors and cell lineage commitment (Mulder et al., 2008), immature neuronal migration and axonal path finding (Berghuis et al., 2005, 2007; Harkany et al., 2008; Mulder et al., 2008), as well as initiation of synaptic communication of neural networks in perinatal rat tissue (Berghuis et al., 2007; Bernard et al., 2005).

A functional eCB system is present in the rat central nervous system as early as gestational days 11–14 as indicated by the presence of CB_1Rs (Berrendero, Sepe, Ramos, Di Marzo, & Fernández-Ruiz, 1999; Rodriguez de Fonseca, Ramos, Bonnin, & Fernández-Ruiz, 1993), AEA and 2-AG (Berrendero et al., 1999; Harkany et al., 2008). In humans, CB_1Rs are detected in fetal tissue as early as the 9th week of gestation (Zurolo et al., 2010). These receptors are apparently functional in the rat and human fetal brain in that application of the cannabinoid agonist, WIN 55,212-2, stimulated $[^{35}S]GTP\gamma S$ binding (Berrendero et al., 1999; Mato, Del Olmo, & Pazos, 2003). Moreover, examination of specific neural structures reveals that high CB_1R expression is limited to the amygdala and hippocampus, with low distributions in the striatum, thalamus, and cerebral cortex during early to mid-gestation in the human fetus (gestational weeks 17–22) and rat neonate (Rodriguez de Fonseca et al., 1993). These findings contrast with reports of higher and wider density of these ubiquitous receptors in the adult rat (Herkenham et al., 1991) and human brain (Jutras-Aswad et al., 2009). The unequal distribution of CB_1R expression

Table 1 Adolescent Corticolimbic Endocannabinoid System Development in the Rat

Endocannabinoid Marker	Sex/Strain	Age	Brain Region	Developmental Trajectory	Reference
CB_1R binding	Wistar (♀,♂)	PNDs 10–70	Limbic system	↑ From PND 10 to PNDs 30–40, then ↓ to PND 70	Rodriguez de Fonseca et al. (1993)
CB_1R mRNA	Sprague-Dawley (♂)	PNDs 25, 40, and 70	PFC	↓ From PNDs 25 to 70	Heng et al. (2011)
AEA	Sprague-Dawley (♀)	PNDs 5, 15, 25, ~30, and 250	Hypothalamus	AEA peaks immediately before pubertal onset (vaginal opening) ~PND 30, then decreases to adult levels	Wenger et al. (2002)
AEA, 2-AG, and CB_1R	Long Evans (♂)	PNDs 29, 38, and 50	PFC	AEA ↑ from PNDs 29 to 50; 2-AG ↓ from PNDs 29 to 38, then ↑ to PND 50; CB_1R ↓ from PNDs 29 to 50	Ellgren et al. (2008)
AEA, FAAH	Sprague-Dawley (♂)	PNDs 25, 35, 45, and 70	PFC, amygdala, hypothalamus, and hippocampus	AEA ↑ from PNDs 25 to 35, ↓ from PNDs 35 to 45, ↑ from PNDs 45 to 70 with corresponding changes in FAAH activity	Lee et al. (2013)
AEA, 2-AG, CB_1R binding, FAAH/MAGL activity	Sprague-Dawley (♀)	PNDs 46, 60, and 75	PFC	AEA ↑ from PNDs 46 to 60, then ↓ by PND 75; CB_1R binding ↑ from PNDs 46 to 60, then ↓ by PND 75; no significant changes in 2-AG, FAAH, and MAGL activity	Rubino et al. (2015)

CB_1R, CB_1 receptor; AEA, anandamide; 2-AG, 2-arachidonoylglyercol; FAAH, fatty acid amide hydrolase; MAGL, monoacylglycerol lipase; PND, postnatal day; PFC, prefrontal cortex.

throughout the immature brain suggests a region-dependent role in ensuring normative neurodevelopment (Campolongo, Trezza, Ratano, Palmery, & Cuomo, 2011; Fride, Suris, Weidenfeld, & Mechoulam, 2005; Harkany et al., 2008; Jutras-Aswad et al., 2009; Marco, Adriani, Llorente, Laviola, & Viveros, 2009; Schneider, 2008).

There are growing reports indicating that adolescent eCB system activity in the brain changes in a highly temporal-specific, dynamic, and sometimes region-dependent manner. Adolescent CB_1R binding in the rat brain appears to be highest just prior to the onset of adolescence (PNDs 25–29), followed by a general linear decline to adult levels within limbic, striatal, and cortical structures (Rodriguez de Fonseca et al., 1993). This is consistent with more recent findings that PFC CB_1R expression declines in pre- to early adolescence (PND 25; Ellgren et al., 2008; Heng, Beverley, Steiner, & Tseng, 2011). More specific examination between limbic/associative and sensorimotor cortical regions has revealed differential rates at which these CB_1R expression declines occur in the PFC; declines in limbic/associative regions occur gradually throughout adolescence, whereas major changes in sensorimotor regions are not exhibited until mid- to late adolescence, with the functionality of these receptors following the same developmental pattern (Heng et al., 2011). In female rodents, CB_1Rs have been reported to increase from mid (PND 46) to late adolescence (PND 60) and decline by adulthood (PND 75; Rubino et al., 2015).

Investigation of the maturational trajectory of the eCB ligands has revealed somewhat divergent results that are likely influenced by strain and sex differences. Earlier reports suggest AEA concentrations in the male rodent PFC exhibit a gradual and progressive increase (i.e., PNDs 29–50) to adult levels, while 2-AG levels were highest very early in adolescence (PND 29), decreased by PND 38, and increased again by late adolescence (PND 50; Ellgren et al., 2008). In contrast, AEA concentration increases from PNDs 46 to 60 but decreases from PNDs 60 to 75 in the PFC of female rats, with no differences in 2-AG concentrations as well as FAAH and MAGL activity (Rubino et al., 2015). Furthermore, hypothalamic AEA content in female rats is observed to increase immediately preceding vaginal opening (as a physical marker of pubertal onset; Wenger et al., 2002). We have also previously reported that throughout the male rodent adolescent period (PNDs 25–70), corticolimbic AEA concentrations fluctuate in a similar pattern across the amygdala, PFC, hippocampus, and hypothalamus (Lee, Hill, Hillard, & Gorzalka, 2013).

Moreover, two other related *N*-acylethanolamines hydrolyzed by FAAH, oleoylethanolamine and palmitoylethanolamine, exhibit the same temporal-specific pattern as AEA, indicating that the corticolimbic AEA fluctuations are at least partly due to corresponding changes in FAAH activity (Lee et al., 2013). Thus, the rodent literature indicates that corticolimbic eCB ligand concentrations, particularly with respect to AEA, fluctuate throughout adolescence, while CB_1R expression peaks at the onset of adolescence before declining to adult levels (Ellgren et al., 2008; Lee et al., 2013; Rubino et al., 2015; Wenger et al., 2002).

In humans, CB_1R expression is reported to increase gradually to adult levels in postmortem brain tissue (Mato et al., 2003). In contrast, a more recent study found that CB_1R expression peaks during infancy (<1 year old) to toddler age (1.5 to 4.5 years old), then gradually decreases to adult levels in human PFC tissue samples (Long, Lind, Webster, & Weickert, 2012). Moreover, expression of enzymes that contribute to AEA synthesis (i.e., NAPE-PLD) and degradation (i.e., FAAH) both increase from infancy to adulthood, indicating greater AEA regulation in adulthood than earlier in life (Long et al., 2012). In contrast, 2-AG and DAGL-α were found to peak between school age and young adulthood, while MAGL followed a similar pattern as CB_1R expression (i.e., expression peaks around toddler age; Long et al., 2012). Interestingly, although a similar general pattern of CB_1R expression occurs in humans and rodents, the maturational trajectories of these receptors exhibit a peak around toddler age in humans, whereas this peak occurs closer to pubertal onset/early adolescence in the rodent.

While the functional consequences of these fluctuations require further investigation, it is possible that the dynamic changes in adolescent rodent eCB signaling suggest a general instability compared to that of adults and could underlie adolescent behavioral profiles that are regulated by corticolimbic circuits, such as relatively low anxiety, high responsivity to reward, and reduced inhibitory control (Casey & Jones, 2010). Given that adolescence is a period when this behavioral profile is maximal and is coincident with these changes in AEA/CB_1R expression and activity, the eCB system may represent a neural substrate of the adolescent phenotype. Moreover, the dynamic nature of adolescent eCB signaling could also be contributing to alterations in stress sensitivity, emotionality, and executive function, serving as a nexus between developmental stage and HPA axis stress responsivity.

8. ADOLESCENT HPA AXIS FUNCTION: A REGULATORY ROLE FOR THE ENDOCANNABINOID SYSTEM?

8.1 Adolescent Stress Exposure Modulates Corticolimbic Endocannabinoid Signaling

To our knowledge, there are relatively few studies that directly explore eCB-related mechanisms subserving interactions between adolescent development and HPA responsivity, particularly in humans. The results of preclinical studies suggest that stress exposure produces age-specific short- and long-term effects on eCB signaling (e.g., Lee & Hill, 2013; Reich, Mihalik, Iskander, Seckler, & Weiss, 2013; Wamsteeker, Kuzmiski, & Bains, 2010b). Similarly, there is some human and rodent evidence that adolescent cannabinoid use/exposure has the capacity to alter HPA axis stress responses (Biscaia et al., 2003; Lee, Hill, Hillard, & Gorzalka, 2015; Lee, Wainwright, Hill, Galea, & Gorzalka, 2014; van Leeuwen et al., 2011). Together, these studies suggest that the adolescent eCB system exhibits a greater general sensitivity to perturbations (e.g., stress exposure and cannabinoid treatment) than adults, possibly due to greater instability in the regulation of eCBs during this age.

Age of stress exposure region dependently modulates CB_1R binding in the corticolimbic stress circuit, both in the short and long term. Specifically, adolescent rats exposed to repeated restraint stress (10 days, 30 min/day) exhibit no immediate effects in the hippocampus, whereas a downregulation in CB_1Rs is observed in adults. In the long term, however, stress exposure produced an upregulation of hippocampal CB_1Rs in adolescent and adult rats (Lee & Hill, 2013). In contrast, a comparable upregulation of PFC CB_1Rs 24 h following the last restraint session is observed in both age groups; however, 40 days later, adolescent rats exposed to the repeated stress exhibited a CB_1R downregulation, whereas adult rats displayed normalized CB_1R-binding levels (Lee & Hill, 2013). These observations are in line with the idea that adolescent eCB signaling is less stable or not as tightly regulated as in adulthood, potentially leaving an organism more susceptible to the long-term effects of stress exposure (Lee & Gorzalka, 2012). In further support of this, prepubertal male rats (PNDs 21–30) exhibit a transient loss of depolarization-induced suppression of inhibition (DSI), in parvocellular neuroendocrine cells of the PVN immediately following 5 consecutive days of 30 min restraint stress (Wamsteeker, Kuzmiski, & Bains, 2010a, 2010b).

These effects are observed in prepubertal, but not adult rats exposed to repeated restraint (Wamsteeker, Kuzmiski, & Bains, 2010a). The loss in eCB signaling (as revealed by reduced DSI) was a result of stress-induced CORT activating intracellular GR receptors to cause a downregulation of presynaptic CB_1R activity (Wamsteeker, Kuzmiski, & Bains, 2010b). Thus, this study provides early evidence of divergent age-dependent downregulation of CB_1Rs in parvocellular neuroendocrine cells of the hypothalamus as a result of repeated stress exposure. Moreover, the fact that repeated restraint exposure fails to produce the neuroendocrine habituation response in adolescent rats (Romeo, 2010a) coupled with downregulation of hypothalamic PVN CB_1Rs (Wamsteeker, Kuzmiski, & Bains, 2010b), suggests that CB_1R signaling plays a significant role in launching this adaptive response and indicates this mechanism is not fully functional in adolescence. Much like repeated restraint exposure (Lee & Hill, 2013), chronic mild stress exposure during adolescence produces a downregulation of CB_1Rs in the hippocampus (Reich, Taylor, & McCarthy, 2009) and has recently been found to increase excitatory neurotransmission in the CA1 region of the hippocampus through the sensitization of CB_1Rs on GABAergic terminals (Reich et al., 2013).

8.2 Adolescent Cannabinoid Exposure Modifies Adult Stress-Induced HPA Axis Activity

Despite several investigations of the short- and long-term impact of adolescent cannabinoid exposure on the adult brain and behavior of rodents, very few have assessed HPA axis stress responsivity. From what is known, adolescent cannabis users demonstrate lower HPA axis stress responses to the Groningen Social Stress Task relative to nonusers (van Leeuwen et al., 2011), and adolescent male cannabis users exhibit region-dependent alterations in PFC thickness relative to healthy controls (Lopez-Larson et al., 2011). Furthermore, preclinical studies of adolescent cannabinoid exposure suggest that cannabinoids have the capacity to alter stress-induced HPA axis activity in the long term, without affecting basal CORT levels. Adolescent (PNDs 35–45) male and female rats were chronically administered the synthetic THC analog, CP 55,940. This treatment reduced emotionality- and anxiety-like behavior in the hole board, open-field tests, and elevated plus maze, but had no long-term effect on CORT levels (Biscaia et al., 2003). Subsequent studies investigating the long-term impact of adolescent CB_1R agonist, HU-210, and CB_1R inverse agonist, AM-251, treatment

(PNDs 35–44) are in line with previous findings that basal CORT and ACTH levels are unaffected in the long term (Lee et al., 2015; Lee, Wainwright, et al., 2014). However, adolescent HU-210 treatment produced greater overall CORT responses to acute restraint stress exposure in adult male, and to a lesser extent, female rats (Lee, Wainwright, et al., 2014). In contrast, adolescent AM-251 treatment failed to alter the acute HPA axis stress response, but did elicit a relatively modest reduction in the overall magnitude of the HPA axis habituation response, increased risk-assessment behavior in the elevated plus maze and increased stress-coping behavior in the forced swim test in male rats (Lee et al., 2015). These moderate alterations in stress responsivity and emotional behavior were also associated with an increase in PFC CB_1R binding, and a modest reduction and increase in AEA within the amygdala and hypothalamus, respectively (Lee et al., 2015).

Long-term compensatory alterations of HPA axis activity to perinatal cannabinoid exposure are relatively well documented (see review Campolongo et al., 2011), and there is a growing body of literature indicating similar effects with adolescent cannabinoid exposure. Perinatal cannabinoid treatment produces a short-term surge in CORT levels and it is believed that this stimulates long-term HPA axis hypoactivity in compensation (Campolongo et al., 2011). This theory is partially supported by findings that adult male rodents exposed to perinatal THC treatment exhibit reductions in medial basal hypothalamic CRH and CORT levels, while adult females exhibit increases in these markers (Rubio et al., 1995). This theory is further supported by reports that male rats exposed to high perinatal doses of HU-210 (CB_1R agonist) exhibit elevated adult basal CORT levels, but reduced responsivity to a subsequent HU-210 injection challenge in adulthood, whereas males exposed to a low perinatal dose exhibit smaller elevations in basal CORT levels with an elevated response to acute HU-210 challenge (del Arco et al., 2000). We postulate that adolescent cannabinoid exposure similarly provokes high or protracted exposure to glucocorticoids in the short term, possibly leading to permanent hypoactivity or dysregulation of HPA axis activity and reduced responsivity to subsequent cannabinoid-induced HPA axis challenge. Furthermore, the modulation of corticolimbic CB_1R density and binding in response to adolescent cannabinoid treatment could represent a long-term compensatory mechanism of the eCB system. However, these hypotheses remain largely unexplored.

8.3 Long-Term Consequences of Adolescent Endocannabinoid System Dysregulation on the Developing Brain and Behavior

Multiple studies (Lee, Wainwright, et al., 2014; Rubino et al., 2008, 2015; Table 2; Rubino, Realini, Braida, Alberio, et al., 2009; Rubino, Realini, Braida, Guidi, et al., 2009; Zamberletti et al., 2014) have demonstrated that chronic pharmacological enhancement of adolescent eCB signaling results in long-term modulation of corticolimbic CB_1R density and binding as well as other long-term neural and behavioral changes similar to those produced by chronic stress (via glucocorticoid hypersecretion; e.g., McEwen, 2005) and prolonged glucocorticoid exposure (Hill et al., 2008). On a behavioral level, this drug regimen reduces stress-coping behavior in the forced swim test in female rats, impairs spatial working memory in males, and leads to anhedonia in both males and females as measured by the sucrose preference test (Rubino et al., 2008). Neural consequences include differential reductions in amygdalar and hippocampal CB_1R expression and $[^{35}S]GTP\gamma S$ binding in males and females (Rubino et al., 2008), decreases in markers of neuroplasticity such as synaptophysin and PSD 95 in the PFC of females (Rubino, Realini, Braida, Alberio, et al., 2009), dendritic atrophy and reduced number of spines in hippocampal neurons in males (Rubino, Realini, Braida, Guidi, et al., 2009), and reductions in cell proliferation (Realini et al., 2011) and survival (Lee, Wainwright, et al., 2014) in the dentate gyrus of the hippocampus. Research from the same laboratory also demonstrated that chronic adolescent administration of the FAAH inhibitor, URB-597, reduces hippocampal CB_1R expression in male rats (Marco, Rubino, et al., 2009) and reverses the THC-induced reduction in stress coping and increase in anxiety behaviors in female rats (Realini et al., 2011). In contrast, adolescent administration of CP 55,940 (PNDs 28–43) has been reported to increase CB_1R activity in the PFC of adult male, but not female rats (Mateos et al., 2011).

Other studies employing cannabinoid agonists during adolescence report long-term deficits in cognitive processes such as working memory and social discrimination (e.g., Romero et al., 2002; Rubino, Realini, Braida, Guidi, et al., 2009; Schneider & Koch, 2003), further supporting the conclusion that adolescents are sensitive to the effects of cannabinoid exposure. Moreover, adolescence appears to be a period of particular susceptibility to the effects of chronic cannabinoid exposure given that adult rats exposed to the same treatments either do not exhibit these behavioral phenotypes or

Table 2 Effects of Adolescent Cannabinoid Exposure on Adult Neural and HPA Axis Functioning, the Endocannabinoid System, and Emotional Behavior in the Rodent

Age of Exposure (PND)	Species/Sex	Cannabinoid Treatment	Age of Testing (PND)	Neural and HPA Axis Function Effects	Endocannabinoid System	Emotional Behavior	Reference
28	Albino Wistar (♀,♂)	CP 55,940; PNDs 28–43; 0.4 mg/kg	70–76	–	↓ CB_1R activity in PFC of adult ♂, but not ♀	No effect on anxiety behavior	Mateos et al. (2011)
30 and 56	Albino Wistar (♀)	Escalating CP 55,940; PNDs 30–50 or 56–76; 0.15, 0.20, and 0.30 mg/kg for 3, 8, and 10 days	70 and 96	–	–	↓ Social interaction with a novel conspecific in adolescents, but not adults	O'Shea, Singh, McGregor, and Mallet (2004)
4, 30, and 56	Albino Wistar (♂)	Escalating CP 55,940; PNDs 4–24, 30–50, or 56–76; 0.15, 0.20, and 0.30 mg/kg for 7 days each	59, 78, and 104	–	–	↑ Anxiety behavior regardless of age of exposure	O'Shea et al. (2006)
35	Albino Wistar (♀,♂)	CP 55,940; PNDs 35–45; 0.4 mg/kg	>75	No alterations to basal CORT levels	–	↓ Anxiety in ♀,♂	Biscaia et al. (2003)
35	Sprague-Dawley (♀,♂)	Escalating THC twice a day; PNDs 35–45; 2.5, 5, and 10 mg/kg	>75	Low CREB activity in hippocampus and PFC of ♀ only	↓ CB_1R and CB_1/G protein coupling in amygdala, PFC, and hippocampus in ♀,♂	No effect on anxiety, ↑ anhedonia in ♀,♂; ↓ stress coping in ♀ only	Rubino et al. (2008)
35	Sprague-Dawley (♂)	Escalating THC twice a day; PNDs 35–45; 2.5, 5, and 10 mg/kg	>75	Dendritic atrophy and ↓ spines in hippocampal neurons; ↓ PSD95 and NMDA receptors in hippocampus	–	No effect on aversive memory; ↓ spatial memory	Rubino, Realini, Braida, Guidi, et al. (2009)
35	Sprague-Dawley (♀)	Escalating THC twice a day; PNDs 35–45; 2.5, 5, and 10 mg/kg	>75	↓ Synaptophysin and PSD95 in PFC, but not hippocampus	–	No effect on aversive memory; ↓ spatial memory	Rubino, Realini, Braida, Alberio, et al. (2009)
35	Sprague-Dawley (♀)	Escalating THC (Rubino, Realini, Braida, Alberio, et al., 2009); URB 597; 3 injections between PNDs 75 and 105; 0.3 mg/kg	>75	THC ↓ hippocampal neurogenesis	–	↑ Anhedonia, ↓ stress coping, ↓ social interaction, ↓ object recognition memory; URB-597 reversed these effects	Realini et al. (2011)
35	Sprague-Dawley (♀)	Escalating THC twice a day; PNDs 35–45; 2.5, 5, and 10 mg/kg	>75	↓ GAD67 in CCK and parvalbumin interneurons and basal GABA levels in PFC	–	–	Zamberletti et al. (2014)

Age of Exposure (PND)	Species/Sex	Cannabinoid Treatment	Age of Testing (PND)	Neural and HPA Axis Function Effects	Endocannabinoid System	Emotional Behavior	Reference
35	Sprague-Dawley (♀,♂)	Escalating HU-210; PNDs 35–46; 25, 50, and 100 μg/kg	>75	↓ Hippocampal neurogenesis and ↑ acute stress response in ♂ more so than ♀	–	–	Lee, Wainwright, et al. (2014)
35	C57BL/6 mice (♀)	Escalating WIN 55,212-2 twice a day; PNDs 35–45; 0.5, 1, and 2 mg/kg	70–170	Deficient eCB- and mGluR2/3-dependent long-term depression in the medial PFC (mPFC)	↓ CB_1R functionality in mPFC; MAGL inhibition ameliorates deficiency in long-term depression	–	Lovelace et al. (2015)
35	Sprague-Dawley (♀)	Escalating THC twice a day; PNDs 35–45; 2.5, 5, and 10 mg/kg	46, 60, and 75	THC ↑ NMDA GluN2B subunits in adulthood, ↓ PFC spine density	CB_1R ↑ PNDs 46–60, ↓ PNDs 60–75; no change in 2-AG, FAAH, or MAGL activity; PFC AEA ↑ PNDs 46–60, ↓ from PNDs 60 to 75. THC ↓ PFC CB_1R binding and AEA	–	Rubino et al. (2015)
35	Sprague-Dawley (♂)	AM-251; PNDs 35–45; 5 mg/kg	>75	AM-251 treatment modified CORT habituation	↑ PFC CB_1R binding; AEA ↓ in amygdala and ↑ in hypothalamus	↑ Stress coping in FST; ↑ risk-assessment behavior in EPM	Lee et al. (2015)
38	Wistar (♂)	URB-597; PNDs 38–43; 0.1 or 0.5 mg/kg	>100	–	0.5 mg/kg dose ↓ CB_1R expression; no effect on PFC and amygdala	–	Marco, Rubino, et al. (2009)
35, 40, 50, and 75	Sprague-Dawley (♂)	WIN 55,212-2; 5 days; 2 mg/kg	65–95	Only early cannabinoid exposure (PNDs 35 and 40) ↓ PFC GABAergic transmission	–	–	Cass et al. (2014)
40	Sprague-Dawley (♂)	Acute injection of CP 55,940 at 0.1, 0.2, 0.4, or 0.6 mg/kg	40	No effect on dopamine concentration in striatum or limbic forebrain; ↑activated CORT	–	–	Romero et al. (2002)
40 and >80	Wistar (♂)	WIN 55,212-2; 20 injections in 25 days; 1.2 mg/kg	41, 65, and 70	–	–	Immediate ↓ social discrimination at all ages, but persistent in adolescents on final day and 15 days after treatment	Schneider, Schömig, and Leweke (2008)

CB_1R, CB_1 receptor; AEA, anandamide; 2-AG, 2-arachidonoylglyercol; FAAH, fatty acid amide hydrolase; MAGL, monoacylglycerol lipase; PND, postnatal day; PFC, prefrontal cortex; CORT, corticosterone.

do not appear to be as greatly affected in the long term (Bambico, Nguyen, Katz, & Gobbi, 2010; Realini et al., 2011; Schneider & Koch, 2003). However, there are some contrasting reports that chronic adolescent exposure to CP 55,940 either decreased or had no effect on anxiety-like behavior as measured in the open-field test, elevated plus maze (Biscaia et al., 2003; Wegener & Koch, 2009), and social interaction test (O'Shea, McGregor, & Mallet, 2006). As others have suggested, this is likely a result of differences in the type of cannabinoid agonist used, dose, strain, sex, and administration schedule (Rubino & Parolaro, 2008; Schneider, 2008).

Recent work has identified that adolescent cannabinoid exposure disrupts several processes in the rodent PFC, which may contribute to development of psychopathological symptoms. Rubino et al. (2015) suggest that the detrimental impact of adolescent cannabinoid exposure treatment is due, at least in part, to disruption of normative adolescent eCB signaling that regulates maturational processes, such as synaptic pruning, in the PFC. Furthermore, in a mouse model of female adolescent cannabinoid exposure, eCB-mediated presynaptic plasticity in glutamatergic synapses of the PFC is impaired in adulthood (Lovelace, Corches, Vieira, Mackie, & Korzus, 2015). These findings indicate a loss in the ability to regulate neural network homeostasis and neural adaptation to changing environments, possibly resulting in vulnerability to psychopathological processes (Lovelace et al., 2015). These findings may be particularly relevant to the development of stress-related diseases given that eCB signaling plays critical roles in constraining HPA axis activity and facilitating HPA axis adaptation. Moreover, a specific window of vulnerability to exogenous cannabinoids is especially evident in early adolescent rats (i.e., PNDs 35–45). WIN 55,212-2 treatment, during PNDs 35–40 and 40–45 but not in late adolescence (PNDs 50–55) or adulthood (PNDs 75–80), disrupts adult GABAergic transmission via a CB_1R-dependent mechanism in the PFC (Cass et al., 2014). Similarly, early adolescent cannabinoid exposure (PNDs 35–40), but not late-adolescent cannabinoid exposure, suppresses cortical oscillations through a CB_1R-mediated mechanism in adult male mice (Raver & Keller, 2014).

9. CONCLUDING REMARKS

A review of the evidence supports the idea that adolescence is a period of heightened susceptibility to disturbances such as stress as well as cannabinoid exposure. This is likely due, at least in part, to age-dependent differences in corticolimbic eCB signaling as well as HPA axis development and functionality. Adolescent neurodevelopment consists of maturational processes that are both time dependent and region specific. This provides the adolescent brain with remarkable plasticity yet may also confer vulnerability to developing mental illness as a result of biological, environmental, and genetic disruptions that exacerbate transient imbalances occurring between these region- and time-dependent maturational processes (Lee, Heimer, et al., 2014). Given the ability of the eCB system to exert regulation over HPA axis activity and neurodevelopment, it is reasonable to propose that this system plays an important role in adolescent vulnerability to mental illness. In this chapter, we presented evidence demonstrating how the eCB system exerts tight regulation over adult HPA axis stress responsivity, that corticolimbic eCB signaling changes dynamically during the adolescent period, that adolescent stress exposure can alter eCB signaling both in the immediate and long term and similarly, that adolescent cannabinoid exposure generally results in behavioral and neural consequences reminiscent of stress exposure. However, there is a significant gap in the literature investigating the role of normative eCB signaling in the corticolimbic circuit and regulation of stress responsivity in the adolescent HPA axis. Uncovering the relationship between eCB mediation of HPA axis functioning and development is of special importance in light of current health concerns that adolescents frequently engage in recreational drug use coupled with the considerable body of literature, indicating that adolescent cannabis consumption can serve as a risk factor for the onset of psychiatric disease (Hill, 2014; Rubino, Zamberletti, & Parolaro, 2012).

ACKNOWLEDGMENTS

The authors would like to thank the following sources of funding support: TTYL—Canadian Institutes of Health Research (CIHR; salary); BBG—Operating grants from the Natural Sciences and Engineering Research Council of Canada and CIHR.

REFERENCES

Aguilera, G., & Rabadan-Diehl, C. (2000). Vasopressinergic regulation of the hypothalamic–pituitary–adrenal axis: Implications for stress adaptation. *Regulatory Peptides*, *96*, 23–29.

Ahn, K., McKinney, M. K., & Cravatt, B. F. (2008). Enzymatic pathways that regulate endocannabinoid signaling in the nervous system. *Chemical Reviews*, *108*, 1687–1707.

Bambico, F., Nguyen, N., Katz, N., & Gobbi, G. (2010). Chronic exposure to cannabinoids during adolescence but not during adulthood impairs emotional behaviour and monoaminergic neurotransmission. *Neurobiology of Disease*, *37*, 641–655.

Barna, I., Zelena, D., Arszovszki, A. C., & Ledent, C. (2004). The role of endogenous cannabinoids in the hypothalamo-pituitary-adrenal axis regulation: In vivo and in vitro studies in CB1 receptor knockout mice. *Life Sciences*, *75*, 2959–2970.

Berghuis, P., Dobszay, M. B., Wang, X., Spano, S., Ledda, F., Sousa, K. M., et al. (2005). Endocannabinoids regulate interneuron migration and morphogenesis by transactivating the TrkB receptor. *Proceedings of the National Academy of Sciences of the United States of America, 102*, 19115–19120.

Berghuis, P., Rajnicek, A. M., Morozov, Y. M., Ross, R. A., Mulder, J., Urbán, G. M., et al. (2007). Hardwiring the brain: Endocannabinoids shape neuronal connectivity. *Science, 316*, 1212–1216.

Bernard, C., Milh, M., Morozov, Y. M., Ben-Ari, Y., Freund, T. F., & Gozlan, H. (2005). Altering cannabinoid signaling during development disrupts neuronal activity. *Proceedings of the National Academy of Sciences of the United States of America, 102*, 9388–9393.

Berrendero, F., Sepe, N., Ramos, J. A., Di Marzo, V., & Fernández-Ruiz, J. J. (1999). Analysis of cannabinoid receptor binding and mRNA expression and endogenous cannabinoid contents in the developing rat brain during late gestation and early postnatal period. *Synapse, 33*, 181–191.

Bingham, B., McFadden, K., Zhang, X., Bhatnagar, S., Beck, S., & Valentino, R. (2011). Early adolescence as a critical window during which social stress distinctly alters behavior and brain norepinephrine activity. *Neuropsychopharmacology, 36*, 896–909.

Biscaia, M., Marín, S., Fernández, B., Marco, E. M., Rubio, M., Guaza, C., et al. (2003). Chronic treatment with CP 55,940 during the peri-adolescent period differentially affects the behavioural responses of male and female rats in adulthood. *Psychopharmacology, 170*, 301–308.

Borcel, E., Pérez-Alvarez, L., de Ceballos, M. L., Ramirez, B. G., Marco, E. M., Fernández, B., et al. (2004). Functional responses to the cannabinoid agonist WIN 55,212-2 in neonatal rats of both genders: Influence of weaning. *Pharmacology, Biochemistry and Behavior, 78*, 593–602.

Cabral, G. A., Raborn, E. S., Griffin, L., Dennis, J., & Marciano-Cabral, F. (2008). CB2 receptors in the brain: Role in central immune function. *British Journal of Pharmacology, 153*, 240–251.

Campolongo, P., Trezza, V., Ratano, P., Palmery, M., & Cuomo, V. (2011). Developmental consequences of perinatal cannabis exposure: Behavioral and neuroendocrine effects in adult rodents. *Psychopharmacology, 214*, 5–15.

Casey, B. J., & Jones, R. M. (2010). Neurobiology of the adolescent brain and behavior. *Journal of the American Academy of Child and Adolescent Psychiatry, 49*, 1189–1285.

Casey, B. J., Jones, R. M., & Somerville, L. H. (2011). Braking and accelerating of the adolescent brain. *Journal of Research on Adolescence, 21*, 21–33.

Cass, D. K., Flores-Barrera, E., Thomases, D. R., Vital, W. F., Caballero, A., & Tseng, K. Y. (2014). CB1 cannabinoid receptor stimulation during adolescence impairs the maturation of GABA function in the adult rat prefrontal cortex. *Molecular Psychiatry, 19*, 536–543.

Chrousos, G. P. (2009). Stress and disorders of the stress system. *Nature Reviews Endocrinology, 5*, 374–381.

Costello, E. J., Egger, H., & Angold, A. (2005). 10-Year research update review: The epidemiology of child and adolescent psychiatric disorders: I. Methods and public health burden. *Journal of the American Academy of Child and Adolescent Psychiatry, 44*, 972–986.

Cota, D., Steiner, M. A., Marsicano, G., Cervino, C., Herman, J. P., Grübler, Y., et al. (2007). Requirement of cannabinoid receptor type 1 for the basal modulation of hypothalamic-pituitary-adrenal axis function. *Endocrinology, 148*, 1574–1581.

Crews, F., He, J., & Hodge, C. (2007). Adolescent cortical development: A critical period of vulnerability for addiction. *Pharmacology, Biochemistry and Behavior, 86*, 189–199.

del Arco, I., Muñoz, R., Rodríguez De Fonseca, F., Escudero, L., Martín-Calderón, J. L., Navarro, M., et al. (2000). Maternal exposure to the synthetic cannabinoid HU-210:

Effects on the endocrine and immune systems of the adult male offspring. *Neuroimmunomodulation*, *7*, 16–26.
Di Luigi, L., Guidetti, L., Baldari, C., Gallotta, M., Sgrò, P., Perroni, F., et al. (2006). Cortisol, dehydroepiandrosterone sulphate and dehydroepiandrosterone sulphate/cortisol ratio responses to physical stress in males are influenced by pubertal development. *Journal of Endocrinological Investigation*, *29*, 796–804.
Di Marzo, V. (2011). Endocannabinoid signaling in the brain: Biosynthetic mechanisms in the limelight. *Nature Neuroscience*, *14*, 9–15.
Di, S., Malcher-Lopes, R., Halmos, K. C., & Tasker, J. G. (2003). Nongenomic glucocorticoid inhibition via endocannabinoid release in the hypothalamus: A fast feedback mechanism. *Journal of Neuroscience*, *23*, 4850–4857.
Doremus-Fitzwater, T. L., Varlinskaya, E. I., & Spear, L. P. (2009). Social and non-social anxiety in adolescent and adult rats after repeated restraint. *Physiology & Behavior*, *97*, 484–494.
Dziedzic, N., Ho, A., Adabi, B., Foilb, A., & Romeo, R. (2014). Shifts in hormonal stress reactivity during adolescence are not associated with changes in glucocorticoid receptor levels in the brain and pituitary of male rats. *Developmental Neuroscience*, *36*, 261–268.
Eiland, L., & Romeo, R. D. (2013). Stress and the developing adolescent brain. *Neuroscience*, *249*, 162–171.
Ellgren, M., Artmann, A., Tkalych, O., Gupta, A., Hansen, H. S., Hansen, S. H., et al. (2008). Dynamic changes of the endogenous cannabinoid and opioid mesocorticolimbic systems during adolescence: THC effects. *European Neuropsychopharmacology*, *18*, 826–834.
Evanson, N. K., Tasker, J. G., Hill, M. N., Hillard, C. J., & Herman, J. P. (2010). Fast feedback inhibition of the HPA axis by glucocorticoids is mediated by endocannabinoid signaling. *Endocrinology*, *151*, 4811–4819.
Evuarherhe, O., Leggett, J., Waite, E., Kershaw, Y., & Lightman, S. (2009). Reversal of the hypothalamo-pituitary-adrenal response to oestrogens around puberty. *Journal of Endocrinology*, *202*, 279–285.
Figueiredo, H. F., Bruestle, A., Bodie, B., Dolgas, C. M., & Herman, J. P. (2003). The medial prefrontal cortex differentially regulates stress-induced c-fos expression in the forebrain depending on type of stressor. *European Journal of Neuroscience*, *18*, 2357–2364.
Finn, D. P. (2010). Endocannabinoid-mediated modulation of stress responses: Physiological and pathophysiological significance. *Immunobiology*, *215*, 629–646.
Foilb, A. R., Lui, P., & Romeo, R. D. (2011). The transformation of hormonal stress responses throughout puberty and adolescence. *Journal of Endocrinology*, *210*, 391–398.
Freund, T. F., Katona, I., & Piomelli, D. (2003). Role of endogenous cannabinoids in synaptic signaling. *Physiological Reviews*, *83*, 1017–1066.
Fride, E., Suris, R., Weidenfeld, J., & Mechoulam, R. (2005). Differential response to acute and repeated stress in cannabinoid CB1 receptor knockout newborn and adult mice. *Behavioural Pharmacology*, *16*, 431–440.
Ganon-Elazar, E., & Akirav, I. (2009). Cannabinoid receptor activation in the basolateral amygdala blocks the effects of stress on the conditioning and extinction of inhibitory avoidance. *Behavioural Pharmacology*, *29*, 11078–11088.
Gaoni, Y., & Mechoulam, R. (1964). Isolation, structure and partial synthesis of an active constituent of hashish. *Journal of the American Chemical Society*, *86*, 1646–1647.
Gogtay, N., Giedd, J. N., Lusk, L., Hayashi, K. M., Greenstein, D., Vaituzis, A. C., et al. (2004). Dynamic mapping of human cortical development during childhood through early adulthood. *Proceedings of the National Academy of Sciences of the United States of America*, *101*, 8174–8179.

Goldman, L., Winget, C., Hollingshead, G. W., & Levine, S. (1973). Postweaning development of negative feedback in the pituitary-adrenal system of the rat. *Neuroendocrinology, 12*, 199–211.
Gorzalka, B. B., & Hill, M. N. (2009). Integration of endocannabinoid signaling into the neural network regulating stress-induced activation of the hypothalamic-pituitary-adrenal axis. *Current Topics in Behavioral Neurosciences, 1*, 289–306.
Gorzalka, B. B., Hill, M. N., & Hillard, C. J. (2008). Regulation of endocannabinoid signalling by stress: Implications for stress-related affective disorders. *Neuroscience and Biobehavioral Reviews, 32*, 1152–1160.
Gray, J. M., Vecchiarelli, H. A., Morena, M., Lee, T. T.-Y., Hermanson, D. J., Kim, A. B., et al. (2015). Corticotropin-releasing hormone drives anandamide hydrolysis in the amygdala to promote anxiety. *Journal of Neuroscience, 35*, 3879–3892.
Grissom, N., & Bhatnagar, S. (2009). Habituation to repeated stress: Get used to it. *Neurobiology of Learning and Memory, 92*, 215–224.
Guerry, J. D., & Hastings, P. D. (2011). In search of HPA axis dysregulation in child and adolescent depression. *Clinical Child and Family Psychology Review, 14*, 135–160.
Haller, J., Bakos, N., Szirmay, M., Ledent, C., & Freund, T. F. (2002). The effects of genetic and pharmacological blockade of the CB1 cannabinoid receptor on anxiety. *European Journal of Neuroscience, 16*, 1395–1398.
Haring, M., Guggenhuber, S., & Lutz, B. (2012). Neuronal populations mediating the effects of endocannabinoids on stress and emotionality. *Neuroscience, 204*, 145–158.
Harkany, T., Keimpema, E., Barabás, K., & Mulder, J. (2008). Endocannabinoid functions controlling neuronal specification during brain development. *Molecular and Cellular Endocrinology, 286*, S84–S90.
Heng, L., Beverley, J. A., Steiner, H., & Tseng, K. Y. (2011). Differential developmental trajectories for CB1 cannabinoid receptor expression in limbic/associative and sensorimotor cortical areas. *Synapse, 65*, 278–286.
Herkenham, M., Lynn, A. B., Johnson, M. R., Melvin, L. S., de Costa, B. R., & Rice, K. C. (1991). Characterization and localization of cannabinoid receptors in rat brain: A quantitative in vitro autoradiographic study. *Journal of Neuroscience, 11*, 563–583.
Herman, J. P., Figueiredo, H., Mueller, N. K., Ulrich-Lai, Y., Ostrander, M. M., & Choi, D. C. (2003). Central mechanisms of stress integration: Hierarchical circuitry controlling hypothalamo-pituitary-adrenocortical responsiveness. *Frontiers in Neuroendocrinology, 24*, 151–180.
Hill, M. N. (2014). Clearing the smoke: What do we know about adolescent cannabis use and schizophrenia? *Journal of Psychiatry and Neuroscience, 39*, 75–77.
Hill, M. N., Carrier, E. J., Ho, W. S., Shi, L., Patel, S., Gorzalka, B. B., et al. (2008). Prolonged glucocorticoid treatment decreases cannabinoid CB1 receptor density in the hippocampus. *Hippocampus, 18*, 221–226.
Hill, M. N., Karatsoreos, I. N., Hillard, C. J., & McEwen, B. S. (2010). Rapid elevations in limbic endocannabinoid content by glucocorticoid hormones in vivo. *Psychoneuroendocrinology, 35*, 1333–1338.
Hill, M. N., & McEwen, B. S. (2009). Involvement of the endocannabinoid system in the neurobehavioural effects of stress and glucocorticoids. *Progress in Neuro-Psychopharmacology & Biological Psychiatry, 34*, 791–797.
Hill, M. N., McLaughlin, R. J., Bingham, B., Shrestha, L., Lee, T. T.-Y., Gray, J. M., et al. (2010). Endogenous cannabinoid signaling is essential for stress adaptation. *Proceedings of the National Academy of Sciences of the United States of America, 107*, 9406–9411.
Hill, M. N., McLaughlin, R. J., Morrish, A. C., Viau, V., Floresco, S. B., Hillard, C. J., et al. (2009). Suppression of amygdalar endocannabinoid signaling by stress contributes to activation of the hypothalamic-pituitary-adrenal axis. *Neuropsychopharmacology, 34*, 2733–2745.

Hill, M. N., McLaughlin, R. J., Pan, B., Fitzgerald, M. L., Roberts, C. J., Lee, T. T., et al. (2011). Recruitment of prefrontal cortical endocannabinoid signaling by glucocorticoids contributes to termination of the stress response. *Journal of Neuroscience, 31*, 10506–10615.

Hill, M. N., Patel, S., Campolongo, P., Tasker, J. G., Wotjak, C. T., & Bains, J. S. (2010). Functional interactions between stress and the endocannabinoid system: From synaptic signaling to behavioral output. *Journal of Neuroscience, 30*, 14980–14986.

Hill, M. N., & Tasker, J. G. (2012). Endocannabinoid signaling, glucocorticoid-mediated negative feedback, and regulation of the hypothalamic-pituitary-adrenal axis. *Neuroscience, 204*, 5–16.

Hillard, C. J. (2000). Biochemistry and pharmacology of the endocannabinoids arachidonylethanolamide and 2-arachidonylglycerol. *Prostaglandins and Other Lipid Mediators, 61*, 3–18.

Isgor, C., Kabbaj, M., Akil, H., & Watson, J. B. (2004). Delayed effects of chronic variable stress during peripubertal-juvenile period on hippocampal morphology and on cognitive and stress axis functions in rats. *Hippocampus, 14*, 636–648.

Jaferi, A., & Bhatnagar, S. (2006). Corticosterone can act at the posterior paraventricular thalamus to inhibit hypothalamic-pituitary-adrenal activity in animals that habituate to repeated stress. *Endocrinology, 147*, 4917–4930.

Jankord, R., Solomon, M. B., Albertz, J., Flak, J. N., Zhang, R., & Herman, J. P. (2011). Stress vulnerability during adolescent development in rats. *Endocrinology, 152*, 629–638.

Johnson, S. B., Blum, R. W., & Giedd, J. N. (2009). Adolescent maturity and the brain: The promise and pitfalls of neuroscience research in adolescent health policy. *Journal of Adolescent Health, 45*, 216–221.

Juraska, J. M., Sisk, C. L., & DonCarlos, L. L. (2013). Sexual differentiation of the adolescent rodent brain: Hormonal influences and developmental mechanisms. *Hormones and Behavior, 64*, 203–210.

Jutras-Aswad, D., DiNieri, J. A., Harkany, T., & Hurd, Y. L. (2009). Neurobiological consequences of maternal cannabis on human fetal development and its neuropsychiatric outcome. *European Archives of Psychiatry and Clinical Neuroscience, 259*, 395–412.

Karst, H., Berger, S., Erdmann, G., Schütz, G., & Joëls, M. (2010). Metaplasticity of amygdalar responses to the stress hormone corticosterone. *Proceedings of the National Academy of Sciences of the United States of America, 107*, 14449–14454.

Kessler, R. C., Demler, O., Frank, R. G., Olfson, M., Pincus, H. A., Walters, E. E., et al. (2005). Prevalence and treatment of mental disorders, 1990 to 2003. *New England Journal of Medicine, 352*, 2515–2523.

Kim-Cohen, J., Caspi, A., Moffitt, T. E., Harrington, H., Milne, B. J., & Poulton, R. (2003). Prior juvenile diagnoses in adults with mental disorder: Developmental follow-back of a prospective-longitudinal cohort. *Archives of General Psychiatry, 60*, 709–717.

Lee, T. T.-Y., & Gorzalka, B. B. (2012). Timing is everything: Evidence for a role of corticolimbic endocannabinoids in modulating hypothalamic-pituitary-adrenal axis activity across developmental periods. *Neuroscience, 204*, 17–30.

Lee, F. S., Heimer, H., Giedd, J. N., Lein, E. S., Šestan, N., Weinberger, D. R., et al. (2014). Mental health. Adolescent mental health—Opportunity and obligation. *Science, 346*, 547–549.

Lee, T. T.-Y., & Hill, M. N. (2013). Age of stress exposure modulates the immediate and sustained effects of repeated stress on corticolimbic cannabinoid CB(1) receptor binding in male rats. *Neuroscience, 249*, 106–114.

Lee, T. T.-Y., Hill, M. N., Hillard, C. J., & Gorzalka, B. B. (2013). Temporal changes in N-acylethanolamine content and metabolism throughout the peri-adolescent period. *Synapse, 67*, 4–10.

Lee, T. T.-Y., Hill, M. N., Hillard, C. J., & Gorzalka, B. B. (2015). Disruption of peri-adolescent endocannabinoid signaling modulates neuroendocrine and behavioural responses to stress in adulthood. *Neuropharmacology*, *99*, 89–97.

Lee, T. T.-Y., Wainwright, S., Hill, M., Galea, L., & Gorzalka, B. (2014). Sex, drugs, and adult neurogenesis: Sex-dependent effects of escalating adolescent cannabinoid exposure on adult hippocampal neurogenesis, stress reactivity, and amphetamine sensitization. *Hippocampus*, *24*, 280–292.

Lewis, E. M., Barnett, J. F., Jr., Freshwater, L., Hoberman, A. M., & Christian, M. S. (2002). Sexual maturation data for CRL Sprague-Dawley rats: Criteria and confounding factors. *Drug and Chemical Toxicology*, *25*, 437–458.

Long, L. E., Lind, J., Webster, M., & Weickert, C. S. (2012). Developmental trajectory of the endocannabinoid system in human dorsolateral prefrontal cortex. *BMC Neuroscience*, *13*, 87.

Lopez-Larson, M. P., Bogorodzki, P., Rogowska, J., McGlade, E., King, J. B., Terry, J., et al. (2011). Altered prefrontal and insular cortical thickness in adolescent marijuana users. *Behavioural Brain Research*, *220*, 164–172.

Lovelace, J. W., Corches, A., Vieira, P. A., Mackie, K., & Korzus, E. (2015). An animal model of female adolescent cannabinoid exposure elicits a long-lasting deficit in presynaptic long-term plasticity. *Neuropharmacology*, *99*, 242–255.

Maccarrone, M., Guzmán, M., Mackie, K., Doherty, P., & Harkany, T. (2014). Programming of neural cells by (endo) cannabinoids: From physiological rules to emerging therapies. *Nature Reviews Neuroscience*, *15*, 786–801.

Mackie, K. (2006). Mechanisms of CB1 receptor signaling: Endocannabinoid modulation of synaptic strength. *International Journal of Obesity*, *30*, S19–S23.

Malcher-Lopes, R., Di, S., Marcheselli, V. S., Weng, F. J., Stuart, C. T., Bazan, N. G., et al. (2006). Opposing crosstalk between leptin and glucocorticoids rapidly modulates synaptic excitation via endocannabinoid release. *Journal of Neuroscience*, *26*, 6643–6650.

Marco, E. M., Adriani, W., Llorente, R., Laviola, G., & Viveros, M. P. (2009). Detrimental psychophysiological effects of early maternal deprivation in adolescent and adult rodents: Altered responses to cannabinoid exposure. *Neuroscience and Biobehavioral Reviews*, *33*, 498–507.

Marco, E. M., Rubino, T., Adriani, W., Viveros, M. P., Parolaro, D., & Laviola, G. (2009). Long-term consequences of URB597 administration during adolescence on cannabinoid CB1 receptor binding in brain areas. *Brain Research*, *1257*, 25–31.

Marshall, W., & Tanner, J. (1969). Variations in patterns of pubertal changes in girls. *Archives of Disease in Childhood*, *44*, 291–303.

Marshall, W., & Tanner, J. (1970). Variations in the pattern of pubertal changes in boys. *Archives of Disease in Childhood*, *45*, 14–23.

Mateos, B., Borcel, E., Loriga, R., Luesu, W., Bini, V., Llorente, R., et al. (2011). Adolescent exposure to nicotine and/or the cannabinoid agonist CP 55,940 induces gender-dependent long-lasting memory impairments and changes in brain nicotinic and CB1 cannabinoid receptors. *Journal of Psychopharmacology*, *25*, 1676–1690.

Mato, S., Del Olmo, E., & Pazos, A. (2003). Ontogenetic development of cannabinoid receptor expression and signal transduction functionality in the human brain. *European Journal of Neuroscience*, *17*, 1747–1754.

McCormick, C. M., Mathews, I. Z., Thomas, C., & Waters, P. (2010). Investigations of HPA function and the enduring consequences of stressors in adolescence in animal models. *Brain and Cognition*, *72*, 73–85.

McEwen, B. S. (1992). Steroid hormones: Effect on brain development and function. *Hormone Research*, *37*, 1–10.

McEwen, B. S. (2005). Glucocorticoids, depression and mood disorders: Structural remodeling in the brain. *Metabolism*, *54*, 20–23.

McEwen, B. S. (2007). Physiology and neurobiology of stress and adaptation: Central role of the brain. *Physiological Reviews*, *87*, 873–904.

Miller, G. E., Chen, E., & Zhou, E. S. (2007). If it goes up, must it come down? Chronic stress and the hypothalamic-pituitary-adrenocortical axis in humans. *Psychological Bulletin*, *133*, 25–45.

Mulder, J., Aguado, T., Keimpema, E., Barabas, K., Ballester Rosado, C. J., Nguyen, L., et al. (2008). Endocannabinoid signaling controls pyramidal cell specification and long-range axon patterning. *Proceedings of the National Academy of Sciences of the United States of America*, *105*, 8760–8765.

Onaivi, E., Ishiguro, H., Gong, J., Patel, S., Perchuk, A., Meozzi, P., et al. (2006). Discovery of the presence and functional expression of cannabinoid CB2 receptors in brain. *Annals of the New York Academy of Sciences*, *1074*, 514–536.

Ordaz, S., & Luna, B. (2012). Sex differences in physiological reactivity to acute psychosocial stress in adolescence. *Psychoneuroendocrinology*, *37*, 1135–1157.

O'Shea, M., McGregor, I. S., & Mallet, P. E. (2006). Repeated cannabinoid exposure during perinatal, adolescent or early adult ages produces similar long lasting deficits in object recognition and reduced social interaction in rats. *Journal of Psychopharmacology*, *20*, 611–621.

O'Shea, M., Singh, M. E., McGregor, I. S., & Mallet, P. E. (2004). Chronic cannabinoid exposure produces lasting memory impairment and increased anxiety in adolescent but not adult rats. *Journal of Psychopharmacology*, *18*, 502–508.

Patel, S., & Hillard, C. J. (2008). Adaptations in endocannabinoid signaling in response to repeated homotypic stress: A novel mechanism for stress habituation. *European Journal of Neuroscience*, *27*, 2821–2829.

Patel, S., Kingsley, P. J., Mackie, K., Marnett, L. J., & Winder, D. G. (2009). Repeated homotypic stress elevates 2-arachidonoylglycerol levels and enhances short-term endocannabinoid signaling at inhibitory synapses in basolateral amygdala. *Neuropsychopharmacology*, *34*, 2699–2709.

Patel, S., Roelke, C. T., Rademacher, D. J., Cullinan, W. E., & Hillard, C. J. (2004). Endocannabinoid signaling negatively modulates stress-induced activation of the hypothalamic-pituitary-adrenal axis. *Endocrinology*, *145*, 5431–5438.

Patel, S., Roelke, C. T., Rademacher, D. J., & Hillard, C. J. (2005). Inhibition of restraint-stress induced neural and behavioural activation by endogenous cannabinoid signalling. *European Journal of Neuroscience*, *21*, 1057–1069.

Paus, T., Keshavan, M., & Giedd, J. N. (2008). Why do many psychiatric disorders emerge during adolescence? *Nature Reviews Neuroscience*, *9*, 947–957.

Pignatelli, D., Xiao, F., Gouveia, A. M., Ferreira, J. G., & Vinson, G. P. (2006). Adrenarche in the rat. *Journal of Endocrinology*, *191*, 301–308.

Pine, D. S., Cohen, P., Gurley, D., Brook, J., & Ma, Y. (1998). The risk for early-adulthood anxiety and depressive disorders in adolescents with anxiety and depressive disorders. *Archives of General Psychiatry*, *55*, 56–64.

Rademacher, D., Meier, S., Shi, L., Ho, W., Jarrahian, A., & Hillard, C. (2008). Effects of acute and repeated restraint stress on endocannabinoid content in the amygdala, ventral striatum, and medial prefrontal cortex in mice. *Neuropharmacology*, *54*, 108–116.

Radley, J., Arias, C., & Sawchenko, P. (2006). Regional differentiation of the medial prefrontal cortex in regulating adaptive responses to acute emotional stress. *Journal of Neuroscience*, *26*, 12967–12976.

Raver, S. M., & Keller, A. (2014). Permanent suppression of cortical oscillations in mice after adolescent exposure to cannabinoids: Receptor mechanisms. *Neuropharmacology*, *86*, 161–173.

Realini, N., Vigano', D., Guidali, C., Zamberletti, E., Rubino, T., & Parolaro, D. (2011). Chronic URB597 treatment at adulthood reverted most depressive-like symptoms induced by adolescent exposure to THC in female rats. *Neuropharmacology*, *60*, 235–243.

Reich, C. G., Mihalik, G. R., Iskander, A. N., Seckler, J. C., & Weiss, M. S. (2013). Adolescent chronic mild stress alters hippocampal CB1 receptor-mediated excitatory neurotransmission and plasticity. *Neuroscience*, *253*, 444–454.

Reich, C. G., Taylor, M. E., & McCarthy, M. M. (2009). Differential effects of chronic unpredictable stress on hippocampal CB1 receptors in male and female rats. *Behavioural Brain Research*, *203*, 264–269.

Rice, D., & Barone, S. J. (2000). Critical periods of vulnerability for the developing nervous system: Evidence from humans and animal models. *Environmental Health Perspectives*, *108*, 511–533.

Rodriguez de Fonseca, F., Ramos, J. A., Bonnin, A., & Fernández-Ruiz, J. J. (1993). Presence of cannabinoid binding sites in the brain from early postnatal stages. *Neuroreport*, *4*, 135–138.

Romeo, R. D. (2010a). Adolescence: A central event in shaping stress reactivity. *Developmental Psychobiology*, *52*, 244–253.

Romeo, R. D. (2010b). Pubertal maturation and programming of hypothalamic–pituitary–adrenal reactivity. *Frontiers in Neuroendocrinology*, *31*, 232–240.

Romeo, R. D., Bellani, R., Karatsoreos, I. N., Chhua, N., Vernov, M., Conrad, C. D., et al. (2006). Stress history and pubertal development interact to shape hypothalamic-pituitary-adrenal axis plasticity. *Endocrinology*, *147*, 1664–1674.

Romeo, R. D., Ferhina, S. A., Karatsoreos, I. N., Bellani, R., Chhua, N., Vernov, M., et al. (2008). Glucocorticoid receptor mRNA expression in the hippocampal formation of male rats before and after pubertal development in response to acute or repeated restraint stress. *Neuroendocrinology*, *87*, 160–167.

Romeo, R. D., Lee, S. J., Chhua, N., McPherson, C. R., & McEwen, B. S. (2004). Testosterone cannot activate an adult-like stress response in prepubertal male rats. *Neuroendocrinology*, *79*, 125–132.

Romeo, R. D., Lee, S. J., & McEwen, B. S. (2004). Differential stress reactivity in intact and ovariectomized prepubertal and adult female rats. *Neuroendocrinology*, *80*, 387–393.

Romeo, R. D., & McEwen, B. S. (2006). Stress and the adolescent brain. *Annals of the New York Academy of Sciences*, *1094*, 202–214.

Romeo, R., Minhas, S., Svirsky, S., Hall, B., Savenkova, M., & Karatsoreos, I. (2014). Pubertal shifts in adrenal responsiveness to stress and adrenocorticotropic hormone in male rats. *Psychoneuroendocrinology*, *42*, 146–152.

Romero, E. M., Fernández, B., Sagredo, O., Gomez, N., Urigüen, L., Guaza, C., et al. (2002). Antinociceptive, behavioural and neuroendocrine effects of CP 55,940 in young rats. *Brain Research. Developmental Brain Research*, *136*, 85–92.

Rubino, T., & Parolaro, D. (2008). Long lasting consequences of cannabis exposure in adolescence. *Molecular and Cellular Endocrinology*, *286*, S108–S113.

Rubino, T., Prini, P., Piscitelli, F., Zamberletti, E., Trusel, M., Melis, M., et al. (2015). Adolescent exposure to THC in female rats disrupts developmental changes in the prefrontal cortex. *Neurobiology of Disease*, *73*, 60–69.

Rubino, T., Realini, N., Braida, D., Alberio, T., Capurro, V., Viganò, D., et al. (2009). The depressive phenotype induced in adult female rats by adolescent exposure to THC is associated with cognitive impairment and altered neuroplasticity in the prefrontal cortex. *Neurotoxicity Research*, *15*, 567–607.

Rubino, T., Realini, N., Braida, D., Guidi, S., Capurro, V., Viganò, D., et al. (2009). Changes in hippocampal morphology and neuroplasticity induced by adolescent THC treatment are associated with cognitive impairment in adulthood. *Hippocampus*, *19*, 363–372.

Rubino, T., Vigano, D., Realini, N., Guidali, C., Braida, D., Capurro, V., et al. (2008). Chronic delta 9-tetrahydrocannabinol during adolescence provokes sex-dependent changes in the emotional profile in adult rats: Behavioral and biochemical correlates. *Neuropsychopharmacology, 33*, 1760–1769.

Rubino, T., Zamberletti, E., & Parolaro, D. (2012). Adolescent exposure to cannabis as a risk factor for psychiatric disorders. *Journal of Psychopharmacology, 26*, 177–188.

Rubio, P., Rodríguez de Fonseca, F., Muñoz, R. M., Ariznavarreta, C., Martín-Calderón,-J. L., & Navarro, M. (1995). Long-term behavioral effects of perinatal exposure to delta 9-tetrahydrocannabinol in rats: Possible role of pituitary-adrenal axis. *Life Sciences, 56*, 2169–2176.

Sapolsky, R. M., Krey, L., & McEwen, B. S. (1984). Glucocorticoid sensitive hippocampal neurons are involved in terminating the adrenocortical stress response. *Proceedings of the National Academy of Sciences of the United States of America, 81*, 6174–6177.

Schneider, M. (2008). Puberty as a highly vulnerable developmental period for the consequences of cannabis exposure. *Addiction Biology, 13*, 253–263.

Schneider, M., & Koch, M. (2003). Chronic pubertal, but not adult chronic cannabinoid treatment impairs sensorimotor gating, recognition memory, and the performance in a progressive ratio task in adult rats. *Neuropsychopharmacology, 28*, 1760–1769.

Schneider, M., Schömig, E., & Leweke, F. M. (2008). Acute and chronic cannabinoid treatment differentially affects recognition memory and social behavior in pubertal and adult rats. *Addiction Biology, 13*, 345–357.

Schoenfeld, N. M., Leadhem, J. H., & Rabii, J. (1980). Maturation of adrenal stress responsiveness in the rat. *Neuroendocrinology, 31*, 101–105.

Sisk, C. L., & Foster, D. L. (2004). The neural basis of puberty and adolescence. *Nature Neuroscience*, 7, 1040–1047.

Sisk, C. L., & Zehr, J. L. (2005). Pubertal hormones organize the adolescent brain and behavior. *Frontiers in Neuroendocrinology, 26*, 163–174.

Somerville, L. H., Jones, R. M., & Casey, B. J. (2010). A time of change: Behavioral and neural correlates of adolescent sensitivity to appetitive and aversive environmental cues. *Brain and Cognition, 72*, 124–133.

Spear, L. P. (2000). The adolescent brain and age-related behavioral manifestations. *Neuroscience and Biobehavioral Reviews, 24*, 417–463.

Stanton, M. E., & Levine, S. (1990). Inhibition of infant glucocorticoid stress response: Specific role of maternal cues. *Developmental Psychobiology, 23*, 411–426.

Steiner, M. A., Marsicano, G., Nestler, E. J., Holsboer, F., Lutz, B., & Wotjak, C. T. (2008). Antidepressant-like behavioral effects of impaired cannabinoid receptor type 1 signaling coincide with exaggerated corticosterone secretion in mice. *Psychoneuroendocrinology, 33*, 54–67.

Sugiura, T., Kobayashi, Y., Oka, S., & Waku, K. (2002). Biosynthesis and degradation of anandamide and 2-arachidonoylglycerol and their possible physiological significance. *Prostaglandins, Leukotrienes and Essential Fatty Acids, 66*, 173–192.

Sullivan, R., & Holman, P. (2011). Transitions in sensitive period attachment learning in infancy: The role of corticosterone. *Neuroscience and Biobehavioral Reviews, 34*, 835–844.

Tasker, J. G., & Herman, J. P. (2011). Mechanisms of rapid glucocorticoid feedback inhibition of the hypothalamic-pituitary-adrenal axis. *Stress, 14*, 398–406.

Ulrich-Lai, Y. M., & Herman, J. P. (2009). Neural regulation of endocrine and autonomic stress responses. *Nature Reviews Neuroscience, 10*, 397–409.

van Leeuwen, A. P., Creemers, H. E., Greaves-Lord, K., Verhulst, F. C., Ormel, J., & Huizink, A. C. (2011). Hypothalamic-pituitary-adrenal axis reactivity to social stress and adolescent cannabis use: The TRAILS study. *Addiction, 106*, 1484–1492.

Van Sickle, M., Duncan, M., Kingsley, P., Mouihate, A., Urbani, P., Mackie, K., et al. (2005). Identification and functional characterization of brainstem cannabinoid CB2 receptors. *Science, 310*, 329–332.
Vázquez, D. (1998). Stress and the developing limbic-hypothalamic-pituitary-adrenal axis. *Psychoneuroendocrinology, 23*, 663–700.
Vázquez, D. M., & Akil, H. (1993). Pituitary-adrenal response to ether vapor in the weanling animal: Characterization of the inhibitory effect of glucocorticoids on adrenocorticotropin secretion. *Pediatric Research, 34*, 646–653.
Walker, E. F., Sabuwalla, Z., & Huot, R. (2004). Pubertal maturation, stress sensitivity and psychopathology. *Development and Psychopathology, 16*, 807–824.
Wamsteeker, J. I., Kuzmiski, J. B., & Bains, J. S. (2010a). Maintenance of retrograde endocannabinoid signaling at synapses onto hypothalamic neuroendocrine cells reflects age-specific adaptation to stress. Program No. 389.18, In *40th annual meeting of the society for neuroscience. Neuroscience meeting planner. Online, San Diego, CA.*
Wamsteeker, J. I., Kuzmiski, J. B., & Bains, J. S. (2010b). Repeated stress impairs endocannabinoid signaling in the paraventricular nucleus of the hypothalamus. *Journal of Neuroscience, 20*, 11188–11196.
Wang, M., Hill, M. N., Zhang, L., Gorzalka, B. B., Hillard, C. J., & Alger, B. E. (2011). Acute restraint stress enhances hippocampal endocannabinoid function via glucocorticoid receptor activation. *Journal of Psychopharmacology, 26*, 56–70.
Wegener, N., & Koch, M. (2009). Behavioural disturbances and altered Fos protein expression in adult rats after chronic pubertal cannabinoid treatment. *Brain Research, 1253*, 81–91.
Weintraub, A., Singaravelu, J., & Bhatnagar, S. (2010). Enduring and sex-specific effects of adolescent social isolation in rats on adult stress reactivity. *Brain Research, 1343*, 83–92.
Wenger, T., Gerendai, I., Fezza, F., González, S., Bisogno, T., Fernandez-Ruiz, J., et al. (2002). The hypothalamic levels of the endocannabinoid, anandamide, peak immediately before the onset of puberty in female rats. *Life Sciences, 70*, 1407–1414.
Zamberletti, E., Beggiato, S., Steardo, L. J., Prini, P., Antonelli, T., Ferraro, L., et al. (2014). Alterations of prefrontal cortex GABAergic transmission in the complex psychotic-like phenotype induced by adolescent delta-9-tetrahydrocannabinol exposure in rats. *Neurobiology of Disease, 63*, 35–47.
Zurolo, E., Iyer, A. M., Spliet, W. G. M., Van Rijen, P. C., Troost, D., Gorter, J. A., et al. (2010). CB1 and CB2 cannabinoid receptor expression during development and in epileptogenic developmental pathologies. *Neuroscience, 170*, 28–41.

CHAPTER THREE

The Endocannabinoid System and Its Role in Regulating the Intrinsic Neural Circuitry of the Gastrointestinal Tract

Samantha M. Trautmann[*,†,‡], **Keith A. Sharkey**[*,†,‡,1]
[*]Hotchkiss Brain Institute, Cumming School of Medicine, University of Calgary, Calgary, Alberta, Canada
[†]Snyder Institute for Chronic Diseases, Cumming School of Medicine, University of Calgary, Calgary, Alberta, Canada
[‡]Department of Physiology and Pharmacology, Cumming School of Medicine, University of Calgary, Calgary, Alberta, Canada
[1]Corresponding author: e-mail address: ksharkey@ucalgary.ca

Contents

1. Introduction 87
2. The Enteric Nervous System and the Brain–Gut Axis 88
   2.1 The Structure of the ENS 89
3. The Endocannabinoid System 91
   3.1 Cannabinoid Receptors and Their Ligands 91
   3.2 Cannabinoid Receptors in the ENS 92
   3.3 Endocannabinoid Synthesis 94
   3.4 2-AG Synthesis and Degradation 95
   3.5 Anandamide Synthesis and Degradation 97
4. Endocannabinoid Signaling Mechanisms 98
   4.1 Endocannabinoid Transport Mechanisms 98
5. Other Receptors, Agonists, and Antagonists 100
   5.1 Phytocannabinoids 100
   5.2 Peptide Endocannabinoids 102
   5.3 Transient Receptor Potential Vanilloid 1 102
   5.4 Virodhamine, an Endogenous CB_1 Antagonist 103
   5.5 G Protein-Coupled Receptor 55 103
   5.6 Oleoylethanolamide and Palmitoylethanolamide 104
6. Endocannabinoid Tone 105
7. Cannabinoid Receptor Metaplasticity in the ENS 107
8. Cannabinoid Receptor Properties 107
   8.1 Selective Coupling of GPCRs 107
   8.2 Dimerization of CB Receptors and Interactions with Other Receptors 108
   8.3 Desensitization of Cannabinoid Receptors 109
9. Enteric Glia 110

International Review of Neurobiology, Volume 125
ISSN 0074-7742
http://dx.doi.org/10.1016/bs.irn.2015.10.002

10. Neurotransmission in the ENS and GI Tract 110
10.1 Neuromuscular Transmission 110
10.2 Secretomotor Transmission 112
10.3 Enteroendocrine Cells 114
11. Brain–Gut Axis 114
12. The Endocannabinoid System of the ENS in GI Inflammation 115
13. Summary and Future Directions 116
Disclosures 116
References 117

Abstract

Endocannabinoids are important neuromodulators in the central nervous system. They regulate central transmission through pre- and postsynaptic actions on neurons and indirectly through effects on glial cells. Cannabinoids (CBs) also regulate neurotransmission in the enteric nervous system (ENS) of the gastrointestinal (GI) tract. The ENS consists of intrinsic primary afferent neurons, interneurons, and motor neurons arranged in two ganglionated plexuses which control all the functions of the gut. Increasing evidence suggests that endocannabinoids are potent neuromodulators in the ENS. In this review, we will highlight key observations on the localization of CB receptors and molecules involved in the synthesis and degradation of endocannabinoids in the ENS. We will discuss endocannabinoid signaling mechanisms, endocannabinoid tone and concepts of CB receptor metaplasticity in the ENS. We will also touch on some examples of enteric neural signaling in relation neuromuscular, secretomotor, and enteroendocrine transmission in the ENS. Finally, we will briefly discuss some key future directions.

ABBREVIATIONS

2-AG 2-arachidonoylglycerol
5-HT serotonin
AEA anandamide
AH after hyperpolarization
CB cannabinoid
CBD cannabidiol
CCK cholecystokinin
ChAT choline acetyltransferase
CNS central nervous system
DAG diacylglycerol
ENS enteric nervous system
GI gastrointestinal
GIP glucose-dependent insulinotropic peptide
GLP-2 glucagon-like peptide 2
GPCR G protein-coupled receptor
LPS lipopolysaccharide
MAGL monoacylglycerol
NAPE *N*-acyl phosphatidylethanolamine

NAPE-PLD *N*-acyl phosphatidylethanolamine phospholipase D
NANC nonadrenergic noncholinergic
NF-κB nuclear factor kappa B
NO nitric oxide
NOS nitric oxide synthase
THC Δ^9-tetrahydrocannabinol
THCV Δ^9-tetrahydrocannabidivarin
TRPV1–4 transient receptor potential vanilloid 1–4 receptor
VIP vasoactive intestinal peptide

1. INTRODUCTION

Extracts of the hemp plant (*Cannabis sativa*) have been used for millennia to relieve pain, as well as many other conditions including intestinal cramping and diarrhea (Zuardi, 2006). Gaoni and Mechoulam (1964) isolated and identified many of the components of *Cannabis*, including Δ^9-tetrahydrocannabinol (THC), the major psychoactive compound of the plant. This discovery led to the subsequent identification of receptors for THC; cannabinoid (CB) receptor 1 (CB_1; Matsuda, Lolait, Brownstein, Young, & Bonner, 1990) in the brain and cannabinoid receptor 2 (CB_2; Munro, Thomas, & Abu-Shaar, 1993) in the periphery. Shortly after the discovery of the CB_1 receptor, endogenous ligands for these receptors, anandamide (AEA) and 2-arachidonoylglycerol (2-AG), together termed endocannabinoids, were isolated and found to be lipid mediators derived from arachidonic acid (Devane et al., 1992; Mechoulam et al., 1995).

The realization that there was a widely distributed "endocannabinoid system" in the brain led investigators to focus on the role of these newly discovered neurotransmitters. In 2001, three laboratories showed independently that endocannabinoids were retrograde synaptic transmitters that presynaptically inhibit neurotransmitter release (Kreitzer & Regehr, 2001; Ohno-Shosaku, Maejima, & Kano, 2001; Wilson & Nicoll, 2001). Endocannabinoids are one of the most widespread transmitter systems acting in almost every region of the brain. More recently, postsynaptic actions of endocannabinoids have been reported and endocannabinoids may also influence central transmission through actions involving glial cells in the brain (Oliveira da Cruz, Robin, Drago, Marsicano, & Metna-Laurent, 2015; Stella, 2010).

CB receptors are very widely distributed throughout the body. They are prominently expressed in the gastrointestinal (GI) tract, which explains some of the therapeutic benefits of *Cannabis* in GI disorders. In early studies, THC was shown to inhibit GI motility through CB_1 receptors (Pertwee, 2001; Pinto, Capasso, Di Carlo, & Izzo, 2002). Endocannabinoids are also found in the GI tract, in fact, 2-AG was originally isolated from the gut (Mechoulam et al., 1995). CBs and endocannabinoids are able to regulate key GI functions including motility and secretion, making the endocannabinoid system an important regulatory element of the GI tract.

In this review, we will focus on how the endocannabinoid system serves to regulate gut function through actions on the enteric nervous system (ENS): the intrinsic nervous system of the GI tract. We will examine how endocannabinoids influence the enteric neural control of GI physiology. We will address endocannabinoid signaling mechanisms, ligands and their receptors and focused aspects of the neural control of the GI tract with an emphasis on the role of endocannabinoids. Interested readers are encouraged to also look at other recent reviews on this subject (Cani, 2012; Cluny, Reimer, & Sharkey, 2012; Izzo & Sharkey, 2010; Maccarrone et al., 2015; Nasser, Bashashati, & Andrews, 2014).

2. THE ENTERIC NERVOUS SYSTEM AND THE BRAIN–GUT AXIS

The GI tract is extensively innervated by the third division of the autonomic nervous system, the ENS. The ENS is found throughout the wall of the gut from the esophagus to the anus. It is essential for regulating motility and secretion, and plays important roles in epithelial barrier, vasomotor and immune functions in the gut (Furness, 2012). Though the ENS receives sympathetic and parasympathetic input from the central nervous system (CNS), it is able to function autonomously. Thus the ENS is more "brain-like" than other components of the autonomic nervous system and is a complex, integrative neural network. The ENS consists of intrinsic primary afferent, interneurons and motor neurons arranged in two ganglionated plexuses. The myenteric plexus lies between the longitudinal and circular muscle layers and the submucosal plexus in the submucosa, between the circular muscle and mucosa. The submucosal plexus contains fewer neurons than the myenteric plexus and is not found in the esophagus or stomach, which only have a myenteric plexus. Both plexuses are found throughout the rest of the length of the small and large intestines (Furness, 2006). The human

ENS has around 200–400 million neurons, which is similar to the number of neurons found in the spinal cord (Furness, 2012).

The ENS is an important component in the brain–gut axis. The brain–gut axis is the system of central, autonomic and enteric nerves that provide bidirectional communication between the gut and the brain. Interoceptive and nociceptive signals from the gut relay critical information about the state of digestion, the luminal environment, and mucosal defense to the brain. In turn the brain, via the autonomic nervous system, can influence GI functions; though the ENS operates autonomously in most circumstances. In health, we are rarely aware of the state of our GI tract except for gastric fullness after a large meal, hunger, the passage of gas and the regular episodes of rectal fullness that precede defecation. In contrast, in GI disorders such as irritable bowel syndrome, pain and bloating occur together with marked alterations in the function of the gut, leading to constipation, diarrhea or both. Irritable bowel syndrome is a disorder of the brain–gut axis, illustrating how alterations in the sensory-motor neural control of the gut can profoundly affect intestinal homeostasis (Furness, Callaghan, Rivera, & Cho, 2014).

Recently, considerable attention has been paid to the presence and function of the enteric microbiota (Rhee, Pothoulakis, & Mayer, 2009). The gut is host to a vast number of bacteria (~100 trillion) consisting of between 500 and 1000 commensal bacterial species (Mayer, Tillisch, & Gupta, 2015; Rhee et al., 2009). These play a critical role in homeostasis, and digestion and are now recognized to signal back and forth between themselves and between their host. Microbial communication to the ENS influences gut motor and secretory functions (Cryan & O'Mahony, 2011; Koboziev, Reinoso Webb, Furr, & Grisham, 2014). Moreover, the composition of the microbiome also alters CNS functions, including higher cognitive functions such as emotion (Mayer et al., 2015). Thus, it is crucial to understand the neural circuitry of the ENS as it not only has an impact on GI functions but also on the communication between the brain and gut that can affect higher brain functions.

2.1 The Structure of the ENS

The ENS consists of a dense neuropil consisting of the unmyelinated axons and dendrites of enteric neurons and the processes of enteric glia, astrocyte-like glial cells that surround enteric neurons (Furness, 2012; Gulbransen & Sharkey, 2012). Enteric glia outnumber enteric neurons and are involved in regulatory functions within the GI tract (Sharkey,

2015). While it is recognized that different populations of enteric glia exist, classifying enteric glia into subpopulations has been difficult (Gulbransen & Sharkey, 2012). Enteric glia not only lie within the ENS but also are found in the mucosa lying beneath the epithelium. These cells have been suggested to be important for regulating barrier function as the ablation of enteric glia *in vivo* induces fulminant jejuno-ileitis due to disruption of the epithelial barrier (Neunlist et al., 2013). Consistent with this idea, it was shown in a coculture model, that enteric glia cells increase barrier resistance and thus reduce barrier permeability (Savidge et al., 2007). Enteric glia also play a role in modulating neurotransmission in the ENS (Gulbransen & Sharkey, 2012). Enteric glia, like astrocytes, are able to regulate several important factors of neurotransmission such as the availability, degradation, and sequestering of neurotransmitters.

The ENS receives autonomic input from sympathetic and parasympathetic nerves that arise from abdominal prevertebral ganglia and the brainstem and spinal cord, respectively (Furness, 2006, 2012) There is a direct sympathetic innervation to the enteric ganglia that is involved in controlling neurotransmission and also to blood vessels, regulating blood flow through vasoconstriction. Parasympathetic nerves innervate the enteric ganglia and modulate the enteric neural programs that control motility and secretion.

Enteric nerves in the mucosa do not penetrate the epithelium lying below it. Hence luminal signaling to the ENS is indirect and is accomplished through the various populations of enteroendocrine cells that lie in the epithelium along the length of the GI tract. These cells sense the luminal environment within the intestine and relay information to the ENS (Bohorquez & Liddle, 2015). The largest population of enteroendocrine cells contain serotonin (5-HT) and are called enterochromaffin cells (Gershon, 2013). 5-HT is an important mediator of peristalsis and other motor and secretory reflexes (Smith, Park, & Hennig, 2014). Other populations of enteroendocrine cells sense luminal nutrients, the products of digestion and bacteria and release peptides that serve as paracrine mediators or hormones. Paracrine mediators include somatostatin, neurotensin, glucagon-like peptide 2 (GLP-2) and cholecystokinin (CCK) (Kairupan et al., 2015; Rehfeld, 2014). CCK is also a hormone and acts alongside such molecules peptide YY, glucose-dependent insulinotropic peptide (GIP), and GLP-1 to regulate glucose homeostasis, food intake, and energy balance (Ellis, Chambers, Gwynne, & Bornstein, 2013; Psichas, Reimann, & Gribble, 2015; Spreckley & Murphy, 2015).

For the rest of this review, we will consider the endocannabinoid system and how endocannabinoids impact the neural control of enteric reflex circuits and the brain–gut axis. This is an emerging area of research and compared to our understanding of the role of endocannabinoids in the brain, relatively little is known about the role of endocannabinoids in the neural control of the GI tract.

3. THE ENDOCANNABINOID SYSTEM

The endocannabinoid system is comprised of two CB receptors (CB_1 and CB_2), their endogenous ligands, 2-AG (Devane et al., 1992), and *N*-arachidonoylethanolamine (AEA; Mechoulam et al., 1995), and a variety of enzymes required for the synthesis and degradation of these ligands. The endocannabinoid system is unique as its ligands are synthesized "on demand" and typically act as retrograde messengers binding to CB receptors on presynaptic terminal (Katona & Freund, 2012; Piomelli, 2003). More recent evidence also supports postsynaptic actions of endocannabinoids in the CNS and actions mediated by glial cells (Navarrete, Diez, & Araque, 2014).

3.1 Cannabinoid Receptors and Their Ligands

CB receptors are G protein-coupled receptors (GPCRs) that primarily couple to $G_{i/o}$ proteins to inhibit adenylate cyclase and promote mitogen-activated protein kinase signaling (Howlett, 2005; Pertwee, 2006; Pertwee et al., 2010). Furthermore, they inhibit N- and P/Q-type calcium channels and couple positively to activate A-type outward potassium channels. Together these actions prevent calcium influx into the cell, block depolarization, and inhibit the release of neurotransmitters into the synaptic cleft (Howlett, 2005).

In the ENS, the primary fast excitatory neurotransmitter is acetylcholine, but unlike in the CNS, there are no fast inhibitory neurotransmitters (Furness, 2006). The effect of activation of CB_1 receptors is to inhibit acetylcholine release from presynaptic terminals (Coutts & Pertwee, 1997; Pertwee, Fernando, Nash, & Coutts, 1996). There have not been detailed studies examining the molecular mechanism of CB_1 receptor activation in the ENS.

There are several endogenous, exogenous, and synthetic ligands that act as agonists of the CB receptors. Ligands that act as CB agonists are grouped into four classes based on their chemical structures: (i) classic CBs, otherwise, exogenous CBs that are plant derived such as THC (ii) nonclassic CBs such as CP55,940, (iii) Eicosanoids, the endogenously produced CBs AEA, 2-AG, and (iv) aminoalkylindoles (WIN55,212-2) (Hudson, Hebert, & Kelly, 2010a; Pertwee et al., 2010). Furthermore, there are several inverse agonists and antagonists that act on these receptors, most of which are synthetically produced. Virodhamine, is the only known endogenous antagonist but only when it binds to the CB_1 receptor (Porter et al., 2002).

CB_2 and CB_1 exhibit only 44% sequence homology between each other allowing room for functional differences between the receptors. For example, virodhamine acts as a full agonist on the CB_2 receptor rather than an antagonist (Porter et al., 2002). Also, there is some evidence suggesting that in the adult hippocampus and brainstem, CB_2 is located postsynaptically rather than presynaptically like CB_1 (Brusco, Tagliaferro, Saez, & Onaivi, 2008; Van Sickle et al., 2005). Unlike CB_1, the CB_2 receptor also appears to be inducible and whereas CB_1 is implicated in control under physiological conditions, CB_2 is thought to be of more importance under inflammatory conditions (Basu & Dittel, 2011; Benito et al., 2008; Patel, Davison, Pittman, & Sharkey, 2010). This idea will be discussed later in the review.

3.2 Cannabinoid Receptors in the ENS

CB_1 receptors were the first CB receptors localized in the ENS. In early studies, CB_1 mRNA was shown in the embryonic rat to be localized to both the myenteric and submucosal plexuses and in the adult guinea pig in the myenteric plexus (Buckley, Hansson, Harta, & Mezey, 1998; Griffin et al., 1997). Griffin et al. (1997) were not able to detect CB_2 mRNA in the guinea pig myenteric plexus, but they were able to show it in whole gut extracts. Using immunohistochemistry, Kulkarni-Narla and Brown showed that CB_1 receptor immunoreactivity colocalized with choline acetyltransferase (ChAT), the biosynthetic enzyme for acetylcholine in the ENS of the porcine ileum and colon (Kulkarni-Narla & Brown, 2000), which was consistent with the functional evidence described above showing CB_1 receptor activation inhibited acetylcholine release. In this study, however, no CB_1 immunoreactivity was seen in either substance P, vasoactive intestinal peptide (VIP) or nitric oxide synthase (NOS) containing myenteric or submucosal neurons. The former are excitatory motor and secretomotor neurons and so this result was somewhat surprising, particularly since substance P and ChAT colocalized to a great extent. That CB_1 receptors were not found on NOS neurons suggested that CB_1 does not influence these inhibitory motor neurons and similarly for VIP, which is a marker of inhibitory motor neurons and secretomotor neurons.

Coutts et al. performed an extensive evaluation of the neuronal subpopulations of myenteric neurons in the guinea pig and rat ileum that expressed CB_1 receptor (Coutts, Irving, Mackie, Pertwee, & Anavi-Goffer, 2002). They determined that CB_1 receptors are located not only on motor neurons but also on intrinsic primary afferent neurons and interneurons within the myenteric ganglia, both of which are also cholinergic in nature. Storr and colleagues confirmed that CB_1 receptor was not colocalized with NOS in the mouse colon (Storr et al., 2004), Abalo et al. showed the majority of myenteric neurons expressed CB_1 receptor in the guinea pig ileum (Abalo, Rivera, Vera, Suardiaz, & Martin, 2005) and Adami and colleagues showed that CB_1 receptor was extensively expressed in the gastric myenteric plexus of the rat, colocalized in cholinergic neurons (Adami et al., 2002).

While one study determined that CB_1 receptors did not colocalize VIP in the porcine ileum or colon (Kulkarni-Narla & Brown, 2000) another found that CB_1 receptor colocalized with VIP in the submucosal plexus of the guinea pig ileum (MacNaughton et al., 2004). It should be noted that this difference may be species-specific or due to the different antibodies used in each study. This study also showed that CB_1 was found on some neuropeptide Y-containing neurons (which are a cholinergic secretomotor population) and extrinsic primary afferent fibers that express TRPV1 in the submucosal plexus (MacNaughton et al., 2004).

While CB_1 receptor localization has been well investigated in the ENS, CB_2 receptor localization has only been investigated to a limited extent. CB_2 receptors are primarily found on immune cells, such as macrophages and neutrophils within the periphery (Galiegue et al., 1995). In the ENS, CB_2 receptors were localized using immunohistochemistry on myenteric neurons, but not on enteric glia of the rat ileum (Duncan, Mouihate, et al., 2008). Similarly to CB_1, CB_2 receptors were not found on NOS neurons. These data were supported by RT-PCR and Western blot and the antibody specificity using CB_2 receptor knockout mice. The expression of CB_2 receptors in the submucosal plexus has not been assessed to date. While this expression was shown under physiological conditions, the evidence for functional CB_2 receptors under these conditions is quite limited (see below), suggesting these receptors, if present, are not coupled to second messengers or are somehow not in an active conformation.

Thus, while studies have been conducted on the localization of both CB_1 and CB_2 receptors in some enteric neuronal populations, there is an incomplete understanding of the neuronal subpopulations that express these receptors throughout the length of the gut. To date, there have been no investigations into the ultrastructural localization of CB receptors in the ENS; however, the immunohistochemical data potentially supports both pre- and postsynaptic localization. Hons and colleagues showed that CB_1 receptor was colocalized with a presynaptic protein, synaptotagmin and was found on cholinergic vesicles as it colocalized with the vesicular acetylcholine transporter (Hons et al., 2012). However, CB_1 receptor was also observed at other cellular sites which could potentially be postsynaptic. The finding that CB_1 was found on intrinsic primary afferent neurons while apparently robust, is interesting in the light of a pharmacological study that showed that WIN55,212-2 had no effects on this neuronal population (Lopez-Redondo, Lees, & Pertwee, 1997). However, further studies are needed to confirm these observations.

Taken together, there is clearly a need for further research that examines the expression of the CB receptors in the ENS, especially in pathophysiological states.

3.3 Endocannabinoid Synthesis

The "on demand" synthesis of endocannabinoids is dependent on an increase of intracellular calcium via an influx of calcium into the cell or a release of calcium from internal stores or a combination of the two (Cascio & Marini, 2015; Di Marzo & Izzo, 2006; Ohno-Shosaku, Hashimotodani, Maejima, & Kano, 2005). Thus, the endocannabinoid system is unique as it is an activity-dependent signaling mechanism since it depends on the depolarization of neurons or activation of other GPCRs such as metabotropic glutamate receptors in order to be produced (Leung, Saghatelian, Simon, & Cravatt, 2006). In general, calcium levels must be in the micromolar range in order to produce endocannabinoids (Ohno-Shosaku et al., 2005). 2-AG and AEA synthesis can occur through multiple pathways, all of which are calcium dependent. However, it may be that different synthesis pathways are activated depending on whether the calcium is released through internal store or through an influx of calcium (Ohno-Shosaku et al., 2005). In Purkinje cells, a calcium concentration of 15 μM is required to mediate the release of endocannabinoids from neurons (Brenowitz & Regehr, 2003). In the ENS, the required calcium concentrations in order to activate the production of endocannabinoids have not been investigated, but are assumed to also be high. Once released into the synapse, endocannabinoids are limited to specific regions as they are quickly degraded (Cascio & Marini, 2015).

3.4 2-AG Synthesis and Degradation

There are three main pathways for the synthesis of 2-AG. Arachidonic acid-containing diacylglycerol (DAG) is the precursor in the primary 2-AG synthesis pathway and is dependent on the activity of PLC-β. In this pathway, DAG is hydrolyzed to 2-AG by DAG lipase (either DAGLα or DAGLβ) (Stella, Schweitzer, & Piomelli, 1997). A second pathway involves the hydrolysis reaction of 2-arachidonoyl-lysophosphatidylinositol to 2-AG via the enzyme lyso-PLC (Higgs & Glomset, 1994). Lastly, Nakane et al. (2002) determined a third pathway to synthesize 2-AG in rat brain. This involves the dephosphorylation of arachidonoyl-lysophosphatidic acid to 2-AG via a LPA phosphatase.

It is generally accepted that the primary pathway for 2-AG synthesis in the brain occurs through DAG lipase but these other two pathways are also involved, as DAGLα inhibitors do not completely prevent 2-AG production (Bisogno et al., 2003; Zhang, Wang, Bisogno, Di Marzo, & Alger, 2011). The DAG pathway was assumed to occur in the gut because of the high content of 2-AG, however, the presence of the biosynthetic enzymes was not shown until recently. DAGLα, but not DAGLβ, involved in the hydrolysis of DAG to 2-AG, was shown to localize primarily to cholinergic neurons in the ileum and colon of the GI tract (Bashashati et al., 2015). However, DAGLα had minimal colocalization to NOS and substance P immunoreactive neurons in the ileum and colon. In another study, DAGLα immunoreactivity was seen in the plasma cells within the lamina propria, the muscularis mucosae, and externa and around ganglion cells within the myenteric plexus of the human colon (Marquez et al., 2009). DAGLα was shown to increase in the colon during conditions of pancolitis in human samples, potentially supporting a protective role of endocannabinoids in conditions of intestinal inflammation.

DAGL is functional in the ENS of the mouse intestine. In normal mice, under baseline conditions inhibition of DAGL has no effect on motility or contractility. DAGL inhibitors did not alter evoked contractility or 2-AG levels under basal conditions in either the ileum or colon *in vitro* and whole gut transit or fecal output *in vivo* in healthy mice. This observation suggests that basal endocannabinoid tone to inhibit gut contractility is instead largely under the control of AEA, as previously suggested by Pinto et al., (2002), and Capasso et al. (2005). When intestinal contractility was pharmacologically reduced with the muscarinic antagonist scopolamine or the μ-opioid receptor agonist loperamide, DAGL inhibitors could partly or fully reverse the effects in both the ileum and colon. DAGL inhibitors can also reverse the pharmacologically induced prolongation of whole gut transit. These data suggest that 2-AG synthesis contributes to the control of GI motility. The effects of DAGL inhibition are via CB_1 receptor activation as CB_1 receptor antagonists reverse it and it is not present in CB_1 knockout mice. These findings suggest that either at the neuromuscular junction and/or in the myenteric plexus, 2-AG biosynthesis can be upregulated over basal levels via GPCRs (including cholinergic muscarinic and μ-opioid receptors) and contribute to constipation by inhibition of cholinergic transmission following CB_1 receptor activation. When 2-AG biosynthesis is inhibited, these mechanisms are also inhibited and motility is increased (Bashashati et al., 2015). The other two 2-AG synthesis pathways have not been examined within the ENS and therefore their presence and involvement in the synthesis of endocannabinoids is unknown at this time.

The degradation of 2-AG is relatively complex. The canonical degradation of 2-AG produces arachidonic acid and glycerol products (Freund, Katona, & Piomelli, 2003). The primary pathway for degradation is monoacylglycerol lipase (MAGL), however, other enzymes such as fatty acid amide hydrolase (FAAH), and the serine hydrolases α-β-hydrolase domain 6 and 12 (ABHD6 and ABHD12) are also involved in 2-AG degradation, but to a lesser extent a cyclooxygenase pathway can alter the molecular composition of 2-AG and AEA (Yu, Ives, & Ramesha, 1997) to produce prostaglandin glyceryl esters and ethanolamides rather than arachidonic acid and glycerol (Kozak et al., 2002). These products are able to mediate effects through signaling pathways such as extracellular signal regulated kinases, p38, inositol triphosphate, and nuclear factor kappa B (NF-κB) rather than through CB_1 and CB_2 (Sang, Zhang, & Chen, 2007). 2-AG is also metabolized to prostaglandin glyceryl esters and ethanolamides by cyclooxygenase 2 (Yang, Zhang, Andreasson, & Chen, 2008).

Similarly to its synthesis, 2-AG degradation and metabolism have not been thoroughly studied within the ENS. 2-AG inhibits motility through CB_1 receptors if its normally rapid degradation is prevented (Bashashati et al., 2015), strongly suggesting that MAGL is present and functionally active in the ENS.

MAGL was shown using immunohistochemistry to be present in the intestinal epithelium and enteric neurons within the GI tract (Duncan, Thomas, et al., 2008). Furthermore, MAGL levels were increased in the distal colon relative to the duodenum, which coincides with decreased levels of 2-AG in the colon relative to the ileum. In patients with ulcerative colitis, MAGL expression is increased (Marquez et al., 2009). Interestingly, MAGL inhibitors, such as JZL184 are able to protect against gastric damage caused by the administration of nonsteroidal anti-inflammatory drugs. The mechanism of action appears to be due to the reduction in proinflammatory cytokines, and the effects are mediated through CB_1 but not CB_2 receptors (Kinsey et al., 2011).

Overall, the primary pathways involved in the synthesis and degradation of 2-AG in the brain have been found in the ENS. Considerable further research is required, especially establishing the cellular sources of 2-AG in the gut, and into the presence and localization of the second and tertiary pathways for its biosynthesis and alternative pathways for its degradation.

3.5 Anandamide Synthesis and Degradation

AEA synthesis begins with *N*-acyl phosphatidylethanolamine (NAPE) as the precursor. There are three different pathways from which AEA can be produced from NAPE. The first pathway, considered to be the primary pathway, is the conversion of NAPE to AEA catalyzed by phospholipase D (NAPE-PLD) (Di Marzo et al., 1994). Secondly, ABHD4 can deacetylate NAPE and then cleave glycerophosphate to yield AEA (Simon & Cravatt, 2006, 2010). Lastly, Phospholipase C can hydrolyze NAPE to produce phosphoanadamide that becomes dephosphorylated by phosphatases to yield AEA (Liu et al., 2006). In NAPE-PLD$^{(-/-)}$ mice AEA levels in the brain were unaltered providing further evidence that other pathways are involved in the synthesis of AEA (Leung et al., 2006). These pathways have not been studied in the GI tract.

AEA is primarily degraded by FAAH, producing ethanolamine and arachidonic acid (Di Marzo & Izzo, 2006). FAAH is considered to be the major enzyme involved in the degradation of AEA, as large increases in AEA are observed in the brain when FAAH is inhibited or genetically knocked out (Cravatt et al., 2001; Di Marzo & Izzo, 2006). Therefore, while other mechanisms may exist, FAAH appears to have the greatest influence on the concentration AEA thus making FAAH a key regulator of AEA levels.

Similarly to the pathways involved in 2-AG synthesis and degradation, AEA pathways have not been thoroughly studied within the GI tract. Immunohistochemistry has been used to examine the presence of FAAH in the myenteric plexus, but it is still not known which subtypes of enteric neurons express this enzyme (Duncan, Davison, & Sharkey, 2005). FAAH inhibitors increase AEA levels in the GI tract (Capasso et al., 2008). NAPE-PLD has been found in enteric nerves in the human colonic myenteric plexus, but the neurochemical nature of these nerves was not established (Marquez et al., 2009; Suarez et al., 2012).

In summary, research of the synthesis and degradation pathways of AEA in the ENS is well behind research in the brain where enzymes such as FAAH have been localized to the postsynaptic terminals (Di Marzo, 2011). Furthermore, NAPE-PLD has been localized both pre- and postsynaptically in Purkinje neurons. However, the localization of NAPE-PLD has not been confirmed in the ENS and thus is an important step to be able to fully understand how the endocannabinoid signaling works within the ENS (Di Marzo, 2011).

4. ENDOCANNABINOID SIGNALING MECHANISMS

4.1 Endocannabinoid Transport Mechanisms

As mentioned above, endocannabinoids are produced postsynaptically, where they bind to the CB receptors on the presynaptic terminal (Katona & Freund, 2012). What is not known is how endocannabinoids are able to pass through the cell membrane to inhibit neurotransmitter release on the presynaptic cell. Mechanisms of transport such as traveling against particular concentration gradients, as well as ATP-dependent transport have largely been excluded (Nicolussi & Gertsch, 2015). Models of cellular uptake and transport of AEA and 2-AG have been proposed (Hermann, Kaczocha, & Deutsch, 2006; Nicolussi & Gertsch, 2015).

The most likely player is an endocannabinoid-mediated transporter; however, its identity is still unknown. Studies show that an endocannabinoid-mediated transporter may exist as specific properties of known membrane transporters are similar to the electrophysiological properties of neurons upon stimulation to produce AEA (Nicolussi & Gertsch, 2015). However, studies have not examined any of these properties in the ENS.

Similarly, the mechanism by which endocannabinoids are transported to their appropriate degradative enzymes are not yet fully resolved (Kaczocha et al., 2014). Fatty acid binding proteins FABP5 and FABP7 act as lipid transport proteins and are able to transport AEA to FAAH controlling the degradation of the endocannabinoid. Fatty acid binding proteins have been found in the monkey cerebellum and hippocampus as well as dorsal root ganglia of the rat and mouse (Boneva et al., 2010, 2011; De Leon et al., 1996). Intestinal-type and liver-type fatty acid binding proteins have been found in the ileum and jejunum of newborn rats and following necrotizing enterocolitis are significantly reduced (Goncalves et al., 2015). A better understanding of these proteins under normal and inflammatory conditions in the GI tract is obviously required. Also, an understanding of the impact it has on endocannabinoid levels will give further insight into how the endocannabinoid signaling system is regulated. The transport of 2-AG to its degradative enzymes is not known and requires further investigation.

Other than the transport of endocannabinoids across a lipid membrane and to appropriate degradative enzymes, research into the signaling mechanism has been well studied in the brain. CB_1 receptors are located in several regions of the CNS including the hippocampus, cerebellum, basal ganglia, and amygdala. When activated, they suppress the release of neurotransmitters including both glutamate and gamma-aminobutryic acid in various brain regions (Katona et al., 2001; Schlicker & Kathmann, 2001). Recently, endocannabinoids have been shown to induce both long-term depression and long-term potentiation in the same cell within the striatum. This was dependent on spike timing and shows that endocannabinoids not only mediate long-term depression but also influence potentiation in cells (Cui et al., 2015). The mechanism by which this occurs is through the inactivation of voltage-gated calcium channels, activation of potassium channels, and negative coupling to adenylyl cyclase (Howlett et al., 2002).

In the ENS, activation of CB_1 receptors has been shown to lead to a reduction of fast and slow excitatory postsynaptic potentials in myenteric neurons, as the release of acetylcholine and other (not determined) slow excitatory neurotransmitters is inhibited (Hons et al., 2012; Lopez-Redondo et al., 1997). Recently, the activity of the myenteric plexus was studied in CB_1 receptor knockout mice (Hons et al., 2012). The increased spontaneous and fast excitatory postsynaptic potentials evoked by prolonged presynaptic stimulation in CB_1 receptor knockout mice combined with the CB_1-mediated depression of fast excitatory postsynaptic potentials and paired-pulse facilitation in wild-type mice, implicate a presynaptic site of action of endocannabinoids. Boesmans, Ameloot, van den Abbeel, Tack, and Vanden Berghe (2009) demonstrated in enteric neurons that CB_1 receptor antagonism increased the percentage of neurons with spontaneous Ca^{2+} waves within the soma. This result is consistent with an increased percentage of myenteric neurons receiving spontaneous fast excitatory postsynaptic potentials in CB_1knockout mice (Hons et al., 2012). It seems most likely that the CB_1 receptor-mediated inhibition of synaptic transmission is due to changes in Ca^{2+} influx to presynaptic neurons in the ENS. As a result, there is a functional decrease in motility within the GI tract. While it is clear that CB_1 receptors inhibit the release of acetylcholine throughout the ENS, there is some evidence that CB_1 may modulate inhibitory transmission in the mouse colon (Storr et al., 2004). This effect was demonstrated using WIN55,212-2, a potent, synthetic agonist for CB_1 and not with the endogenous ligand, AEA, but this might have been due to the rapid degradation of AEA. However, since CB_1 receptors are not localized on inhibitory neurons that release NO, the physiology of endocannabinoid-mediated effects on inhibitory neurotransmission remain to be fully understood.

More recently, it has been suggested that the CB_2 receptor may be implicated in modulating an increased release of VIP, an important inhibitory neurotransmitter (Grider & Rivier, 1990). With the addition either CB_1 agonist, AEA or WIN55,212-2, there was an increase in VIP release that was prevented using a CB_2 antagonist (Kurjak, Hamel, Allescher, Schusdziarra, & Storr, 2008). Both AEA and WIN55,212-2 also increased NOS activity in synaptosomal membrane preparations of the intestine, which was not blocked by either a CB_1 or CB_2 antagonist (Kurjak et al., 2008).

While it is widely accepted that CB_1 prevents the release of acetylcholine throughout the GI tract, its involvement in the prevention of inhibitory neurons within the GI tract still requires further investigation. CB_2 may have a role regulating VIP and thus modulating inhibitory neurotransmission, but the conditions under which this occurs requires further investigation. Lastly, the mechanism by which WIN55,212-2 and AEA increase NOS activity requires research as it is not mediated by CB_1 or CB_2 but perhaps by another receptor that is able to bind the two ligands. In summary, endocannabinoid signaling in the brain is far ahead of research into the GI tract. The role that CB_1 plays in the inhibition of excitatory neurotransmission has been well established. However, the control of endocannabinoids in modulating inhibitory neurotransmission is not well understood.

5. OTHER RECEPTORS, AGONISTS, AND ANTAGONISTS

5.1 Phytocannabinoids

Over 70 plant-derived CBs, "phytocannabinoids," have been identified from *C. sativa* (Elsohly & Slade, 2005). The most common, THC, is a highly abundant psychoactive CB, but others have been investigated and we will discuss some examples of those that have been studied in relation to the ENS of the GI tract.

Cannabidiol (CBD) is a phytocannabinoid that has been reported to have medicinal benefits, notably by reducing inflammation and neuronal excitability (Izzo, Borrelli, Capasso, Di Marzo, & Mechoulam, 2009). Rats treated with 2,4,6-trinitrobenzene sulphonic acid, a chemical used to induce acute colitis and then treated with either CBD, THC, or both showed reductions in the degree of inflammation and intestinal damage and improvement of colitis. THC was more efficacious than CBD, but lower doses of THC combined with CBD were highly effective at reducing colitis (Jamontt, Molleman, Pertwee, & Parsons, 2010). Associated with the colitis was a marked reduction in neurally evoked contraction and relaxation of colonic segments studies *in vitro*. Both CBs also improved the degree of contractility and enhanced the extent of relaxation, presumably by reducing the degree of inflammation (Jamontt et al., 2010). However, there may also be direct actions on the ENS itself. De Filippis and colleagues reported that CBD was able to reduce the degree of inflammation in mice treated with lipopolysaccharide (LPS) (De Filippis et al., 2011). This model of intestinal inflammation was associated with gliosis of the enteric glia, as demonstrated by elevated levels of S-100B expression (a glial-specific protein). CBD was able to also reverse this change as well as reduce TNF-α and inducible NOS expression. They also studied rectal biopsy samples from patients with ulcerative colitis *in vitro*, and again CBD was able to reduce the LPS + interferon γ-stimulated samples. The mechanism of action of CBD in these processes is unclear but suspected to be via PPAR activation.

De Petrocellis and colleagues studied 12 different CBs and their effect on different transient receptor potential vanilloid (TRPV) channels in a model of ileal inflammation induced with croton oil (De Petrocellis et al., 2012). Some of these include the degradation product of THC, cannabinol (CBN), and a propyl homolog of THC, Δ^9-tetrahydrocannabidivarin (THCV) and a structural analog cannabichromene (CBC) among several others. In this study, the response of HEK-293 cells with various TRP channels was measured via a calcium response when a particular agonist was given. Depending on the specific channel, different CBs would elicit a response while others would not. For example, only CBD and THCV elicited a response on the four TRPV channels under study, one of which was TRPV2, the only channel that THC is able to activate. Furthermore, CBC reduced the expression of TRPV1, TRPV3, and TRPV4 mRNA in the jejunum, and TRPV3 and TRPV4 mRNA in the ileum of croton oil-treated mice. TRPV1 and

TRPV4 channels in particular have been implicated in pain sensation from the gut and TRPV4 also in the control of motility through actions in the ENS (Akbar et al., 2010; Fichna et al., 2015; Lapointe et al., 2015; Vergnolle, 2014). This study illustrates that phytocannabinoids may help relieve symptoms of GI disorders such as inflammatory bowel disease through actions at the level of the ENS. It also points to the importance of a deeper understanding of the entire CB system in control of intestinal functions through actions on the ENS.

5.2 Peptide Endocannabinoids

The CB receptor ligands were always thought to be of a lipid nature. However, recently, it has been found that peptide endocannabinoids, coined "pepcans" (Hofer et al., 2015) are also able to interact with the CB_1 receptor (Heimann et al., 2007). Bauer et al. (2012) determined that pepcans are able to bind with high affinity to the CB_1 receptor and Hofer et al. (2015) showed that the brainstem and adrenal glands produce pepcans. However, the physiological relevance of pepcans are unknown and still require further investigation in both the CNS and ENS.

5.3 Transient Receptor Potential Vanilloid 1

For over a decade, there has been evidence that AEA is an agonist of the transient receptor potential vanilloid receptor 1 (TRPV1), and thus modulates the receptor's activation (Ross, 2003). Mang and colleagues studied the release of acetylcholine from the myenteric plexus of the guinea pig ileum. In their preparations which were studied under basal conditions and stimulated at very low frequency (0.1 Hz), AEA facilitated acetylcholine release under basal conditions but inhibited it when the enteric nerves were stimulated. Both were due to nervous mechanisms as they were sensitive to tetrodotoxin. The basal facilitation was shown to be due to TRPV1 receptor activation, but not involve CB_1 or CB_2 receptors, whereas the action on evoked release, as expected was CB_1 receptor mediated (Mang, Erbelding, & Kilbinger, 2001). TRPV1 receptors in the ENS are located on both enteric neurons and extrinsic primary afferent nerve (Buckinx et al., 2013). This facilitatory role of AEA was suggested to be mediated by extrinsic primary afferent nerves which are sensitive to TRPV1 blockade and release substance P, because the effect of AEA under baseline conditions was also attenuated by neurokinin antagonists (Mang et al., 2001). This is interesting because AEA is also thought to contribute to endocannabinoid "tone" in the ENS, but this is CB_1 mediated and is likely due to an action on enteric neurons. These data possibly suggest a dual site and mode of action of AEA in the ENS when it is in a resting state, something that requires further study.

5.4 Virodhamine, an Endogenous CB_1 Antagonist

Virodhamine is an endocannabinoid that acts as an antagonist at the CB_1 receptor and a full agonist at the CB_2 receptor (Porter et al., 2002). Virodhamine has not been study profoundly in either the CNS or ENS. However, one study looked at pain-related responses to colorectal distensions (Brusberg et al., 2009). Colorectal distensions were used to induce visceral pain in mice. Different groups of mice had different CB ligands including WIN55,212-2, AEA, and virodhamine. While WIN55,212-2 had analgesic effects that were blocked by a CB_1 antagonist, virodhamine did not. However, the authors of this study considered virodhamine as an agonist of CB_1 rather than an antagonist and found that it had no effect on pain-related responses, results which are hard to interpret given the nature of this endocannabinoid.

5.5 G Protein-Coupled Receptor 55

G protein-coupled receptor 55 (GPR55) is an orphan GPCR that binds many CB and endocannabinoid ligands, including agonists and antagonists (Gasperi, Dainese, Oddi, Sabatucci, & Maccarrone, 2013; Ryberg et al., 2007). Because it is not structurally related to the other CB receptors it is not currently considered a CB receptor, but nevertheless is a potential target of endocannabinoids in the GI tract. The localization of the receptor in the GI tract is still under debate as Li et al. (2013) report GPR55 immunoreactivity to be more abundant in the mouse colon myenteric plexus over the ileum whereas Lin et al. (2011) reports the highest mRNA expression in the mouse ileum. Nevertheless, it seems that GPR55 is strongly expressed in the ENS.

In mice, similar to CB_1, GPR55 agonists are able to inhibit neurogenic contractions in the ileum and colon elicited *in vitro* largely through a prejunctional site of action (Li et al., 2013; Ross, Lichtman, Dewey, & Akbarali, 2012). The GPR55 agonist O-1602 concentration-dependently reduced evoked contractions in muscle strips from the colon ($\sim$60%) and weakly ($\sim$25%) from the ileum. O-1602 also slowed whole gut transit and colonic bead expulsion; effects that were not present in GPR55 knockout mice.

Under inflammatory conditions, GPR55 protein and gene levels were upregulated (Lin et al., 2011), an occurrence that is also seen with CB_1 and CB_2 receptors as well (Izzo et al., 2001). Using two models of colitis, Schicho et al. showed that administration of O-1602 reduced the degree of inflammation, probably by inhibiting neutrophil infiltration (Schicho et al., 2011). Using an LPS model of sepsis, CBD, a GPR55 agonist, was found to attenuate the hypomotility seen under these conditions (Lin et al., 2011). The function of GPR55 in the ENS requires considerable further investigation as little is known about which neurons it is expressed in and what its endogenous ligand is.

5.6 Oleoylethanolamide and Palmitoylethanolamide

Two other fatty acid ethanolamides, oleoylethanolamide (OEA), and palmitoylethanolamide (PEA) are substrates for FAAH and are found in the GI tract. Though structurally similar to AEA these two compounds are unable to bind to CB_1 and CB_2 receptors. However, there is evidence that PEA may directly or indirectly stimulate CB_2 receptors (Re, Barbero, Miolo, & Di Marzo, 2007). There is some evidence that PEA may also bind to CB_1 receptors but this has yet to be confirmed (Lin, Lu, Wu, Huang, & Wang, 2015). PEA and OEA exert their effects through the proliferator-activated receptor alpha (PPARα) or GPR119 (Hansen & Artmann, 2008).

OEA inhibits gastric emptying and intestinal motility, but its mechanism of action is not yet determined (Aviello et al., 2008; Capasso et al., 2005; Cluny, Keenan, Lutz, Piomelli, & Sharkey, 2009). PPARα immunoreactivity was found to be present throughout the ENS, in neurons in the myenteric and submucosal plexuses along the length of the GI tract. However, OEA-inhibited upper GI transit was still present in PPARα, CB_1, and CB_2 knockout mice and in the presence of the PPARα antagonist GW6471, the TRPV 1 antagonist SB366791 and the GLP-1 antagonist exendin-3 (9–39) amide, suggesting neither PPARα nor the CBs and other likely receptors are involved in mediating the effects of OEA (Cluny et al., 2009).

Recently, PEA has gained considerable interest because of its anti-inflammatory and antinociceptive actions (Re et al., 2007). Exogenous PEA inhibited the inflammation-induced increases in intestinal transit and tended to increase AEA levels in the gut (Capasso et al., 2014). Inhibition of transit by PEA was blocked by the CB_1 receptor antagonist rimonabant, but not a CB_2 receptor antagonist. Interestingly, motility was further increased by 5′-iodoresiniferatoxin, a TRPV1 antagonist, but was not significantly altered one way or another by the PPARα antagonist GW6471. Whether this is a direct CB_1 receptor-mediated effect or is indirect, occurring at the level of the ENS remains to be determined. Further support for an action of PEA on the ENS was recently presented when it was shown to improve colitis through an effect on enteric glia (Esposito et al., 2014). Using mouse models of colitis, colonic biopsies from patients with ulcerative colitis and primary cultures of mouse and human enteric glial cells, the effects of PEA, alone or in the presence of specific PPARα or PPARγ antagonists were studied. PEA treatment was shown to improve all macroscopic signs of colitis in mice and reduce proinflammatory cytokines. The effects of PEA were mediated by PPARα, but not PPARγ, through the selective targeting of the S-100B/toll-like receptor 4 on enteric glia that limited the downstream inhibition of NF-κB-dependent inflammation (Esposito et al., 2014). These exciting results also reveal the complexity of this system, since PEA, like AEA, appears to act on various targets in the ENS—both glial and neuronal—and via different receptors on each cell type. Further detailed investigations are required since to date, PPARα has not been directly demonstrated on enteric glial cells.

6. ENDOCANNABINOID TONE

The concept of endocannabinoid tone comes from observations that CB_1 receptor antagonists either applied to preparations *in vitro* or administered *in vivo* enhance contractility of the gut or accelerate transit and that CB_1 receptor knockout mice have accelerated transit compared to wild-type controls (Storr et al., 2010). There are two explanations for these observations. The first relates to the receptor itself. When inverse agonists, such as rimonabant, bind to the CB_1 receptor they lower the receptors basal activity by inactivating it (Howlett et al., 2011). This can reduce second messenger coupling and so reduce the effect of this constitutive receptor activity. The second is that there is ongoing release of endocannabinoids in the tissue which is never "at rest" even under basal conditions. Storr et al. have examined this problem using a combination of inverse agonists/antagonists and neutral antagonists which have no activity at the receptor in their own right (Storr et al., 2010). They showed that rimonabant enhanced electrically stimulated contractility in a concentration-dependent manner, with its greatest effect at a stimulation frequency of 4 Hz. In contrast, the inverse agonist/antagonist AM251 had no effect on contractility over the same concentration range, and neither did the neutral antagonist AM4113 (Storr et al., 2010). As all the compounds were completely effective at reversing the actions of WIN55,212-2, it suggests that their ability to act as antagonists is separable from other actions and that while there may be constitutive receptor activity it does not completely explain the effects of CB receptor antagonism in the ENS, which is thus presumably due to endocannabinoid production. These observations were extended by studying AM251 and AM4113 on upper GI transit *in vivo*. Both compounds increased upper GI transit in a dose-dependent manner. As the effect of the neutral antagonist AM4113 cannot be explained by inverse agonist activity and it mirrors what is observed in the CB_1 receptor knockout mice and the *in vitro* studies, these results strongly support the idea that under physiological conditions *in vivo* endogenous CB tone is importantly involved in the regulation of upper GI transit through the production of endocannabinoids (Storr et al., 2010). Endocannabinoid production is also supported by other work (Boesmans et al., 2009; Hons et al., 2012; Pertwee, 2001; Pinto, Capasso, et al., 2002; Pinto, Izzo, et al., 2002).

Endocannabinoid tone is relevant in patients with altered intestinal motility. Patients with slow transit constipation have decreased FAAH activity in the colon and therefore an increased endocannabinoid tone (Zhang, Wang, Su, Jiang, & Yuan, 2014). In contrast, in patients with diarrhea predominant irritable bowel syndrome, a specific FAAH CA/AA genotype was associated with the condition (Camilleri et al., 2008). Camilleri et al. (2008) suggest that the reduced expression of FAAH prevents the release of inhibitory neurotransmitters and thus enhanced motility.

Interestingly, Russo (2008) reviewed the previously mentioned concept of a clinical endocannabinoid deficiency. This concept relies on the fact that an endocannabinoid tone exists in normal patients, and that in patients with irritable bowel syndrome, for example, this tone is altered. In the CNS, researchers found that AEA levels are able to predict stress-induced anxiety in mice (Bluett et al., 2014). Mice that were given a FAAH inhibitor after the stress were observed to reverse their anxiety-like behavior. Interestingly, the FAAH inhibitor had no effect on control mice that were in nonstressed conditions. In the CNS, these results support the notion that endocannabinoid deficiencies could possibly play a role in disorders such as anxiety and potentially irritable bowel syndrome. Considering endocannabinoids are able to influence GI motility, and irritable bowel syndrome is characterized by diarrhea and constipation, this prediction is quite likely. Of course, clinical trials need to be conducted in order to confirm the theory that individuals with irritable bowel syndrome may have an endocannabinoid deficiency, and if so, whether it is a particular endocannabinoid such as 2-AG or AEA, or if it is the combination of the two.

7. CANNABINOID RECEPTOR METAPLASTICITY IN THE ENS

Endocannabinoid receptor metaplasticity in the gut was first described by Hons et al. (2012) who found that retrogradely released endocannabinoids are not only able to regulate the synaptic strength of cholinergic neurotransmission in the ENS but that they do so together with a retrogradely released purinergic transmitter. In the absence of the inhibitory influence of the CB_1 receptor potent activity-dependent purinergic facilitation of fast EPSPs was observed (Hons et al., 2012). These data indicate that retrograde endocannabinoid and purinergic transmitters interact to regulate vesicle release probability and control synaptic communication in the myenteric plexus. Removal of endocannabinoid signaling in CB_1 receptor knockout mice results in unopposed purinergic synaptic facilitation and increased vesicle release, ultimately culminating in the increased excitatory neurotransmission in the ENS and this may lead to the accelerated GI transit seen in these animals (Hons et al., 2012). In the absence of endocannabinoid modulation a purinergic signal is revealed and is involved in synaptic regulation by facilitating transmitter release in an activity-dependent manner. These observations are consistent with other studies that show CB regulation of ATP in the control of ileal contractility (Baldassano, Zizzo, Serio, & Mule, 2009). Taken together, these findings suggest a novel form of metaplasticity through the balance of endocannabinoid and purinergic signaling at the enteric synapse and have potentially important implications for our understanding of enteric neurophysiology.

8. CANNABINOID RECEPTOR PROPERTIES

8.1 Selective Coupling of GPCRs

The GPCRs, CB_1 and CB_2, are primarily coupled to GPCR $G_{i/o}$, however, in some cases the G protein coupling is altered. One example is through agonist trafficking where the agonist is able to promote the receptor to couple to one G protein versus another, even within the same G protein subtype (Mukhopadhyay & Howlett, 2005). Mukhopadhyay and Howlett (2001) determined that different loop regions within the CB_1 receptor configuration would change the G protein subtype coupling. This suggests that different agonists could direct G protein signaling depending where they bind to the CB_1 receptor. Furthermore, it has been shown that ligands can direct signaling of multiple G proteins. For example, WIN55,2-212 is able to act as an agonist and activate all G protein pathways (Hudson et al., 2010a). Much of these signaling pathway studies have been conducted in either cell lines or in the brain but none of which have yet been confirmed in the ENS. This is an area open to further investigation.

8.2 Dimerization of CB Receptors and Interactions with Other Receptors

Some GPCR proteins are capable of forming stable dimers but the functional role of this remains uncertain (Gurevich & Gurevich, 2008). In 2002, studies that examined whether the CB_1 receptor was able to dimerize indicated this was the case as two molecular weight forms of the CB_1 receptor were identified by Western blot, a fast migrating form at 53–64 kDa and a much slower form between 160 and 200 kDa. This higher molecular weight suggested the potential for the CB_1 receptor to dimerize (Wager-Miller, Westenbroek, & Mackie, 2002). Since then, CB_1 has been shown to exist as a dimer with the dopamine D_2 receptors in cell lines. The presence of a CB_1-D_2 heterodimer causes a change in the G protein preference from G_i to G_s (Pertwee et al., 2010). In the presence of a CB_1 agonist, such as CP55940, an increase in the K_D value of the D_2 receptors was observed in rat striatal membranes (Marcellino et al., 2008). CB_1 receptors also dimerize with the β2-adrenergic receptor (Hudson, Hebert, & Kelly, 2010b). When CB_1 and β2-adrenergic receptor were coexpressed in HEK 293 cells, CB_1 localization was primarily on the cell surface whereas, when CB_1 is expressed alone, the receptor is internalized. This suggests a physical interaction between the receptors. Hudson et al. (2010b) also found that the β2- adrenergic receptor attenuated constituent activity of the CB_1 receptor suggesting an important interaction between the two GPCRs.

CB_1 receptors have also been shown to form heterodimers with μ opioid receptors in HEK-293 cells using bioluminescence resonance energy transfer assays (Rios, Gomes, & Devi, 2006). Fluorescence resonance energy transfer analysis further confirms the formation of this heterodimer in baby hamster kidney cells (Hojo et al., 2008). Rios et al. (2006) showed that both the CB_1 and the μ-opioid receptor can modulate the other upon ligand binding in both cell lines and endogenous tissue. In cells expressing both receptors DAMGO and CP55940 elicited calcium responses thus showing that these receptors act through G_{qi5} coupled protein signaling (Hojo et al., 2008). This study also confirmed that the receptors function on the plasma membrane to transmit these signals through the G_{qi5} pathway and not G proteins expressed in oocytes. Functionally, it was determined that coactivation of these receptors attenuates the response of a neuritogenesis relative to the activation of either receptor alone in Neuro-2A cells (Rios et al., 2006). The functional relevance of this interaction has yet to be determined. Similarly, the formation of CB_1 receptor heterodimers in the ENS has yet to be discovered. However, a potential interaction between κ-opioid receptors and CB receptors has been found in the ileum of the mouse GI tract, though the formation of a heterodimer has not been confirmed (Capasso et al., 2008).

Overall, the understanding of the interactions between CB_1 and other receptors is becoming better understood, but research needs to be conducted to see if this occurs *in vivo* and ultimately to determine if this happens in the ENS and if so, what implications it may have on the control of the GI tract.

8.3 Desensitization of Cannabinoid Receptors

GPCRs are able to be internalized into endosomes upon activation leading to their desensitization (Kennedy & Marchese, 2015). One of the earliest reports of CB_1 desensitization and internalization was in 1999 where it was found that GPCR kinase and β-arrestin would cause internalization of the CB_1 receptor in *Xenopus* oocytes (Jin et al., 1999). More importantly, distal C terminal residues 418–439 of the CB_1 receptor was found to be an important site for desensitization as mutations in the residues S426 or S430 would eliminate receptor desensitization but not internalization. In 2012, a similar study was conducted where the CB_1 receptors were expressed in hippocampal CB_1 knockout neurons (Straiker, Wager-Miller, & Mackie, 2012). As such, a similar result was seen where residues in the distal C terminus were important for desensitization caused by WIN55,212-2 treatment. Whether this is important in the ENS remains to be determined but it may form part of the mechanism of tolerance to CB agonists that has been previously described (reviewed in Izzo & Sharkey, 2010).

9. ENTERIC GLIA

In the canonical CB signaling mechanism, astrocytes were not thought to play a role. However, their function and participation in endocannabinoid signaling is becoming better understood. Astrocytes have been shown to respond to neurotransmitters by an increased calcium concentration and subsequently release of gliotransmitters (Navarrete et al., 2014). Thus, glia are able to modulate the synaptic signaling between neurons. Importantly, Navarrete and Araque (2008) showed that the CB_1 receptor is localized on astrocytes and that with the application of CB_1 receptor agonist, WIN55,212-2 there is an elevated intracellular calcium concentration. This was caused by the mobilization of calcium from intracellular stores, thus activating the $G_{q/11}$ pathway. Importantly, the increase of intracellular calcium led to the subsequent release of glutamate from astrocytes allowing the glutamate to act at the level of the neurons.

In the ENS, enteric glia and their potential involvement in the control of synaptic signaling have not been fully evaluated (Gulbransen & Sharkey, 2012). It is unknown whether CB receptors localize to enteric glia as they do on astrocytes in the CNS. However, Duncan, Mouihate, et al. (2008) showed that CBs do have an effect on enteric glia, as well as enteric neurons, by demonstrating that in LPS-induced inflammation Fos expression in enteric glia, is significantly reduced by a CB_2 receptor antagonist. Thus, implicating the role of CB_2 receptors on enteric glia (and neurons) in regulating cellular activity. Functionally, the addition of a CB_2 agonist reverses the increased intestinal transit associated with LPS. This implies that at the very least, an indirect pathway exists where endocannabinoids can alter the activity of enteric glia. In this particular circumstance, the CB_2 receptor was implicated in mediating the functional response through either a direct or indirect pathway on enteric glia and neurons.

10. NEUROTRANSMISSION IN THE ENS AND GI TRACT

10.1 Neuromuscular Transmission

The regulation of neuromuscular transmission by endocannabinoids in the GI tract is important for the control of GI motility. The control of peristalsis by the myenteric plexus has been extensively studied and CB receptors are involved in signaling involving intrinsic primary afferents, ascending excitatory pathways and descending inhibitory pathways. In the rat colon, Grider and colleagues examined the effect of CBs on the peristaltic reflex using a three-compartment preparation that allows separation of ascending contraction, descending relaxation, and the sensory components of the reflex (Grider et al., 2009). They showed that AEA decreased and that AM251 increased ascending contractions and the release of substance P orally. Similarly, AEA reduced descending relaxation and the release of VIP. In the central sensory compartment, AEA increased both ascending contraction and SP release orally and descending relaxation and VIP release caudally, suggesting a role for CB_1 receptors in modulation of primary afferent transmission in the myenteric plexus. This was further supported by the observation that calcitonin gene-related peptide (CGRP) release was inhibited by AEA in a CB_1 dependent manner (Grider et al., 2009). CGRP is a peptide found in intrinsic primary afferent neurons of the ENS (Furness, 2006). These findings are largely consistent with work discussed above, but it is of interest to note that using electrophysiological approaches López-Redondo et al. were unable to record any effects of a CB_1 agonist (WIN55,212-2) on myenteric after hyperpolarization (AH) neurons in the guinea pig ileum (Lopez-Redondo et al., 1997). AH neurons in the guinea pig ileum are CGRP immunoreactive intrinsic primary afferent neurons. There are many reasons why there might be a difference between these studies comparing the effects on intrinsic primary afferents in rat colon and guinea pig ileum. However, the work by Grider and colleagues is supported by the immunohistochemical localization studies described above.

The effects of AEA have also been investigated in segments of rat ileum in which contractions of the circular smooth muscle and the ascending myenteric component of the peristaltic reflex was studied (Sibaev et al., 2014). AEA significantly reduced cholinergic twitch contractions of ileal smooth muscle and as expected this was CB_1 receptor mediated. AEA also reduced the ascending peristaltic contraction by affecting ganglionic and neuromuscular neurotransmission in the rat ileum, as it is in the guinea pig ileum (Heinemann, Shahbazian, & Holzer, 1999). AEA reduced excitatory and inhibitory junction potentials, whereas intestinal slow waves were not affected, suggesting an action on both acetylcholine and NO release (Sibaev et al., 2014). The latter remains puzzling in the light of observations on the localization of CB_1 receptors (discussed above), but is somewhat consistent with observations made using synaptosomal preparations (Kurjak et al., 2008). However, in this case the stimulatory effect of AEA on NO synthase was not antagonized by CB_1, CB_2, or TRPV1 antagonists.

The effects of endocannabinoids on nonadrenergic noncholinergic (NANC) excitatory and inhibitory neurotransmission have been examined in the mouse colon (Mule, Amato, Baldassano, & Serio, 2007). Under

NANC conditions, electrically evoked responses are characterized by a nitrergic relaxation followed by a tachykinergic contraction. WIN55,212-2 and AEA produced concentration-dependent CB_1 receptor-mediated reductions of the NANC contractile responses, without affecting the NANC relaxation. Interestingly, the FAAH inhibitor URB597 was without any effect on the NANC evoked responses, perhaps suggesting that endogenous ligands do not tonically regulate NANC transmission in the mouse colon.

Taken together these data suggest that endocannabinoids acting at CB_1 receptors regulate the intestinal peristaltic reflex and these actions involve intrinsic primary afferent neurons, interneurons and motor neurons. There is limited understanding of the role of endogenously released endocannabinoids, but AEA when exogenously administered mediates these effects by activation of CB_1 receptors.

There have also been some investigations in other regions of the GI tract. In the rat gastric fundus, AEA and WIN55,212-2 do not affect the baseline tone of muscle contraction (Storr, Gaffal, Saur, Schusdziarra, & Allescher, 2002). However, AEA and WIN55,212-2 reduced electrically stimulated twitch contractions. The effect of AEA was partially reversed by the CB_2 receptor antagonist AM630, which did not have an effect on twitch contractions alone. This result is surprising given the overwhelming evidence that CB_2 receptors are not involved elsewhere in the gut under baseline conditions and is worthy of further investigation. Storr et al. (2002) found that endocannabinoids can also regulate electrically stimulated relaxations. Similarly to their effect on contractions, AEA and WIN55,212-2 suppressed the gastric fundic relaxations. However, this time, AM630 caused an increase in induced relaxations whereas it had no effect on the induced contractions.

Overall, along the length of the GI tract, endocannabinoids reduce contractility and propulsive motility through inhibition of acetylcholine release and that of other neurotransmitters through actions on myenteric neurons. This effect is largely mediated by CB_1 receptors however, there is evidence that the GPR55 receptor also inhibits colonic muscle contraction, similar to CB_1 (Ross et al., 2012).

10.2 Secretomotor Transmission

There have been few studies that examine the control of epithelial secretions by endocannabinoids. Storr et al. (2010) studied the effects of a variety of CB_1 receptor antagonists to ascertain if endogenous CB ligands were involved. They assessed the response to electrical stimulation by measuring short-circuit current *in vitro* using Ussing chambers and assessing stool fluid content *in vivo* in mouse colon. Interestingly, neither the CB_1 receptor inverse agonists/antagonist AM251 nor the neutral CB_1 receptor antagonist AM4113 had any effect on fecal stool water content, suggesting that the increased whole gut transit in animals treated with AM251 is a motor and not a secretomotor effect. To address whether there may be a physiological regulation of secretomotor function a variety of stimuli, including electrical field stimulation, carbachol and forskolin, which stimulate secretion indirectly via the ENS and directly at the epithelium were used (Storr et al., 2010). However, no effects of the CB_1 receptor antagonists on ion transport, tissue conductance or mucosal permeability were seen. However, in other studies a role of CB_1 receptors has been observed in the control of ion transport in the guinea pig ileum and in the control of intestinal barrier function.

MacNaughton and colleagues studied preparations of the submucosa and mucosa of the guinea pig ileum *in vitro* where short-circuit current was measured as an indicator of net electrogenic ion transport in Ussing chambers (MacNaughton et al., 2004). They stimulated extrinsic primary afferent nerves with capsaicin and enteric nerves with electrical field stimulation. The responses to capsaicin and EFS were reduced by $47\% \pm 12\%$ and $30\% \pm 14\%$, respectively, by the CB_1 receptor agonist WIN55,212-2. The inhibitory effect of WIN55,212-2 on electrically evoked secretion was not observed in extrinsically denervated segments of ileum which taken together, show that CB_1 receptors on extrinsic primary afferent nerves, are capable of inhibiting the release of transmitters that act on cholinergic secretomotor pathways in the submucosal plexus and so regulating secretion. These data were supported by immunohistochemical findings of CB_1 localization in this region of the ENS (MacNaughton et al., 2004). These data once again point to the potential specificity of CB actions in the ENS, and highlight the need for further studies on the submucosal plexus in other species and other regions of the gut.

The control of barrier function is critical to homeostasis. The first report showing that CB_1 receptors controlled barrier function suggested that they played a detrimental role, i.e., that blocking CB_1 receptors reduced permeability of the gut, and that there was a unique crosstalk between endocannabinoids, and the gut microbiota that controlled aspects of metabolism (Muccioli et al., 2010). Here, it was thought that CB_1 receptors on the epithelium were mediating these effects. Recently, however, using CB_1 receptor knockout mice, Zoppi and colleagues showed that these animals had a degree of colonic barrier dysfunction including a significantly lower IgA secretion, higher paracellular permeability and a greater degree of bacterial translocation, both under basal conditions and after a period of psychological stress (Zoppi et al., 2012). Some of these effects were also observed after pharmacological antagonism with rimonabant. They concluded that in the colon at least, CB_1 receptors exert a protective role through the regulation of intestinal secretion of IgA and paracellular permeability. It remains to be determined whether this is due to epithelial or enteric neural receptors in the colon.

10.3 Enteroendocrine Cells

As noted above, the initiation of GI reflexes occurs by the release of peptides and 5-HT from enteroendocrine cells. Enteroendocrine cells lie in the epithelium of the GI tract and secrete varying hormones that are able to either act locally or enter into the bloodstream (Sternini, Anselmi, & Rozengurt, 2008). Specialized endocrine cells, I-cells, which release CCK contain mRNA for CB_1 receptor suggesting that if the protein is expressed they may be able to respond to endocannabinoid signals (Sykaras, Demenis, Case, McLaughlin, & Smith, 2012). CB_1 receptor expression on vagal afferent neurons of the nodose ganglion is regulated by CCK (Burdyga et al., 2004). This is part of the mechanism of peripheral control of energy balance and hence the possibility exists that this may be an autocrine control mechanism used to regulate secretion in response to nutrients, but further work is required to test this notion.

Another population of enteroendocrine cells called K cells release GIP to control insulin secretion. These cells also express a functional CB_1 receptor that inhibits GIP release, however, the physiological role of endocannabinoid regulation of GIP has yet to be determined (Moss et al., 2012). Clearly, much more work is required in this area and especially to understand if peristaltic and other GI reflexes are regulated by the endocannabinoid system at the level of their transduction.

11. BRAIN–GUT AXIS

Though the ENS is able to function independently from the CNS, there is extensive bidirectional communication as described above. The endocannabinoid system of the GI tract is able to regulate satiety and energy balance, dependent on the brain–gut axis (Cluny et al., 2012; DiPatrizio & Piomelli, 2015; Dockray, 2014). Further discussion of this interesting topic is beyond the scope of the current review, but it is worth noting the important role of the endocannabinoid system and the GI tract in terms of the development of obesity.

In animal models of obesity, endocannabinoid tone is upregulated due to an increase in CB_1 expression and intestinal barrier permeability is increased (Muccioli et al., 2010). This provides a mechanisms for the GI tract play a role in the development of obesity, particularly involving the translocation of commensal bacteria or bacterial products.

The impact that the gut microbiota has on the brain is currently a very active area of research (Cryan & O'Mahony, 2011; Mayer, 2011; Mayer et al., 2015). Gut microbiota are able to influence central neurotransmitters and act on the brain–gut axis to modulate energy balance and food intake, thus playing an important role in obesity (Manco, 2012; Rhee et al., 2009). Interestingly, individuals with obesity have a different microbiota than those who are not obese and that an alteration of gut microbiota of obese mice can alter the endocannabinoid tone in that subject (Cani, Geurts, Matamoros, Plovier, & Duparc, 2014). When the gut microbiota of mice were changed to those of lean animals, the levels of both AEA and CB_1 receptor were reduced in adipose tissues. This suggests a profound link between gut microbiota and the endocannabinoid signaling systems in the body. In addition to the gut microbiota changing endocannabinoid tone, it has also been shown that the microbiota is able to change the expression of antiobesity and pro-obesity peptides which can act on the gut–brain axis (Schele et al., 2013). Together, this shows that the brain–gut axis and endocannabinoid system are critically linked to the development of obesity and further research linking this to the ENS control of gut function is warranted.

12. THE ENDOCANNABINOID SYSTEM OF THE ENS IN GI INFLAMMATION

In pathophysiological states, both CB1 and CB2 receptor activation reduces enhanced GI motility, with CB2 receptor activation normalizing motility and serving as a braking mechanism to limit abnormal motility (Duncan, Mouihate, et al., 2008; Mathison, Ho, Pittman, Davison, & Sharkey, 2004). Interestingly, inhibition of FAAH, which elevates the levels of endocannabinoids, also normalizes accelerated GI motility in pathophysiological states, but not in normal animals (Bashashati et al., 2012). Studies of the ENS in intestinal inflammation have focused on neural control mechanisms (Hons et al., 2012), but to date the functional role of the endocannabinoid system has not been evaluated in any detail.

13. SUMMARY AND FUTURE DIRECTIONS

The ENS of the GI tract is the only part of the peripheral nervous able to function autonomously without any input from the CNS (Furness, 2012). The effect of marijuana on the GI tract has been reported for centuries, but the mechanisms of how these effects are exerted remain to be fully understood. CB_1 and CB_2 receptors expressed on enteric nerves and throughout the intestinal mucosa including on enteroendocrine cells (CB_1) and on epithelial cells (CB_1 and CB_2). Under physiological conditions, the dominant receptor controlling the ENS is CB_1, but in pathophysiological conditions CB_2 receptors play a role. However, how these receptors interact, under what conditions and whether they are functionally linked remains to be determined.

The production of endocannabinoids in the ENS has only recently been investigated as have the degradative enzyme systems for endocannabinoids. Interesting results described above suggest that these enzymes are valuable targets that can be utilized to control the ENS in pathophysiological conditions, potentially conferring benefits to patients with GI motor disorders. Increasing our understanding of these enzymes systems is also a key priority for future research.

Though endocannabinoids have been implicated in relation to motor and secretomotor control, further research need so to be conducted to determine the downstream signaling pathways of CBs and what processes lead to the coupling of a preferred G protein. In the GI tract specifically, a better understanding of the localization of CB receptors will lead to knowledge of how the endocannabinoid system is able to interact with other receptor systems such as opioid receptors, 5-HT receptors and muscarinic and nicotinic cholinergic receptors that are widely distributed in the ENS. Emphasis has been placed on the CB_1 receptor as it appears to play the more dominant role under normal conditions; however, more focus needs to be placed on the CB_2 receptor as it becomes extremely important under inflammatory conditions which would aid in understanding the etiology common GI diseases such as inflammatory bowel disease and irritable bowel syndrome.

DISCLOSURES

The authors have no competing interests.

REFERENCES

Abalo, R., Rivera, A. J., Vera, G., Suardiaz, M., & Martin, M. I. (2005). Evaluation of the effect of age on cannabinoid receptor functionality and expression in guinea-pig ileum longitudinal muscle-myenteric plexus preparations. *Neuroscience Letters, 383*, 176–181.

Adami, M., Frati, P., Bertini, S., Kulkarni-Narla, A., Brown, D. R., de Caro, G., et al. (2002). Gastric antisecretory role and immunohistochemical localization of cannabinoid receptors in the rat stomach. *British Journal of Pharmacology, 135*, 1598–1606.

Akbar, A., Yiangou, Y., Facer, P., Brydon, W. G., Walters, J. R., Anand, P., et al. (2010). Expression of the TRPV1 receptor differs in quiescent inflammatory bowel disease with or without abdominal pain. *Gut, 59*, 767–774.

Aviello, G., Matias, I., Capasso, R., Petrosino, S., Borrelli, F., Orlando, P., et al. (2008). Inhibitory effect of the anorexic compound oleoylethanolamide on gastric emptying in control and overweight mice. *Journal of Molecular Medicine (Berlin, Germany), 86*, 413–422.

Baldassano, S., Zizzo, M. G., Serio, R., & Mule, F. (2009). Interaction between cannabinoid CB1 receptors and endogenous ATP in the control of spontaneous mechanical activity in mouse ileum. *British Journal of Pharmacology, 158*, 243–251.

Bashashati, M., Nasser, Y., Keenan, C. M., Ho, W., Piscitelli, F., Nalli, M., et al. (2015). Inhibiting endocannabinoid biosynthesis: A novel approach to the treatment of constipation. *British Journal of Pharmacology, 172*, 3099–3111.

Bashashati, M., Storr, M. A., Nikas, S. P., Wood, J. T., Godlewski, G., Liu, J., et al. (2012). Inhibiting fatty acid amide hydrolase normalizes endotoxin-induced enhanced gastrointestinal motility in mice. *British Journal of Pharmacology, 165*, 1556–1571.

Basu, S., & Dittel, B. N. (2011). Unraveling the complexities of cannabinoid receptor 2 (CB2) immune regulation in health and disease. *Immunologic Research, 51*, 26–38.

Bauer, M., Chicca, A., Tamborrini, M., Eisen, D., Lerner, R., Lutz, B., et al. (2012). Identification and quantification of a new family of peptide endocannabinoids (Pepcans) showing negative allosteric modulation at CB1 receptors. *Journal of Biological Chemistry, 287*, 36944–36967.

Benito, C., Tolon, R. M., Pazos, M. R., Nunez, E., Castillo, A. I., & Romero, J. (2008). Cannabinoid CB2 receptors in human brain inflammation. *British Journal of Pharmacology, 153*, 277–285.

Bisogno, T., Howell, F., Williams, G., Minassi, A., Cascio, M. G., Ligresti, A., et al. (2003). Cloning of the first sn1-DAG lipases points to the spatial and temporal regulation of endocannabinoid signaling in the brain. *Journal of Cell Biology, 163*, 463–468.

Bluett, R. J., Gamble-George, J. C., Hermanson, D. J., Hartley, N. D., Marnett, L. J., & Patel, S. (2014). Central anandamide deficiency predicts stress-induced anxiety: Behavioral reversal through endocannabinoid augmentation. *Translational Psychiatry, 4*, e408.

Boesmans, W., Ameloot, K., van den Abbeel, V., Tack, J., & Vanden Berghe, P. (2009). Cannabinoid receptor 1 signalling dampens activity and mitochondrial transport in networks of enteric neurones. *Neurogastroenterology and Motility, 21*. 958-e977.

Bohorquez, D. V., & Liddle, R. A. (2015). The gut connectome: Making sense of what you eat. *Journal of Clinical Investigation, 125*, 888–890.

Boneva, N. B., Kaplamadzhiev, D. B., Sahara, S., Kikuchi, H., Pyko, I. V., Kikuchi, M., et al. (2011). Expression of fatty acid-binding proteins in adult hippocampal neurogenic niche of postischemic monkeys. *Hippocampus, 21*, 162–171.

Boneva, N. B., Mori, Y., Kaplamadzhiev, D. B., Kikuchi, H., Zhu, H., Kikuchi, M., et al. (2010). Differential expression of FABP 3, 5, 7 in infantile and adult monkey cerebellum. *Neuroscience Research, 68*, 94–102.

Brenowitz, S. D., & Regehr, W. G. (2003). Calcium dependence of retrograde inhibition by endocannabinoids at synapses onto Purkinje cells. *Journal of Neuroscience, 23*, 6373–6384.

Brusberg, M., Arvidsson, S., Kang, D., Larsson, H., Lindstrom, E., & Martinez, V. (2009). CB1 receptors mediate the analgesic effects of cannabinoids on colorectal distension-induced visceral pain in rodents. *Journal of Neuroscience, 29*, 1554–1564.

Brusco, A., Tagliaferro, P., Saez, T., & Onaivi, E. S. (2008). Postsynaptic localization of CB2 cannabinoid receptors in the rat hippocampus. *Synapse, 62*, 944–949.

Buckinx, R., Van Nassauw, L., Avula, L. R., Alpaerts, K., Adriaensen, D., & Timmermans, J. P. (2013). Transient receptor potential vanilloid type 1 channel (TRPV1) immunolocalization in the murine enteric nervous system is affected by the targeted C-terminal epitope of the applied antibody. *Journal of Histochemistry and Cytochemistry, 61*, 421–432.

Buckley, N. E., Hansson, S., Harta, G., & Mezey, E. (1998). Expression of the CB1 and CB2 receptor messenger RNAs during embryonic development in the rat. *Neuroscience, 82*, 1131–1149.

Burdyga, G., Lal, S., Varro, A., Dimaline, R., Thompson, D. G., & Dockray, G. J. (2004). Expression of cannabinoid CB1 receptors by vagal afferent neurons is inhibited by cholecystokinin. *Journal of Neuroscience, 24*, 2708–2715.

Camilleri, M., Carlson, P., McKinzie, S., Grudell, A., Busciglio, I., Burton, D., et al. (2008). Genetic variation in endocannabinoid metabolism, gastrointestinal motility, and sensation. *American Journal of Physiology Gastrointestinal and Liver Physiology, 294*, G13–G19.

Cani, P. D. (2012). Crosstalk between the gut microbiota and the endocannabinoid system: Impact on the gut barrier function and the adipose tissue. *Clinical Microbiology and Infection, 18*(Suppl. 4), 50–53.

Cani, P. D., Geurts, L., Matamoros, S., Plovier, H., & Duparc, T. (2014). Glucose metabolism: Focus on gut microbiota, the endocannabinoid system and beyond. *Diabetes & Metabolism, 40*, 246–257.

Capasso, R., Borrelli, F., Cascio, M. G., Aviello, G., Huben, K., Zjawiony, J. K., et al. (2008). Inhibitory effect of salvinorin A, from Salvia divinorum, on ileitis-induced hypermotility: Cross-talk between κ-opioid and cannabinoid CB(1) receptors. *British Journal of Pharmacology, 155*, 681–689.

Capasso, R., Matias, I., Lutz, B., Borrelli, F., Capasso, F., Marsicano, G., et al. (2005). Fatty acid amide hydrolase controls mouse intestinal motility in vivo. *Gastroenterology, 129*, 941–951.

Capasso, R., Orlando, P., Pagano, E., Aveta, T., Buono, L., Borrelli, F., et al. (2014). Palmitoylethanolamide normalizes intestinal motility in a model of post-inflammatory accelerated transit: Involvement of CB(1) receptors and TRPV1 channels. *British Journal of Pharmacology, 171*, 4026–4037.

Cascio, M. G., & Marini, P. (2015). Biosynthesis and fate of endocannabinoids. *Handbook of Experimental Pharmacology, 231*, 39–58.

Cluny, N. L., Keenan, C. M., Lutz, B., Piomelli, D., & Sharkey, K. A. (2009). The identification of peroxisome proliferator-activated receptor alpha-independent effects of oleoylethanolamide on intestinal transit in mice. *Neurogastroenterology and Motility, 21*, 420–429.

Cluny, N. L., Reimer, R. A., & Sharkey, K. A. (2012). Cannabinoid signalling regulates inflammation and energy balance: The importance of the brain-gut axis. *Brain, Behavior, and Immunity, 26*, 691–698.

Coutts, A. A., Irving, A. J., Mackie, K., Pertwee, R. G., & Anavi-Goffer, S. (2002). Localisation of cannabinoid CB(1) receptor immunoreactivity in the guinea pig and rat myenteric plexus. *Journal of Comparative Neurology, 448*, 410–422.

Coutts, A. A., & Pertwee, R. G. (1997). Inhibition by cannabinoid receptor agonists of acetylcholine release from the guinea-pig myenteric plexus. *British Journal of Pharmacology, 121*, 1557–1566.

Cravatt, B. F., Demarest, K., Patricelli, M. P., Bracey, M. H., Giang, D. K., Martin, B. R., et al. (2001). Supersensitivity to anandamide and enhanced endogenous cannabinoid signaling in mice lacking fatty acid amide hydrolase. *Proceedings of the National Academy of Sciences of the United States of America, 98*, 9371–9376.

Cryan, J. F., & O'Mahony, S. M. (2011). The microbiome-gut-brain axis: From bowel to behavior. *Neurogastroenterology and Motility, 23*, 187–192.

Cui, Y., Paille, V., Xu, H., Genet, S., Delord, B., Fino, E., et al. (2015). Endocannabinoids mediate bidirectional striatal spike-timing-dependent plasticity. *Journal of Physiology, 593*, 2833–2849.

De Filippis, D., Esposito, G., Cirillo, C., Cipriano, M., De Winter, B. Y., Scuderi, C., et al. (2011). Cannabidiol reduces intestinal inflammation through the control of neuroimmune axis. *PLoS One, 6*, e28159.

De Leon, M., Welcher, A. A., Nahin, R. H., Liu, Y., Ruda, M. A., Shooter, E. M., et al. (1996). Fatty acid binding protein is induced in neurons of the dorsal root ganglia after peripheral nerve injury. *Journal of Neuroscience Research, 44*, 283–292.

De Petrocellis, L., Orlando, P., Moriello, A. S., Aviello, G., Stott, C., Izzo, A. A., et al. (2012). Cannabinoid actions at TRPV channels: Effects on TRPV3 and TRPV4 and their potential relevance to gastrointestinal inflammation. *Acta Physiologica (Oxford, England), 204*, 255–266.

Devane, W. A., Hanus, L., Breuer, A., Pertwee, R. G., Stevenson, L. A., Griffin, G., et al. (1992). Isolation and structure of a brain constituent that binds to the cannabinoid receptor. *Science, 258*, 1946–1949.

Di Marzo, V. (2011). Endocannabinoid signaling in the brain: Biosynthetic mechanisms in the limelight. *Nature Neuroscience, 14*, 9–15.

Di Marzo, V., Fontana, A., Cadas, H., Schinelli, S., Cimino, G., Schwartz, J. C., et al. (1994). Formation and inactivation of endogenous cannabinoid anandamide in central neurons. *Nature, 372*, 686–691.

Di Marzo, V., & Izzo, A. A. (2006). Endocannabinoid overactivity and intestinal inflammation. *Gut, 55*, 1373–1376.

DiPatrizio, N. V., & Piomelli, D. (2015). Intestinal lipid-derived signals that sense dietary fat. *Journal of Clinical Investigation, 125*, 891–898.

Dockray, G. J. (2014). Gastrointestinal hormones and the dialogue between gut and brain. *Journal of Physiology, 592*, 2927–2941.

Duncan, M., Davison, J. S., & Sharkey, K. A. (2005). Review article: Endocannabinoids and their receptors in the enteric nervous system. *Alimentary Pharmacology & Therapeutics, 22*, 667–683.

Duncan, M., Mouihate, A., Mackie, K., Keenan, C. M., Buckley, N. E., Davison, J. S., et al. (2008). Cannabinoid CB2 receptors in the enteric nervous system modulate gastrointestinal contractility in lipopolysaccharide-treated rats. *American Journal of Physiology Gastrointestinal and Liver Physiology, 295*, G78–G87.

Duncan, M., Thomas, A. D., Cluny, N. L., Patel, A., Patel, K. D., Lutz, B., et al. (2008). Distribution and function of monoacylglycerol lipase in the gastrointestinal tract. *American Journal of Physiology Gastrointestinal and Liver Physiology, 295*, G1255–G1265.

Ellis, M., Chambers, J. D., Gwynne, R. M., & Bornstein, J. C. (2013). Serotonin and cholecystokinin mediate nutrient-induced segmentation in guinea pig small intestine. *American Journal of Physiology Gastrointestinal and Liver Physiology, 304*, G749–G761.

Elsohly, M. A., & Slade, D. (2005). Chemical constituents of marijuana: The complex mixture of natural cannabinoids. *Life Sciences, 78*, 539–548.

Esposito, G., Capoccia, E., Turco, F., Palumbo, I., Lu, J., Steardo, A., et al. (2014). Palmitoylethanolamide improves colon inflammation through an enteric glia/toll like receptor 4-dependent PPAR-alpha activation. *Gut, 63*, 1300–1312.

Fichna, J., Poole, D. P., Veldhuis, N., MacEachern, S. J., Saur, D., Zakrzewski, P. K., et al. (2015). Transient receptor potential vanilloid 4 inhibits mouse colonic motility by activating NO-dependent enteric neurotransmission. *Journal of Molecular Medicine (Berlin, Germany).*

Freund, T. F., Katona, I., & Piomelli, D. (2003). Role of endogenous cannabinoids in synaptic signaling. *Physiological Reviews, 83*, 1017–1066.

Furness, J. B. (2006). *The enteric nervous system.* Massachusetts: Blackwell Publishing.

Furness, J. B. (2012). The enteric nervous system and neurogastroenterology. *Nature Reviews Gastroenterology & Hepatology, 9*, 286–294.

Furness, J. B., Callaghan, B. P., Rivera, L. R., & Cho, H. J. (2014). The enteric nervous system and gastrointestinal innervation: Integrated local and central control. *Advances in Experimental Medicine and Biology, 817*, 39–71.

Galiegue, S., Mary, S., Marchand, J., Dussossoy, D., Carriere, D., Carayon, P., et al. (1995). Expression of central and peripheral cannabinoid receptors in human immune tissues and leukocyte subpopulations. *European Journal of Biochemistry, 232*, 54–61.

Gaoni, Y., & Mechoulam, R. (1964). Isolation, structure, and partial synthesis of an active constituent of Hashish. *Journal of the American Chemical Society, 86*, 1646–1647.

Gasperi, V., Dainese, E., Oddi, S., Sabatucci, A., & Maccarrone, M. (2013). GPR55 and its interaction with membrane lipids: Comparison with other endocannabinoid-binding receptors. *Current Medicinal Chemistry, 20*, 64–78.

Gershon, M. D. (2013). 5-Hydroxytryptamine (serotonin) in the gastrointestinal tract. *Current Opinion in Endocrinology, Diabetes, and Obesity, 20*, 14–21.

Goncalves, F. L., Soares, L. M., Figueira, R. L., Simoes, A. L., Gallindo, R. M., & Sbragia, L. (2015). Evaluation of the expression of I-FABP and L-FABP in a necrotizing enterocolitis model after the use of Lactobacillus acidophilus. *Journal of Pediatric Surgery, 50*, 543–549.

Grider, J. R., Mahavadi, S., Li, Y., Qiao, L. Y., Kuemmerle, J. F., Murthy, K. S., et al. (2009). Modulation of motor and sensory pathways of the peristaltic reflex by cannabinoids. *American Journal of Physiology Gastrointestinal and Liver Physiology, 297*, G539–G549.

Grider, J. R., & Rivier, J. R. (1990). Vasoactive intestinal peptide (VIP) as transmitter of inhibitory motor neurons of the gut: Evidence from the use of selective VIP antagonists and VIP antiserum. *Journal of Pharmacology and Experimental Therapeutics, 253*, 738–742.

Griffin, G., Fernando, S. R., Ross, R. A., McKay, N. G., Ashford, M. L., Shire, D., et al. (1997). Evidence for the presence of CB2-like cannabinoid receptors on peripheral nerve terminals. *European Journal of Pharmacology, 339*, 53–61.

Gulbransen, B. D., & Sharkey, K. A. (2012). Novel functional roles for enteric glia in the gastrointestinal tract. *Nature Reviews Gastroenterology & Hepatology, 9*, 625–632.

Gurevich, V. V., & Gurevich, E. V. (2008). How and why do GPCRs dimerize? *Trends in Pharmacological Sciences, 29*, 234–240.

Hansen, H. S., & Artmann, A. (2008). Endocannabinoids and nutrition. *Journal of Neuroendocrinology, 20*(Suppl. 1), 94–99.

Heimann, A. S., Gomes, I., Dale, C. S., Pagano, R. L., Gupta, A., de Souza, L. L., et al. (2007). Hemopressin is an inverse agonist of CB(1) cannabinoid receptors. *Proceedings of the National Academy of Sciences of the United States of America, 104*, 20588–20593.

Heinemann, A., Shahbazian, A., & Holzer, P. (1999). Cannabinoid inhibition of guinea-pig intestinal peristalsis via inhibition of excitatory and activation of inhibitory neural pathways. *Neuropharmacology, 38*, 1289–1297.

Hermann, A., Kaczocha, M., & Deutsch, D. G. (2006). 2-Arachidonoylglycerol (2-AG) membrane transport: History and outlook. *AAPS Journal, 8*, E409–E412.

Higgs, H. N., & Glomset, J. A. (1994). Identification of a phosphatidic acid-preferring phospholipase A1 from bovine brain and testis. *Proceedings of the National Academy of Sciences of the United States of America, 91*, 9574–9578.

Hofer, S. C., Ralvenius, W. T., Gachet, M. S., Fritschy, J. M., Zeilhofer, H. U., & Gertsch, J. (2015). Localization and production of peptide endocannabinoids in the rodent CNS and adrenal medulla. *Neuropharmacology*, In press.

Hojo, M., Sudo, Y., Ando, Y., Minami, K., Takada, M., Matsubara, T., et al. (2008). mu-Opioid receptor forms a functional heterodimer with cannabinoid CB1 receptor: Electrophysiological and FRET assay analysis. *Journal of Pharmacological Sciences, 108*, 308–319.

Hons, I. M., Storr, M. A., Mackie, K., Lutz, B., Pittman, Q. J., Mawe, G. M., et al. (2012). Plasticity of mouse enteric synapses mediated through endocannabinoid and purinergic signaling. *Neurogastroenterology and Motility, 24*, e113–e124.

Howlett, A. C. (2005). Cannabinoid receptor signaling. *Handbook of Experimental Pharmacology, 168*, 53–79.

Howlett, A. C., Barth, F., Bonner, T. I., Cabral, G., Casellas, P., Devane, W. A., et al. (2002). International Union of Pharmacology. XXVII. Classification of cannabinoid receptors. *Pharmacological Reviews, 54*, 161–202.

Howlett, A. C., Reggio, P. H., Childers, S. R., Hampson, R. E., Ulloa, N. M., & Deutsch, D. G. (2011). Endocannabinoid tone versus constitutive activity of cannabinoid receptors. *British Journal of Pharmacology, 163*, 1329–1343.

Hudson, B. D., Hebert, T. E., & Kelly, M. E. (2010a). Ligand- and heterodimer-directed signaling of the CB(1) cannabinoid receptor. *Molecular Pharmacology, 77*, 1–9.

Hudson, B. D., Hebert, T. E., & Kelly, M. E. (2010b). Physical and functional interaction between CB1 cannabinoid receptors and beta2-adrenoceptors. *British Journal of Pharmacology, 160*, 627–642.

Izzo, A. A., Borrelli, F., Capasso, R., Di Marzo, V., & Mechoulam, R. (2009). Non-psychotropic plant cannabinoids: New therapeutic opportunities from an ancient herb. *Trends in Pharmacological Sciences, 30*, 515–527.

Izzo, A. A., Fezza, F., Capasso, R., Bisogno, T., Pinto, L., Iuvone, T., et al. (2001). Cannabinoid CB1-receptor mediated regulation of gastrointestinal motility in mice in a model of intestinal inflammation. *British Journal of Pharmacology, 134*, 563–570.

Izzo, A. A., & Sharkey, K. A. (2010). Cannabinoids and the gut: New developments and emerging concepts. *Pharmacology & Therapeutics, 126*, 21–38.

Jamontt, J. M., Molleman, A., Pertwee, R. G., & Parsons, M. E. (2010). The effects of delta-tetrahydrocannabinol and cannabidiol alone and in combination on damage, inflammation and in vitro motility disturbances in rat colitis. *British Journal of Pharmacology, 160*, 712–723.

Jin, W., Brown, S., Roche, J. P., Hsieh, C., Celver, J. P., Kovoor, A., et al. (1999). Distinct domains of the CB1 cannabinoid receptor mediate desensitization and internalization. *Journal of Neuroscience, 19*, 3773–3780.

Kaczocha, M., Rebecchi, M. J., Ralph, B. P., Teng, Y. H., Berger, W. T., Galbavy, W., et al. (2014). Inhibition of fatty acid binding proteins elevates brain anandamide levels and produces analgesia. *PLoS One, 9*, e94200.

Kairupan, T. S., Amitani, H., Cheng, K. C., Runtuwene, J., Asakawa, A., & Inui, A. (2015). Role of gastrointestinal hormones in feeding behavior and obesity treatment. *Journal of Gastroenterology*, In press.

Katona, I., & Freund, T. F. (2012). Multiple functions of endocannabinoid signaling in the brain. *Annual Review of Neuroscience, 35*, 529–558.

Katona, I., Rancz, E. A., Acsady, L., Ledent, C., Mackie, K., Hajos, N., et al. (2001). Distribution of CB1 cannabinoid receptors in the amygdala and their role in the control of GABAergic transmission. *Journal of Neuroscience, 21*, 9506–9518.

Kennedy, J. E., & Marchese, A. (2015). Regulation of GPCR trafficking by ubiquitin. *Progress in Molecular Biology and Translational Science, 132*, 15–38.

Kinsey, S. G., Nomura, D. K., O'Neal, S. T., Long, J. Z., Mahadevan, A., Cravatt, B. F., et al. (2011). Inhibition of monoacylglycerol lipase attenuates nonsteroidal anti-inflammatory drug-induced gastric hemorrhages in mice. *Journal of Pharmacology and Experimental Therapeutics*, *338*, 795–802.

Koboziev, I., Reinoso Webb, C., Furr, K. L., & Grisham, M. B. (2014). Role of the enteric microbiota in intestinal homeostasis and inflammation. *Free Radical Biology & Medicine*, *68*, 122–133.

Kozak, K. R., Crews, B. C., Morrow, J. D., Wang, L. H., Ma, Y. H., Weinander, R., et al. (2002). Metabolism of the endocannabinoids, 2-arachidonylglycerol and anandamide, into prostaglandin, thromboxane, and prostacyclin glycerol esters and ethanolamides. *Journal of Biological Chemistry*, *277*, 44877–44885.

Kreitzer, A. C., & Regehr, W. G. (2001). Retrograde inhibition of presynaptic calcium influx by endogenous cannabinoids at excitatory synapses onto Purkinje cells. *Neuron*, *29*, 717–727.

Kulkarni-Narla, A., & Brown, D. R. (2000). Localization of CB1-cannabinoid receptor immunoreactivity in the porcine enteric nervous system. *Cell and Tissue Research*, *302*, 73–80.

Kurjak, M., Hamel, A. M., Allescher, H. D., Schusdziarra, V., & Storr, M. (2008). Differential stimulatory effects of cannabinoids on VIP release and NO synthase activity in synaptosomal fractions from rat ileum. *Neuropeptides*, *42*, 623–632.

Lapointe, T. K., Basso, L., Iftinca, M. C., Flynn, R., Chapman, K., Dietrich, G., et al. (2015). TRPV1 sensitization mediates postinflammatory visceral pain following acute colitis. *American Journal of Physiology Gastrointestinal and Liver Physiology*, *309*, G87–G99.

Leung, D., Saghatelian, A., Simon, G. M., & Cravatt, B. F. (2006). Inactivation of N-acyl phosphatidylethanolamine phospholipase D reveals multiple mechanisms for the biosynthesis of endocannabinoids. *Biochemistry*, *45*, 4720–4726.

Li, K., Fichna, J., Schicho, R., Saur, D., Bashashati, M., Mackie, K., et al. (2013). A role for O-1602 and G protein-coupled receptor GPR55 in the control of colonic motility in mice. *Neuropharmacology*, *71*, 255–263.

Lin, T. Y., Lu, C. W., Wu, C. C., Huang, S. K., & Wang, S. J. (2015). Palmitoylethanolamide inhibits glutamate release in rat cerebrocortical nerve terminals. *International Journal of Molecular Sciences*, *16*, 5555–5571.

Lin, X. H., Yuece, B., Li, Y. Y., Feng, Y. J., Feng, J. Y., Yu, L. Y., et al. (2011). A novel CB receptor GPR55 and its ligands are involved in regulation of gut movement in rodents. *Neurogastroenterology and Motility*, *23*. 862-e342.

Liu, J., Wang, L., Harvey-White, J., Osei-Hyiaman, D., Razdan, R., Gong, Q., et al. (2006). A biosynthetic pathway for anandamide. *Proceedings of the National Academy of Sciences of the United States of America*, *103*, 13345–13350.

Lopez-Redondo, F., Lees, G. M., & Pertwee, R. G. (1997). Effects of cannabinoid receptor ligands on electrophysiological properties of myenteric neurones of the guinea-pig ileum. *British Journal of Pharmacology*, *122*, 330–334.

Maccarrone, M., Bab, I., Biro, T., Cabral, G. A., Dey, S. K., Di Marzo, V., et al. (2015). Endocannabinoid signaling at the periphery: 50 years after THC. *Trends in Pharmacological Sciences*, *36*, 277–296.

MacNaughton, W. K., Van Sickle, M. D., Keenan, C. M., Cushing, K., Mackie, K., & Sharkey, K. A. (2004). Distribution and function of the cannabinoid-1 receptor in the modulation of ion transport in the guinea pig ileum: Relationship to capsaicin-sensitive nerves. *American Journal of Physiology Gastrointestinal and Liver Physiology*, *286*, G863–G871.

Manco, M. (2012). Gut microbiota and developmental programming of the brain: From evidence in behavioral endophenotypes to novel perspective in obesity. *Frontiers in Cellular and Infection Microbiology*, *2*, 109.

Mang, C. F., Erbelding, D., & Kilbinger, H. (2001). Differential effects of anandamide on acetylcholine release in the guinea-pig ileum mediated via vanilloid and non-CB1 cannabinoid receptors. *British Journal of Pharmacology, 134*, 161–167.

Marcellino, D., Carriba, P., Filip, M., Borgkvist, A., Frankowska, M., Bellido, I., et al. (2008). Antagonistic cannabinoid CB1/dopamine D2 receptor interactions in striatal CB1/D2 heteromers. A combined neurochemical and behavioral analysis. *Neuropharmacology, 54*, 815–823.

Marquez, L., Suarez, J., Iglesias, M., Bermudez-Silva, F. J., Rodriguez de Fonseca, F., & Andreu, M. (2009). Ulcerative colitis induces changes on the expression of the endocannabinoid system in the human colonic tissue. *PLoS One, 4*, e6893.

Mathison, R., Ho, W., Pittman, Q. J., Davison, J. S., & Sharkey, K. A. (2004). Effects of cannabinoid receptor-2 activation on accelerated gastrointestinal transit in lipopolysaccharide-treated rats. *British Journal of Pharmacology, 142*, 1247–1254.

Matsuda, L. A., Lolait, S. J., Brownstein, M. J., Young, A. C., & Bonner, T. I. (1990). Structure of a cannabinoid receptor and functional expression of the cloned cDNA. *Nature, 346*, 561–564.

Mayer, E. A. (2011). Gut feelings: The emerging biology of gut–brain communication. *Nature Reviews Neuroscience, 12*, 453–466.

Mayer, E. A., Tillisch, K., & Gupta, A. (2015). Gut/brain axis and the microbiota. *Journal of Clinical Investigation, 125*, 926–938.

Mechoulam, R., Ben-Shabat, S., Hanus, L., Ligumsky, M., Kaminski, N. E., Schatz, A. R., et al. (1995). Identification of an endogenous 2-monoglyceride, present in canine gut, that binds to cannabinoid receptors. *Biochemical Pharmacology, 50*, 83–90.

Moss, C. E., Marsh, W. J., Parker, H. E., Ogunnowo-Bada, E., Riches, C. H., Habib, A. M., et al. (2012). Somatostatin receptor 5 and cannabinoid receptor 1 activation inhibit secretion of glucose-dependent insulinotropic polypeptide from intestinal K cells in rodents. *Diabetologia, 55*, 3094–3103.

Muccioli, G. G., Naslain, D., Backhed, F., Reigstad, C. S., Lambert, D. M., Delzenne, N. M., et al. (2010). The endocannabinoid system links gut microbiota to adipogenesis. *Molecular Systems Biology, 6*, 392.

Mukhopadhyay, S., & Howlett, A. C. (2001). CB1 receptor-G protein association. Subtype selectivity is determined by distinct intracellular domains. *European Journal of Biochemistry, 268*, 499–505.

Mukhopadhyay, S., & Howlett, A. C. (2005). Chemically distinct ligands promote differential CB1 cannabinoid receptor-Gi protein interactions. *Molecular Pharmacology, 67*, 2016–2024.

Mule, F., Amato, A., Baldassano, S., & Serio, R. (2007). Evidence for a modulatory role of cannabinoids on the excitatory NANC neurotransmission in mouse colon. *Pharmacological Research, 56*, 132–139.

Munro, S., Thomas, K. L., & Abu-Shaar, M. (1993). Molecular characterization of a peripheral receptor for cannabinoids. *Nature, 365*, 61–65.

Nakane, S., Oka, S., Arai, S., Waku, K., Ishima, Y., Tokumura, A., et al. (2002). 2-Arachidonoyl-sn-glycero-3-phosphate, an arachidonic acid-containing lysophosphatidic acid: Occurrence and rapid enzymatic conversion to 2-arachidonoyl-sn-glycerol, a cannabinoid receptor ligand, in rat brain. *Archives of Biochemistry and Biophysics, 402*, 51–58.

Nasser, Y., Bashashati, M., & Andrews, C. N. (2014). Toward modulation of the endocannabinoid system for treatment of gastrointestinal disease: FAAHster but not "higher" *Neurogastroenterology and Motility, 26*, 447–454.

Navarrete, M., & Araque, A. (2008). Endocannabinoids mediate neuron-astrocyte communication. *Neuron, 57*, 883–893.

Navarrete, M., Diez, A., & Araque, A. (2014). Astrocytes in endocannabinoid signalling. *Philosophical Transactions of the Royal Society of London Series B, Biological Sciences, 369*, 20130599.

Neunlist, M., Van Landeghem, L., Mahe, M. M., Derkinderen, P., des Varannes, S. B., & Rolli-Derkinderen, M. (2013). The digestive neuronal-glial-epithelial unit: A new actor in gut health and disease. *Nature Reviews Gastroenterology & Hepatology, 10*, 90–100.

Nicolussi, S., & Gertsch, J. (2015). Endocannabinoid transport revisited. *Vitamins and Hormones, 98*, 441–485.

Ohno-Shosaku, T., Hashimotodani, Y., Maejima, T., & Kano, M. (2005). Calcium signaling and synaptic modulation: Regulation of endocannabinoid-mediated synaptic modulation by calcium. *Cell Calcium, 38*, 369–374.

Ohno-Shosaku, T., Maejima, T., & Kano, M. (2001). Endogenous cannabinoids mediate retrograde signals from depolarized postsynaptic neurons to presynaptic terminals. *Neuron, 29*, 729–738.

Oliveira da Cruz, J. F., Robin, L. M., Drago, F., Marsicano, G., & Metna-Laurent, M. (2015). Astroglial type-1 cannabinoid receptor (CB): A new player in the tripartite synapse. *Neuroscience*, In press.

Patel, K. D., Davison, J. S., Pittman, Q. J., & Sharkey, K. A. (2010). Cannabinoid CB(2) receptors in health and disease. *Current Medicinal Chemistry, 17*, 1393–1410.

Pertwee, R. G. (2001). Cannabinoids and the gastrointestinal tract. *Gut, 48*, 859–867.

Pertwee, R. G. (2006). The pharmacology of cannabinoid receptors and their ligands: An overview. *International Journal of Obesity, 30*(Suppl. 1), S13–S18.

Pertwee, R. G., Fernando, S. R., Nash, J. E., & Coutts, A. A. (1996). Further evidence for the presence of cannabinoid CB1 receptors in guinea-pig small intestine. *British Journal of Pharmacology, 118*, 2199–2205.

Pertwee, R. G., Howlett, A. C., Abood, M. E., Alexander, S. P., Di Marzo, V., Elphick, M. R., et al. (2010). International Union of Basic and Clinical Pharmacology. LXXIX. Cannabinoid receptors and their ligands: Beyond CB(1) and CB(2). *Pharmacological Reviews, 62*, 588–631.

Pinto, L., Capasso, R., Di Carlo, G., & Izzo, A. A. (2002). Endocannabinoids and the gut. *Prostaglandins, Leukotrienes, and Essential Fatty Acids, 66*, 333–341.

Pinto, L., Izzo, A. A., Cascio, M. G., Bisogno, T., Hospodar-Scott, K., Brown, D. R., et al. (2002). Endocannabinoids as physiological regulators of colonic propulsion in mice. *Gastroenterology, 123*, 227–234.

Piomelli, D. (2003). The molecular logic of endocannabinoid signalling. *Nature Reviews Neuroscience, 4*, 873–884.

Porter, A. C., Sauer, J. M., Knierman, M. D., Becker, G. W., Berna, M. J., Bao, J., et al. (2002). Characterization of a novel endocannabinoid, virodhamine, with antagonist activity at the CB1 receptor. *Journal of Pharmacology and Experimental Therapeutics, 301*, 1020–1024.

Psichas, A., Reimann, F., & Gribble, F. M. (2015). Gut chemosensing mechanisms. *Journal of Clinical Investigation, 125*, 908–917.

Re, G., Barbero, R., Miolo, A., & Di Marzo, V. (2007). Palmitoylethanolamide, endocannabinoids and related cannabimimetic compounds in protection against tissue inflammation and pain: Potential use in companion animals. *Veterinary Journal, 173*, 21–30.

Rehfeld, J. F. (2014). Gastrointestinal hormones and their targets. *Advances in Experimental Medicine and Biology, 817*, 157–175.

Rhee, S. H., Pothoulakis, C., & Mayer, E. A. (2009). Principles and clinical implications of the brain-gut-enteric microbiota axis. *Nature Reviews Gastroenterology & Hepatology, 6*, 306–314.

Rios, C., Gomes, I., & Devi, L. A. (2006). mu opioid and CB1 cannabinoid receptor interactions: Reciprocal inhibition of receptor signaling and neuritogenesis. *British Journal of Pharmacology, 148*, 387–395.

Ross, R. A. (2003). Anandamide and vanilloid TRPV1 receptors. *British Journal of Pharmacology, 140*, 790–801.

Ross, G. R., Lichtman, A., Dewey, W. L., & Akbarali, H. I. (2012). Evidence for the putative cannabinoid receptor (GPR55)-mediated inhibitory effects on intestinal contractility in mice. *Pharmacology, 90*, 55–65.

Russo, E. B. (2008). Clinical endocannabinoid deficiency (CECD): Can this concept explain therapeutic benefits of cannabis in migraine, fibromyalgia, irritable bowel syndrome and other treatment-resistant conditions? *Neuro Endocrinology Letters, 29*, 192–200.

Ryberg, E., Larsson, N., Sjogren, S., Hjorth, S., Hermansson, N. O., Leonova, J., et al. (2007). The orphan receptor GPR55 is a novel cannabinoid receptor. *British Journal of Pharmacology, 152*, 1092–1101.

Sang, N., Zhang, J., & Chen, C. (2007). COX-2 oxidative metabolite of endocannabinoid 2-AG enhances excitatory glutamatergic synaptic transmission and induces neurotoxicity. *Journal of Neurochemistry, 102*, 1966–1977.

Savidge, T. C., Newman, P., Pothoulakis, C., Ruhl, A., Neunlist, M., Bourreille, A., et al. (2007). Enteric glia regulate intestinal barrier function and inflammation via release of S-nitrosoglutathione. *Gastroenterology, 132*, 1344–1358.

Schele, E., Grahnemo, L., Anesten, F., Hallen, A., Backhed, F., & Jansson, J. O. (2013). The gut microbiota reduces leptin sensitivity and the expression of the obesity-suppressing neuropeptides proglucagon (Gcg) and brain-derived neurotrophic factor (Bdnf) in the central nervous system. *Endocrinology, 154*, 3643–3651.

Schicho, R., Bashashati, M., Bawa, M., McHugh, D., Saur, D., Hu, H. M., et al. (2011). The atypical cannabinoid O-1602 protects against experimental colitis and inhibits neutrophil recruitment. *Inflammatory Bowel Diseases, 17*, 1651–1664.

Schlicker, E., & Kathmann, M. (2001). Modulation of transmitter release via presynaptic cannabinoid receptors. *Trends in Pharmacological Sciences, 22*, 565–572.

Sharkey, K. A. (2015). Emerging roles for enteric glia in gastrointestinal disorders. *Journal of Clinical Investigation, 125*, 918–925.

Sibaev, A., Yuece, B., Allescher, H. D., Saur, D., Storr, M., & Kurjak, M. (2014). The endocannabinoid anandamide regulates the peristaltic reflex by reducing neuro-neuronal and neuro-muscular neurotransmission in ascending myenteric reflex pathways in rats. *Pharmacological Reports, 66*, 256–263.

Simon, G. M., & Cravatt, B. F. (2006). Endocannabinoid biosynthesis proceeding through glycerophospho-N-acyl ethanolamine and a role for alpha/beta-hydrolase 4 in this pathway. *Journal of Biological Chemistry, 281*, 26465–26472.

Simon, G. M., & Cravatt, B. F. (2010). Characterization of mice lacking candidate N-acyl ethanolamine biosynthetic enzymes provides evidence for multiple pathways that contribute to endocannabinoid production in vivo. *Molecular BioSystems, 6*, 1411–1418.

Smith, T. K., Park, K. J., & Hennig, G. W. (2014). Colonic migrating motor complexes, high amplitude propagating contractions, neural reflexes and the importance of neuronal and mucosal serotonin. *Journal of Neurogastroenterology and Motility, 20*, 423–446.

Spreckley, E., & Murphy, K. G. (2015). The L-cell in nutritional sensing and the regulation of appetite. *Frontiers in Nutrition, 2*, 23.

Stella, N. (2010). Cannabinoid and cannabinoid-like receptors in microglia, astrocytes, and astrocytomas. *Glia, 58*, 1017–1030.

Stella, N., Schweitzer, P., & Piomelli, D. (1997). A second endogenous cannabinoid that modulates long-term potentiation. *Nature, 388*, 773–778.

Sternini, C., Anselmi, L., & Rozengurt, E. (2008). Enteroendocrine cells: A site of 'taste' in gastrointestinal chemosensing. *Current Opinion in Endocrinology, Diabetes, and Obesity, 15*, 73–78.

Storr, M. A., Bashashati, M., Hirota, C., Vemuri, V. K., Keenan, C. M., Duncan, M., et al. (2010). Differential effects of CB(1) neutral antagonists and inverse agonists on gastrointestinal motility in mice. *Neurogastroenterology and Motility, 22*, 787–796. e223.

Storr, M., Gaffal, E., Saur, D., Schusdziarra, V., & Allescher, H. D. (2002). Effect of cannabinoids on neural transmission in rat gastric fundus. *Canadian Journal of Physiology and Pharmacology*, *80*, 67–76.

Storr, M., Sibaev, A., Marsicano, G., Lutz, B., Schusdziarra, V., Timmermans, J. P., et al. (2004). Cannabinoid receptor type 1 modulates excitatory and inhibitory neurotransmission in mouse colon. *American Journal of Physiology Gastrointestinal and Liver Physiology*, *286*, G110–G117.

Straiker, A., Wager-Miller, J., & Mackie, K. (2012). The CB1 cannabinoid receptor C-terminus regulates receptor desensitization in autaptic hippocampal neurones. *British Journal of Pharmacology*, *165*, 2652–2659.

Suarez, J., Romero-Zerbo, Y., Marquez, L., Rivera, P., Iglesias, M., Bermudez-Silva, F. J., et al. (2012). Ulcerative colitis impairs the acylethanolamide-based anti-inflammatory system reversal by 5-aminosalicylic acid and glucocorticoids. *PLoS One*, *7*, e37729.

Sykaras, A. G., Demenis, C., Case, R. M., McLaughlin, J. T., & Smith, C. P. (2012). Duodenal enteroendocrine I-cells contain mRNA transcripts encoding key endocannabinoid and fatty acid receptors. *PLoS One*, *7*, e42373.

Van Sickle, M. D., Duncan, M., Kingsley, P. J., Mouihate, A., Urbani, P., Mackie, K., et al. (2005). Identification and functional characterization of brainstem cannabinoid CB2 receptors. *Science*, *310*, 329–332.

Vergnolle, N. (2014). TRPV4: New therapeutic target for inflammatory bowel diseases. *Biochemical Pharmacology*, *89*, 157–161.

Wager-Miller, J., Westenbroek, R., & Mackie, K. (2002). Dimerization of G protein-coupled receptors: CB1 cannabinoid receptors as an example. *Chemistry and Physics of Lipids*, *121*, 83–89.

Wilson, R. I., & Nicoll, R. A. (2001). Endogenous cannabinoids mediate retrograde signalling at hippocampal synapses. *Nature*, *410*, 588–592.

Yang, H., Zhang, J., Andreasson, K., & Chen, C. (2008). COX-2 oxidative metabolism of endocannabinoids augments hippocampal synaptic plasticity. *Molecular and Cellular Neuroscience*, *37*, 682–695.

Yu, M., Ives, D., & Ramesha, C. S. (1997). Synthesis of prostaglandin E2 ethanolamide from anandamide by cyclooxygenase-2. *Journal of Biological Chemistry*, *272*, 21181–21186.

Zhang, L., Wang, M., Bisogno, T., Di Marzo, V., & Alger, B. E. (2011). Endocannabinoids generated by Ca2+ or by metabotropic glutamate receptors appear to arise from different pools of diacylglycerol lipase. *PLoS One*, *6*, e16305.

Zhang, S. C., Wang, W. L., Su, P. J., Jiang, K. L., & Yuan, Z. W. (2014). Decreased enteric fatty acid amide hydrolase activity is associated with colonic inertia in slow transit constipation. *Journal of Gastroenterology and Hepatology*, *29*, 276–283.

Zoppi, S., Madrigal, J. L., Perez-Nievas, B. G., Marin-Jimenez, I., Caso, J. R., Alou, L., et al. (2012). Endogenous cannabinoid system regulates intestinal barrier function in vivo through cannabinoid type 1 receptor activation. *American Journal of Physiology Gastrointestinal and Liver Physiology*, *302*, G565–G571.

Zuardi, A. W. (2006). History of cannabis as a medicine: A review. *Revista Brasileira de Psiquiatria*, *28*, 153–157.

CHAPTER FOUR

Endocannabinoid Mechanisms Influencing Nausea

Martin A. Sticht*,†, Erin M. Rock*, Cheryl L. Limebeer*, Linda A. Parker*,[1]

*Department of Psychology, Collaborative Neuroscience Graduate Program, University of Guelph, Guelph, Ontario, Canada
†Department of Physiology and Pharmacology, Hotchkiss Brain Institute, University of Calgary, Calgary, Alberta, Canada
[1]Corresponding author: e-mail address: parkerl@uoguelph.ca

Contents

1. Endocannabinoids in Acute Nausea 132
   1.1 Cannabinoids and Chemotherapy-Induced Nausea in Humans 132
   1.2 Endocannabinoid Levels During the Experience of Nausea in Humans 134
   1.3 Animal Models of Nausea 137
   1.4 Endocannabinoids Reduce Acute Nausea-Induced Conditioned Gaping 140
   1.5 Brainstem Circuitry Involved in Nausea 142
   1.6 Insular Cortex: A Critical Forebrain Area for Nausea 142
   1.7 Motion Sickness in Rodents 144
2. Endocannabinoids in Anticipatory Nausea 145
   2.1 Human Studies of Anticipatory Nausea 145
   2.2 Cannabinoid Treatments for Anticipatory Nausea in Humans 146
   2.3 Animal Studies of Anticipatory Nausea: Contextually Elicited Conditioned Gaping 149
   2.4 Cannabinoid Treatments Reduce Anticipatory Nausea in Rodents 150
   2.5 Endocannabinoid Manipulations Reduced Anticipatory Nausea in Rodents 151
3. Conclusions 152
References 152

Abstract

One of the first recognized medical uses of Δ^9-tetrahydrocannabinol was treatment of chemotherapy-induced nausea and vomiting. Although vomiting is well controlled with the currently available non-cannabinoid antiemetics, nausea continues to be a distressing side effect of chemotherapy and other disorders. Indeed, when nausea becomes conditionally elicited by the cues associated with chemotherapy treatment, known as anticipatory nausea (AN), currently available antiemetics are

International Review of Neurobiology, Volume 125
ISSN 0074-7742
http://dx.doi.org/10.1016/bs.irn.2015.09.001

largely ineffective. Considerable evidence demonstrates that the endocannabinoid system regulates nausea in humans and other animals. In this review, we describe recent evidence suggesting that cannabinoids and manipulations that enhance the functioning of the natural endocannabinoid system are promising treatments for both acute nausea and AN.

For several centuries, cannabis has been used for therapeutic purposes (Mechoulam, 2005), including the attenuation of nausea and vomiting. Ineffective treatment of chemotherapy-induced nausea and vomiting (CINV) prompted oncologists to examine the antiemetic properties of cannabinoids in the late 1970s and early 1980s. The first cannabinoid agonist, nabilone (Cesamet), which is a synthetic analog of Δ^9-tetrahydrocannabinol (THC) was specifically licensed for the treatment of CINV in cancer patients. As well, synthetic THC, dronabinol, entered the clinic as Marinol in 1985 as an antiemetic (Pertwee, 2009).

A major discovery in the control of acute vomiting in chemotherapy treatment came in the late 1980s with the development of 5-hydroxytryptamine 3 (5-HT_3) receptor antagonists, such as ondansetron (e.g., Costall et al., 1986; Miner & Sanger, 1986). However, as the use of these 5-HT_3 antagonists in the clinic progressed, it became clear that they were much more effective in reducing vomiting than they were in suppressing nausea (see Hickok et al., 2003). Indeed, in the case of anticipatory nausea (AN) experienced as a conditioned response, when the patient returns to the clinic in which the treatment occurred, the 5-HT_3 antagonists are completely ineffective (see Morrow & Dobkin, 1987). Therefore, more effective antinausea treatments are still required. Recent evidence based upon the development of new animal models for preclinical evaluation of novel treatments suggests that cannabinoids may offer therapeutic potential in reducing both acute nausea and AN. Here we review this evidence, first for cannabinoid suppression of acute nausea and, second, for the potential of cannabinoids to reduce AN. A list of treatments/manipulations and their efficacy in humans and animal models is presented in Table 1. The neurocircuitry and proposed mechanism by which the endocannabinoid system regulates nausea is presented in Fig. 1.

Table 1 Efficacy of Various Cannabinoid Compounds to Alleviate Acute Nausea in Humans or in Animal Models

Compound	Efficacy in Human Acute Nausea	Efficacy in Animals Models of Acute Nausea
Cannabinoid treatments		
Dronabinol (Marinol)	Effective compared to placebo (Chang et al., 1979; Frytak et al., 1979; Orr, McKernan, & Bloome, 1980; Sallan, Zinberg, & Frei, 1975; Sweet, Miller, Weddington, Senay, & Sushelsky, 1981) Effective compared to D2 antagonists (Ekert, Waters, Jurk, Mobilia, & Loughnan, 1979; Orr & McKernan, 1981) Ineffective compared to D2 antagonists (Carey, Burish, & Brenner, 1983; Crawford & Buckman, 1986; Frytak et al., 1979; Tramèr et al., 2001; Ungerleider, Andrysiak, Fairbanks, Tesler, & Parker, 1984) Equally effective as a 5-HT3 antagonist for nausea; greater reduction in nausea intensity (Meiri et al., 2007)	Not evaluated
Nabilone (Cesamet)	Effective compared to D2 antagonists (Ahmedzai, Carlyle, Calder, & Moran, 1983; Dalzell, Bartlett, & Lilleyman, 1986; Herman et al., 1979; Pomeroy, Fennelly, & Towers, 1986)	Not evaluated
Sativex (1:1 THC:CBD)	Ineffective compared to placebo (Duran et al., 2010) when added to standard emetic treatment	Not evaluated

Continued

Table 1 Efficacy of Various Cannabinoid Compounds to Alleviate Acute Nausea in Humans or in Animal Models—cont'd

Compound	Efficacy in Human Acute Nausea	Efficacy in Animals Models of Acute Nausea
WIN55212-2	Not evaluated	Ineffective for chronic cisplatin-induced pica (Vera et al., 2007)
THC	Not evaluated	Reduced cyclophosphamide (Limebeer & Parker, 1999) and LiCl-induced (Parker & Mechoulam, 2003; Parker et al., 2003) gaping in rats
THCA	Not evaluated	Reduced LiCl-induced gaping in rats (Rock, Kopstick, Limebeer, & Parker, 2013)
HU-210	Not evaluated	Reduced LiCl-induced gaping in rats (Parker et al., 2003)
CBD	Not evaluated	Reduced LiCl-induced gaping in rats (Parker et al., 2002; Rock et al., 2012)
CBDA	Not evaluated	Reduced LiCl-induced gaping in rats (Bolognini et al., 2013)
Endocannabinoid manipulations		
Anandamide	Not evaluated	Reduced LiCl-induced aversive reactions in rats when combined with a FAAH inhibitor; ineffective alone (Cross-Mellor, Ossenkopp, Piomelli, & Parker, 2007) Intra-VIC administration reduced LiCl-induced gaping in rats when combined with an FAAH inhibitor (Sticht et al., submitted); ineffective alone when delivered into VIC (Sticht, Limebeer, Rafla, & Parker, 2015)

Table 1 Efficacy of Various Cannabinoid Compounds to Alleviate Acute Nausea in Humans or in Animal Models—cont'd

Compound	Efficacy in Human Acute Nausea	Efficacy in Animals Models of Acute Nausea
2-AG	Not evaluated	Reduced LiCl-induced gaping in rats (Sticht et al., 2012) Intra-VIC administration reduced LiCl-induced gaping in rats (Sticht et al., 2015)
FAAH inhibition	Not evaluated	PF3845 reduced LiCl-induced gaping in rats (Rock et al., 2015); but ineffective when administered into the VIC in rats (Sticht et al., submitted) URB597 attenuated LiCl-induced aversive reactions in rats; greater effect when combined with exogenous anandamide (Cross-Mellor et al., 2007) URB597 reduced LiCl-induced gaping in rats when administered into the VIC with exogenous AEA; intra-VIC URB597 ineffective alone (Sticht et al., submitted)
MAGL inhibition	Not evaluated	JZL184 reduced LiCl-induced gaping in rats when combined with exogenous 2-AG; ineffective alone (Sticht et al., 2012) MJN110 reduced LiCl-induced gaping in rats (Parker, Rock, Sticht, Wills, & Limebeer, 2015); also effective when delivered into the VIC (Sticht et al., submitted)
Dual FAAH/MAGL inhibition	Not evaluated	JZL195 reduced LiCl-induced gaping in rats when administered into the VIC (Sticht et al., submitted)

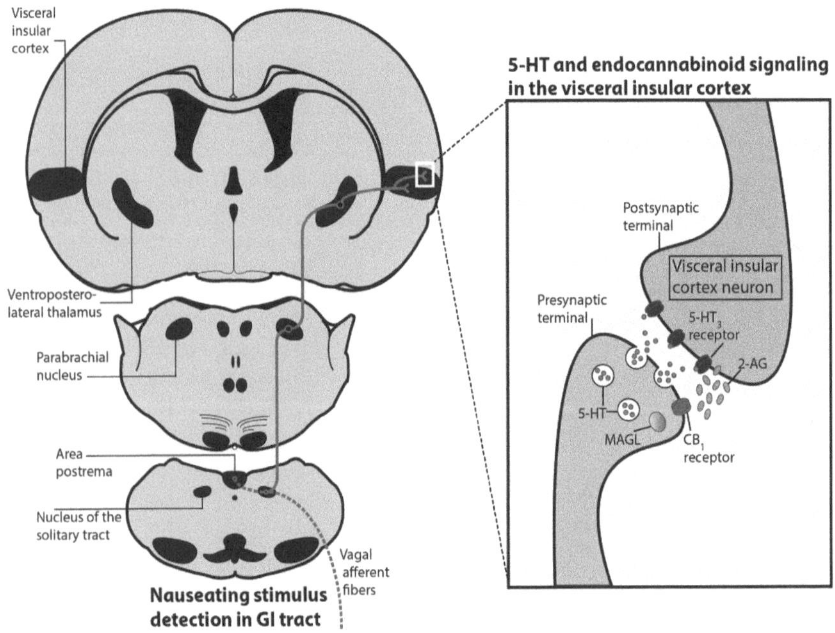

Figure 1 Brain neurocircuitry involved in nausea. The circuits and projections involved in nausea are shown, including the proposed mechanism by which the endocannabinoid system modulates nausea sensation in the insular cortex. Illness-inducing (emetic) stimuli activate key components of the gut–brain axis to elicit the sensation of nausea and vomiting (in emetic-capable species). Toxins or pharmacological treatments are detected in the gastrointestinal tract and relayed to the brain through vagal afferents terminating in the nucleus of the solitary tract (NTS); alternatively, emetogens (toxins) also gain access to the CNS through the area postrema (AP). Ultimately, nausea processing occurs in the visceral insular cortex, which receives visceral information from the brainstem dorsal vagal complex (DVC; and parabrachial nucleus) and is relayed through the ventroposterolateral parvicellular (VPLpc) thalamic nucleus before reaching the posterior granular layer of the IC. Within the VIC, activation of postsynaptic 5-HT_3 receptors produces nausea, resulting in biosynthesis of the endocannabinoid, 2-arachidonoylglycerol (2-AG), which reduces its sensation. We hypothesize that the 'on-demand' synthesis of 2-AG in the VIC functions to reduce subsequent 5-HT release via inhibitory presynaptic CB_1 signaling, thereby reducing nausea. *Coronal sections were adapted from Paxinos and Watson (2007).*

1. ENDOCANNABINOIDS IN ACUTE NAUSEA

1.1 Cannabinoids and Chemotherapy-Induced Nausea in Humans

Early studies in humans investigating the potential for cannabinoid compounds to reduce nausea were focused on assessing the effects of marihuana to alleviate CINV (Cotter, 2009). The most well-studied phytocannabinoid has been THC (Gaoni & Mechoulam, 1964), which is the major psychoactive constituent in cannabis (Gaoni & Mechoulam, 1964; Mechoulam, Shani, Edery, & Grunfeld, 1970). THC is a high-affinity agonist for both Cannabinoid 1 (CB_1) and Cannabinoid 2 (CB_2) receptors; thus, these studies provided early insights into the important role played by the endocannabinoid system in regulating nausea. As presented in Table 1, an orally administered form of THC, known as dronabinol (Marinol), has been shown to effectively alleviate chemotherapy-induced nausea (and vomiting) when compared to a placebo control (Chang et al., 1979; Frytak et al., 1979; Orr et al., 1980; Sallan et al., 1975; Sweet et al., 1981), or compared to common dopamine 2 (D_2) receptor antagonists prescribed at the time (Ekert et al., 1979; Orr & McKernan, 1981). In some cases, however, dronabinol was no more effective than the D_2 antagonist in alleviating these symptoms of chemotherapy treatment (Carey et al., 1983; Crawford & Buckman, 1986; Frytak et al., 1979; Tramèr et al., 2001; Ungerleider et al., 1984). More recently, Meiri and colleagues compared the antinausea potential of dronabinol with the currently prescribed antiemetic, ondansetron (a 5-HT_3 antagonist), and found that either treatment was effective in alleviating both nausea and vomiting (Meiri et al., 2007); however, when assessing nausea severity, treatment with dronabinol resulted in a greater reduction of nausea intensity compared to ondansetron (Meiri et al., 2007). In this case, it appears that dronabinol may have had greater effect in reducing nausea intensity resulting from mild to moderately severe emetogenic treatment, but not highly emetogenic treatments (Meiri et al., 2007).

Another orally active and synthetic analog of THC, nabilone (Cesamet), was licensed for treatment-resistant nausea and vomiting in 1985, but only after conventional treatments were deemed ineffective. Although nabilone has only been assessed in comparison with dopamine (D_2) antagonists (as with Dronabinol), chemotherapy patients reported fewer episodes of nausea when receiving the cannabinoid (Ahmedzai et al., 1983; Dalzell et al., 1986; Herman et al., 1979; Pomeroy et al., 1986). Although Pomeroy and colleagues (1986) reported that nabilone produced a marginal decrease in nausea when patients received highly emetogenic chemotherapy, others have found nabilone to be no more effective than a D_2 antagonist (Crawford & Buckman, 1986); thus, as with dronabinol, nabilone may be no more effective in reducing nausea when receiving a highly emetogenic drug than conventional antiemetics.

Recently, an oromucosal cannabis-based medicine, Sativex (1:1 THC: cannabidiol (CBD)) has been evaluated in phase II clinical trials for its potential to reduce CINV (Duran et al., 2010). In this case, Sativex or placebo was administered to patients receiving concomitant treatment with standard antiemetic drugs including a 5-HT_3 antagonist and a corticosteroid. Duran and colleagues (2010) found that Sativex was effective at relieving delayed nausea (and vomiting) among patients receiving moderately emetogenic cancer chemotherapy. In particular, approximately half (57%) of the Sativex-treated patients experienced no delayed nausea and an even greater number (71%) experienced no delayed emesis compared to patients receiving placebo (Duran et al., 2010). However, given the known antinausea/antiemetic effects of CBD demonstrated in animal models (Parker et al., 2015; Sharkey, Darmani, & Parker, 2014), it is unclear how much of the antinausea effects of Sativex are attributable to either THC or CBD; nonetheless, this study demonstrates the therapeutic potential of combining these cannabinoids in treating delayed CINV.

Despite the overwhelming evidence for the antinausea potential of cannabinoids in humans, it is worthwhile to note that chronic marihuana use has also been reported to produce the sensation of nausea (and vomiting). This paradoxical effect of marihuana, known as cannabinoid hyperemesis syndrome (Sullivan, 2010), has been documented in numerous case reports in recent years (Allen, de Moore, Heddle, & Twartz, 2004; Roca-Pallín, López-Pelayo, Sugranyes, & Balcells-Oliveró, 2013; Simonetto, Oxentenko, Herman, & Szostek, 2012). Although the cause of hyperemesis is not known, it can be speculated that changes in CB_1 receptor expression and relative concentrations of THC (and other marijuana constituents) may play a role in this paradoxical effect. Further research is undoubtedly needed to fully understand the mechanism(s) underlying this peculiar effect of cannabinoids that may lead to increases in nausea in vomiting following chronic marihuana consumption. Nevertheless, these cases highlight the dynamic nature played by the endocannabinoid system in regulating nausea and suggest a possible consequence of endocannabinoid system dysregulation, which may lead to undesirable results such as an increased sensation of nausea and emesis.

1.2 Endocannabinoid Levels During the Experience of Nausea in Humans

Endogenous, or endocannabinoids, are lipid messengers that bind to either of the metabotropic cannabinoid receptors (CB_1 or CB_2). Synthesized in an activity-dependent manner from membrane phospholipids, these endogenous ligands consist of *N*-arachidonoylethanolamide (anandamide) and 2-arachidonoylglycerol (2-AG). Following release, anandamide and 2-AG are hydrolysed primarily by fatty acid amide hydrolase (FAAH) and monoacylglycerol lipase (MAGL), respectively. Both of these enzymes are targets of pharmacological manipulations in which blockade of FAAH and MAGL results in respective increases in anandamide and 2-AG levels. This approach is widely used to explore the role of the endocannabinoid system in physiology and behavior (Blankman & Cravatt, 2013) and offers tremendous therapeutic potential in alleviating nausea as discussed throughout this chapter. Moreover, given that endocannabinoids are synthesized on-demand, manipulations that target endocannabinoid hydrolysis result in a much more localized increase in levels of 2-AG and anandamide compared to administration of endocannabinoids or cannabinoid receptor agonists. Therefore, a focused increase in endocannabinoid levels is less likely to produce unwanted side effects attributable to a global effect and is, thus, preferable for selectively reducing nausea.

Although at present there have not been any investigations assessing the potential for pro-endocannabinoid drugs to reduce nausea—that is, manipulations that directly increase the biological action(s) of anandamide and 2-AG—studies have assessed endocannabinoid levels in response to nausea-inducing manipulations. These investigations have provided important insight into the role of the endocannabinoid system, through the actions of anandamide and 2-AG, to modulate the experience of nausea. To this end, Schelling and colleagues (2006) investigated the effects of general anesthesia on anandamide levels using the volatile inhalant, sevoflurane, or the intravenously administered anesthetic, propofol. Although anandamide levels were found to remain unchanged among patients receiving propofol, Schelling et al. (2006) reported that sevoflurane administration resulted in a significant decrease in anandamide levels from the point of induction until 40 min after intubation (measurements taken at 0, 10, 20, and 40 min post-intubation). This pattern of findings is consistent with reports that anesthesia with propofol, a drug that inhibits FAAH and increases anandamide levels in rodents (Patel et al., 2003), results in less postoperative nausea compared to sevoflurane (Kumar, Stendall, Mistry, Gurusamy, & Walker, 2014) and may be the result of increased anandamide signaling; although, increases in other CB_1-inactive FAAH substrates such as oleoylethanolamide (OEA) and palmitoylethanolamide (PEA) should also be considered, and thus it is not clear whether other fatty acid ethanolamides mediate some of the effects of propofol, as well. Nonetheless, recent findings by Jarzimski and colleagues are in contrast to those previously reported, such that both sevoflurane and propofol decreased plasma anandamide levels similarly (Jarzimski et al., 2012). Further research is necessary to clarify the effects of anesthetics on endocannabinoid levels and the particular effects of propofol on anandamide, in particular. At the very least, however, propofol appears to avoid the undesirable side effect of nausea following general anesthesia, and, thus, may be a more advantageous anesthetic treatment.

A recent study by Sadhasivam et al. (2015) also investigated the role of anandamide in postoperative nausea and vomiting (PONV) by assessing the extent to which common genetic variants of FAAH, or FAAH polymorphisms, are associated with PONV occurrence. Such FAAH polymorphisms have previously been studied with respect to psychiatric disorders and these genetic changes likely contribute to endocannabinoid dysfunction underlying addiction, depression, and other behaviors (Hillard, Weinlander, & Stuhr, 2012). It is not surprising, therefore, that FAAH polymorphisms, which ultimately modify endocannabinoid system function, may also play a role in the experience of nausea. Specifically, Sadhasivam and colleagues (2015) found five specific FAAH single nucleotide polymorphisms that were associated with an increased risk for PONV among child-patients undergoing a common surgical procedure, tonsillectomy. Because these patients were anesthetized with a combination of drugs including morphine, which often leads to unwanted side effects of respiratory depression and PONV, it is no surprise that some patients experienced both of these symptoms; however, it appears that differences in anandamide signaling among individuals with different FAAH polymorphisms underlies the extent to which patients experienced PONV and respiratory depression symptoms. As mentioned before, it is important to consider the possibility that other FAAH substrates (e.g., PEA/OEA) may also be involved given that, like anandamide, they are regulated by FAAH. Nonetheless, it appears that genetic variants of FAAH may underlie similar postoperative outcomes of respiratory depression and PONV in children (Sadhasivam et al., 2015). The extent to which these FAAH polymorphisms play a role in PONV and respiratory depression among adult patients remains unknown and should be an aim of future research.

The modulatory role of the endocannabinoid system in humans has also been demonstrated in a model of motion-induced nausea. Schelling and colleagues (Choukèr et al., 2010) analyzed endocannabinoid levels from participants undergoing parabolic flight maneuvers (PFs) and this revealed that blood anandamide and 2-AG levels were lower among those who experienced motion sickness. Moreover, while blood concentrations of anandamide dropped among participants experiencing motion sickness, it was found to increase in volunteers who did not report nausea. Another noticeable difference between volunteers was that levels of 2-AG increased following the end of PFs among those that did not experience nausea, whereas blood 2-AG concentrations remained steady (but lower) throughout the entire test for those that reported nausea. Thus, blood endocannabinoid levels were higher among individuals that were less susceptible to motion-induced nausea, albeit with changes in endocannabinoid levels occurring at different time points during an experience of nausea. Interestingly, it was also found that CB_1 receptor mRNA, an index of CB_1 receptor expression, in isolated leukocytes found in blood was significantly decreased following PFs among participants that experienced motion sickness, whereas there was no change in expression levels for CB_1 among those that did not report nausea. On the other hand, CB_2 expression did not change regardless of whether participants experience nausea or not. Taken together, therefore, this study suggests that the experience of motion-induced nausea is associated with reduced endocannabinoid signaling, which is evident through a reduction in anandamide and 2-AG levels, as well as reduced CB_1 receptor expression (Choukèr et al., 2010).

1.3 Animal Models of Nausea

Animal models of nausea are indispensible preclinical tools to evaluate potential pharmacological treatments for this distressing symptom. Moreover, such models provide an opportunity to explore the neurobiological mechanisms of nausea, which ultimately leads to the development of more effective therapies.

1.3.1 Pica

One established measure of nausea in rodents is the consumption of nonnutritive kaolin clay, a behavior known as pica. The eating of a nonfood substance serves a functional purpose in alleviating gastrointestinal malaise to prevent toxin digestion, and this is particularly important among nonemetic species such as rats and mice. Emetic drugs such as lithium chloride (LiCl) or the chemotherapy drug, cyclophosphamide, result in kaolin consumption in a dose-dependent manner (Mitchell et al., 1976), and this can be blocked with conventional antiemetic treatments (Rudd, Yamamoto, Yamatodani, & Takeda, 2002). Thus, it has been argued that pica is analogous to vomiting in emetic subjects (Takeda, Hasegawa, Morita, & Matsunaga, 1993), albeit it is often measured in species that are incapable of this behavior. Although pica may be a useful *unconditioned* measure of nausea (in contrast to models of *conditioned* nausea discussed later in this chapter), it is important to note, however, that pica occurs in response to manipulations that do not necessarily produce nausea such as stress (Burchfield, Elich, & Woods, 1977). Moreover, pica is absent in the emetic shrew species, *Suncus murinus*, following treatment with emetogenic compounds, despite the fact that such drugs produce vomiting in this species (Yamamoto, Ngan, Takeda, Yamatodani, & Rudd, 2004). Therefore, pica is neither selectively produced by nausea-inducing manipulations nor is it observed in all species. Nonetheless, Vera and colleagues (2007) assessed whether the synthetic cannabinoid, WIN55212-2, would affect pica in response to chronic cisplatin administration and found that kaolin ingestion was not modified by cannabinoid treatment. To date, there have not been any investigations of endocannabinoid manipulations on pica.

1.3.2 Conditioned Taste Avoidance

In addition to the measures of nausea just described, rodents also display several conditioned behaviors to stimuli that are present at the time of illness. Although rats are not capable of producing an emetic reaction in response to nausea-inducing stimuli, they come to reject a taste or flavor that has been associated with its occurrence following consumption. This phenomenon, known as conditioned taste avoidance (aversion), was initially investigated by Garcia and colleagues (Garcia, Hankins, & Rusiniak 1974), who described a type of associative learning in which a particular taste comes to predict toxin-induced illness following conditioning. As this associative learning is directly related to regulation of the internal homeostatic environment, Garcia and colleagues (1974) argued that the palatability of a taste is modified by internal effects subsequent to consumption. This type of learning is particularly critical for nonemetic species such as rats, which are neophobic and generally consume only small amounts of novel food upon initial exposure; hence, taste aversion learning is so prevalent that it results after a single taste–illness pairing following any change in homeostatic state. Interestingly, even rewarding drugs—those that are self-administered by rats in operant conditioning paradigms—produce taste avoidance (Parker, 1995). Moreover, antiemetic treatments do not prevent the establishment of conditioned taste avoidance, either. Therefore, these findings suggest that nausea is not a requirement for conditioned taste avoidance, and, thus, it is not surprising that cannabinoids (and pro-endocannabinoid manipulations) do not prevent its establishment.

1.3.3 Nausea-Induced Conditioned Gaping

Rats not only avoid consuming any food associated with illness but also display conditioned disgust (rejection) reactions when reexposed to the illness-paired taste. This finding was initially reported by Garcia et al. (1974) but was subsequently replicated by Grill and Norgren using the taste-reactivity test, a measure that assesses palatability of intraorally infused taste solutions (Grill & Norgren, 1978). In the event that a palatable solution such as sucrose was delivered intraorally, rats were observed to display ingestive reactions such as tongue protrusions; alternatively, bitter quinine solution elicited disgust reactions in the form of gaping, as well as a series of body responses (e.g., chin rubs, pawtreads) all of which were considered to be indicative of an aversion to the taste. Interestingly, upon reexposure to a sweet solution that was previously paired with a nauseating drug such as LiCl, the same disgust responses were observed, of which gaping is the most reliable aversive reaction (Breslin, Spector, & Grill, 1992). In fact, electromyographic analysis in rats revealed a similarity between the facial musculature involved in gaping and those involved in vomiting responses among emetic species (Travers & Norgren, 1986), indicating that gaping reactions may represent an incomplete emetic response.

Numerous studies have since demonstrated that only emetic drugs (including withdrawal from morphine (McDonald, Parker, & Siegel, 1997) or cocaine (Wheeler et al., 2008, 2011) result in conditioned gaping behavior in rats, the establishment of which is subsequently blocked by antiemetic treatments including cannabinoids. For example, THC reduced the establishment of acute nausea-induced conditioned gaping to the chemotherapy drug, cyclophosphamide (Limebeer & Parker, 1999) and LiCl (Parker & Mechoulam, 2003; Parker et al., 2003), and the antinausea effects were reversed following concomitant pretreatment with a CB_1 receptor antagonist/inverse agonist (AM-251 or rimonabant). Similarly, the acidic precursor of THC (tetrahydrocannabinolic acid—THCA) has also been shown to reduce conditioned gaping, and with greater potency than THC (Rock et al., 2013). Also, the synthetic cannabinoid agonist, HU-210, was found to exert antinausea effects through a CB_1 receptor-dependent mechanism of action, as well (Parker et al., 2003). Thus, CB_1 agonism leads to antinausea effects in the conditioned gaping model.

It is worthwhile to mention that, although CB_1 antagonists/inverse agonists reverse the antiemetic effects of the aforementioned cannabinoid drugs, high doses of these compounds alone are sufficient to produce nausea-induced gaping. For example, McLaughlin et al. (2005) demonstrated that

AM-251 not only reduced the consumption of palatable food, but it also resulted in conditioned gaping and taste avoidance when paired with an infusion of intraoral saccharin (McLaughlin et al., 2005). Lower doses of AM-251 have also been found to potentiate LiCl-induced conditioned gaping (Limebeer et al., 2010) and a similar effect was also demonstrated with rimonabant (Parker & Mechoulam, 2003). It is important to note that the mere blockade of CB_1 receptors is not sufficient to produce nausea on its own, however, as the neutral antagonist, AM4113, did not produce conditioned gaping when paired with intraoral saccharin (Sink et al., 2008). Likewise, the respective centrally and peripherally active CB_1 neutral antagonists, AM6527 and AM6545, were also ineffective in potentiating LiCl-induced gaping, as well (Limebeer et al., 2010). Therefore, whereas CB_1 antagonism alone does not produce conditioned gaping in rats, inverse agonism of CB_1 receptors leads to the sensation of nausea.

Although THC has received much interest for its therapeutic potential in alleviating nausea, another well-studied constituent of cannabis, CBD, has also been shown to reduce nausea-induced conditioned gaping (Parker & Mechoulam, 2003). However, this nonpsychoactive cannabinoid does not bind to CB_1 or CB_2 receptors, but rather exerts antinausea effects through activation of somatodendritic 5-HT_{1A} autoreceptors of the dorsal raphe nucleus (Rock et al., 2012), thereby limiting 5-HT release in terminal forebrain regions underlying the experience of nausea. Furthermore, the carboxylic precursor of CBD, cannabidiolic acid (CBDA), exhibits even greater potency compared to CBD in its ability to reduce LiCl-induced conditioned gaping in rats (Bolognini et al., 2013). Taken together, cannabis constituents suppress nausea through cannabinoid receptor-dependent and independent mechanisms, and the acid precursors of these compounds may have an even greater therapeutic potential in alleviating the sensation of nausea.

1.4 Endocannabinoids Reduce Acute Nausea-Induced Conditioned Gaping

Early evidence that the endocannabinoid system is involved in the regulation of nausea was largely the result of experiments assessing the effects of THC, and animal studies were instrumental in demonstrating the ability of CB_1 antagonists to modify these antinausea effects. Therefore, these studies demonstrated the importance of CB_1 signaling in modulating the experience of nausea. As such, it should come as no surprise that recent studies have focused on the role of endogenous ligands for CB_1 receptors and their ability to modify nausea. With the development of compounds that increase levels of endogenously released anandamide and 2-AG (via inhibition of FAAH and MAGL), these studies have provided important insights into endocannabinoid regulation of nausea.

Inhibition of anandamide hydrolysis with the FAAH inhibitor, URB597, has been shown to attenuate LiCl-induced aversive reactions (but not gaping specifically) in the taste-reactivity test, with an even greater suppressive effect following concomitant pretreatment with exogenous anandamide (Cross-Mellor et al., 2007). More recently, the novel FAAH inhibitor, PF3845, has also been shown to be considerably more effective than URB597 in reducing acute nausea-induced conditioned gaping (Rock et al., 2015). However, rimonabant only reversed the suppressive effects of URB597 on aversive taste reactivity (Cross-Mellor et al., 2007), whereas the peroxisome proliferator-activated receptor-α (PPARα) antagonist, MK886, blocked the antinausea effect of PF3845 (Rock et al., 2015). The fatty acid ethanolamides, PEA and OEA, although structurally related to anandamide, do not bind to cannabinoid receptors; instead, these lipids are known to activate a particular subtype of nuclear transcription factors known as PPARs (Issemann & Green, 1990). Specifically, OEA (Fu et al., 2003) and PEA (Verme et al., 2005) bind to the PPARα subtype to exert much of their biological effects. In the case of nausea, the suppressive effects of these lipids, indeed, appear to be through PPARα, and this finding is consistent with the robust increases in OEA and PEA following PF3845 administration (Ahn et al., 2009).

Manipulations that increase levels of 2-AG have also been shown to reduce acute nausea in rats. For example, administration of exogenous 2-AG (which is rapidly deactivated by MAGL) has been found to reduce LiCl-induced conditioned gaping (Sticht et al., 2012), and inhibition of MAGL-mediated hydrolysis of 2-AG for up to 24 h with the selective inhibitor, MJN110, has also been shown to reduce LiCl-induced conditioned gaping (Parker et al., 2014). The antinausea effects of MJN110 were blocked following CB_1 antagonism, however, the suppressive effects of 2-AG were not attenuated following administration of a CB_1 or CB_2 antagonist, likely because it had been metabolized by MAGL (Parker et al., 2014; Sticht et al., 2012). In the latter case, administration of the cyclooxygenase (COX) inhibitor, indomethacin, blocked the ability of 2-AG to reduce nausea-induced conditioned gaping, as well as its metabolic product, arachidonic acid (Sticht et al., 2012). This finding suggests that downstream COX-derived metabolites of 2-AG (or arachidonic acid) may also play a role in reducing nausea, as well (Sticht et al., 2012, 2015). Taken together, these findings demonstrated that various mechanisms underlie endocannabinoid suppression of nausea in rats; however, the particular brain structures that mediate these actions remain less clear.

1.5 Brainstem Circuitry Involved in Nausea

Illness-inducing (emetic) stimuli activate key components of the gut–brain axis to elicit the sensation of nausea and vomiting (in emetic-capable species). Specifically, ingested toxins or pharmacological treatments stimulate the release of serotonin from enterochromaffin cells of the gastrointestinal tract (Andrews & Bhandari, 1993), which in turn activates 5-HT_3 or 5-HT_4 receptors located on vagal afferents (Nemoto et al., 2001) terminating in the dorsal vagal complex (DVC) of the brainstem (Andrews & Horn, 2006; Horn, 2014; Hornby, 2001). Alternatively, emetogens (toxins) also gain access to the CNS through the area postrema (AP), a circumventricular organ lying outside the blood brain barrier that functions as an important chemoreceptor for circulating toxins in the blood and cerebrospinal fluid (Miller & Leslie, 1994). In either case, these signals activate neurons in the DVC circuit (Leslie, 1985), and visceral information is relayed from the nucleus of the solitary tract to the insular cortex (IC) via the parabrachial nucleus and thalamus (Cechetto & Saper, 1987). Although endocannabinoids have not been assessed for their potential to regulate nausea through a direct action in the DVC, cannabinoid compounds have been shown to reduce toxin-induced vomiting in ferrets through CB_1 (Van Sickle et al., 2001, 2003) and CB_2 (Van Sickle et al., 2005) receptors in this region; therefore, it is likely that endocannabinoids could exert antinausea effects through similar mechanisms in the DVC and, thereby, inhibit the relay of visceral malaise information to the forebrain.

1.6 Insular Cortex: A Critical Forebrain Area for Nausea

The earliest reports that the IC may be involved in nausea were from the study of patients undergoing treatment for temporal lobe epilepsy. Upon stimulation of the IC, visceral sensory phenomena were reported by patients as the feeling of nausea and nausea-like gastric sensations (Penfield & Faulk, 1955). These findings have been reported in more recent studies in which patients underwent a similar procedure that included stimulation of the IC (Catenoix et al., 2008; Isnard, Guénot, Sindou, & Mauguière, 2004; Ostrowsky et al., 2000). Recently, the brain circuitry underlying nausea has been explored in functional neuroimaging studies in humans

(Napadow et al., 2013; Sclocco et al., 2014), revealing an important role for the IC. Specifically, Napadow et al. (2013) demonstrated that a strong sensation of nausea resulted in sustained activation of the interoceptive insula, as well as limbic and subcortical regions. Importantly, autonomic activation, which is essential for nausea perception (LaCount et al., 2011), appears to be modulated by the IC (Sclocco et al., 2014). The IC also plays a central role in the processing of disgust-related facial expressions (see Calder, Lawrence, & Young, 2001) and the experience of disgust in response to nauseating stimuli (Calder et al., 2007). Therefore, these studies demonstrate the importance of the IC in several aspects of nausea and disgust. In rats, the IC likely plays a similar role, because ablation of this entire region has been shown to selectively interfere with the establishment of LiCl-induced conditioned gaping, but not taste avoidance (Kiefer & Orr, 1992).

The rat IC is composed of three distinct cellular layers along the rostrocaudal axis (Krieg, 1946), extending from a ventral agranular layer through dysgranular and granular layers located more dorsally (Cechetto & Saper, 1987; Krieg, 1946). These cortical layers correspond to rostrally located taste-responsive neurons contained in the agranular and dysgranular layers, while general visceral stimulation comprises the neurons in the granular layer known as the visceral insular cortex (VIC) (Cechetto & Saper, 1987). In the case of the VIC, this subregion receives visceral information from the brainstem DVC (and parabrachial nucleus) and is relayed through the ventroposterolateral parvicellular (VPLpc) thalamic nucleus before reaching the posterior granular layer of the IC (Cechetto & Saper, 1987). This area appears to be critically involved in the sensation of nausea as determined using a number of different approaches and techniques. Contreras et al. (2007) demonstrated that temporary inactivation of this area attenuated LiCl-induced lying-on-belly behavior. Moreover, LiCl administration results in VIC Fos immunoreactivity (Contreras et al., 2007; Sticht et al., submitted), an index of neural activation during an episode of nausea. Recently, a 5-HT_3 agonist, mCPBG, was found to elicit conditioned gaping when delivered directly to the VIC (Tuerke, Limebeer, Fletcher, & Parker, 2012), while the conventional 5-HT_3 antagonist, ondansetron or the synthetic cannabinoid, HU-210, reduced nausea-induced conditioned gaping (Limebeer, Rock, Mechoulam, & Parker, 2012; Tuerke et al., 2012). Thus, activation of the VIC by serotonin appears to underlie an experience of nausea in rats and inactivation of this region via CB_1-signaling serves to reduce its sensation.

Recently, the role of the VIC endocannabinoid system has been further investigated in which the effects of exogenous anandamide and 2-AG were assessed following localized administration. Specifically, intra-VIC 2-AG, but not anandamide, effectively suppressed conditioned gaping; however, intra-VIC-treated rats still displayed conditioned taste avoidance to LiCl-paired saccharin regardless of the endocannabinoid treatment (Sticht et al., 2015). Most likely because of rapid metabolism by MAGL, the anti-nausea effects of 2-AG within the VIC were not reversed following CB_1 antagonism, but were blocked with the COX inhibitor, indomethacin (Sticht et al., 2015). This pattern was observed in a previous study with systemic administration of 2-AG, in which its antinausea effects were also reversed following COX enzyme inhibition (Sticht et al., 2012).

In a separate study, inhibition of endocannabinoid catabolic enzymes within the VIC was also found to reduce conditioned gaping, as well. Consistent with the effects of systemic MAGL inhibition, intra-VIC MJN110 administration blocked the establishment of conditioned gaping without modifying taste avoidance (Sticht et al., submitted). However, unlike exogenous 2-AG in the VIC, localized MAGL inhibition reduced nausea via CB_1 receptor signaling because the effects of MJN110 were reversed by AM-251, but not the COX inhibitor, indomethacin. This finding suggests that under conditions of reduced 2-AG metabolism, the antinausea effects are mediated by CB_1 receptors; yet, downstream (COX) metabolic products of 2-AG also play a role to reduce nausea, as well (Sticht et al., 2012, 2015). Interestingly, neither of the FAAH inhibitors URB597 nor PF3845 were effective in reducing conditioned gaping in rats following intra-VIC administration, although the combination of URB597 together with exogenous anandamide was effective when delivered into the VIC (Sticht et al., submitted). This finding is particularly noteworthy given that URB597 did not suppress conditioned gaping when administered systemically (Cross-Mellor et al., 2007; Rock et al., 2015). Altogether, these behavioral results are consistent with the finding that LiCl administration results in a selective increase in 2-AG levels over anandamide within the VIC (Sticht et al., submitted); thus, it appears that the effects of the VIC endocannabinoid system during an experience of acute nausea are mediated by the multiple actions of 2-AG—either through CB_1 receptors or downstream metabolic products.

1.7 Motion Sickness in Rodents

The animal models discussed thus far have all assessed the effects of pharmacologically induced illness. However, several published investigations have utilized a measure of motion-induced illness that is a purported index of motion sickness in rodents (Wei et al., 2011; Yu, Cai, Liu, Chu, & Su, 2007). Specifically, this model assesses the extent to which rotation-induced illness elicits the symptoms of piloerection, tremble, and urinal/fecal incontinence. The classic anticholinergic drug (muscarinic antagonist), scopolamine, has been shown to reduce this motion sickness index in rats (Yu et al., 2007) and mice (Wei et al., 2011; Yu et al., 2007), as did administration of other antimotion sickness drugs such as the antihistamine and diphenhydramine (Wei et al., 2011). Bilateral labyrinthectomy (disruption of the vestibular system) in mice reduced the index of motion sickness (Wei et al., 2011), and, interestingly, a labyrinthectomy procedure has also been shown to disrupt the establishment of conditioned gaping in rats to a flavor paired with motion-induced illness (Ossenkopp et al., 2003). Although no direct endocannabinoid manipulations have been assessed for their potential to reduce motion sickness in rodents, Zheng, Wang, Mo, and Li (2014) demonstrated that dexamethasone treatment, which directly increases levels of anandamide and 2-AG (Di, Malcher-Lopes, Marcheselli, Bazan, & Tasker, 2005), resulted in fewer motion sickness-related symptoms and, surprisingly, the CB_1 antagonist, AM-251, prevented the suppressive effects of dexamethasone on motion-induced illness (Zheng et al., 2014). Moreover, whereas anandamide levels remained unchanged following rotation and dexamethasone treatment, 2-AG levels appeared to decline following rotation yet remained steady among rats receiving dexamethasone (Zheng et al., 2014). And, similar to human participants experiencing motion sickness (Choukèr et al., 2010), rats also displayed a reduction in CB_1 mRNA and CB_1 protein observed in the stomach and brainstem DVC in response to rotation, whereas dexamethasone-pretreated rats did not display a decrease in CB_1 receptors Zheng et al., 2014. Altogether, it appears that dexamethasone, a drug that is prescribed to reduce nausea and vomiting among chemotherapy patients (and has been shown to increase 2-AG and anandamide), reduces the severity of motion sickness through a CB_1 receptor-dependent mechanism.

2. ENDOCANNABINOIDS IN ANTICIPATORY NAUSEA

2.1 Human Studies of Anticipatory Nausea

Cancer patients undergoing outpatient chemotherapy treatment can develop AN as a result of the association between the contextual cues of the chemotherapy clinic, with the subsequent nausea experienced from their treatment (Nesse, Carli, Curtis, & Kleinman, 1980). Initially, clinic staff classified this phenomenon as a type of neurosis until Nesse et al. (1980) described it as a form of Pavlovian conditioning defined as AN (Burish & Carey, 1986; Eckert, 2001; Hickok, & Morrow, 2002; Kamen et al., 2014; Matteson, Roscoe, Nesse et al., 1980; Stockhorst, Steingrueber, Enck, & Klosterhalfen, 2006). Consistent with predictions based on conditioning theory, across eight chemotherapy infusions, patients experiencing AN indicated that they had greater increases in the severity of AN closer to scheduled infusions relative to more distal times (Montgomery & Bovbjerg, 1997). These findings may perhaps inform clinical practices, such that patients could administer prophylactic treatments prior to their arrival at the clinic to ameliorate AN severity.

Some 25–59% of chemotherapy patients experience AN during their treatment course (Akechi et al., 2010; Bovbjerg et al., 1992; Hickok, Roscoe, & Morrow, 2001; Stockhorst, Klosterhalfen, Klosterhalfen, Winkelmann, & Steingrueber, 1993; Tyc, Mulhern, & Bieberich, 1997; Watson, McCarron, & Law, 1992; Zachariae et al., 2007). The risk of developing AN increases with the number of chemotherapy cycles a patient undergoes during which nausea has not been properly managed, and can occur in 25–39% of chemotherapy patients after their fourth treatment cycle (Aapro, Molassiotis, & Olver, 2005; Hickok et al., 2003; Janelsins et al., 2013; Morrow et al., 1998; Morrow, Roscoe, Kirshner, Hynes, & Rosenbluth, 1998; Roscoe, Morrow, Aapro, Molassiotis, & Olver, 2011). Once AN does develop, it is refractive to pharmacological treatment by antiemetics such as the classic 5-HT_3 receptor antagonist ondansetron (Aapro et al., 2005; Foubert & Vaessen, 2005; Morrow et al., 1998), and is currently treated with nonspecific antianxiety drugs (benzodiazepines, such as lorazepam), which have unpleasant sedative side effects (Malik et al., 1995; Razavi et al., 1993).

It is clear from the literature that, currently, AN is not properly managed in the clinic. This portion of the chapter will focus on current animal models of AN and the potential of endocannabinoids as treatments for this distressing symptom of chemotherapy, for which there is currently no specific treatment. The efficacy of these treatments/manipulations in animal models is presented in Table 2.

2.2 Cannabinoid Treatments for Anticipatory Nausea in Humans

Cannabinoid compounds such as THC have been shown to be effective in managing acute nausea in human patients (as discussed above) and to reduce

Table 2 Efficacy of Various Compounds Tested for Alleviation of Anticipatory Nausea for Humans or in Animal Models

Compound	Efficacy in Human AN	Efficacy in Animal Models of AN
Traditional pharmacological agents		
$5\text{-}HT_3$ receptor antagonists	Ineffective (reviewed in Aapro et al., 2005; Eckert, 2001; Figueroa-Moseley et al., 2007; Kamen et al., 2014; Morrow & Hickok, 1993; Morrow & Rosenthal, 1996; Morrow et al., 1998; Mustian et al., 2011; Stewart, 1990)	Ineffective in LiCl-induced rat gaping (Limebeer et al., 2006; Rock et al., 2014) Ineffective in LiCl-induced shrew gaping (Parker et al., 2006)
Benzodiazepines	Effective (Malik et al., 1995; Razavi et al., 1993)	Reduced LiCl-induced rat gaping (Rock et al., 2014)
Cannabinoid treatments		
Dronabinol (Marinol)	Less effective than D_2 receptor antagonist alone (Lane et al., 1991) or D_2 receptor antagonist + dronabinol	Not evaluated
THC	Not evaluated	Reduced LiCl-induced rat gaping (Limebeer et al., 2006; Rock et al., 2014) Reduced LiCl-induced shrew gaping (Parker & Kemp, 2001; Parker et al., 2006)
THCA	Not evaluated	Reduced LiCl-induced rat gaping (Rock et al., 2013)
CBD	Not evaluated	Reduced LiCl-induced rat gaping (Rock et al., 2008)
CBDA	Not evaluated	Reduced LiCl-induced rat gaping (Bolognini et al., 2013; Rock et al., 2014)
Endocannabinoid manipulations		
FAAH inhibition	Not evaluated	Reduced LiCl-induced rat gaping (Rock et al., 2008, 2015)

Continued

Table 2 Efficacy of Various Compounds Tested for Alleviation of Anticipatory Nausea for Humans or in Animal Models—cont'd

Compound	Efficacy in Human AN	Efficacy in Animal Models of AN
MAGL inhibition	Not evaluated	Reduced LiCl-induced rat gaping and LiCl-induced shrew gaping (Parker et al., 2015)
FAAH and MAGL inhibition	Not evaluated	Reduced LiCl-induced rat gaping (Limebeer et al., 2014)

AN in animal models (discussed further below). Therefore, there is a need for human trials investigating the ability of the endocannabinoid system to reduce AN.

To our knowledge, only one published human clinical trial has examined the effects of a CB_1 receptor agonist in human AN patients. When compared with prochlorperazine (a dopamine D_2 receptor antagonist) alone or in combination, dronabinol (Marinol) was less effective at reducing AN (AN occurred in 30% of dronabinol group, 0% in prochlorperazine group, 26% in combination group, Lane et al., 1991). Of interest is the fact that 86% of the patients evaluated were receiving highly emetogenic chemotherapeutic agents. Therefore, it is possible that dronabinol may not be as effective for AN that has developed as a result of nausea and/or vomiting in response to highly emetogenic chemotherapeutics.

As proper management of acute nausea is crucial in the prevention of AN development, the ability of THC and its synthetic derivatives to manage acute nausea (as discussed earlier) should decrease the risk of developing AN. Clinical trials need to specifically evaluate these compounds' efficacies in reducing AN. Unfortunately, no published reports have directly compared THC (or its synthetic analogs) with the current first-line treatment of 5-HT_3 receptor antagonist/dexamethasone/NK1 receptor antagonist, on acute, or anticipatory chemotherapy-induced nausea in human clinical trials, or even in animal models.

Investigations surrounding the role of the endocannabinoid system in nausea have relied mainly on animal models. To our knowledge, no human clinical trials have been conducted with FAAH or MAGL inhibitors in AN; thus, there is a clear need for these pharmacological tools to be moved into the clinic.

2.3 Animal Studies of Anticipatory Nausea: Contextually Elicited Conditioned Gaping

The lack of reliable rodent preclinical models of AN has limited the development of an effective pharmacological treatment for AN. The most parsimonious explanation for AN in chemotherapy patients is the development of a conditional association between nausea-inducing chemotherapy treatment and the contextual cues of the clinic where treatment is dispensed. This type of association is seen in the rat and supports the validity of a rodent model of AN.

As previously discussed, rats display conditioned gaping (rapid, large-amplitude opening of the mouth and simultaneous retraction at the corners of the mouth; Grill & Norgren, 1978) in response to the infusion of a flavored solution (such as sweet saccharin) that has previously been paired with an emetic drug such as LiCl. If suppressed consumption in a LiCl-paired context is a measure of AN, as suggested by Rodriguez et al. (2000), then the intraoral infusion of a novel solution in that environment should elicit gaping in rats. This hypothesis was tested by Limebeer et al. (2006) such that at test, rats were placed in the LiCl-paired context and infused with a novel saccharin solution for one min every five min during a 30-min test. Rats gaped to the infusion of saccharin, thus supporting the hypothesis. Most interestingly, gaping was also displayed during the interinfusion interval, suggesting conditioned nausea was also elicited by the context. To rule out the possibility that this gaping may have simply been an artifact of saccharin entering/remaining in the oral cavity between infusions, Limebeer et al. (2008) explored whether a LiCl-paired context could elicit conditioned gaping in the absence of a flavored solution. Indeed, exposure to the LiCl-paired context elicited conditioned gaping responses, as did exposure to a context where provocative motion stimulation was experienced. These results indicated that alternative nausea-inducing conditioned stimuli (such as contextual cues) are capable of eliciting gaping responses in rats. Conditioned gaping to a LiCl-paired context in rats is analogous to the phenomenon of AN experience by outpatient chemotherapy patients upon returning to the clinic context where they have received illness-inducing chemotherapy treatment. Interestingly, the *Suncus murinus* (house musk shrew), which vomits in response to an emetic treatment, also displays a conditioned gaping reaction when returned to a LiCl-paired context (Parker & Kemp, 2001; Parker, Kwiatkowska, & Mechoulam, 2006; Parker et al., 2015). Interestingly, as seen with human chemotherapy patients, ondansetron does not interfere with the expression of contextually elicited conditioned gaping in either rats or shrews (Limebeer, Hall, & Parker, 2006; Parker & Kemp, 2001; Parker et al., 2006; Rock et al., 2014).

2.4 Cannabinoid Treatments Reduce Anticipatory Nausea in Rodents

In animal models, THC (0.5–10 mg/kg, ip), but not ondansetron, attenuated contextually elicited gaping in rats (Limebeer et al., 2006; Rock et al., 2014). These results indicate that just as in human studies, ondansetron is ineffective in conditioned gaping models of AN, while THC is effective. Currently, human patients are treated with benzodiazepines (Razavi et al., 1993) for AN; indeed the benzodiazepine, chlordiazepoxide, also suppressed contextually elicited gaping in rats, but unlike THC, only at a dose that also impaired locomotor activity (Rock et al., 2014). In addition, THC (3 mg/kg) also suppresses contextually elicited gaping in shrews (Parker & Kemp, 2001; Parker et al., 2006) without altering locomotor activity. Therefore, THC may be a more desirable therapeutic over benzodiazepines in treating AN.

THC's acidic precursor, THCA (0.05 mg/kg, ip), which is thought to be devoid of psychoactive properties (Grunfeld & Edery, 1969), potently attenuated contextually elicited gaping in rats (Rock et al., 2013). The suppression of contextually elicited gaping by THCA was via CB_1 receptors as administration of SR141716 blocked this effect (Rock et al., 2013). There is, however, the possibility that THCA could exhibit CB_1 receptor-mediated abuse liability problems (see Oliere, Joliette-Riopel, Potvin, & Jutras-Aswad, 2013, for a recent review), perhaps making it a less desirable therapeutic agent for AN over that of other phytocannabinoids.

As mentioned earlier in the chapter, there is an increasing interest in the therapeutic potential of another well-studied cannabis constituent, CBD, which has also been shown to reduce contextually elicited gaping in rats (Parker & Mechoulam, 2003). Although it does not bind to CB_1 or CB_2 receptors, CBD is likely to reduce forebrain 5-HT release through an action at somatodendritric $5\text{-}HT_{1A}$ autoreceptors of the dorsal raphe nucleus (Rock et al., 2012). Furthermore, as with suppression of acute nausea, CBDA exhibits even greater potency compared to CBD in its ability to reduce AN in rats, as well (Bolognini et al., 2013). As such, CBDA appears to be a promising treatment for AN that is devoid of CB_1-related psychotropic side effects.

2.5 Endocannabinoid Manipulations Reduced Anticipatory Nausea in Rodents

In addition to managing acute nausea, the endogenous cannabinoid system has also been implicated in the control of AN (see Parker et al., 2015 for review). FAAH and/or MAGL inhibitors extend and enhance the effects of the endogenous CB_1 and/or PPARα ligands when they are produced selectively in the relevant brain region. Therefore, FAAH and/or MAGL inhibition might be a preferred therapeutic option, over that of direct exogenous CB_1 agonists like THC, which "flood" the system. As there are no specific treatments for AN, inhibition of the enzymes that deactivate endocannabinoids may be an effective treatment option for these patients. Additionally, as FAAH or MAGL inhibition suppresses chemotherapy-induced neuropathic pain in animal models (Guindon, Lai, Takacs, Bradshaw, & Hohmann, 2013), manipulation of the endocannabinoid system may be dually beneficial for chemotherapy patients who suffer from AN and neuropathic pain.

2.5.1 FAAH Inhibition

Inhibition of FAAH effectively reduces contextually elicited gaping in rats, presumably through elevation of AEA. The FAAH inhibitor URB597 (0.3 mg/kg, ip) effectively suppressed the expression of contextually elicited gaping (Rock, Limebeer, Mechoulam, Piomelli, & Parker, 2008). This effect was blocked by the administration of SR141617, indicating a CB_1 receptor-mediated mechanism of action, likely due to AEA elevation (Rock et al., 2008). Most recently, in comparison to URB597, the more selective FAAH inhibitor PF-3845 (Ahn et al., 2009) reduced AN when administered at 10 or 20 mg/kg (ip, Rock et al., submitted). This effect was also CB_1 receptor mediated, but not PPARα mediated (as compared to acute nausea), as SR141716, but not MK886 reversed this effect (Rock et al., 2015). These results are intriguing, as it seems that PF-3845 and URB597 suppressed AN via CB_1 receptor activation (but not PPARα activation), presumably by elevating AEA; whereas PF-3845 (but not URB597) suppressed acute nausea via PPARα activation (but not CB_1 receptor activation), presumably by elevating OEA and PEA.

2.5.2 MAGL Inhibition

Inhibition of MAGL also effectively reduces contextually elicited gaping in rats and shrews. The selective MAGL inhibitor, MJN110 (10, 20 mg/kg, ip), interfered with the expression of contextually elicited gaping in rats, and this effect was blocked by pretreatment with SR141716 (Parker et al., 2014). In addition, MJN110 (20 mg/kg, ip) also suppressed contextually elicited conditioned gaping in shrews to a LiCl-paired context.

2.5.3 Dual FAAH and MAGL Inhibition

Dual inhibition of FAAH and MAGL also effectively reduced contextually elicited gaping in rats. The dual FAAH–MAGL inhibitor JZL195 (10 mg/kg, ip) suppressed contextually elicited gaping and elevated AEA, PEA, and OEA (Limebeer et al., 2014). This effect was blocked by SR141716 (but not by AM630) indicating a CB_1 receptor-mediated effect. The suppressive effect of JZL195 on gaping, along with the corresponding elevation of AEA and 2-AG, was augmented by pretreatment with either AEA or 2-AG. Furthermore, AEA alone (but not 2-AG) also suppressed gaping; an effect that was blocked by CB_1 receptor antagonism. These results indicate that JZL195 reduced AN primarily by inhibiting FAAH, but inhibition of MAGL is also indicated. Further work should clarify the interaction between AEA and 2-AG, through the use of dual FAAH and MAGL inhibition, in the suppression of contextually elicited gaping in rats.

3. CONCLUSIONS

Since the discovery of the endocannabinoid system, our understanding of the mechanism(s) by which cannabinoids reduce nausea and vomiting has been greatly improved. Indeed, animal models demonstrate that direct CB_1 agonists reduce both acute nausea and AN. As well, since novel FAAH and MAGL inhibitors have been developed, these pro-endocannabinoid manipulations have also been shown to possess great potential in the treatment of both acute nausea and AN. Since nausea is much more resistant to conventional treatments such as $5\text{-}HT_3$ antagonists, manipulations of the endocannabinoid system have great promise for reducing this distressing side effect of cancer treatment. Determining the generality of the potential for endocannabinoid manipulations to reduce all forms of nausea (in response to various nauseating stimuli) is important in understanding the mechanisms underlying its sensation and experience.

REFERENCES

Aapro, M. S., Molassiotis, A., & Olver, I. (2005). Anticipatory nausea and vomiting. *Supportive Care in Cancer*, *13*(2), 117–121.

Ahmedzai, S., Carlyle, D. L., Calder, I. T., & Moran, F. (1983). Anti-emetic efficacy and toxicity of nabilone, a synthetic cannabinoid, in lung cancer chemotherapy. *British Journal of Cancer*, *48*(5), 657–663.

Ahn, K., Johnson, D. S., Mileni, M., Beidler, D., Long, J. Z., McKinney, M. K., et al. (2009). Discovery and characterization of a highly selective FAAH inhibitor that reduces inflammatory pain. *Chemistry & Biology*, *16*(4), 411–420. http://doi.org/10.1016/j.chembiol.2009.02.013.

Akechi, T., Okuyama, T., Endo, C., Sagawa, R., Uchida, M., Nakaguchi, T., et al. (2010). Anticipatory nausea among ambulatory cancer patients undergoing chemotherapy: Prevalence, associated factors, and impact on quality of life. *Cancer Science*, *101*(12), 2596–2600. http://doi.org/10.1111/j.1349-7006.2010.01718.x.

Allen, J. H., de Moore, G. M., Heddle, R., & Twartz, J. C. (2004). Cannabinoid hyperemesis: Cyclical hyperemesis in association with chronic cannabis abuse. *Gut*, *53*(11), 1566–1570. http://doi.org/10.1136/gut.2003.036350.

Andrews, P. L. R., & Bhandari, P. (1993). The 5-hydroxytryptamine receptor antagonists as antiemetics: Preclinical evaluation and mechanism of action. *European Journal of Cancer*, *29A*(Suppl.), S11–S16. http://doi.org/10.1016/S0959-8049(05)80254-0.

Andrews, P. L. R., & Horn, C. C. (2006). Signals for nausea and emesis: Implications for models of upper gastrointestinal diseases. *Autonomic Neuroscience: Basic and Clinical*, *125*(1–2), 100–115. http://doi.org/10.1016/j.autneu.2006.01.008.

Blankman, J. L., & Cravatt, B. F. (2013). Chemical probes of endocannabinoid metabolism. *Pharmacological Reviews*, *65*(2), 849–871. http://doi.org/10.1124/pr.112.006387.

Bolognini, D., Rock, E. M., Cluny, N. L., Cascio, M. G., Limebeer, C. L., Duncan, M., et al. (2013). Cannabidiolic acid prevents vomiting in *Suncus murinus* and nausea-induced behaviour in rats by enhancing 5-HT1A receptor activation. *British Journal of Pharmacology*, *168*(6), 1456–1470. http://doi.org/10.1111/bph.12043.

Bovbjerg, D. H., Redd, W. H., Jacobsen, P. B., Manne, S. L., Taylor, K. L., Surbone, A., et al. (1992). An experimental analysis of classically conditioned nausea during cancer chemotherapy. *Psychosomatic Medicine*, *54*(6), 623–637. Retrieved from, http://www.ncbi.nlm.nih.gov/pubmed/1454956.

Breslin, P. A., Spector, A. C., & Grill, H. J. (1992). A quantitative comparison of taste reactivity behaviors to sucrose before and after lithium chloride pairings: A unidimensional account of palatability. *Behavioral Neuroscience*, *106*(5), 820–836. http://doi.org/10.1037/0735-7044.106.5.820.

Burchfield, S. R., Elich, M. S., & Woods, S. C. (1977). Geophagia in response to stress and arthritis. *Physiology and Behavior*, *19*(2), 265–267. http://doi.org/10.1016/0031-9384(77)90337-7.

Burish, T. G., & Carey, M. P. (1986). Conditioned aversive responses in cancer chemotherapy patients: Theoretical and developmental analysis. *Journal of Consulting and Clinical Psychology*, *54*(5), 593–600. Retrieved from, http://www.ncbi.nlm.nih.gov/pubmed/3534029.

Calder, A. J., Lawrence, A. D., & Young, A. W. (2001). Neuropsychology of fear and loathing. *Nature Reviews. Neuroscience*, *2*(5), 352–363. http://doi.org/10.1028/35072584.

Calder, A. J., Beaver, J. D., Davis, M. H., Van Ditzhuijzen, J., Keane, J., & Lawrence, A. D. (2007). Disgust sensitivity predicts the insula and pallidal response to pictures of disgusting foods. *European Journal of Neuroscience*, *25*(11), 3422–3428. http://doi.org/10.1111/j.1460-9568.2007.05604.x.

Carey, M. P., Burish, T. G., & Brenner, D. E. (1983). Delta-9-tetrahydrocannabinol in cancer chemotherapy: Research problems and issues. *Annals of Internal Medicine*, *99*(1), 106–114.

Catenoix, H., Isnard, J., Guénot, M., Petit, J., Remy, C., & Mauguière, F. (2008). The role of the anterior insular cortex in ictal vomiting: A stereotactic electroencephalography study. *Epilepsy and Behavior*, *13*(3), 560–563. http://doi.org/10.1016/j.yebeh.2008.06.019.

Cechetto, D. F., & Saper, C. B. (1987). Evidence for a viscerotopic sensory representation in the cortex and thalamus in the rat. *The Journal of Comparative Neurology*, *262*(1), 27–45. http://doi.org/10.1002/cne.902620104.

Chang, A. E., Shiling, D. J., Stillman, R. C., Goldberg, N. H., Seipp, C. A., Barofsky, I., et al. (1979). Delata-9-tetrahydrocannabinol as an antiemetic in cancer patients receiving high-dose methotrexate. A prospective, randomized evaluation. *Annals of Internal Medicine*, *91*(6), 819–824.

Choukèr, A., Kaufmann, I., Kreth, S., Hauer, D., Feuerecker, M., Thieme, D., et al. (2010). Motion sickness, stress and the endocannabinoid system. *PLoS One*, *5*(5), 1–7. http://doi.org/10.1371/journal.pone.0010752.

Contreras, M., Ceric, F., & Torrealba, F. (2007). Inactivation of the interoceptive insula disrupts drug craving and malaise induced by lithium. *Science*, *318*(5850), 655–658. http://doi.org/10.1126/science.1145590.

Costall, B., Domeney, A. M., Naylor, R. J., & Tattersall, F. D. (1986). 5-Hydroxytryptamine M-receptor antagonism to prevent cisplatin-induced emesis. *Neuropharmacology*, *25*(8), 959–961. Retrieved from, http://www.ncbi.nlm.nih.gov/pubmed/3774121.

Cotter, J. (2009). Efficacy of crude marijuana and synthetic delta-9-tetrahydrocannabinol as treatment for chemotherapy-induced nausea and vomiting: A systematic literature review. *Oncology Nursing Forum*, *36*(3), 345–352. http://doi.org/10.1188/09.ONF.345-352.

Crawford, S. M., & Buckman, R. (1986). Nabilone and metoclopramide in the treatment of nausea and vomiting due to cisplatinum: A double blind study. *Medical Oncology and Tumor Pharmacotherapy*, *3*(1), 39–42. http://doi.org/10.1007/BF02934575.

Cross-Mellor, S. K., Ossenkopp, K.-P., Piomelli, D., & Parker, L. A. (2007). Effects of the FAAH inhibitor, URB597, and anandamide on lithium-induced taste reactivity responses: A measure of nausea in the rat. *Psychopharmacology*, *190*(2), 135–143. http://doi.org/10.1007/s00213-006-0589-7.

Dalzell, A. M., Bartlett, H., & Lilleyman, J. S. (1986). Nabilone: An alternative antiemetic for cancer chemotherapy. *Archives of Disease in Childhood*, *61*(5), 502–505. http://doi.org/10.1136/adc.61.5.502.

Di, S., Malcher-Lopes, R., Marcheselli, V. L., Bazan, N. G., & Tasker, J. G. (2005). Rapid glucocorticoid-mediated endocannabinoid release and opposing regulation of glutamate and γ-aminobutyric acid inputs to hypothalamic magnocellular neurons. *Endocrinology*, *146*(10), 4292–4301. http://doi.org/10.1210/en.2005-0610.

Duran, M., Pérez, E., Abanades, S., Vidal, X., Saura, C., Majem, M., et al. (2010). Preliminary efficacy and safety of an oromucosal standardized cannabis extract in chemotherapy-induced nausea and vomiting. *British Journal of Clinical Pharmacology*, *70*(5), 656–663. http://doi.org/10.1111/j.1365-2125.2010.03743.x.

Eckert, R. M. (2001). Understanding anticipatory nausea. *Oncology Nursing Forum*, *28*(10), 1553–1558. quiz 1559–1560. Retrieved from, http://www.ncbi.nlm.nih.gov/pubmed/11759303.

Ekert, H., Waters, K. D., Jurk, I. H., Mobilia, J., & Loughnan, P. (1979). Amelioration of cancer chemotherapy-induced nausea and vomiting by delta-9-tetrahydrocannabinol. *Medical Journal of Australia*, *2*(12), 657–659.

Figueroa-Moseley, C., Jean-Pierre, P., Roscoe, J. A., Ryan, J. L., Kohli, S., Palesh, O. G., et al. (2007). Behavioral interventions in treating anticipatory nausea and vomiting. *Journal of the National Comprehensive Cancer Network*, *5*(1), 44–50. Retrieved from, http://www.ncbi.nlm.nih.gov/pubmed/17239325.

Foubert, J., & Vaessen, G. (2005). Nausea: The neglected symptom? *European Journal of Oncology Nursing: The Official Journal of European Oncology Nursing Society*, *9*(1), 21–32. http://doi.org/10.1016/j.ejon.2004.03.006.

Frytak, S., Moertel, C. G., OFallon, J. R., Rubin, J., Creagan, E. T., OConnell, M. J., et al. (1979). Delta-9-tetrahydrocannabinol as an antiemetic for patients receving cancer chemotherapy. A comparison with prochlorperazine and a placebo. *Annals of Internal Medicine*, *91*(6), 825–830.

Fu, J., Gaetani, S., Oveisi, F., Lo Verme, J., Serrano, A., Rodríguez De Fonseca, F., et al. (2003). Oleylethanolamide regulates feeding and body weight through activation of the nuclear receptor PPAR-alpha. *Nature*, *425*(6953), 90–93. http://doi.org/10.1038/nature01921.

Gaoni, Y., & Mechoulam, R. (1964). Isolation, structure, and partial synthesis of an active constituent of hashish. *Journal of the American Chemical Society*, *86*, 1646–1647. http://doi.org/10.1021/ja01062a046.

Garcia, J., Hankins, W. G., & Rusiniak, K. W. (1974). Behavioral regulation of the milieu interne in man and rat. *Science (New York)*, *185*(4154), 824–831. http://doi.org/10.1126/science.185.4154.824.

Grill, H. J., & Norgren, R. (1978). The taste reactivity test. I. Mimetic responses to gustatory stimuli in neurologically normal rats. *Brain Research*, *143*(2), 263–279. http://doi.org/10.1016/0006-8993(78)90568-1.

Grunfeld, Y., & Edery, H. (1969). Psychopharmacological activity of the active constituents of hashish and some related cannabinoids. *Psychopharmacologia*, *14*(3), 200–210. Retrieved from, http://www.ncbi.nlm.nih.gov/pubmed/4981896.

Guindon, J., Lai, Y., Takacs, S. M., Bradshaw, H. B., & Hohmann, A. G. (2013). Alterations in endocannabinoid tone following chemotherapy-induced peripheral neuropathy: Effects of endocannabinoid deactivation inhibitors targeting fatty-acid amide hydrolase and monoacylglycerol lipase in comparison to reference analgesics. *Pharmacological Research*, *67*(1), 94–109. http://doi.org/10.1016/j.phrs.2012.10.013.

Herman, T. S., Einhorn, L. H., Jones, S. E., Nagy, C., Chester, A. B., Dean, J. C., et al. (1979). Superiority of nabilone over prochlorperazine as an antiemetic in patients receiving cancer chemotherapy. *The New England Journal of Medicine*, *300*(23), 1295–1297.

Hickok, J. T., Roscoe, J. A., & Morrow, G. R. (2001). The role of patients' expectations in the development of anticipatory nausea related to chemotherapy for cancer. *Journal of Pain and Symptom Management*, *22*(4), 843–850. Retrieved from, http://www.ncbi.nlm.nih.gov/pubmed/11576801.

Hickok, J. T., Roscoe, J. A., Morrow, G. R., King, D. K., Atkins, J. N., & Fitch, T. R. (2003). Nausea and emesis remain significant problems of chemotherapy despite prophylaxis with 5-hydroxytryptamine-3 antiemetics: A University of Rochester James P. Wilmot Cancer Center Community Clinical Oncology Program Study of 360 cancer patients treated in t. *Cancer*, *97*(11), 2880–2886. http://doi.org/10.1002/cncr.11408.

Hillard, C. J., Weinlander, K. M., & Stuhr, K. L. (2012). Contributions of endocannabinoid signaling to psychiatric disorders in humans: Genetic and biochemical evidence. *Neuroscience*, *204*, 207–229. http://doi.org/10.1016/j.neuroscience.2011.11.020.

Horn, C. C. (2014). Measuring the nausea-to-emesis continuum in non-human animals: Refocusing on gastrointestinal vagal signaling. *Experimental Brain Research*, *232*(8), 2471–2481. http://doi.org/10.1007/s00221-014-3985-y.

Hornby, P. J. (2001). Central neurocircuitry associated with emesis. *The American Journal of Medicine*, *111*(Suppl.), 106S–112S. http://doi.org/10.1016/S0002-9343(01)00849-X.

Isnard, J., Guénot, M., Sindou, M., & Mauguière, F. (2004). Clinical manifestations of insular lobe seizures: A stereo-electroencephalographic study. *Epilepsia*, *45*(9), 1079–1090. http://doi.org/10.1016/S1388-2457(08)60529-5.

Issemann, I., & Green, S. (1990). Activation of a member of the steroid hormone receptor superfamily by peroxisome proliferators. *Nature*, *347*(6294), 645–650. http://doi.org/10.1038/347645a0.

Janelsins, M. C., Tejani, M. A., Kamen, C., Peoples, A. R., Mustian, K. M., & Morrow, G. R. (2013). Current pharmacotherapy for chemotherapy-induced nausea and vomiting in cancer patients. *Expert Opinion on Pharmacotherapy*, *14*(6), 757–766. http://doi.org/10.1517/14656566.2013.776541.

Jarzimski, C., Karst, M., Zoerner, A. A., Rakers, C., May, M., Suchy, M. T., et al. (2012). Changes of blood endocannabinoids during anaesthesia: A special case for fatty acid amide hydrolase inhibition by propofol? *British Journal of Clinical Pharmacology, 74*(1), 54–59. http://doi.org/10.1111/j.1365-2125.2012.04175.x.

Kamen, C., Tejani, M. A., Chandwani, K., Janelsins, M., Peoples, A. R., Roscoe, J. A., et al. (2014). Anticipatory nausea and vomiting due to chemotherapy. *European Journal of Pharmacology, 722*, 172–179. http://doi.org/10.1016/j.ejphar.2013.09.071.

Kiefer, S. W., & Orr, M. R. (1992). Taste avoidance, but not aversion, learning in rats lacking gustatory cortex. *Behavioral Neuroscience, 106*(1), 140–146. http://doi.org/10.1037/0735-7044.106.1.140.

Krieg, W. J. S. (1946). Connections of the cerebral cortex. *The Journal of Comparative Neurology, 84*(3), 277–323.

Kumar, G., Stendall, C., Mistry, R., Gurusamy, K., & Walker, D. (2014). A comparison of total intravenous anaesthesia using propofol with sevoflurane or desflurane in ambulatory surgery: Systematic review and meta-analysis. *Anaesthesia, 69*(10), 1138–1150. http://doi.org/10.1111/anae.12713. 1–13.

LaCount, L. T., Barbieri, R., Park, K., Kim, J., Brown, E. N., Kuo, B., et al. (2011). Static and dynamic autonomic response with increasing nausea perception. *Aviation, Space and Environmental Medicine, 82*(4), 424–433. http://doi.org/10.3357/ASEM.2932.2011.

Lane, M., Vogel, C. L., Ferguson, J., Krasnow, S., Saiers, J. L., Hamm, J., et al. (1991). Dronabinol and prochlorperazine in combination for treatment of cancer chemotherapy-induced nausea and vomiting. *Journal of Pain and Symptom Management, 6.*

Leslie, R. A. (1985). Neuroactive substances in the dorsal vagal complex of the medulla oblongata: Nucleus of the tractus solitarius, area postrema, and dorsal motor nucleus of the vagus. *Neurochemistry International,* 7(2), 191–211. http://doi.org/10.1016/0197-0186(85)90106-8.

Limebeer, C. L., & Parker, L. A. (1999). Delta-9-tetrahydrocannabinol interferes with the establishment and the expression of conditioned rejection reactions produced by cyclophosphamide: A rat model of nausea. *Neuroreport, 10*(18), 3769–3772. http://doi.org/10.1097/00001756-199912160-00009.

Limebeer, C. L., Hall, G., & Parker, L. A. (2006). Exposure to a lithium-paired context elicits gaping in rats: A model of anticipatory nausea. *Physiology and Behavior, 88*(4–5), 398–403. http://doi.org/10.1016/j.physbeh.2006.04.014.

Limebeer, C. L., Krohn, J. P., Cross-Mellor, S. K., Litt, D. E., Ossenkopp, K.-P., & Parker, L. A. (2008). Exposure to a context previously associated with nausea elicits conditioned gaping in rats: A model of anticipatory nausea. *Behavioural Brain Research, 187*(1), 33–40. http://doi.org/10.1016/j.bbr.2007.08.024.

Limebeer, C. L., Abdullah, R. A., Rock, E. M., Imhof, E., Wang, K., Lichtman, A. H., et al. (2014). Attenuation of anticipatory nausea in a rat model of contextually elicited conditioned gaping by enhancement of the endocannabinoid system. *Psychopharmacology, 231*(3), 603–612. http://doi.org/10.1007/s00213-013-3282-7.

Limebeer, C. L., Vemuri, V. K., Bedard, H., Lang, S. T., Ossenkopp, K.-P., Makriyannis, A., et al. (2010). Inverse agonism of cannabinoid CB 1 receptors potentiates LiCl-induced nausea in the conditioned gaping model in rats. *British Journal of Pharmacology, 161*(2), 336–349. http://doi.org/10.1111/j.1476-5381.2010.00885.x.

Limebeer, C. L., Rock, E. M., Mechoulam, R., & Parker, L. A. (2012). The anti-nausea effects of CB1 agonists are mediated by an action at the visceral insular cortex. *British Journal of Pharmacology, 167*(5), 1126–1136. http://doi.org/10.1111/j.1476-5381.2012.02066.x.

Malik, I. A., Khan, W. A., Qazilbash, M., Ata, E., Butt, A., & Khan, M. A. (1995). Clinical efficacy of lorazepam in prophylaxis of anticipatory, acute, and delayed nausea and vomiting induced by high doses of cisplatin. A prospective randomized trial. *American*

Journal of Clinical Oncology, 18(2), 170–175. Retrieved from, http://www.ncbi.nlm.nih.gov/pubmed/7900711.

Matteson, S., Roscoe, J., Hickok, J., & Morrow, G. R. (2002). The role of behavioral conditioning in the development of nausea. *American Journal of Obstetrics and Gynecology, 186*(5 Suppl. Understanding), S239–S243. Retrieved from, http://www.ncbi.nlm.nih.gov/pubmed/12011893.

McDonald, R. V., Parker, L. A., & Siegel, S. (1997). Conditioned sucrose aversions produced by naloxone-precipitated withdrawal from acutely administered morphine. *Pharmacology, Biochemistry, and Behavior, 58*(4), 1003–1008. Retrieved from, http://www.ncbi.nlm.nih.gov/pubmed/9408206.

McLaughlin, P. J., Winston, K. M., Limebeer, C. L., Parker, L. A., Makriyannis, A., & Salamone, J. D. (2005). The cannabinoid CB1 antagonist AM 251 produces food avoidance and behaviors associated with nausea but does not impair feeding efficiency in rats. *Psychopharmacology, 180*(2), 286–293. http://doi.org/10.1007/s00213-005-2171-0.

Mechoulam, R., Shani, A., Edery, H., & Grunfeld, Y. (1970). Chemical basis of hashish activity. *Science, 169*(945), 611–612. http://doi.org/10.1126/science.169.3945.611.

Meiri, E., Jhangiani, H., Vredenburgh, J. J., Barbato, L. M., Carter, F. J., Yang, H.-M., et al. (2007). Efficacy of dronabinol alone and in combination with ondansetron versus ondansetron alone for delayed chemotherapy-induced nausea and vomiting. *Current Medical Research and Opinion, 23*(3), 533–543. http://doi.org/10.1185/030079907X167525.

Miller, A. D., & Leslie, R. A. (1994). The area postrema and vomiting. *Frontiers in Neuroendocrinology, 15*(4), 301–320. http://doi.org/10.1006/frne.1994.1012. S0091-3022(84)71012-0 [pii].

Miner, W. D., & Sanger, G. J. (1986). Inhibition of cisplatin-induced vomiting by selective 5-hydroxytryptamine M-receptor antagonism. *British Journal of Pharmacology, 88*(3), 497–499. Retrieved from, http://www.ncbi.nlm.nih.gov/pubmed/3742146.

Mitchell, D., Wells, C., Hoch, N., Lind, K., Woods, S. C., & Mitchell, L. K. (1976). Poison induced pica in rats. *Physiology & Behavior, 17*(4), 691–697. http://doi.org/10.1016/0031-9384(76)90171-2.

Montgomery, G. H., & Bovbjerg, D. H. (1997). The development of anticipatory nausea in patients receiving adjuvant chemotherapy for breast cancer. *Physiology & Behavior, 61*(5), 737–741. Retrieved from, http://www.ncbi.nlm.nih.gov/pubmed/9145945.

Morrow, G. R., & Dobkin, P. L. (1987). Behavioral approaches for the management of adversive side effects of cancer treatment. *Psychiatric Medicine, 5*(4), 299–314. Retrieved from, http://www.ncbi.nlm.nih.gov/pubmed/3329349.

Morrow, G. R., & Hickok, J. T. (1993). Behavioral treatment of chemotherapy-induced nausea and vomiting. *Oncology (Williston Park, N.Y.)*, 7(12), 83–89. discussion 93–94, 97. Retrieved from, http://www.ncbi.nlm.nih.gov/pubmed/8292509.

Morrow, G. R., & Rosenthal, S. N. (1996). Models, mechanisms and management of anticipatory nausea and emesis. *Oncology*, (53 Suppl. 1), 4–7. Retrieved from, http://www.ncbi.nlm.nih.gov/pubmed/8692550.

Morrow, G. R., Roscoe, J. A., Hickok, J. T., Stern, R. M., Pierce, H. I., King, D. B., et al. (1998). Initial control of chemotherapy-induced nausea and vomiting in patient quality of life. *Oncology (Williston Park, N.Y.), 12*(3 Suppl. 4), 32–37. Retrieved from, http://www.ncbi.nlm.nih.gov/pubmed/9556781.

Morrow, G. R., Roscoe, J. A., Kirshner, J. J., Hynes, H. E., & Rosenbluth, R. J. (1998). *Anticipatory nausea and vomiting in the era of 5-HT3 antiemetics*. Supportive Care in Cancer. http://doi.org/10.1007/s005200050161.

Mustian, K. M., Devine, K., Ryan, J. L., Janelsins, M. C., Sprod, L. K., Peppone, L. J., et al. (2011). Treatment of nausea and vomiting during chemotherapy. *US Oncology & Hematology*, 7(2), 91–97. Retrieved from, http://www.ncbi.nlm.nih.gov/pubmed/24466408.

Napadow, V., Sheehan, J. D., Kim, J., Lacount, L. T., Park, K., Kaptchuk, T. J., et al. (2013). The brain circuitry underlying the temporal evolution of nausea in humans. *Cerebral Cortex*, *23*(4), 806–813. http://doi.org/10.1093/cercor/bhs073.

Nemoto, M., Endo, T., Minami, M., Yoshioka, M., Ito, H., & Saito, H. (2001). 5-Hydroxytryptamine (5-HT)-induced depolarization in isolated abdominal vagus nerves in the rat: Involvement of 5-HT3 and 5-HT4 receptors. *Research Communications in Molecular Pathology and Pharmacology*, *109*(3-4), 217–230.

Nesse, R. M., Carli, T., Curtis, G. C., & Kleinman, P. D. (1980). Pretreatment nausea in cancer chemotherapy: A conditioned response? *Psychosomatic Medicine*, *42*(1), 33–36.

Olière, S., Joliette-Riopel, A., Potvin, S., & Jutras-Aswad, D. (2013). Modulation of the endocannabinoid system: Vulnerability factor and new treatment target for stimulant addiction. *Frontiers in Psychiatry*, *4*, 109. http://doi.org/10.3389/fpsyt.2013.00109.

Orr, L. E., & McKernan, J. F. (1981). Antiemetic effect of delta 9-tetrahydrocannabinol in chemotherapy-associated nausea and emesis as compared to placebo and compazine. *Journal of Clinical Pharmacology*, *21*(8-9 supplement), 76S–80S.

Orr, L. E., McKernan, J. F., & Bloome, B. (1980). Antiemetic effect of tetrahydrocannabinol. Compared with placebo and prochlorperazine in chemotherapy-associated nausea and emesis. *Archives of Internal Medicine*, *140*(11), 1431–1433.

Ossenkopp, K.-P., Parker, L. A., Limebeer, C. L., Burton, P., Fudge, M. A., & Cross-Mellor, S. K. (2003). Vestibular lesions selectively abolish body rotation-induced, but not lithium-induced, conditioned taste aversions (oral rejection responses) in rats. *Behavioral Neuroscience*, *117*(1), 105–112. http://doi.org/10.1037/0735-7044.117.1.105.

Ostrowsky, K., Isnard, J., Ryvlin, P., Guénot, M., Fischer, C., & Mauguière, F. (2000). Functional mapping of the insular cortex: Clinical implication in temporal lobe epilepsy. *Epilepsia*, *41*(6), 681–686. http://doi.org/10.1111/j.1528-1157.2000.tb00228.x.

Parker, L. A. (1995). Rewarding drugs produce taste avoidance, but not taste aversion. *Neuroscience and Biobehavioral Reviews*, *19*(1), 143–151. http://doi.org/10.1016/0149-7634(94)00028-Y.

Parker, L. A., & Kemp, S. (2001). Tetrahydrocannabinol (THC) interferes with conditioned retching in Suncus murinus: An animal model of anticipatory nausea and vomiting (ANV). *Neuroreport*, *12*(4), 749–752.

Parker, L. A., & Mechoulam, R. (2003). Cannabinoid agonists and antagonists modulate lithium-induced conditioned gaping in rats. *Integrative Physiological and Behavioral Science*, *38*(2), 133–145.

Parker, L. A., Mechoulam, R., & Schlievert, C. (2002). Cannabidiol, a non-psychoactive component of cannabis and its synthetic dimethylheptyl homolog suppress nausea in an experimental model with rats. *Neuroreport*, *13*(5), 567–570. http://doi.org/10.1097/00001756-200204160-00006.

Parker, L. A., Kwiatkowska, M., & Mechoulam, R. (2006). Delta-9-tetrahydrocannabinol and cannabidiol, but not ondansetron, interfere with conditioned retching reactions elicited by a lithium-paired context in Suncus murinus: An animal model of anticipatory nausea and vomiting. *Physiology and Behavior*, *87*(1), 66–71. http://doi.org/10.1016/j.physbeh.2005.08.045.

Parker, L. A., Mechoulam, R., Schlievert, C., Abbott, L., Fudge, M. L., & Burton, P. (2003). Effects of cannabinoids on lithium-induced conditioned rejection reactions in a rat model of nausea. *Psychopharmacology*, *166*(2), 156–162. http://doi.org/10.1007/s00213-002-1329-2.

Parker, L. A., Niphakis, M. J., Downey, R., Limebeer, C. L., Rock, E. M., Sticht, M. A., et al. (2014). Effect of selective inhibition of monoacylglycerol lipase (MAGL) on acute nausea, anticipatory nausea, and vomiting in rats and *Suncus murinus*. *Psychopharmacology*, *232*(3), 583–593. http://doi.org/10.1007/s00213-014-3696-x.

Parker, L. A., Rock, E. M., Sticht, M. A., Wills, K. L., & Limebeer, C. L. (2015). Cannabinoids suppress acute and anticipatory nausea in pre-clinical rat models of conditioned gaping. *Clinical Pharmacology & Therapeutics*, *97*(6) 559–561. http://doi.org/10.1002/cpt.98.

Patel, S., Wohlfeil, E. R., Rademacher, D. J., Carrier, E. J., Perry, L. J., Kundu, A., et al. (2003). The general anesthetic propofol increases brain N-arachidonylethanolamine (anandamide) content and inhibits fatty acid amide hydrolase. *British Journal of Pharmacology*, *139*(5), 1005–1013. http://doi.org/10.1038/sj.bjp.0705334.

Paxinos, G., & Watson, C. (2007). *The rat brain in stereotaxic coordinates* (6th ed.). New York: Elsevier/Academic Press.

Penfield, W., & Faulk, M. E. (1955). The insula: Further observations on its function. *Brain*, *78*(4), 445–470. http://doi.org/10.1093/brain/78.4.445.

Pertwee, R. G. (2009). Emerging strategies for exploiting cannabinoid receptor agonists as medicines. *British Journal of Pharmacology*, *156*(3), 397–411. http://doi.org/10.1111/j.1476-5381.2008.00048.x.

Pomeroy, M., Fennelly, J. J., & Towers, M. (1986). Prospective randomized double-blind trial of nabilone versus domperidone in the treatment of cytotoxic-induced emesis. *Cancer Chemotherapy and Pharmacology*, *17*(3), 285–288.

Razavi, D., Delvaux, N., Farvacques, C., De Brier, F., Van Heer, C., Kaufman, L., et al. (1993). Prevention of adjustment disorders and anticipatory nausea secondary to adjuvant chemotherapy: A double-blind, placebo-controlled study assessing the usefulness of alprazolam. *Journal of Clinical Oncology*, *11*(7), 1384–1390. Retrieved from, http://www.ncbi.nlm.nih.gov/pubmed/8315437.

Roca-Pallín, J. M., López-Pelayo, H., Sugranyes, G., & Balcells-Oliveró, M. M. (2013). Cannabinoid hyperemesis syndrome. *CNS Neuroscience and Therapeutics*, *19*(12), 994–995. http://doi.org/10.1111/cns.12207.

Rock, E. M., Limebeer, C. L., Mechoulam, R., Piomelli, D., & Parker, L. A. (2008). The effect of cannabidiol and URB597 on conditioned gaping (a model of nausea) elicited by a lithium-paired context in the rat. *Psychopharmacology*, *196*(3), 389–395. http://doi.org/10.1007/s00213-007-0970-1.

Rock, E. M., Limebeer, C. L., Navaratnam, R., Sticht, M.a., Bonner, N., Engeland, K., et al. (2014). A comparison of cannabidiolic acid with other treatments for anticipatory nausea using a rat model of contextually elicited conditioned gaping. *Psychopharmacology*, *231*(16), 3207–3215. http://doi.org/10.1007/s00213-014-3498-1.

Rock, E. M., Bolognini, D., Limebeer, C. L., Cascio, M. G., Anavi-Goffer, S., Fletcher, P. J., et al. (2012). Cannabidiol, a nonpsychotropic component of cannabis, attenuates vomiting and nausea-like behaviour via indirect agonism of 5-HT 1A somatodendritic autoreceptors in the dorsal raphe nucleus. *British Journal of Pharmacology*, *165*(8), 2620–2634. http://doi.org/10.1111/j.1476-5381.2011.01621.x.

Rock, E. M., Kopstick, R. L., Limebeer, C. L., & Parker, L. A. (2013). Tetrahydrocannabinolic acid reduces nausea-induced conditioned gaping in rats and vomiting in *Suncus murinus*. *British Journal of Pharmacology*, *170*(3), 641–648. http://doi.org/10.1111/bph.12316.

Rock, E. M., Limebeer, C. L., Ward, J. M., Cohen, A., Grove, K., Niphakis, M. J., et al. (2015). Interference with acute nausea and anticipatory nausea in rats by fatty acid amide hydrolase (FAAH) inhibition through a PPARα and CB1 receptor mechanism, respectively: A double dissociation. *Psychopharmacology*, *232*(20), 3841–3848. http://doi.org/10.1007/s00213-015-4050-7.

Rodriguez, M., Lopez, M., Symonds, M., & Hall, G. (2000). Lithium-induced context aversion in rats as a model of anticipatory nausea in humans. *Physiology and Behavior*, *71*(5), 571–579. http://dx.doi.org/10.1016/S0031-9384(00)00376–0.

Roscoe, J. A., Morrow, G. R., Aapro, M. S., Molassiotis, A., & Olver, I. (2011). Anticipatory nausea and vomiting. *Supportive Care in Cancer : Official Journal of the Multinational Association of Supportive Care in Cancer, 19*(10), 1533–1538. http://doi.org/10.1007/s00520-010-0980-0.

Rudd, J. A., Yamamoto, K., Yamatodani, A., & Takeda, N. (2002). Differential action of ondansetron and dexamethasone to modify cisplatin-induced acute and delayed kaolin consumption ("pica") in rats. *European Journal of Pharmacology, 454*(1), 47–52. http://doi.org/10.1016/S0014-2999(02)02472-X.

Sadhasivam, S., Zhang, X., Chidambaran, V., Mavi, J., Pilipenko, V., Mersha, T. B., et al. (2015). Novel associations between FAAH genetic variants and postoperative central opioid-related adverse effects. *The Pharmacogenomics Journal, 15*(5), 436–442. http://doi.org/10.1038/tpj.2014.79.

Sallan, S. E., Zinberg, N. E., & Frei, E. (1975). Antiemetic effect of delta-9-tetrahydrocannabinol in patients receiving cancer chemotherapy. *The New England Journal of Medicine, 293*, 795–797.

Schelling, G., Hauer, D., Azad, S. C., Schmoelz, M., Choukèr, A., Schmidt, M., et al. (2006). Effects of general anesthesia on anandamide blood levels in humans. *Anesthesiology, 104*(2), 273–277. http://doi.org/10.1097/00000542-200602000-00012.

Sclocco, R., Kim, J., Garcia, R. G., Sheehan, J. D., Beissner, F., Bianchi, A. M., et al. (2014). Brain circuitry supporting multi-organ autonomic outflow in response to nausea. *Cerebral Cortex*, In press. http://doi.org/10.1093/cercor/bhu172.

Sharkey, K. A., Darmani, N. A., & Parker, L. A. (2014). Regulation of nausea and vomiting by cannabinoids and the endocannabinoid system. *European Journal of Pharmacology, 722*(1), 134–146. http://doi.org/10.1016/j.ejphar.2013.09.068.

Simonetto, D. A., Oxentenko, A. S., Herman, M. L., & Szostek, J. H. (2012). Cannabinoid hyperemesis: A case series of 98 patients. *Mayo Clinic Proceedings, 87*(2), 114–119. http://doi.org/10.1016/j.mayocp.2011.10.005.

Sink, K. S., McLaughlin, P. J., Wood, J. A. T., Brown, C., Fan, P., Vemuri, V. K., et al. (2008). The novel cannabinoid CB1 receptor neutral antagonist AM4113 suppresses food intake and food-reinforced behavior but does not induce signs of nausea in rats. *Neuropsychopharmacology, 33*(4), 946–955. http://doi.org/10.1038/sj.npp.1301476.

Stewart, D. J. (1990). Cancer therapy, vomiting, and antiemetics. *Canadian Journal of Physiology and Pharmacology, 68*(2), 304–313. Retrieved from, http://www.ncbi.nlm.nih.gov/pubmed/2178754.

Sticht, M. A., Long, J. Z., Rock, E. M., Limebeer, C. L., Mechoulam, R., Cravatt, B. F., et al. (2012). Inhibition of monoacylglycerol lipase attenuates vomiting in *Suncus murinus* and 2-arachidonoyl glycerol attenuates nausea in rats. *British Journal of Pharmacology, 165*(8), 2425–2435. http://doi.org/10.1111/j.1476-5381.2011.01407.x.

Sticht, M. A., Limebeer, C. L., Rafla, B., & Parker, L. A. (2015). Intra-visceral insular cortex 2-arachidonoylglycerol, but not N-arachidonoylethanolamide, suppresses acute nausea-induced conditioned gaping in rats. *Neuroscience, 286*, 338–344. http://doi.org/10.1016/j.neuroscience.2014.11.058.

Stockhorst, U., Klosterhalfen, S., Klosterhalfen, W., Winkelmann, M., & Steingrueber, H. J. (1993). Anticipatory nausea in cancer patients receiving chemotherapy: Classical conditioning etiology and therapeutical implications. *Integrative Physiological and Behavioral Science, 28*(2), 177–181. Retrieved from, http://www.ncbi.nlm.nih.gov/pubmed/8318445.

Stockhorst, U., Steingrueber, H.-J., Enck, P., & Klosterhalfen, S. (2006). Pavlovian conditioning of nausea and vomiting. *Autonomic Neuroscience : Basic & Clinical, 129*(1–2), 50–57. http://doi.org/10.1016/j.autneu.2006.07.012.

Sullivan, S. (2010). Cannabinoid hyperemesis. *Canadian Journal of Gastroenterology, 24*(5), 284–285. http://doi.org/10.1111/j.1445-5994.2009.02047.x.
Sweet, D. L., Miller, N. J., Weddington, W., Senay, E., & Sushelsky, L. (1981). delta 9-Tetrahydrocannabinol as an antiemetic for patients receiving cancer chemotherapy. A pilot study. *Journal of Clinical Pharmacology, 21*(8–9 Suppl.), 70S–75S.
Takeda, N., Hasegawa, S., Morita, M., & Matsunaga, T. (1993). Pica in rats is analogous to emesis: An animal model in emesis research. *Pharmacology, Biochemistry and Behavior, 45*(4), 817–821. http://doi.org/10.1016/0091-3057(93)90126-E.
Tramèr, M. R., Carroll, D., Campbell, F. A., Reynolds, D. J., Moore, R. A., & McQuay, H. J. (2001). Cannabinoids for control of chemotherapy induced nausea and vomiting: Quantitative systematic review. *British Medical Journal, 323*(7303), 16–21. http://doi.org/10.1136/bmj.323.7303.16.
Travers, J. B., & Norgren, R. (1986). Electromyographic analysis of the ingestion and rejection of sapid stimuli in the rat. *Behavioral Neuroscience, 100*(4), 544–555. http://doi.org/10.1037/0735-7044.100.4.544.
Tuerke, K. J., Limebeer, C. L., Fletcher, P. J., & Parker, L. A. (2012). Double dissociation between regulation of conditioned disgust and taste avoidance by serotonin availability at the 5-HT3 receptor in the posterior and anterior insular cortex. *Journal of Neuroscience, 32*(40), 13709–13717. http://doi.org/10.1523/JNEUROSCI.2042-12.2012.
Tyc, V. L., Mulhern, R. K., & Bieberich, A. A. (1997). Anticipatory nausea and vomiting in pediatric cancer patients: An analysis of conditioning and coping variables. *Journal of Developmental and Behavioral Pediatrics, 18*(1), 27–33. Retrieved from, http://www.ncbi.nlm.nih.gov/pubmed/9055147.
Ungerleider, J. T., Andrysiak, T. A., Fairbanks, L. A., Tesler, A. S., & Parker, R. G. (1984). Tetrahydrocannabinol vs. prochlorperazine. The effects of two antiemetics on patients undergoing radiotherapy. *Radiology, 150*(2), 598–599.
Van Sickle, M. D., Oland, L. D., Ho, W., Hillard, C. J., Mackie, K., Davison, J. S., et al. (2001). Cannabinoids inhibit emesis through CB1 receptors in the brainstem of the ferret. *Gastroenterology, 121*(4), 767–774. S0016508501038586 [pii].
Van Sickle, M. D., Oland, L. D., Mackie, K., Davison, J. S., Sharkey, K. A., Sickle, V., et al. (2003). Delta 9-tetrahydrocannabinol selectively acts on CB 1 receptors in specific regions of dorsal vagal complex to inhibit emesis in ferrets. *American Journal of Physiology. Gastrointestinal and Liver Physiology, 285*(3), G566–G576. http://doi.org/10.1152/ajpgi.00113.2003.
Van Sickle, M. D., Duncan, M., Kingsley, P. J., Mouihate, A., Urbani, P., Mackie, K., et al. (2005). Identification and functional characterization of brainstem cannabinoid CB2 receptors. *Science, 310*(5746), 329–332. http://doi.org/10.1126/science.1115740.
Vera, G., Chiarlone, A., Cabezos, P. A., Pascual, D., Martín, M. I., & Abalo, R. (2007). WIN 55,212-2 prevents mechanical allodynia but not alterations in feeding behaviour induced by chronic cisplatin in the rat. *Life Sciences, 81*(6), 468–479. http://doi.org/10.1016/j.lfs.2007.06.012.
Verme, J. Lo, Fu, J., Astarita, G., Rana, G. La, Russo, R., Calignano, A., et al. (2005). The nuclear receptor peroxisome proliferator-activated receptor-α mediates the anti-inflammatory actions of palmitoylethanolamide. *Molecular Pharmacology, 67*(1), 15–19. http://doi.org/10.1124/mol.104.006353.dyshev.
Watson, M., McCarron, J., & Law, M. (1992). Anticipatory nausea and emesis, and psychological morbidity: Assessment of prevalence among out-patients on mild to moderate chemotherapy regimens. *British Journal of Cancer, 66*(5), 862–866. Retrieved from, http://www.ncbi.nlm.nih.gov/pubmed/1419628.
Wei, X., Wang, Z.-B., Zhang, L.-C., Liu, W.-Y., Su, D.-F., & Li, L. (2011). Verification of motion sickness index in Mice. *CNS Neuroscience & Therapeutics, 17*(6), 790–792. http://doi.org/10.1111/j.1755-5949.2011.00272.x.

Wheeler, R. A., Twining, R. C., Jones, J. L., Slater, J. M., Grigson, P. S., & Carelli, R. M. (2008). Behavioral and electrophysiological indices of negative affect predict cocaine self-administration. *Neuron*, *57*(5), 774–785. http://doi.org/10.1016/j.neuron.2008.01.024.

Wheeler, R. A., Aragona, B. J., Fuhrmann, K. A., Jones, J. L., Day, J. J., Cacciapaglia, F., et al. (2011). Cocaine cues drive opposing context-dependent shifts in reward processing and emotional state. *Biological Psychiatry*, *69*(11), 1067–1074. http://doi.org/10.1016/j.biopsych.2011.02.014.

Yamamoto, K., Ngan, M. P., Takeda, N., Yamatodani, A., & Rudd, J. A. (2004). Differential activity of drugs to induce emesis and pica behavior in *Suncus murinus* (house musk shrew) and rats. *Physiology and Behavior*, *83*(1 SPEC. ISS), 151–156. http://doi.org/10.1016/j.physbeh.2004.08.006.

Yu, X. H., Cai, G. J., Liu, A. J., Chu, Z. X., & Su, D. F. (2007). A novel animal model for motion sickness and its first application in rodents. *Physiology and Behavior*, *92*(4), 702–707. http://doi.org/10.1016/j.physbeh.2007.05.067.

Zachariae, R., Paulsen, K., Mehlsen, M., Jensen, A. B., Johansson, A., & von der Maase, H. (2007). Anticipatory nausea: The role of individual differences related to sensory perception and autonomic reactivity. *Annals of Behavioral Medicine : A Publication of the Society of Behavioral Medicine*, *33*(1), 69–79. http://doi.org/10.1207/s15324796abm3301_8.

Zheng, Y., Wang, X. L., Mo, F. F., & Li, M. (2014). Dexamethasone alleviates motion sickness in rats in part by enhancing the endocannabinoid system. *European Journal of Pharmacology*, *727*(1), 99–105. http://doi.org/10.1016/j.ejphar.2014.01.047.

CHAPTER FIVE

Endocannabinoid Regulation of Neuroendocrine Systems

Jeffrey G. Tasker*[,†,1], Chun Chen*, Marc O. Fisher[†], Xin Fu[†], Jennifer R. Rainville*, Grant L. Weiss*

*Department of Cell and Molecular Biology, Tulane University, New Orleans, Louisiana, USA
[†]Neuroscience Program, Tulane University, New Orleans, Louisiana, USA
[1]Corresponding author: e-mail address: tasker@tulane.edu

Contents

1. The Endocannabinoids and Their Receptors 164
2. Cannabinoid Regulation of the Neuroendocrine Stress Response 166
   2.1 Inhibitory Endocannabinoid Tone on the HPA Axis 167
   2.2 Acute Stress Regulation of Endocannabinoids and the HPA Axis 168
   2.3 Chronic Stress Regulation of Endocannabinoids and the HPA Axis 170
3. Endocannabinoid Regulation of Energy Homeostasis 172
   3.1 Endocannabinoids and Cannabinoid Receptors in Energy Homeostasis 173
   3.2 Ghrelin and Endocannabinoids in Energy Homeostasis 174
   3.3 CRH, TRH, and Endocannabinoid Regulation of Energy Homeostasis 175
   3.4 Leptin and Endocannabinoid Regulation of Energy Homeostasis 176
4. Endocannabinoids and Fluid Homeostasis 177
   4.1 Receptor Distribution in Osmoregulatory Tissues 177
   4.2 Glucocorticoid Feedback on Vasopressin Secretion Via Endocannabinoids 180
   4.3 Oxytocin and Atrial Natriuretic Peptide 182
5. Endocannabinoid Regulation of Reproduction 183
   5.1 Endocannabinoid Regulation of the HPG Axis 183
   5.2 Ghrelin Inhibits the HPG Axis Through Endocannabinoid Release 185
   5.3 Stress and Glucocorticoid Modulation of Reproductive Behavior Via Endocannabinoid Actions 186
References 189

Abstract

The hypothalamus is a part of the brain that is critical for sustaining life through its homeostatic control and integrative regulation of the autonomic nervous system and neuroendocrine systems. Neuroendocrine function in mammals is mediated mainly through the control of pituitary hormone secretion by diverse neuroendocrine cell groups in the hypothalamus. Cannabinoid receptors are expressed throughout the hypothalamus, and endocannabinoids have been found to exert pronounced regulatory effects on neuroendocrine function via modulation of the outputs of several

International Review of Neurobiology, Volume 125
ISSN 0074-7742
http://dx.doi.org/10.1016/bs.irn.2015.09.003

neuroendocrine systems. Here, we review the physiological regulation of neuroendocrine function by endocannabinoids, focusing on the role of endocannabinoids in the neuroendocrine regulation of the stress response, food intake, fluid homeostasis, and reproductive function.

Cannabis sativa (marijuana) has a long history of recreational and/or medicinal use dating back to ancient times. It was used as an analgesic, anesthetic, and antianxiety herb as early as 2600 B.C. The hedonic, anxiolytic, and mood-elevating properties of cannabis have also been cited in ancient records from different cultures. However, it was not until 1964 that the psychoactive constituent of cannabis, Δ^9-tetrahydrocannabinol, was isolated and its chemical structure determined (Gaoni & Mechoulam, 1964).

1. THE ENDOCANNABINOIDS AND THEIR RECEPTORS

The type 1 cannabinoid (CB_1) receptor was first identified as a heterotrimeric GTP-binding protein-coupled receptor and as the predominant cannabinoid receptor in the brain in 1990 (Matsuda, Lolait, Brownstein, Young, & Bonner, 1990). Radiolabelled cannabinoid-binding studies revealed that the CB_1 receptor is one of the most abundant G protein-coupled receptors in the brain (Herkenham et al., 1991) The type 2 cannabinoid (CB_2) receptor was subsequently identified in 1993 and is expressed at highest levels in peripheral immune tissues (Munro, Thomas, & Abu-Shaar, 1993), but was also subsequently identified in the central nervous system, albeit at significantly lower levels (Gong et al., 2006; Van Sickle et al., 2005). Both of the known cannabinoid receptors are coupled to $G\alpha_{i/o}$, which is negatively coupled to adenylyl cyclase activity and downregulates cAMP production. CB_1 receptors are localized primarily on the presynaptic terminals of glutamatergic and GABAergic synapses, where they reduce the release probabilities of glutamate and GABA and modulate excitatory and inhibitory neurotransmission (Di, Malcher-Lopes, Halmos, & Tasker, 2003; Hirasawa et al., 2004; Melis et al., 2004; Ohno-Shosaku et al., 2002; Wilson & Nicoll, 2001), although they have also been reported to modulate the release of other neurotransmitters, such as norepinephrine (Carvalho & Van Bockstaele, 2012). Activation of presynaptic CB_1 receptors has been shown to suppress calcium influx via presynaptic voltage-gated calcium channels (Wilson, Kunos, & Nicoll, 2001), and to facilitate the opening of presynaptic voltage-gated potassium channels (Mackie, Lai, Westenbroek, & Mitchell, 1995), which lead in both cases to a decrease in the probability of neurotransmitter release.

The endogenous ligands for the CB_1 and CB_2 receptors, the endocannabinoids (eCBs), were isolated in the early 1990s and belong to a family of long-chain polyunsaturated fatty acids. The two best characterized and widely studied eCBs are *N*-arachidonoylethanolamine (AEA), also known as anandamide, and 2-arachidonoylglycerol (2-AG) (Hillard, 2000). Both AEA and 2-AG are produced from lipid precursors in the plasma membrane, albeit under different conditions. Because of their hydrophobic properties, AEA and 2-AG are not stored in vesicles, but are released upon synthesis by metabolic enzymes. Although both are synthesized and released via enzymatic activation, recent evidence suggests that AEA may be constitutively synthesized and released tonically (Di, Popescu, & Tasker, 2013; Kim & Alger, 2010), whereas 2-AG synthesis is stimulated phasically by electrical activity-, GPCR-, or steroid receptor-dependent mechanisms (Di et al., 2013; Hashimotodani et al., 2013; Kim & Alger, 2010), which are often associated with an increase in intracellular calcium. Both AEA and 2-AG are thought to work in the brain primarily as retrograde messengers, synthesized and released by postsynaptic neurons and act at presynaptic cannabinoid receptors to suppress neurotransmitter release. However, eCB actions have also been reported to modulate postsynaptic membrane properties via putative autocrine actions (Bacci, Huguenard, & Prince, 2004; Marinelli et al., 2008). While both AEA and 2-AG bind to CB_1 and CB_2 receptors, AEA also binds to and activates the transient receptor potential vanilloid type 1 (TRPV1) receptor, which can increase calcium influx and stimulate neurotransmitter release, the opposite effect of cannabinoid receptor activation (Di Marzo et al., 1998).

The biosynthesis of AEA has not been fully characterized, however one pathway has been described in which phospholipase D (NAPE-PLD) cleaves the lipid precursor *N*-arachidonoylphosphatidylethano lamine (NAPE) to synthesize AEA (Di Marzo et al., 1994). Other pathways also have been proposed that utilize phospholipase A2 (Sun et al., 2004) or phospholipase C and protein phosphatase activity (Liu et al., 2006). Interestingly, genetic deletion of NAPE-PLD does not reduce AEA levels in the mouse brain (Leung, Saghatelian, Simon, & Cravatt, 2006), which suggests that there are multiple alternative routes to AEA synthesis (Malcher-Lopes, Franco, & Tasker, 2009). The 2-AG synthetic pathway is induced by activation of diacylglycerol (DAG) and the downstream activation of a DAG lipase (Bisogno et al., 2003) that acts on membrane phosphatidic acid to stimulate 2-AG production (Bisogno, Melck, Petrocellis, & Marzo, 2008). 2-AG and AEA are degraded by different enzymes at synapses in the brain. 2-AG is acted on primarily by monoacylglycerol lipase to produce arachidonic acid and glycerol (Di Marzo et al., 1998; Saario, Savinainen, Laitinen, Järvinen, & Niemi, 2004). AEA is hydrolyzed by fatty acid amide hydrolase (FAAH) into arachidonic acid and ethanolamine (Di Marzo et al., 1994).

Endocannabinoid synthesis and release can be induced by the electrical activation of neurons. Thus, depolarization during action potential generation leads to calcium influx that can induce the synthesis of an eCB, likely to be 2-AG, which is released retrogradely and acts at presynaptic cannabinoid receptors to suppress transmitter release. This activity-dependent eCB release is referred to as depolarization-induced suppression of inhibition (DSI) if it occurs at GABA synapses and as depolarization-induced suppression of excitation (DSE) at glutamate synapses. Several G protein-coupled receptors that signal via $G\alpha_q$ are also capable of stimulating eCB synthesis and release. These include muscarinic cholinergic receptors (Kim, Isokawa, Ledent, & Alger, 2002), dopamine receptors (Giuffrida et al., 1999), and metabotropic glutamate receptors (Varma, Carlson, Ledent, & Alger, 2001). The $G\alpha_q$-coupled receptors can either elicit eCB release on their own or they can facilitate the activity-dependent release of eCB.

Steroids have also been reported by us and others to induce the synthesis and retrograde release of eCBs via nongenomic signaling mechanisms at glutamate and/or GABA synapses in different parts of the brain. For example, the synthesis and release of an eCB, likely to be 2-AG, is induced by glucocorticoids at glutamate synapses in the hypothalamus (Di et al., 2003; Di, Malcher-Lopes, Marcheselli, Bazan, & Tasker, 2005; Malcher-Lopes et al., 2006), at GABA synapses in the prefrontal cortex (Hill et al., 2008, 2011), and at glutamate and GABA synapses in the basolateral amygdala (Karst, Berger, Erdmann, Schütz, & Joëls, 2010; Di, Itoga, Fisher, Solomonow, Gilpin, et al., unpublished data). Estrogen causes a rapid suppression of inhibition by activating metabotropic glutamate receptors that induce eCB release at GABA synapses onto hippocampal CA1 neurons of female, but not male rats (Huang & Woolley, 2012). We discuss this form of steroid-induced eCB release in the context of modulation of neuroendocrine circuits later in the review.

2. CANNABINOID REGULATION OF THE NEUROENDOCRINE STRESS RESPONSE

The hypothalamic–pituitary–adrenal (HPA) axis is the main neuroendocrine system activated as part of the stress response. During exposure to a stressor, corticotropin-releasing hormone (CRH) neurons in the paraventricular nucleus of the hypothalamus (PVN) release CRH from their axons in the median eminence into the portal capillaries feeding the anterior pituitary gland. Portal CRH activates ACTH release from corticotropes of the anterior pituitary into the general blood circulation. Circulating adrenocorticotropic hormone (ACTH) then acts on the adrenal cortex to trigger the synthesis and systemic release of the final hormone products of the HPA axis, corticosteroids. The hypothalamus and limbic system are critical regulators of the HPA axis and sites of feedback regulation by corticosteroids (Evanson, Tasker, Hill, Hillard, & Herman, 2010; Hill et al., 2011; Sapolsky, Armanini, Packan, Sutton, & Plotsky, 1990). Immunohistochemical studies have shown that CB_1 receptors are expressed within cortical-limbic and hypothalamic circuits in rats, including in the PVN, the amygdala, the hippocampus, the bed nucleus of the stria terminalis (BNST), the anterior cingulate cortex, and the prefrontal cortex (Marco et al., 2004). The eCB system generally has been found to suppress activation of the HPA axis, although opposing effects of cannabinoids on HPA output have been reported that may be related to dose. Application of exogenous cannabinoid agonists has been shown to regulate activation of the HPA axis, with high doses of cannabinoids stimulating an HPA response and low doses constraining HPA activity (Marco et al., 2004; Martin-Calderon et al., 1998; Romero et al., 2002; Weidenfeld, Feldman, & Mechoulam, 1994). The activation of the HPA axis by high-dose cannabinoid administration may contribute to the dysphoric response that includes increased anxiety and decreased locomotor activity and exploration (Haller, Varga, Ledent, & Freund, 2004). Blockade of CRH receptors centrally attenuates cannabinoid receptor-mediated anxiogenic behavior (Chaperon & Thiebot, 1999). These studies demonstrate a close interaction between eCB signaling and HPA axis activation. In this section, we discuss how eCB signaling regulates the HPA axis under basal conditions and following acute and chronic stress exposure.

2.1 Inhibitory Endocannabinoid Tone on the HPA Axis

Behavioral studies in mice with genetic deletion of CB_1 receptors show that deficiency in CB_1 receptor signaling generally leads to anxiety-like behavior, although this may be dependent on the context in which the stressor is presented (Haller, Bakos, Szirmay, Ledent, & Freund, 2002; Haller, Varga, Ledent, Barna, & Freund, 2004; Haller, Varga, Ledent, & Freund, 2004). This suggests a tonic activation of CB_1 receptors by basal eCB levels that is anxiolytic under context-dependent conditions. Similarly, pharmacological blockade of CB_1 receptors with systemic injection of the CB_1 receptor antagonist rimonabant (SR141716) elicited anxiogenic behavior, which was accompanied by elevated levels of serotonin and dopamine in the hypothalamus, and may have resulted from HPA axis activation (Arevalo, de Miguel, & Hernandez-Tristan, 2001). Consistent with this, systemic administration of CB_1 antagonists acutely increased c-Fos expression within the PVN and elevated peripheral corticosterone levels (Atkinson et al., 2010; Doyon et al., 2006; Wade, Degroot, & Nomikos, 2006), although the site of action of the CB_1 antagonist in these studies could have been anywhere in the central stress circuitry. Somewhat surprisingly, microinjection of SR141716 into the PVN had no effect on basal HPA output (Evanson et al., 2010), but injection into the basolateral complex of the amygdala (BLA) increased HPA axis activation, indicated by an increase in circulating corticosterone (Hill et al., 2009). The lack of effect of CB_1 receptor blockade in the PVN is not totally unexpected, since a tonic suppression of GABA synaptic inputs to PVN neurons by eCB has also been found (Oliet, Baimoukhametova, Piet, & Bains, 2007), which is mediated by AEA (Di et al., 2013), and blocking a tonic eCB suppression of GABA release onto CRH neurons would be expected to inhibit, and not activate, the HPA axis. Thus, the BLA has a tonic inhibitory influence on the HPA axis and, therefore, it may suppress HPA activation under baseline, unstressed conditions, and activation of the HPA axis may include a lifting of this tonic inhibitory input from the BLA (Hill & Tasker, 2012) (Fig. 1).

2.2 Acute Stress Regulation of Endocannabinoids and the HPA Axis

The AEA content of the amygdala decreases upon stress exposure (Patel et al., 2005; Rademacher et al., 2008) (Fig. 1) due to a rapid induction of the AEA degradative enzyme fatty acid amide hydrolase (FAAH) and FAAH-mediated AEA hydrolysis, which suggests that the tonic eCB inhibition of the BLA is exerted by AEA (Hill et al., 2009). Intra-BLA administration of a FAAH inhibitor attenuates activation of the HPA axis via increased CB_1 receptor activation, since it was blocked by inhibiting CB_1 receptors (Bedse et al., 2014). It was recently shown that this rapid induction of FAAH in the BLA is mediated by a stress-induced increase in CRH levels (Gray et al., 2015). The CRH-induced decrease in AEA inhibitory tone in the BLA, therefore, may contribute to the acute stress initiation of the HPA response (Hill & Tasker, 2012).

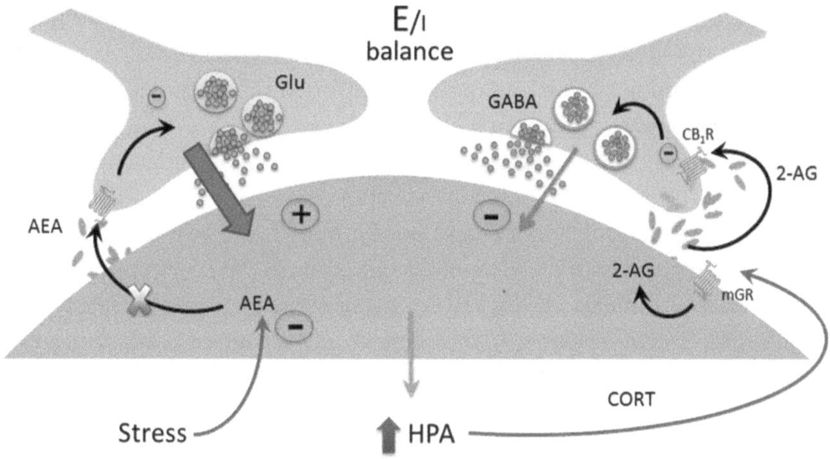

Figure 1 Tonic versus phasic endocannabinoid regulation excitatory and inhibitory synapses. Tonic AEA release in the BLA is thought to tonically constrain excitatory synapses onto BLA principle neurons. Lifting of the tonic eCB suppression of synaptic excitation activates the BLA neurons, which contributes to the initiation of an HPA response (Gray et al., 2015; Patel, Roelke, Rademacher, & Hillard, 2005). HPA activation leads to a surge in circulating glucocorticoids, which feedback on the BLA and activate a membrane-associated glucocorticoid receptor (mGR) on BLA principle neurons. This elicits a phasic 2-AG release at GABA synapses (Di et al., in revision; Hill et al., 2010), which further disinhibits the BLA neurons, leading to increased BLA outputs.

While AEA has been found to be released tonically in the basolateral amygdala (Di et al., in revision), and hypothalamus (Di et al., 2013), and to exert a tonic inhibitory influence on the HPA axis, 2-AG appears to act more as a phasic signal that is evoked during the stress response to attenuate or terminate the stress-induced activation of the HPA axis. Following stress exposure, increased 2-AG levels have been observed in the PVN (Di et al., 2013), hippocampus, and medial prefrontal cortex (Di, Malcher-Lopes, et al., 2005; Evanson et al., 2010; Hill et al., 2011), but not in the amygdala (Hill et al., 2009; Rademacher et al., 2008). While the stress-induced decrease in AEA levels occurs rapidly upon stress presentation, the stress-induced increase in 2-AG is comparatively delayed with respect to the onset of the acute stress (Dubreucq et al., 2012). Systemic corticosterone administration in the unstressed rat or to rat brain slices causes a rapid increase in 2-AG content in the PVN and hippocampus (Di, Malcher-Lopes, et al., 2005; Hill et al., 2011; Wang et al., 2012), indicating that glucocorticoids can induce 2-AG directly via a nongenomic mechanism and suggesting that stress-induced glucocorticoid release activates 2-AG synthesis. In turn, 2-AG suppresses glutamate release within the PVN, thus decreasing the activity of PVN CRH neurons and the HPA axis. Electrophysiological studies in rat and mouse acute brain slices have demonstrated that glucocorticoids induce a decrease in the excitatory synaptic inputs to CRH neurons through the retrograde release of 2-AG and activation of presynaptic CB_1 receptors at glutamate synapses (Di et al., 2003, 2013; Nahar et al., 2015; Tasker, Di, & Malcher-Lopes, 2005). This 2-AG-mediated suppression of excitatory drive to the CRH neurons in the PVN contributes to the feedback inhibition of the HPA axis by elevated glucocorticoid levels during acute stress exposure (Evanson et al., 2010). Bilateral administration of glucocorticoids into the PVN inhibits ACTH and corticosterone release in response to acute stress, whereas intra-PVN CB_1 receptor antagonist application enhances stress activation of the HPA axis. The glucocorticoid-induced 2-AG actions in the PVN are directed specifically to glutamate synapses by astrocytes, while it is thought that tonic eCB actions are mediated by AEA at GABA synapses and are independent of glial regulation (Di et al., 2013).

Glucocorticoids also induce 2-AG release at GABA synapses in the medial prefrontal cortex following stress, albeit on a slower time scale and via a putative genomic mechanism (Hill et al., 2011). The medial prefrontal cortex plays an inhibitory role in the regulation of the HPA axis by activating a GABAergic inhibitory relay in the bed nucleus of the stria terminalis (Radley, Gosselink, & Sawchenko, 2009). Glucocorticoid-induced eCB suppression of synaptic inhibition and the resulting disinhibition of the prefrontal cortical principal neurons, therefore, lead to increased BNST GABAergic input to the PVN and suppression of the HPA axis. Together, these glucocorticoid-2-AG interactions in the PVN and medial prefrontal cortex indicate a combined role for glucocorticoid-induced 2-AG in limbic-hypothalamic circuits in the negative feedback mechanism regulating HPA axis activation.

2.3 Chronic Stress Regulation of Endocannabinoids and the HPA Axis

In general, chronic stress animal models are designed to exhibit either habituating or nonhabituating responses. Repeated exposure to homotypic stressors models a habituating chronic stress, while exposure to variable unpredictable stressors provides a model of nonhabituating chronic stress, as the unpredictability of the successive stressors in the paradigm preclude the development of coping strategies over the course of the chronic stress presentation. Similar to the effect of acute stress, both repeated homotypic stress exposure (Hill et al., 2010; Patel et al., 2005) and chronic unpredictable stress exposure (Hill et al., 2008) result in a reduction in AEA levels in the cortical-limbic circuitry regulating the HPA axis, including in the prefrontal cortex, amygdala, hippocampus, and hypothalamus (Hill et al., 2010; Patel et al., 2005). Chronic unpredictable stress leads to a reduction in CB_1 receptor binding and function in the nucleus accumbens (Wang et al., 2010), CB_1 binding and expression in the hippocampus (Hill et al., 2005), and CB_1 expression and function in the PVN of juvenile rats (Wamsteeker, Kuzmiski, & Bains, 2010). The CB_1 receptor desensitization may be due to direct actions of glucocorticoids, as chronic corticosterone treatment can cause a similar reduction in CB_1 binding and/or protein content and the stress-induced eCB desensitization is reversed by blocking glucocorticoid receptor (Hill et al., 2008; Hong et al., 2011; Rossi et al., 2008). A reduction in tonic AEA levels and a downregulation of CB_1 receptor signaling would be expected to result in the hypersensitivity of the HPA axis, since it should cause a reduction in the baseline AEA constraint on the activation of stress circuits.

Both acute and chronic stressors have been reported to increase 2-AG levels in stress circuits, although chronic stress generally induces more robust changes. Increased 2-AG levels are more consistent following repeated homotypic stress exposure (e.g., acute restraint or social defeat stress) compared to chronic unpredictable stress. Generally, an increased 2-AG level following a chronic homotypic stress has been observed in the amygdala, hippocampus, medial prefrontal cortex, and hypothalamus (Dubreucq et al., 2012; Rademacher et al., 2008). A possible mechanism underlying the upregulation of 2-AG levels by stress exposure may be the reduced expression of the 2-AG hydrolase, monoacylglycerol lipase, as is seen in the amygdala (Sumislawski, Ramikie, & Patel, 2011).

Converging data from multiple studies, therefore, reveal a tight control by eCBs of synaptic signaling in cortical-limbic and hypothalamic circuits regulating the HPA axis (Tasker & Herman, 2011), a control that is regulated by acute and chronic stress. A hypothetical model emerges from these studies of endocannabinoid regulation of the stress circuitry, in which the tonic AEA inhibitory constraint on stress circuits is lifted upon acute stress exposure to facilitate the initiation of the HPA response to the stressor. Following acute stress activation of the HPA axis, elevated circulating glucocorticoids feedback onto the brain and evoke 2-AG synthesis and release within stress-related circuits, which contributes to the negative feedback regulation of the HPA axis. Chronic stress should lead to a sustained depression of the AEA inhibitory tone in stress circuits, causing the disinhibition and resulting hypersensitivity of these circuits (Hill et al., 2010; Patel et al., 2005). Chronic stress is characterized by a hypersensivity of the HPA axis (Herman, Adams, & Prewitt, 1995). Additionally, chronic stress may downregulate cannabinoid receptor mechanisms within stress circuits, which would be expected to desensitize these circuits to the feedback inhibitory regulation by glucocorticoids. Together, these plastic changes in the cannabinoid regulation of the stress circuitry induced by chronic stress should render these circuits hypersensitive to stress activation and less sensitive to the negative feedback actions of glucocorticoids, resulting in a sustained elevation of HPA activity, as is seen in certain stress-associated disorders. We recently found that the HPA hypersecretion during chronic stress is likely to be mediated by a sensitization of the HPA axis to excitatory inputs, rather than to a desensitization to rapid glucocorticoid feedback inhibition (Franco, Chen, Scullen, Zsombok, Salahudeen, et al., unpublished data).

3. ENDOCANNABINOID REGULATION OF ENERGY HOMEOSTASIS

Cannabis has been used for its antiemetic and appetite-inducing properties for centuries, however, it was only recently that its effects were attributed to THC (Gaoni & Mechoulam, 1964). Since then, multiple investigations have focused on identifying and understanding the eCBs that THC mimics, AEA, and 2-AG. Central administration of THC and eCBs stimulates eating (Hao, Avraham, Mechoulam, & Berry, 2000; Williams & Kirkham, 1999). Specifically, THC administration can induce eating in satiated animals (Williams, Rogers, & Kirkham, 1998). Not surprisingly, eCBs mediate these effects by acting on some of the commonly studied metabolic pathways in the brain. One area of the brain heavily implicated in metabolism is the arcuate nucleus of the hypothalamus. The arcuate nucleus contains distinct subpopulations of neurons that express neuropeptide Y (NPY), agouti-related protein (AgRP), and proopiomelanocortin (POMC). NPY and AgRP colocalize in the same population of arcuate neurons and serve as appetite stimulators, while POMC is expressed in a separate arcuate subpopulation and is an appetite inhibitor (Denis, John, & Michael, 1999; Hahn, Breininger, Baskin, & Schwartz, 1998). These neurons are able to sense blood-borne appetitive, or orexigenic, signals such as ghrelin, and satiety, or anorexic, signals such as leptin. This is thought to be due to the close proximity of the arcuate nucleus to a circumventricular organ, the median eminence, and the expression of transporters within the arcuate nucleus that actively pump these signals across the blood–brain barrier (Balland et al., 2014; Norsted, Gömüç, & Meister, 2008). The metabolic state-sensing neurons in the arcuate nucleus project to second-order food regulation centers in the hypothalamus, including the PVN and the lateral hypothalamus. Neurons in the PVN inhibit, whereas neurons in the lateral hypothalamus stimulate food intake (Atasoy, Betley, Su, & Sternson, 2012; Schwartz, Woods, Porte, Seeley & Baskin, 2000). These second-order nuclei send their projections to the nucleus of the tractus solitarius (NTS), an area of the brainstem that integrates food intake-related sensory information from the periphery (Travers, Travers, & Norgren, 1987). As mentioned above, these hypothalamic circuits respond to blood-borne metabolic signals. Leptin is a potent metabolic hormone synthesized by white adipose tissue that works to inhibit appetite; ghrelin, which is synthesized primarily by the stomach, is also a potent metabolic hormone, but it works to stimulate appetite. These molecules, along with the hormones peptide YY, insulin, and glucose, allow the periphery to communicate information on the metabolic state of the organism to the central nervous system (Elmquist, Coppari, Balthasar, Ichinose, & Lowell, 2005).

3.1 Endocannabinoids and Cannabinoid Receptors in Energy Homeostasis

As described above, CB_1 receptors are the primary physiological target for cannabinoids in the central nervous system. CB_1 receptors are widely expressed in regions associated with metabolism, including the hypothalamus (Di Marzo & Matias, 2005). Genetic deletion of the CB_1 receptor generates a lean body phenotype, indicating the receptor's importance in energy metabolism (Cota et al., 2003). Additionally, in food-deprived mice, blocking CB_1 receptors with the CB_1 receptor antagonist SR141716, rimonabant, causes a reduction in food intake to a similar extent as CB_1 deletion (Di Marzo et al., 2001). These findings indicate the important role eCBs can play in the regulation of appetite.

The mechanisms by which eCBs regulate food intake and energy metabolism have been the focus of intensive study in recent years because of the relevance to the emerging societal problem of obesity. The administration of peripheral or intrahypothalamic eCBs or CB_1 agonists leads to an increase in food intake in rodents; this effect is blocked by pretreatment with a CB_1 antagonist (Cota, 2007). Within the hypothalamus, the levels of 2-AG are inversely correlated to the animal's energy status, increase during fasting and decreased during refeeding (Kirkham, Williams, Fezza, & Di Marzo, 2002). At the level of the lateral hypothalamus, depolarization of lateral hypothalamic neurons results in the release of eCBs that bind presynaptically to CB_1 receptors at GABAergic axon terminals, suppressing inhibitory synaptic inputs and thus further exciting the lateral hypothalamic neurons (Jo, Chen, Chua, Talmage, & Role, 2005). Depolarization of neuroendocrine cells in the PVN causes 2-AG release at excitatory synapses, which suppresses excitatory synaptic inputs and decreases the excitatory drive to PVN neurons (Di Marzo & Matias, 2005; Di, Boudaba, et al., 2005; Di et al., 2013; Yoshida et al., 2009). Together, the excitatory regulation of lateral hypothalamic neurons and inhibitory regulation of PVN neurons by eCBs should activate feeding-related autonomic and neuroendocrine outputs.

It has been shown that peripheral administration of AEA results in c-Fos expression, an immediate early gene marker for neuronal activation, in the PVN in as little as 45 minutes (Wenger, Jamali, Juanéda, Léonardelli, & Tramu, 1997). More recently, it was shown that the actions of eCBs within the PVN are not so straight-forward, and are actually dependent on the feeding state of the animal. Specifically, in fasted animals the actions of eCBs were found to be anorexic, while in free-fed animals the eCBs were orexigenic (Soria-Gómez, Massa, et al., 2014).

3.2 Ghrelin and Endocannabinoids in Energy Homeostasis

Ghrelin is a 28-amino acid-peptide hormone that is synthesized primarily in the stomach (Kojima et al., 1999). Ghrelin's receptor, the growth-hormone secretagogue receptor 1a (GHS-R1a), is a G protein-coupled receptor that induces intracellular calcium signaling upon homodimerization following ligand binding (Kaiya, Kangawa, & Miyazato, 2013). Ghrelin exists in two forms, acyl-ghrelin and des-acyl-ghrelin, with acyl-ghrelin being acylated at its serine-3 residue. This acylation allows acyl-ghrelin to activate the GHS-R1a, while des-acyl-ghrelin is inactive at GHS-R1a. The acylation of ghrelin is a unique biochemical process that is mediated by ghrelin *O*-acyltransferase, which is a membrane-bound enzyme that is expressed mainly in the stomach, but also in the hypothalamus (Cowley et al., 2003; Gutierrez et al., 2008; Yang, Brown, Liang, Grishin, & Goldstein, 2008). When ghrelin is directly delivered into the CNS via intracerebroventricular injection, feeding behavior is increased along with body weight gain. Also, central ghrelin administration induces c-Fos expression in NPY and AgRP neurons, both known for their importance in the control of feeding behavior (Nakazato et al., 2001). Ghrelin is thought to act through eCB release in the hypothalamus, since the CB_1 receptor antagonist SR141716 blocks the orexigenic effects of intra-PVN ghrelin application (Tucci, Rogers, Korbonits, & Kirkham, 2004). The effect of central ghrelin administration was lost in CB_1-KO mice, and inhibition of the 2-AG synthetic enzyme diacylglycerol lipase blocked ghrelin's effect. Ghrelin was also found to increase the eCB concentration in the hypothalamus in wild-type mice (Kola et al., 2008). Specifically, it appears that ghrelin's ability in the hypothalamus to induce feeding behavior is mediated by CB_1 receptors, and not CB_2 receptors (Ting, Chi, Li, & Chen, 2015). Another link between ghrelin and eCBs in the hypothalamus is their mutual capacity to stimulate AMP-activated protein kinase activity, which stimulates appetite (Kola et al., 2005). Not all of ghrelin's actions in the hypothalamus, however, are eCB-dependent. Ghrelin also stimulates vasopressin neurons in the PVN via a retrograde neuronal–glial circuit, which is likely to couple the energy state of the organism to fluid homeostasis (Haam, Halmos, Di, & Tasker, 2014).

3.3 CRH, TRH, and Endocannabinoid Regulation of Energy Homeostasis

CRH is a 41-amino acid-peptide hormone expressed in the PVN that plays a key role as mediator of both the endocrine and behavioral stress responses. CRH has also been shown to influence food intake, which may be related to its role as a stress hormone. In rats, chronic intracerebroventricular administration of CRH suppressed food intake (Hotta et al., 1991) and inhibited fasting-induced feeding (Pelleymounter et al., 2000). Investigation into the mechanism of the CRH regulation of feeding behavior revealed that it acts in the PVN to inhibit NPY-stimulated food intake (Heinrichs et al., 1993). It was recently shown that CRH acts in the amygdala to elicit an increase in the catabolic enzyme responsible for degrading AEA, FAAH, which leads to a reduction in amygdalar AEA levels (Gray et al., 2015). Although this has not been demonstrated in the hypothalamus, CRH in the PVN is a known regulator of feeding behavior and may work through a similar mechanism to alter feeding behavior by regulating eCB levels in the PVN. Interestingly, FAAH knockout mice display reduced energy expenditure (Brown et al., 2012), which indicates that FAAH may play a direct role in energy metabolism. Pregastric neurons in the PVN that control gastric motility are subject to control by eCBs, which is likely to be mediated by AEA since both CB_1 and TRPV1 receptors are implicated (Boychuk, Zsombok, Tasker, & Smith, 2013) and AEA targets both receptors, whereas 2-AG binds only to CB_1 receptors.

Thyrotropin-releasing hormone (TRH) is synthesized in neurons of the PVN and acts as the main central regulator of the hypothalamic–pituitary–thyroid axis by controlling the amount of thyroid-stimulating hormone released from the pituitary and thyroid hormone secretion from the thyroid (Fekete & Lechan, 2007). Thyroid hormones increase metabolic rate. Under conditions of food deprivation, TRH neurons in the PVN are inhibited, which reduces thyroid hormone secretion and levels in the systemic circulation to maintain homeostasis and conserve energy (Lechan & Fekete, 2006). Central administration of TRH reduces feeding and drinking behavior in rats (Vettor, Fabris, Pagano, & Federspil, 2002). CB_1 receptors have been localized to the presynaptic terminals of axons that innervate the TRH neurons of the PVN (Deli et al., 2009), which suggests a role for eCBs in the regulation of the hypothalamic–pituitary–thyroid axis and, therefore, in another important facet of energy homeostasis.

3.4 Leptin and Endocannabinoid Regulation of Energy Homeostasis

Leptin is produced primarily in adipose tissue and regulates both short-term and long-term energy homeostasis by acting as one of the main signals of satiety (Ahima & Osei, 2004). One of the ways leptin signals satiety is by reducing both AEA and 2-AG levels in the hypothalamus; leptin achieves this not by working on the enzymes that directly synthesize the eCBs, but rather by modulating the enzymes that make the precursors for the eCBs, NAPE-PLD for AEA, and phosphatidylinositol-specific PLC for 2-AG (Di Marzo et al., 2001). Leptin blocks the eCB-mediated suppression of GABA release in the lateral hypothalamus, causing disinhibition and increased excitability of the lateral hypothalamic neurons. Leptin suppresses eCB release by inhibiting voltage-gated calcium currents, which blocks eCB synthesis (Jo et al., 2005). Interestingly, leptin signals through phosphatidylinositide-3kinase (PI3K), though reciprocally in POMC and AgRP neurons; in POMC neurons, the presence of leptin activates PI3K, whereas leptin's absence leads to PI3K activation in AgRP neurons (Xu et al., 2005). Glucocorticoids rapidly induce eCB synthesis in PVN neuroendocrine cells, including in CRH neurons, via a rapid, nongenomic pathway (Di et al., 2003; Tasker, 2006), which inhibits glutamatergic excitatory synaptic inputs. Leptin blocks eCB synthesis via phosphodiesterase-induced inhibition of cAMP activity, and thus lifts the eCB suppression of excitatory synaptic inputs to the PVN neurons (Malcher-Lopes et al., 2009). These mechanisms allow leptin to exert a rapid regulation of eCB release in the lateral hypothalamus and PVN, and to regulate feeding behavior elicited by hypothalamic eCB actions downstream from the arcuate nucleus.

4. ENDOCANNABINOIDS AND FLUID HOMEOSTASIS

Water retention and excretion are tightly regulated in order to achieve fluid homeostasis. Stable fluid volume and blood pressure are required for healthy cardiovascular function, and isotonic blood osmolality is needed to avoid cellular lysis. Fluid homeostasis is controlled by a variety of neuroendocrine systems, many of which signal through endocannabinoids. Early studies on the effects of cannabis revealed that endocannabinoids must be involved in fluid intake, excretion, and neuroendocrine regulation. Abel's early review of the effects of cannabis on hunger and thirst reported a decrease in water intake and increase in food intake, both of which affect fluid homeostasis (Abel, 1975). Early human studies (Ames, 1958) as well as studies in the rat (Sofia, Dixit, & Barry, 1977) implicated THC as a diuretic agent, while a more recent study showed that THC decreased urine volume and urinary frequency in patients suffering from multiple sclerosis (Paronis et al., 2013). Cannabinoids were also found to directly affect vasoconstriction, as systemic administration of anandamide caused hypotension in rats (Varga, Lake, Martin, & Kunos, 1995). And, while cannabinoids have no direct effect on the antidiuretic hormone, vasopressin, an inverse agonist of the CB_1 receptor was shown to potentiate vasopressin secretion (Ruginsk, Uchoa, Elias, & Antunes-Rodrigues, 2012), suggesting a tonic inhibitory eCB regulation of vasopressin secretion. Thus, systemic administration of cannabinoids has complex, sometimes opposing, effects on fluid homeostasis. These complexities arise because eCBs are produced in discrete locations, commonly at excitatory and/or inhibitory synapses, and act as signals within specific circuits that regulate different physiological processes. The distribution of the cannabinoid receptors in fluid regulating organs is crucial to the effects of cannabis on fluid homeostasis.

4.1 Receptor Distribution in Osmoregulatory Tissues

The TRPV1 receptors form ionotropic cation channels and bind the eCB AEA, but not 2-AG. TRPV1 receptors are found along with CB_1 and CB_2

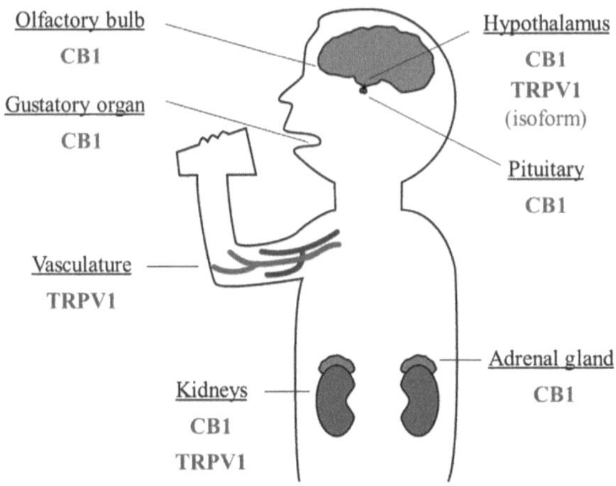

Figure 2 Cannabinoid receptors regulate fluid homeostatic mechanisms throughout the organism. Cannabinoids influence water and salt intake behaviors by enhancing the perception of olfactory (Soria-Gómez, Bellocchio, et al., 2014) and gustatory (Yoshida et al., 2009) senses. Cannabinoids also act on osmosensitive neuroendocrine systems. Excitatory inputs to vasopressin-, oxytocin-, and CRH-synthesizing cells in the hypothalamus are downregulated by an eCB retrograde signal (Di et al., 2003). Cannabinoid receptors are also found on neuroendocrine target tissues. Cannabinoids act at CB_1 receptors in the pituitary to facilitate ACTH release (Pagotto et al., 2001), in the kidney nephron to increase sodium and potassium ATPase activity (Soria-Gómez, Massa, et al., 2014), and at TRPV1 receptors in the adrenal cortex to increase aldosterone production (Ziegler et al., 2010). Macrophage-induced hypotension, mediated by cannabinoid actions at TRPV1 receptors, could have indirect effects on fluid homeostasis (Varga et al., 1995).

receptors in the kidney nephron (Caterina et al., 1997) (Fig. 2), and cannabinoids have been shown to increase salt and fluid retention by upregulating sodium and potassium ATPase activity in these cells (Sampaio et al., 2014). While some studies have shown TRPV1 in the hypothalamus (Menigoz & Boudes, 2011), their association with endocannabinoids and their role in osmoregulation remain unclear. One study suggests that TRPV1 is expressed in vasopressin neurons, but the TRPV1 is truncated and lacks the carboxy terminus, which is thought to make it responsive to changes in osmolarity, but insensitive to the TRPV1 agonist capsaicin (Sudbury, Ciura, Sharif-Naeini, & Bourque, 2010). There is similar evidence of TRPV1 presence in the anteroventricular-third ventricular (AV3V) region of the basal forebrain that projects to the hypothalamus (Fig. 2). Here TRPV1 also acts as an osmoreceptor that senses hyperosmolarity and projects to the vasopressin neurons in the hypothalamus (Ciura & Bourque, 2006). While further studies may determine that TRPV1 in the neuronal circuitry regulating vasopressin release is the mechanism by which AEA induces hypotension, it is more likely that the receptors expressed in mesenteric resistance arteries signal hypotension (Wang, Kaminski, & Wang, 2005). Interestingly, TRPV1 mesenteric expression is upregulated by a high-salt diet, which could make AEA protective against the development of hypertension.

The CB_1 and CB_2 cannabinoid receptors respond to both AEA and 2AG and are found in osmoregulatory tissues (Fig. 2). CB_2 is prominently expressed in tissues of the immune system, where it indirectly influences fluid homeostasis by mediating macrophage-induced hypotension (Varga et al., 1995). CB_1 receptors are found widely distributed in several osmoregulatory tissues, including the gustatory organ, the olfactory system, the hypothalamus, and the pituitary (Mackie, 2005). While cannabinoids have well-known effects on food intake and energy metabolism, their effects on water and salt intake are less clear.

Taste cells in the rat express CB_1 receptors and administration of AEA or 2-AG enhance cellular, neuronal, and behavioral responses to sweet taste, however, eCBs did not induce a preference for salt (Yoshida et al., 2009), indicating that the role of eCBs in fluid homeostasis may not have a gustatory component. CB_1 receptors in the olfactory bulb could play a larger role in the eCB-induced increase in food intake by increasing odor detection (Soria-Gómez, Bellocchio, et al., 2014). While there are no studies directly linking olfactory CB_1 receptors to fluid intake, the olfactory bulb is required for normal fluid intake and salt aversion (Bell, Dennis, & Sly, 1979). Therefore, eCBs affect water and food intake behaviors by modulating sensory information, although the mechanisms are not fully understood.

There is increasing evidence that endocannabinoids are involved in the neuroendocrine control of fluid homeostasis. CB_1 expression is found in the PVN (Castelli et al., 2007; Herkenham et al., 1991), where osmoregulatory neurons that produce the neurohormones vasopressin and CRH are located. CRH is released from the median eminence at the base of the hypothalamus and acts on corticotropic cells of the anterior pituitary to elicit ACTH secretion into the blood. ACTH stimulates the production and secretion of the corticosteroid hormones corticosterone (cortisol in primates) and aldosterone by the adrenal cortex. As described above, corticosterone is involved in the neuroendocrine stress response. Aldosterone acts on the kidneys to inhibit sodium transfer to the urine, or natriuresis. Endocannabinoids are involved in the rapid modulation by corticosterone of the excitatory synaptic inputs to the PVN CRH neurons and play an important role in the negative feedback regulation of the HPA axis (Di et al., 2003; Nahar et al., 2015), as discussed previously. CB_1 receptors are also found in the pituitary, where cannabinoid agonists facilitate CRH-induced secretion of ACTH (Pagotto et al., 2001). Additionally, there are CB_1 receptors in the adrenal cortex that have been shown to directly increase adrenocortical steroidogenesis, including aldosterone synthesis (Ziegler et al., 2010). Thus, eCBs appear to exert an influence over every tier of the HPA axis (Fig. 2). The involvement of eCBs in the control of vasopressin release has been studied and the eCB retrograde control of synaptic signaling to vasopressin neurons will be discussed.

4.2 Glucocorticoid Feedback on Vasopressin Secretion Via Endocannabinoids

Vasopressin is a nine-amino acid-peptide hormone produced by magnocellular neuroendocrine cells in the hypothalamic PVN and supraoptic nucleus (SON). These neurons project to the posterior pituitary, where they release vasopressin directly into the systemic bloodstream. Vasopressin travels to the kidneys, where it stimulates water retention by causing trafficking of aquaporin water channels to the apical membrane of the kidney tubule lumen (Hozawa, Holtzman, & Ausiello, 1996). Aquaporins increase renal cell permeability to water, which allows the kidney to reabsorb fluids. Homeostatic failure of the vasopressin system can result in diabetes insipidus (Bichet et al., 1993). While eCBs can influence ion transport in the kidney (Caterina et al., 1997), there is no evidence that eCBs are involved in aquaporin trafficking. Instead, eCBs appear to exert their influence on fluid homeostasis and diuresis, at least in part, by modulating synaptic inputs to the hypothalamic magnocellular neuroendocrine cells that release vasopressin (Di et al., 2003).

Vasopressin neurons in the hypothalamus are innervated by glutamatergic neurons that include osmoreceptive cells in the hypothalamic anteroventral third ventricular (AV3V) region and a cholinoceptive area on the ventral brain surface (Bisset & Chowdrey, 1988; Ciura & Bourque, 2006). Adrenalectomy leads to an increase in vasopressin expression in CRH neurons of the PVN, indicating that glucocorticoids feedback onto the hypothalamus to downregulate vasopressin expression (Keller-Wood, Shinsako, & Dallman, 1984). Recently, glucocorticoids were shown to rapidly inhibit vasopressin release via an eCB retrograde signaling mechanism (Di et al., 2003). Glucocorticoids induce the synthesis of eCBs in both vasopressin- and oxytocin-secreting magnocellular neurons of the PVN and

SON (Di, Boudaba, et al., 2005; Malcher-Lopes et al., 2006). The glucocorticoid-induced eCB is 2-AG (Di et al., 2013), which serves as a retrograde messenger by activating presynaptic CB_1 receptors at glutamate synapses. Signaling from the CB_1 receptors causes a decrease in synaptic glutamate release and an overall decrease in the synaptic excitation of the magnocellular neurons. While the identity of the membrane-associated glucocorticoid receptor that triggers this response is unknown, the glucocorticoid-induced production of eCBs is dependent on the activation of a $G\alpha$(s)-cAMP-protein kinase A signaling mechanism (Malcher-Lopes et al., 2006). As described above, leptin, a satiety hormone secreted peripherally by white adipose tissue, has been found to inhibit the glucocorticoid-induced eCB synthesis in magnocellular neurons, allowing for crosstalk between food and fluid homeostasis signaling mechanisms. 2-AG is also produced by periods of high electrical activity in magnocellular neurons and at acts at glutamate synapses to suppress synaptic excitation (Di, Boudaba, et al., 2005). Interestingly, CB_1 receptors are present at both excitatory and inhibitory synapses on magnocellular neurons, but the glucocorticoid- and activity-induced 2-AG actions occurs only at excitatory synapses under baseline physiological conditions (Di et al., 2013). This synapse specificity is controlled by the spatial restriction of endocannabinoids by astrocytes. If astrocytic coverage of synapses onto the magnocellular neurons is impaired by dehydration-induced glial retraction or by inhibiting glial metabolism with a gliotoxin, glucocorticoid-, and activity-induced 2-AG signaling can also impact neighboring inhibitory synapses via an apparent "spillover" of the eCB from glutamate synapses onto GABA synapses. This observation suggests an important role for astrocytes in the synapse-specific actions of evoked eCB release. Interestingly, inhibitory synapses on magnocellular neurons of the PVN are subject to a tonic eCB suppression of GABA release (Oliet et al., 2007). The eCB responsible for the tonic modulation of GABA synapses was shown to not be 2-AG, since it was not inhibited by blocking diacylglycerol lipase activity, and is therefore likely to be AEA. Additionally, it is insensitive to manipulation of the astrocytic coverage of magnocellular neurons or of glial metabolism (Di et al., 2013), which suggests that, unlike 2-AG, it is transported to CB_1 receptors on presynaptic GABA terminals without diffusing through the extracellular space.

Despite the role of eCBs in the synaptic signaling to vasopressin neurons, intracerebroventricular administration of AEA does not affect peripheral vasopressin secretion, even under conditions of increased vasopressin secretion induced by blood volume expansion (Ruginsk, Uchoa, Elias, & Antunes-Rodrigues, 2013). Similarly, glucocorticoids administered systemically also have no effect on vasopressin release (Ruginsk et al., 2012). However, administration of a CB_1 receptor antagonist facilitates vasopressin release, and increasing endogenous AEA levels with a FAAH inhibitor causes a decrease in vasopressin release. One study demonstrated that glucocorticoids were required to suppress high levels of vasopressin release caused by hemorrhage (Darlington, Chew, Ha, Keil, & Dallman, 1990). In that study, rats that had been deprived of peripheral glucocorticoids by prior adrenalectomy died within 4 h of hemorrhage, however 50% of animals treated with prior systemic corticosterone survived the hemorrhage. Since the lethality of the hemorrhage was caused by the hypotensive actions of excessive vasopressin release, the protective effect of glucocorticoids was thought to be due to the glucocorticoid suppression of vasopressin release to sublethal levels, which may be meditated by the glucocorticoid-induced eCB suppression of excitatory synaptic inputs activated by the hemorrhage. Together, these findings suggest that while glucocorticoids may not have a strong effect on vasopressin release, eCBs do exert control over circulating vasopressin, albeit by a subtle mechanism. Endocannabinoids have stronger effects, however, on the other main hypothalamic neurohormone involved in fluid homeostasis, oxytocin.

4.3 Oxytocin and Atrial Natriuretic Peptide

Oxytocin is similar to vasopressin in both structure and release by magnocellular neurons of the PVN/SON. Both systems are also subject to the same glucocorticoid-induced depression of excitation via retrograde synaptic endocannabinoid signaling. After being released from the posterior pituitary, oxytocin travels to the right atrium to stimulate the release by the heart of atrial natriuretic peptide (ANP) (Haanwinckel et al., 1995). The ANP travels to the kidney, where it induces natriuresis and diuresis. Although CB_1 receptors have been shown to regulate sodium transport in the kidney (Sampaio et al., 2014), currently, none of the natriuretic effects of eCBs have been attributed to direct modulation of ANP action in the kidney, although this remains a possibility. Instead, eCBs seem to strongly regulate circulating oxytocin levels during glucocorticoid-induced suppression of oxytocin release (Ruginsk, Uchoa, Elias, & Antunes-Rodrigues, 2010). Rats subjected to extracellular volume expansion show an increase in circulating oxytocin, which can be potentiated by treatment with the CB_1

antagonist SR141716 (rimonabant), indicating that eCBs lower oxytocin release *in vivo*. Glucocorticoid administration caused a decrease in circulating oxytocin (Lauand et al., 2007), but the CB_1 receptor antagonist reversed the inhibitory glucocorticoid effect on oxytocin release. These observations *in vivo* are consistent with our *in vitro* findings showing glucocorticoid-induced eCB suppression of excitatory synaptic inputs to oxytocin neurons (Di et al., 2003; Di, Malcher-Lopes, et al., 2005). The strong involvement of cannabinoids in the regulation of oxytocin release suggests that the effects of cannabis on fluid homeostasis may rely largely on oxytocin signaling.

5. ENDOCANNABINOID REGULATION OF REPRODUCTION

Endocannabinoids have also been implicated in the hypothalamic regulation of reproductive behavior, parturition, and lactation through the modulation of gonadotropin-releasing hormone (GnRH) and oxytocin neurons. The hypothalamic–pituitary–gonadal (HPG) axis controls reproductive behavior in humans and lower mammals. The GnRH neurons located in the arcuate nucleus and preoptic area of the hypothalamus secrete GnRH in a pulsatile manner into the portal vessels to drive the rhythmic release of luteinizing hormone (LH) and follicle-stimulating hormone (FSH) from the anterior pituitary into the general circulation, which in turn stimulate the gonads to synthesize and secrete gonadal steroid hormones and promote gametogenesis. Oxytocin neurosecretion plays an important role in parturition and postnatal milk ejection. During parturition and lactation, oxytocin is secreted in a pulsatile fashion from the posterior pituitary by magnocellular neuroendocrine cells of the PVN and SON of the hypothalamus, and acts on oxytocin receptors in the uterus to stimulate uterine contractions during childbirth and in the mammary gland to stimulate milk letdown in response to suckling. In addition, central release of oxytocin also plays a critical role in postnatal maternal behavior.

5.1 Endocannabinoid Regulation of the HPG Axis

The activation of cannabinoid receptors can have significant effects on reproductive behavior and on the physiological function of the HPG axis. In human studies, acute and chronic consumption of marijuana decreases serum testosterone and LH levels in males (Cone, Johnson, Moore, & Roache, 1986; Kolodny, Masters, Kolodner, & Toro, 1974). In females, a single marijuana cigarette significantly suppresses LH levels (Mendelson et al., 1986). Adverse effects of cannabinoids on the HPG axis also are observed in animal studies. Administration of cannabinoids decreases serum LH and testosterone, and reduces sperm count in male rats (Bloch, Thysen, Morrill, Gardner, & Fujimoto, 1978; Gorzalka & Dang, 2012; Kumar & Chen, 1983; Murphy, Steger, Smith, & Bartke, 1990; Wenger, Ledent, Csernus, & Gerendai, 2001). In female rats, chronic cannabinoid exposure was found to delay sexual maturation, disrupt the estrous cycle, and depress follicular maturation (Bloch et al., 1978; Wenger, Croix, & Tramu, 1988). In ovariectomized rats, acute administration of THC suppressed pulsatile LH secretion (Tyrey, 1978).

Recent evidence indicates that eCBs regulate HPG function primarily by modulating the activity of the hypothalamic GnRH neurons. Although acute systemic administration of THC suppresses pulsatile LH secretion from the pituitary, GnRH application still evokes LH secretion by the anterior pituitary gonadotropes and reverses the THC-induced LH suppression (Tyrey, 1978). Additionally, cannabinoid exposure has no effect on basal or GnRH-induced LH release in cultured anterior pituitary cells (Chakravarty, Shah, Sheth, & Ghosh, 1979; Wenger, Rettori, Snyder, Dalterio, & McCann, 1987). These findings indicate that cannabinoids suppress the function of the HPG axis primarily through actions on the hypothalamic GnRH neurons, and not at the pituitary.

Consistent with this, THC exposure lowers the GnRH concentration in the hypothalamus of female rats (Chakravarty et al., 1979). Similarly, the CB_1 receptor agonist WIN55,212 decreases depolarization-induced GnRH secretion and spontaneous pulsatile GnRH secretion from an immortalized GnRH-expressing cell line, GT1-7 cells (Gammon, Freeman, Xie, Petersen & Wetsel, 2005). Although GT1-7 cells are shown to both functionally respond to and release endocannabinoids, *in situ* hybridization findings suggest that GnRH neurons *in vivo* show very low or no CB_1 receptor expression. In contrast, neurons in the vicinity of the GnRH neurons express high levels of the CB_1 receptor (Gammon et al., 2005), indicating that, as in PVN neurons, endocannabinoid released from GnRH neurons may act as a retrograde messenger to regulate inputs from presynaptic neurons. GABAergic transmission, which is excitatory in many of the GnRH neurons because of their high intracellular chloride concentration, plays an important role in the regulation of GnRH neuron activity (Watanabe, Fukuda, & Nabekura, 2014). CB_1 receptor-positive axon terminals are found in close contact with the GnRH neurons, and a subset of these synapses are symmetric, a characteristic of GABAergic synapses (Farkas et al., 2010).

Bath application of the CB_1 receptor agonist WIN55,212 decreases the firing rate of GnRH neurons in extracellular recordings with glutamatergic receptors blocked. Furthermore, whole-cell recordings showed that application of WIN55,212 decreased the miniature postsynaptic current (mPSC) frequency, while the CB_1 receptor antagonist AM251 significantly increased mPSC frequency. These data collectively suggest that eCBs tonically inhibit excitatory GABAergic inputs to GnRH neurons. Finally, blockade of 2-AG synthesis by inhibiting diacylglycerol lipase activity abolished the AM251-induced mPSC frequency increase, indicating that the tonic inhibition of GABAergic inputs to GnRH neurons is mediated by 2-AG release.

GnRH neurons can also modulate their own activity through retrograde endocannabinoid signaling. Thus, strong repeated depolarization of GnRH neurons, which mimics high frequency action potential generation in these neurons, resulted in a short-term suppression of GABAergic transmission and neural activation via retrograde eCB release (Glanowska & Moenter, 2011). The depolarization-induced suppression of GABA transmission was abolished by a diacylglycerol lipase inhibitor. These results suggest that GnRH neurons synthesize and release endocannabinoids to regulate their inhibitory synaptic inputs in a negative feedback manner.

5.2 Ghrelin Inhibits the HPG Axis Through Endocannabinoid Release

Ghrelin has also been found to suppress the function of the HPG axis (Repaci, Gambineri, Pagotto, & Pasquali, 2011). Intracerebroventricular infusion of ghrelin *in vivo* rapidly suppressed the pulsatile secretion of LH in ovariectomized rats (Furuta, Funabashi, & Kimura, 2001). An *in vitro* study in hypothalamic explants from prepubertal and sexually mature male rats also showed that ghrelin incubation significantly prolongs the interpulse interval between GnRH pulses (Lebrethon et al., 2007). Consistent with this, GnRH secretion was inhibited by ghrelin in hypothalamic fragments from ovariectomized female rats (Fernandez-Fernandez et al., 2005). Human studies also support the ghrelin suppression of the HPG axis, as systemic ghrelin administration significantly decreased LH and FSH secretion in women and LH secretion in young men (Kluge, Schussler, Schmidt, Uhr, & Steiger, 2012; Lanfranco et al., 2008).

Recently, Farkas et al. reported that ghrelin acts on the GHS-R1a to regulate GnRH neuron activity via endocannabinoid signaling in an estrous cycle-dependent manner (Farkas, Vastagh, Sarvari, & Liposits, 2013).

Ghrelin rapidly increased the intracellular free calcium concentration in immortalized GT1-7 GnRH cell. This effect was blocked by application of a GHS-R1a antagonist or estradiol, indicating that the ghrelin effect on calcium mobilization is mediated by GHS-R activation and is estrogen dependent. Using single cell RT-PCR from identified GnRH neurons expressing green fluorescent protein, they found that GnRH neurons express GHS-R1a mRNA. Furthermore, they found that ghrelin decreased the mean firing rate and instantaneous firing frequency of GnRH neurons with loose-patch recordings in slices from female mice in metestrus, but had no effect in slice from mice in proestrous. Ghrelin also was shown to decrease the firing rate of GnRH neurons in male mice. Pretreatment with a GHS-R1a antagonist or the CB_1 receptor inverse agonist AM251 abolished the ghrelin-induced decrease in firing rate of GnRH neurons, indicating that ghrelin acts at GHS-Rs to suppress the activity of GnRH neurons in an eCB-dependent manner.

Ghrelin also significantly reduced the frequency of $GABA_A$ receptor-mediated mPSCs in whole-cell recordings in slices from metestrus mice (Farkas et al., 2013). Pretreatment of the slices with AM251 blocked the ghrelin-triggered decrease in mPSC frequency, suggesting an eCB dependence of the ghrelin effect. In addition, the ghrelin-induced decrease in mPSC frequency was abolished by a diacylglycerol lipase inhibitor in the patch solution, suggesting that ghrelin acting at GHS-Rs on GnRH neurons stimulates 2-AG release, which suppresses excitatory GABAergic inputs and inhibits GnRH neurons. Estrogen has been found to exert rapid, nongenomic effects on inhibitory synaptic inputs to hippocampal pyramidal neurons via the retrograde release of an endocannabinoid at GABA synapses (Huang, Chandra, & Rastinejad, 2010). Therefore, as in CA1 neurons, estrogen may modulate eCB mobilization at GABA synapses on GnRH neurons to facilitate ghrelin-induced eCB suppression of GABA release. Ghrelin also has been reported to induce eCB release at GABA synapses onto parvocellular neuroendocrine cells in the PVN and to suppress inhibitory synaptic inputs to these cells (Kola et al., 2005), similar to its actions in GnRH neurons.

5.3 Stress and Glucocorticoid Modulation of Reproductive Behavior Via Endocannabinoid Actions

Maternal behavior improves the survival and well-being of the offspring, and thus is important for the success of the species. In rodents, maternal behaviors include nest building, licking, arched-back nursing, lactation, and maternal aggression. Two peptide hormones are crucial for lactation. Prolactin, secreted from the anterior pituitary, promotes milk production by stimulating the alveoli of the mammary glands to secrete milk. Oxytocin, synthesized by the magnocellular neurons in the PVN and SON, activates smooth muscle cells in the mammary glands to stimulate milk letdown. In addition to stimulating milk letdown, oxytocin has been found to play an important role in maternal behavior and social interactions through central actions of the neuropeptide (Bosch, Meddle, Beiderbeck, Douglas, & Neumann, 2005; Insel, 2010; Numan & Insel, 2003; Pedersen & Boccia, 2002). Central administration of oxytocin elicits complete maternal behavior in virgin rats (Pedersen, Ascher, Monroe, & Prange, 1982), while disruption of the PVN, where many of the OT neurons reside, prevents the activation of maternal behavior (Insel & Harbaugh, 1989). In addition, mothers that display high pup licking/grooming also show increased oxytocin expression in the medial preoptic area and PVN (Shahrokh, Zhang, Diorio, Gratton, & Meaney, 2010). In prairie voles, virgin females that display maternal behavior exhibit a higher oxytocin receptor density in the nucleus accumbens (Olazabal & Young, 2006), an important source of dopamine involved in the brain reward circuitry. These studies indicate that the activity of oxytocin neurons may be modulated by environmental factors to regulate maternal behavior.

Stress and elevated circulating glucocorticoid levels have significant effects on maternal behavior. In lactating rats, the synthetic glucocorticoid dexamethasone reduces maternal behavior in dams, as evidenced by an increased latency to build nesting and to retrieve the first pup, decreased pup weight gain, and reduced time spent in the arched-back nursing position and licking the pups (Vilela & Giusti-Paiva, 2011; Vilela, Ruginsk, de Melo, & Giusti-Paiva, 2013). In addition, dexamethasone treatment also reduced maternal aggression and increased maternal anxiety, evidenced respectively by an increased latency to attack a male intruder and an increase in anxiety-like behavior in the elevated plus maze and open field test (Vilela et al., 2013). Dexamethasone administration also reduced oxytocin and prolactin secretion during lactation (Vilela & Giusti-Paiva, 2011).

As discussed above, glucocorticoid administration reduced the parameters of maternal care in lactating females (Vilela & Giusti-Paiva, 2011; Vilela et al., 2013). However, pretreatment of the female with the CB_1 receptor antagonist AM251 reversed the glucocorticoid-triggered reduction of maternal behavior (Vilela et al., 2013), suggesting that the glucocorticoid effect on maternal behavior is mediated by endocannabinoid release. To investigate the effect of glucocorticoids on synaptic inputs to oxytocin neurons, we recorded the excitatory and inhibitory synaptic currents in putative magnocellular neuroendocrine cells in the PVN and SON from male rats (Di, Malcher-Lopes, et al., 2005). We found that glucocorticoids significantly decreased excitatory synaptic inputs and increased inhibitory synaptic inputs to magnocellular neurons by the activation of an unknown membrane-associated glucocorticoid receptor. The rapid glucocorticoid effect was insensitive to classical type I and type II corticosteroid receptor antagonists, but was blocked by inhibiting postsynaptic G-protein activity, suggesting it acted via a postsynaptic G protein signaling mechanism and synthesis of a retrograde messenger. Glucocorticoids caused a significant increase in both AEA and 2-AG levels in hypothalamic slices, and the rapid modulation of excitatory synaptic inputs by dexamethasone was blocked by CB_1 receptor antagonists/inverse agonists and mimicked and occluded by CB_1 agonists. These results suggested that a glucocorticoid-induced retrograde release of eCB was responsible for the inhibitory effect of glucocorticoid on glutamatergic synaptic transmission. On the other hand, glucocorticoids facilitated GABAergic transmission by activating nitric oxide production at GABAergic synapses (Di, Maxson, Franco, & Tasker, 2009), resulting in an overall inhibitory effect on the activity of oxytocin neurons by glucocorticoids in adult male rats. Future studies are required to determine the mechanism of glucocorticoid modulation of oxytocin neurons in lactating females.

Glucocorticoids were also found to rapidly suppress reproductive clasping behavior in the male salamander (Taricha) by the activation of eCB signaling (Coddington, Lewis, Rose, & Moore, 2007). Acute confinement stress or corticosterone administration suppressed clasping behavior in male Taricha (Moore & Miller, 1984). The suppressive effect of glucocorticoid was rapid, taking place 5–7 min after systemic injection of corticosterone (Orchinik, Murray, & Moore, 1991). In addition, the firing activities of neurons in the rostromedial medulla that respond to sensory stimulation during courtship are rapidly inhibited by corticosterone (Rose, Marrs, & Moore, 1998; Rose, Moore, & Orchinik, 1993). Intraperitoneal injection of a CB_1 receptor agonist caused a similar suppression of clasping behavior in male Taricha (Soderstrom, Leid, Moore, & Murray, 2000), and pretreatment with the CB_1 receptor antagonist abolished the stress- and corticosterone-induced suppression of clasping behavior, suggesting that eCBs may mediate stress-induced suppression of reproductive behavior (Coddington et al., 2007). The corticosterone suppression of neuronal firing in response to cloacal stimulation was blocked by a CB_1 receptor antagonist. These results together suggest that the rapid suppression of sexual behavior in Taricha by stress and glucocorticoid is mediated by eCB release.

REFERENCES

Abel, E. L. (1975). Cannabis: Effects on hunger and thirst. *Behavioral Biology, 15*(3), 255–281.

Ahima, R. S., & Osei, S. Y. (2004). Leptin signaling. *Physiology and Behavior, 81*(2), 223–241. http://doi.org/10.1016/j.physbeh.2004.02.014.

Ames, F. (1958). A clinical and metabolic study of acute intoxication with Cannabis sativa and its role in the model psychoses. *The Journal of Mental Science, 104*(437), 972–999.

Arevalo, C., de Miguel, R., & Hernandez-Tristan, R. (2001). Cannabinoid effects on anxiety-related behaviours and hypothalamic neurotransmitters. *Pharmacology, Biochemistry, and Behavior, 70*(1), 123–131. Retrieved from, http://www.ncbi.nlm.nih.gov/pubmed/11566149.

Atasoy, D., Betley, J. N., Su, H. H., & Sternson, S. M. (2012). Deconstruction of a neural circuit for hunger. *Nature, 488*(7410), 172–177. http://doi.org/10.1038/nature11270.

Atkinson, H. C., Leggett, J. D., Wood, S. A., Castrique, E. S., Kershaw, Y. M., & Lightman, S. L. (2010). Regulation of the hypothalamic-pituitary-adrenal axis circadian rhythm by endocannabinoids is sexually diergic. *Endocrinology, 151*(8), 3720–3727. http://doi.org/10.1210/en.2010-0101.

Bacci, A., Huguenard, J. R., & Prince, D. A. (2004). Long-lasting self-inhibition of neocortical interneurons mediated by endocannabinoids. *Nature, 431*(7006), 312–316. Retrieved from, http://www.ncbi.nlm.nih.gov/pubmed/15372034.

Balland, E., Dam, J., Langlet, F., Caron, E., Steculorum, S., Messina, A., et al. (2014). Hypothalamic tanycytes are an ERK-gated conduit for leptin into the brain. *Cell Metabolism, 19*(2), 293–301. http://doi.org/10.1016/j.cmet.2013.12.015.

Bedse, G., Colangeli, R., Lavecchia, A. M., Romano, A., Altieri, F., Cifani, C., et al. (2014). Role of the basolateral amygdala in mediating the effects of the fatty acid amide hydrolase inhibitor URB597 on HPA axis response to stress. *European Neuropsychopharmacology, 24*(9), 1511–1523. http://doi.org/10.1016/j.euroneuro.2014.07.005.

Bell, F. R., Dennis, B., & Sly, J. (1979). A study of olfaction and gustatory senses in sheep after olfactory bulbectomy. *Physiology & Behavior, 23*(5), 919–924.

Bichet, D. G., Arthus, M. F., Lonergan, M., Hendy, G. N., Paradis, A. J., Fujiwara, T. M., et al. (1993). X-linked nephrogenic diabetes insipidus mutations in North America and the Hopewell hypothesis. *The Journal of Clinical Investigation, 92*(3), 1262–1268. http://doi.org/10.1172/JCI116698.

Bisogno, T., Howell, F., Williams, G., Minassi, A., Cascio, M. G., Ligresti, A., et al. (2003). Cloning of the first sn1-DAG lipases points to the spatial and temporal regulation of endocannabinoid signaling in the brain. *The Journal of Cell Biology, 163*(3), 463–468. http://doi.org/10.1083/jcb.200305129.

Bisogno, T., Melck, D., Petrocellis, L., & Marzo, V. (2008). Phosphatidic acid as the biosynthetic precursor of the endocannabinoid 2-arachidonoylglycerol in intact mouse neuroblastoma cells stimulated with ionomycin. *Journal of Neurochemistry, 72*(5), 2113–2119. Retrieved from, http://doi.wiley.com/10.1046/j.1471-4159.1999.0722113.x.

Bisset, G. W., & Chowdrey, H. S. (1988). Control of release of vasopressin by neuroendocrine reflexes. *Quarterly Journal of Experimental Physiology (Cambridge, England), 73*(6), 811–872.

Bloch, E., Thysen, B., Morrill, G. A., Gardner, E., & Fujimoto, G. (1978). Effects of cannabinoids on reproduction and development. *Vitamins and Hormones, 36*, 203–258.

Bosch, O. J., Meddle, S. L., Beiderbeck, D. I., Douglas, A. J., & Neumann, I. D. (2005). Brain oxytocin correlates with maternal aggression: Link to anxiety. *The Journal of Neuroscience*, *25*(29), 6807–6815. http://doi.org/10.1523/jneurosci.1342-05.2005.

Boychuk, C. R., Zsombok, A., Tasker, J. G., & Smith, B. N. (2013). Rapid glucocorticoid-induced activation of TRP and CB1 receptors causes biphasic modulation of glutamate release in gastric-related hypothalamic preautonomic neurons. *Frontiers in Neuroscience*, 7, 3. http://doi.org/10.3389/fnins.2013.0000.

Brown, W. H., Gillum, M. P., Lee, H.-Y., Camporez, J. P. G., Zhang, X.-M., Jeong, J. K., et al. (2012). Fatty acid amide hydrolase ablation promotes ectopic lipid storage and insulin resistance due to centrally mediated hypothyroidism. *Proceedings of the National Academy of Sciences*, *109*(37), 14966–14971. http://doi.org/10.1073/pnas.1212887109.

Carvalho, A. F., & Van Bockstaele, E. J. (2012). Cannabinoid modulation of noradrenergic circuits: Implications for psychiatric disorders. *Progress in Neuro-Psychopharmacology & Biological Psychiatry*, *38*(1), 59–67. Retrieved from, http://www.pubmedcentral.nih.gov/articlerender.fcgi?artid=3351574&tool=pmcentrez&rendertype=abstract.

Castelli, M. P., Piras, A. P., Melis, T., Succu, S., Sanna, F., Melis, M. R., et al. (2007). Cannabinoid CB1 receptors in the paraventricular nucleus and central control of penile erection: Immunocytochemistry, autoradiography and behavioral studies. *Neuroscience*, *147*(1), 197–206. Retrieved from, http://www.sciencedirect.com/science/article/pii/S0306452207002497.

Caterina, M. J., Schumacher, M. A., Tominaga, M., Rosen, T. A., Levine, J. D., & Julius, D. (1997). The capsaicin receptor: A heat-activated ion channel in the pain pathway. *Nature*, *389*(6653), 816–824. http://doi.org/10.1038/39807.

Chakravarty, I., Shah, P. G., Sheth, A. R., & Ghosh, J. J. (1979). Mode of action of delta-9-tetrahydrocannabinol on hypothalamo-pituitary function in adult female rats. *Journal of Reproduction and Fertility*, *57*(1), 113–115.

Chaperon, F., & Thiebot, M. H. H. (1999). Behavioral effects of cannabinoid agents in animals. *Critical Reviews in Neurobiology*, *13*(3), 243–281. Retrieved from, http://www.ncbi.nlm.nih.gov/pubmed/10803637.

Ciura, S., & Bourque, C. W. (2006). Transient receptor potential vanilloid 1 is required for intrinsic osmoreception in organum vasculosum lamina terminalis neurons and for normal thirst responses to systemic hyperosmolality. *The Journal of Neuroscience*, *26*(35), 9069–9075. http://doi.org/10.1523/JNEUROSCI.0877-06.2006.

Coddington, E., Lewis, C., Rose, J. D., & Moore, F. L. (2007). Endocannabinoids mediate the effects of acute stress and corticosterone on sex behavior. *Endocrinology*, *148*(2), 493–500. http://doi.org/10.1210/en.2006-0740.

Cone, E. J., Johnson, R. E., Moore, J. D., & Roache, J. D. (1986). Acute effects of smoking marijuana on hormones, subjective effects and performance in male human subjects. *Pharmacology, Biochemistry, and Behavior*, *24*(6), 1749–1754.

Cota, D. (2007). CB1 receptors emerging evidence for central and peripheral mechanisms that regulate energer balance. *Diabetes/Metabolism Research and Reviews*, *26*(6), 446–447. http://doi.org/10.1002/dmrr.

Cota, D., Cota, D., Marsicano, G., Marsicano, G., Tschöp, M., Tschöp, M., et al. (2003). The endogenous cannabinoid. *Internal Medicine*, *112*(3), 423–431. http://doi.org/10.1172/JCI200317725.

Cowley, M. A., Smith, R. G., Diano, S., Tschöp, M., Pronchuk, N., Grove, K. L., et al. (2003). The distribution and mechanism of action of ghrelin in the CNS demonstrates a novel hypothalamic circuit regulating energy homeostasis. *Neuron*, *37*(4), 649–661. http://doi.org/10.1016/S0896-6273(03)00063-1.

Darlington, D. N., Chew, G., Ha, T., Keil, L. C., & Dallman, M. F. (1990). Corticosterone, but not glucose, treatment enables fasted adrenalectomized rats to survive moderate hemorrhage. *Endocrinology*, *127*(2), 766–772. http://doi.org/10.1210/endo-127-2-766.

Deli, L., Wittmann, G., Kalló, I., Lechan, R. M., Watanabe, M., Liposits, Z., et al. (2009). Type 1 cannabinoid receptor-containing axons innervate hypophysiotropic thyrotropin-releasing hormone-synthesizing neurons. *Endocrinology*, *150*(1), 98–103. http://doi.org/10.1210/en.2008-0330.

Denis, G., John, F., & Michael, W. (1999). Leptin receptor mRNA identifies a subpopulation of neuropeptide Y neurons activated by fasting in rat hypothalamus. *Diabetes*, *48*, 828–833.

Di Marzo, V., Bisogno, T., Melck, D., Ross, R., Brockie, H., Stevenson, L., et al. (1998). Interactions between synthetic vanilloids and the endogenous cannabinoid system. *FEBS Letters*, *436*(3), 449–454. Retrieved from, http://www.ncbi.nlm.nih.gov/pubmed/9801167.

Di Marzo, V., Fontana, A., Cadas, H., Schinelli, S., Cimino, G., Schwartz, J. C., et al. (1994). Formation and inactivation of endogenous cannabinoid anandamide in central neurons. *Nature*, *372*(6507), 686–691. http://doi.org/10.1038/372686a0.

Di Marzo, V., Goparaju, S. K., Wang, L., Liu, J., Bátkai, S., Járai, Z., et al. (2001). Leptin-regulated endocannabinoids are involved in maintaining food intake. *Nature*, *410*(6830), 822–825. http://doi.org/10.1038/35071088.

Di Marzo, V., & Matias, I. (2005). Endocannabinoid control of food intake and energy balance. *Nature Neuroscience*, *8*(5), 585–589. http://doi.org/10.1038/nn1457.

Di, S., Boudaba, C., Popescu, I. R., Weng, F.-J., Harris, C., Marcheselli, V. L., et al. (2005). Activity-dependent release and actions of endocannabinoids in the rat hypothalamic supraoptic nucleus. *The Journal of Physiology*, *569*(Pt. 3), 751–760. http://doi.org/10.1113/jphysiol.2005.097477.

Di, S., Malcher-Lopes, R., Halmos, K. C., & Tasker, J. G. (2003). Nongenomic glucocorticoid inhibition via endocannabinoid release in the hypothalamus: A fast feedback mechanism. *The Journal of Neuroscience*, *23*(12), 4850–4857. http://doi.org/23/12/4850.

Di, S., Malcher-Lopes, R., Marcheselli, V. L., Bazan, N. G., & Tasker, J. G. (2005). Rapid glucocorticoid-mediated endocannabinoid release and opposing regulation of glutamate and gamma-aminobutyric acid inputs to hypothalamic magnocellular neurons. *Endocrinology*, *146*(10), 4292–4301. http://doi.org/10.1210/en.2005-0610.

Di, S., Maxson, M. M., Franco, A., & Tasker, J. G. (2009). Glucocorticoids regulate glutamate and GABA synapse-specific retrograde transmission via divergent nongenomic signaling pathways. *The Journal of Neuroscience*, *29*(2), 393–401. http://doi.org/10.1523/JNEUROSCI.4546-08.2009.

Di, S., Popescu, I. R., & Tasker, J. G. (2013). Glial control of endocannabinoid heterosynaptic modulation in hypothalamic magnocellular neuroendocrine cells. *The Journal of Neuroscience*, *33*(46), 18331–18342. http://doi.org/10.1523/JNEUROSCI.2971-12.2013.

Doyon, C., Denis, R. G., Baraboi, E. D., Samson, P., Lalonde, J., Deshaies, Y., et al. (2006). Effects of rimonabant (SR141716) on fasting-induced hypothalamic-pituitary-adrenal axis and neuronal activation in lean and obese Zucker rats. *Diabetes*, *55*(12), 3403–3410. http://doi.org/10.2337/db06-0504.

Dubreucq, S., Matias, I., Cardinal, P., Haring, M., Lutz, B., Marsicano, G., et al. (2012). Genetic dissection of the role of cannabinoid type-1 receptors in the emotional consequences of repeated social stress in mice. *Neuropsychopharmacology*, *37*(8), 1885–1900. http://doi.org/10.1038/npp.2012.36.

Elmquist, J. K., Coppari, R., Balthasar, N., Ichinose, M., & Lowell, B. B. (2005). Identifying hypothalamic pathways controlling food intake, body weight, and glucose homeostasis. *Journal of Comparative Neurology*, *493*(1), 63–71. http://doi.org/10.1002/cne.20786.

Evanson, N. K., Tasker, J. G., Hill, M. N., Hillard, C. J., & Herman, J. P. (2010). Fast feedback inhibition of the HPA axis by glucocorticoids is mediated by endocannabinoid signaling. *Endocrinology*, *151*(10), 4811–4819. http://doi.org/10.1210/en.2010-0285.

Farkas, I., Kallo, I., Deli, L., Vida, B., Hrabovszky, E., Fekete, C., et al. (2010). Retrograde endocannabinoid signaling reduces GABAergic synaptic transmission to gonadotropin-releasing hormone neurons. *Endocrinology*, *151*(12), 5818–5829. http://doi.org/10.1210/en.2010-0638.

Farkas, I., Vastagh, C., Sarvari, M., & Liposits, Z. (2013). Ghrelin decreases firing activity of gonadotropin-releasing hormone (GnRH) neurons in an estrous cycle and endocannabinoid signaling dependent manner. *PLoS One*, *8*(10), e78178. http://doi.org/10.1371/journal.pone.0078178.

Fekete, C., & Lechan, R. M. (2007). Negative feedback regulation of hypophysiotropic thyrotropin-releasing hormone (TRH) synthesizing neurons: Role of neuronal afferents and type 2 deiodinase. *Frontiers in Neuroendocrinology*, *28*(2–3), 97–114. http://doi.org/10.1016/j.yfrne.2007.04.002.

Fernandez-Fernandez, R., Tena-Sempere, M., Navarro, M., Barreiro, M. L., Castellano, J. M., Aguilar, E., et al. (2005). Effects of ghrelin upon gonadotropin-releasing hormone and gonadotropin secretion in adult female rats: In vivo and in vitro studies. *Neuroendocrinology*, *82*(5–6), 245–255. http://doi.org/10.1159/000092753.

Furuta, M., Funabashi, T., & Kimura, F. (2001). Intracerebroventricular administration of ghrelin rapidly suppresses pulsatile luteinizing hormone secretion in ovariectomized rats. *Biochemical and Biophysical Research Communications*, *288*(4), 780–785. http://doi.org/10.1006/bbrc.2001.5854.

Gammon, C. M., Freeman, G. M., Jr., Xie, W., Petersen, S. L., & Wetsel, W. C. (2005). Regulation of gonadotropin-releasing hormone secretion by cannabinoids. *Endocrinology*, *146*(10), 4491–4499. http://doi.org/10.1210/en.2004-1672.

Gaoni, Y., & Mechoulam, R. (1964). Isolation, structure, and partial synthesis of an active constituent of hashish. *Journal of the American Chemical Society*, *86*(8), 1646–1647. http://doi.org/10.1021/ja01062a046.

Giuffrida, A., Parsons, L. H., Kerr, T. M., Rodríguez de Fonseca, F., Navarro, M., & Piomelli, D. (1999). Dopamine activation of endogenous cannabinoid signaling in dorsal striatum. *Nature Neuroscience*, *2*(4), 358–363. http://doi.org/10.1038/7268.

Glanowska, K. M., & Moenter, S. M. (2011). Endocannabinoids and prostaglandins both contribute to GnRH neuron-GABAergic afferent local feedback circuits. *Journal of Neurophysiology*, *106*(6), 3073–3081. http://doi.org/10.1152/jn.00046.2011.

Gong, J.-P., Onaivi, E. S., Ishiguro, H., Liu, Q.-R., Tagliaferro, P. A., Brusco, A., et al. (2006). Cannabinoid CB2 receptors: Immunohistochemical localization in rat brain. *Brain Research*, *107*(1), 10–23. Retrieved from, http://www.ncbi.nlm.nih.gov/pubmed/16472786.

Gorzalka, B. B., & Dang, S. S. (2012). Minireview: Endocannabinoids and gonadal hormones: Bidirectional interactions in physiology and behavior. *Endocrinology*, *153*(3), 1016–1024. http://doi.org/10.1210/en.2011-1643.

Gray, J. M., Vecchiarelli, H. A., Morena, M., Lee, T. T. Y., Hermanson, D. J., Kim, A. B., et al. (2015). Corticotropin-releasing hormone drives anandamide hydrolysis in the amygdala to promote anxiety. *The Journal of Neuroscience*, *35*(9), 3879–3892. http://doi.org/10.1523/JNEUROSCI.2737-14.2015.

Gutierrez, J. A., Solenberg, P. J., Perkins, D. R., Willency, J. A., Knierman, M. D., Jin, Z., et al. (2008). Ghrelin octanoylation mediated by an orphan lipid transferase. *Proceedings of the National Academy of Sciences of the United States of America*, *105*(17), 6320–6325. http://doi.org/10.1073/pnas.0800708105.

Haam, J., Halmos, K. C., Di, S., & Tasker, J. G. (2014). Nutritional state-dependent ghrelin activation of vasopressin neurons via retrograde trans-neuronal-glial stimulation of excitatory GABA circuits. *Journal of Neuroscience*, *34*(18), 6201–6213. http://doi.org/10.1523/JNEUROSCI.3178-13.2014.

Haanwinckel, M. A., Elias, L. L., Favaretto, A. L., Gutkowska, J., McCann, S. M., & Antunes-Rodrigues, J. (1995). Oxytocin mediates atrial natriuretic peptide release and natriuresis after volume expansion in the rat. *Proceedings of the National Academy of Sciences of the United States of America, 92*(17), 7902–7906.

Hahn, T. M., Breininger, J. F., Baskin, D. G., & Schwartz, M. W. (1998). Coexpression of Agrp and NPY in fasting-activated hypothalamic neurons. *Nature Neuroscience, 1*(4), 271–272. http://doi.org/10.1038/1082.

Haller, J., Bakos, N., Szirmay, M., Ledent, C., & Freund, T. F. (2002). The effects of genetic and pharmacological blockade of the CB1 cannabinoid receptor on anxiety. *The European Journal of Neuroscience, 16*(7), 1395–1398. Retrieved from, http://www.ncbi.nlm.nih.gov/pubmed/12405999.

Haller, J., Varga, B., Ledent, C., Barna, I., & Freund, T. F. (2004a). Context-dependent effects of CB1 cannabinoid gene disruption on anxiety-like and social behaviour in mice. *European Journal of Neuroscience, 19*(7), 1906–1912. http://doi.org/10.1111/j.1460-9568.2004.03293.x.

Haller, J., Varga, B., Ledent, C., & Freund, T. F. (2004b). CB1 cannabinoid receptors mediate anxiolytic effects: Convergent genetic and pharmacological evidence with CB1-specific agents. *Behavioural Pharmacology, 15*(4), 299–304. http://doi.org/10.1097/01.fbp.0000135704.56422.40.

Hao, S., Avraham, Y., Mechoulam, R., & Berry, E. M. (2000). Low dose anandamide affects food intake, cognitive function, neurotransmitter and corticosterone levels in diet-restricted mice. *European Journal of Pharmacology, 392*(3), 147–156. http://doi.org/10.1016/S0014-2999(00)00059-5.

Hashimotodani, Y., Ohno-Shosaku, T., Tanimura, A., Kita, Y., Sano, Y., Shimizu, T., et al. (2013). Acute inhibition of diacylglycerol lipase blocks endocannabinoid-mediated retrograde signalling: Evidence for on-demand biosynthesis of 2-arachidonoylglycerol. *The Journal of Physiology, 591*(Pt. 19), 4765–4776. Retrieved from, http://www.pubmedcentral.nih.gov/articlerender.fcgi?artid=3800453&tool=pmcentrez&rendertype=abstract.

Heinrichs, S. C., Menzaghi, F., Pich, E. M., Hauger, R. L., & Koob, G. F. (1993). Corticotropin-releasing factor in the paraventricular nucleus modulates feeding induced by neuropeptide Y. *Brain Research, 611*(619), 18–24.

Herkenham, M., Lynn, A. B., Johnson, M. R., Melvin, L. S., de Costa, B. R., & Rice, K. C. (1991). Characterization and localization of cannabinoid receptors in rat brain: A quantitative in vitro autoradiographic study. *The Journal of Neuroscience, 11*(2), 563–583. Retrieved from, http://www.ncbi.nlm.nih.gov/pubmed/1992016.

Herman, J. P., Adams, D., & Prewitt, C. (1995). Regulatory changes in neuroendocrine stress-integrative circuitry produced by a variable stress paradigm. *Neuroendocrinology, 61*(2), 180–190. Retrieved from, http://www.ncbi.nlm.nih.gov/pubmed/7753337.

Hill, M. N., Carrier, E. J., McLaughlin, R. J., Morrish, A. C., Meier, S. E., Hillard, C. J., et al. (2008). Regional alterations in the endocannabinoid system in an animal model of depression: Effects of concurrent antidepressant treatment. *Journal of Neurochemistry, 106*(6), 2322–2336. http://doi.org/10.1111/j.1471-4159.2008.05567.x.

Hill, M. N., McLaughlin, R. J., Morrish, A. C., Viau, V., Floresco, S. B., Hillard, C. J., et al. (2009). Suppression of amygdalar endocannabinoid signaling by stress contributes to activation of the hypothalamic-pituitary-adrenal axis. *Neuropsychopharmacology, 34*(13), 2733–2745. http://doi.org/10.1038/npp.2009.114.

Hill, M. N., McLaughlin, R. J., Pan, B., Fitzgerald, M. L., Roberts, C. J., Lee, T. T. Y., et al. (2011). Recruitment of prefrontal cortical endocannabinoid signaling by glucocorticoids contributes to termination of the stress response. *The Journal of Neuroscience, 31*(29), 10506–10515. http://doi.org/10.1523/JNEUROSCI.0496-11.2011.

Hill, M. N., Patel, S., Campolongo, P., Tasker, J. G., Wotjak, C. T., & Bains, J. S. (2010). Functional interactions between stress and the endocannabinoid system: From synaptic signaling to behavioral output. *The Journal of Neuroscience*, *30*(45), 14980–14986. http://doi.org/10.1523/JNEUROSCI.4283-10.2010.

Hill, M. N., Patel, S., Carrier, E. J., Rademacher, D. J., Ormerod, B. K., Hillard, C. J., et al. (2005). Downregulation of endocannabinoid signaling in the hippocampus following chronic unpredictable stress. *Neuropsychopharmacology*, *30*(3), 508–515. Retrieved from, http://www.ncbi.nlm.nih.gov/pubmed/15525997.

Hill, M. N., & Tasker, J. G. (2012). Endocannabinoid signaling, glucocorticoid-mediated negative feedback, and regulation of the hypothalamic-pituitary-adrenal axis. *Neuroscience*, *204*, 5–16. http://doi.org/10.1016/j.neuroscience.2011.12.030.

Hillard, C. J. (2000). Biochemistry and pharmacology of the endocannabinoids arachidonylethanolamide and 2-arachidonylglycerol. *Prostaglandins & Other Lipid Mediators*, *61*(1–2), 3–18. Retrieved from, http://www.ncbi.nlm.nih.gov/pubmed/10785538.

Hirasawa, M., Schwab, Y., Natah, S., Hillard, C. J., Mackie, K., Sharkey, K. A., et al. (2004). Dendritically released transmitters cooperate via autocrine and retrograde actions to inhibit afferent excitation in rat brain. *The Journal of Physiology*, *559*(Pt. 2), 611–624. Retrieved from, http://www.pubmedcentral.nih.gov/articlerender.fcgi?artid=1665137&tool=pmcentrez&rendertype=abstract.

Hong, S., Zheng, G., Wu, X., Snider, N. T., Owyang, C., & Wiley, J. W. (2011). Corticosterone mediates reciprocal changes in CB 1 and TRPV1 receptors in primary sensory neurons in the chronically stressed rat. *Gastroenterology*. *140*(2). http://dx.doi.org/10.1053/j.gastro.2010.11.003 http://doi.org/10.1053/j.gastro.2010.11.003, 627–637 e4.

Hotta, M., Shibasaki, T., Yamauchi, N., Ohno, H., Benoit, R., Ling, N., et al. (1991). The effects of chronic central administration of corticotropin-releasing factor on food intake, body weight, and hypothalamic-pituitary-adrenocortical hormones. *Life Sciences*, *48*(15), 1483–1491. Retrieved from, http://www.ncbi.nlm.nih.gov/pubmed/1849215.

Hozawa, S., Holtzman, E. J., & Ausiello, D. A. (1996). cAMP motifs regulating transcription in the aquaporin 2 gene. *The American Journal of Physiology*, *270*(6 Pt. 1), C1695–C1702.

Huang, P., Chandra, V., & Rastinejad, F. (2010). Structural overview of the nuclear receptor superfamily: Insights into physiology and therapeutics. *Annual Review of Physiology*, *72*, 247–272. http://doi.org/10.1146/annurev-physiol-021909-135917.

Huang, G., & Woolley, C. S. (2012). Estradiol acutely suppresses inhibition in the hippocampus through a sex-specific endocannabinoid and mGluR-dependent mechanism. *Neuron*, *74*(5), 801–808. http://doi.org/10.1016/j.neuron.2012.03.035.

Insel, T. R. (2010). The challenge of translation in social neuroscience: A review of oxytocin, vasopressin, and affiliative behavior. *Neuron*, *65*(6), 768–779. http://doi.org/10.1016/j.neuron.2010.03.005.

Insel, T. R., & Harbaugh, C. R. (1989). Lesions of the hypothalamic paraventricular nucleus disrupt the initiation of maternal behavior. *Physiology and Behavior*, *45*(5), 1033–1041.

Jo, Y. H., Chen, Y. J. J., Chua, S. C., Talmage, D. A., & Role, L. W. (2005). Integration of endocannabinoid and leptin signaling in an appetite-related neural circuit. *Neuron*, *48*(6), 1055–1066. http://doi.org/10.1016/j.neuron.2005.10.021.

Kaiya, H., Kangawa, K., & Miyazato, M. (2013). Molecular evolution of GPCRs: Ghrelin/ghrelin receptors. *Journal of Molecular Endocrinology*, *52*(3), 1–41. http://doi.org/10.1530/JME-13-0175.

Karst, H., Berger, S., Erdmann, G., Schütz, G., & Joëls, M. (2010). Metaplasticity of amygdalar responses to the stress hormone corticosterone. *Proceedings of the National Academy of Sciences of the United States of America*, *107*(32), 14449–14454. Retrieved from, http://www.pubmedcentral.nih.gov/articlerender.fcgi?artid=2922581&tool=pmcentrez&rendertype=abstract.

Keller-Wood, M., Shinsako, J., & Dallman, M. F. (1984). Interaction between stimulus intensity and corticosteroid feedback in control of ACTH. *The American Journal of Physiology*, *247*(4 Pt. 1), E489–E494.

Kim, J., & Alger, B. E. (2010). Reduction in endocannabinoid tone is a homeostatic mechanism for specific inhibitory synapses. *Nature Neuroscience*, *13*(5), 592–600. http://doi.org/10.1038/nn.2517.

Kim, J., Isokawa, M., Ledent, C., & Alger, B. E. (2002). Activation of muscarinic acetylcholine receptors enhances the release of endogenous cannabinoids in the hippocampus. *The Journal of Neuroscience*, *22*(23), 10182–10191. Retrieved from, http://www.ncbi.nlm.nih.gov/pubmed/12451119.

Kirkham, T. C., Williams, C. M., Fezza, F., & Di Marzo, V. (2002). Endocannabinoid levels in rat limbic forebrain and hypothalamus in relation to fasting, feeding and satiation: Stimulation of eating by 2-arachidonoyl glycerol. *British Journal of Pharmacology*, *136*(4), 550–557. http://doi.org/10.1038/sj.bjp.0704767.

Kluge, M., Schussler, P., Schmidt, D., Uhr, M., & Steiger, A. (2012). Ghrelin suppresses secretion of luteinizing hormone (LH) and follicle-stimulating hormone (FSH) in women. *The Journal of Clinical Endocrinology and Metabolism*, *97*(3), E448–E451. http://doi.org/10.1210/jc.2011-2607.

Kojima, M., Hosoda, H., Date, Y., Nakazato, M., Matsuo, H., & Kangawa, K. (1999). Ghrelin is a growth-hormone-releasing acylated peptide from stomach. *Nature*, *402*(6762), 656–660. http://doi.org/10.1038/45230.

Kola, B., Farkas, I., Christ-Crain, M., Wittmann, G., Lolli, F., Amin, F., et al. (2008). The orexigenic effect of ghrelin is mediated through central activation of the endogenous cannabinoid system. *PLoS One*, *3*(3), e1797. http://doi.org/10.1371/journal.pone.0001797.

Kola, B., Hubina, E., Tucci, S., Kirkham, T. C., Garcia, E. A., Mitchell, S. E., et al. (2005). Cannabinoids and ghrelin have both central and peripheral metabolic and cardiac effects via AMP-activated protein kinase. *Journal of Biological Chemistry*, *280*(26), 25196–25201. http://doi.org/10.1074/jbc.C500175200.

Kolodny, R. C., Masters, W. H., Kolodner, R. M., & Toro, G. (1974). Depression of plasma testosterone levels after chronic intensive marihuana use. *The New England Journal of Medicine*, *290*(16), 872–874. http://doi.org/10.1056/nejm197404182901602.

Kumar, M. S., & Chen, C. L. (1983). Effect of an acute dose of delta 9-THC on hypothalamic luteinizing hormone releasing hormone and met-enkephalin content and serum levels of testosterone and corticosterone in rats. *Substance and Alcohol Actions/Misuse*, *4*(1), 37–43.

Lanfranco, F., Bonelli, L., Baldi, M., Me, E., Broglio, F., & Ghigo, E. (2008). Acylated ghrelin inhibits spontaneous luteinizing hormone pulsatility and responsiveness to naloxone but not that to gonadotropin-releasing hormone in young men: Evidence for a central inhibitory action of ghrelin on the gonadal axis. *The Journal of Clinical Endocrinology and Metabolism*, *93*(9), 3633–3639. http://doi.org/10.1210/jc.2008-0049.

Lauand, F., Ruginsk, S. G., Rodrigues, H. L. P., Reis, W. L., de Castro, M., Elias, L. L., et al. (2007). Glucocorticoid modulation of atrial natriuretic peptide, oxytocin, vasopressin and Fos expression in response to osmotic, angiotensinergic and cholinergic stimulation. *Neuroscience*, *147*(1), 247–257. http://doi.org/10.1016/j.neuroscience.2007.04.021.

Lebrethon, M. C., Aganina, A., Fournier, M., Gerard, A., Parent, A. S., & Bourguignon, J. P. (2007). Effects of in vivo and in vitro administration of ghrelin, leptin and neuropeptide mediators on pulsatile gonadotrophin-releasing hormone secretion from male rat hypothalamus before and after puberty. *Journal of Neuroendocrinology*, *19*(3), 181–188. http://doi.org/10.1111/j.1365-2826.2006.01518.x.

Lechan, R. M., & Fekete, C. (2006). The TRH neuron: A hypothalamic integrator of energy metabolism. *Progress in Brain Research, 153*, 209–235. http://doi.org/10.1016/S0079-6123(06)53012-2.

Leung, D., Saghatelian, A., Simon, G. M., & Cravatt, B. F. (2006). Inactivation of N-acyl phosphatidylethanolamine phospholipase D reveals multiple mechanisms for the biosynthesis of endocannabinoids. *Biochemistry, 45*(15), 4720–4726. http://doi.org/10.1021/bi0601631.

Liu, J., Wang, L., Harvey-White, J., Osei-Hyiaman, D., Razdan, R., Gong, Q., et al. (2006). A biosynthetic pathway for anandamide. *Proceedings of the National Academy of Sciences of the United States of America, 103*(36), 13345–13350. http://doi.org/10.1073/pnas.0601832103.

Mackie, K. (2005). Distribution of cannabinoid receptors in the central and peripheral nervous system. *Handbook of Experimental Pharmacology, 168*, 299–325.

Mackie, K., Lai, Y., Westenbroek, R., & Mitchell, R. (1995). Cannabinoids activate an inwardly rectifying potassium conductance and inhibit Q-type calcium currents in AtT20 cells transfected with rat brain cannabinoid receptor. *The Journal of Neuroscience, 15*(10), 6552–6561. Retrieved from, http://www.ncbi.nlm.nih.gov/pubmed/7472417.

Malcher-Lopes, R., Di, S., Marcheselli, V. L., Weng, F.-J., Stuart, C. T., Bazan, N. G., et al. (2006). Opposing crosstalk between leptin and glucocorticoids rapidly modulates synaptic excitation via endocannabinoid release. *The Journal of Neuroscience, 26*(24), 6643–6650. http://doi.org/10.1523/JNEUROSCI.5126-05.2006.

Malcher-Lopes, R., Franco, A., & Tasker, J. G. (2009). Glucocorticoids shift arachidonic acid metabolism toward endocannabinoi synthesis: A non genomic anti-inflammatory switch. *European Journal of Pharmacology, 583*(2–3), 322–339. http://doi.org/10.1016/j.ejphar.2007.12.033.Glucocorticoids.

Marco, E. M., Perez-Alvarez, L., Borcel, E., Rubio, M., Guaza, C., Ambrosio, E., et al. (2004). Involvement of 5-HT1A receptors in behavioural effects of the cannabinoid receptor agonist CP 55,940 in male rats. *Behavioral Pharmacology, 15*(1), 21–27. Retrieved from, http://www.ncbi.nlm.nih.gov/pubmed/15075623.

Marinelli, S., Pacioni, S., Bisogno, T., Di Marzo, V., Prince, D. A., Huguenard, J. R., et al. (2008). The endocannabinoid 2-arachidonoylglycerol is responsible for the slow self-inhibition in neocortical interneurons. *The Journal of Neuroscience, 28*(50), 13532–13541. Retrieved from, http://www.pubmedcentral.nih.gov/articlerender.fcgi?artid=2615383&tool=pmcentrez&rendertype=abstract.

Martin-Calderon, J. L., Munoz, R. M., Villanua, M. A., del Arco, I., Moreno, J. L., de Fonseca, F. R., et al. (1998). Characterization of the acute endocrine actions of (−)-11-hydroxy-delta8-tetrahydrocannabinol-dimethylheptyl (HU-210), a potent synthetic cannabinoid in rats. *European Journal of Pharmacology, 344*(1), 77–86. Retrieved from, http://www.ncbi.nlm.nih.gov/pubmed/9580419.

Matsuda, L. A., Lolait, S. J., Brownstein, M. J., Young, A. C., & Bonner, T. I. (1990). Structure of a cannabinoid receptor and functional expression of the cloned cDNA. *Nature, 346*(6284), 561–564. http://doi.org/10.1038/346561a0.

Melis, M., Pistis, M., Perra, S., Muntoni, A. L., Pillolla, G., & Gessa, G. L. (2004). Endocannabinoids mediate presynaptic inhibition of glutamatergic transmission in rat ventral tegmental area dopamine neurons through activation of CB1 receptors. *The Journal of Neuroscience, 24*(1), 53–62. Retrieved from, http://www.ncbi.nlm.nih.gov/pubmed/14715937.

Mendelson, J. H., Mello, N. K., Ellingboe, J., Skupny, A. S., Lex, B. W., & Griffin, M. (1986). Marihuana smoking suppresses luteinizing hormone in women. *The Journal of Pharmacology and Experimental Therapeutics, 237*(3), 862–866.

Menigoz, A., & Boudes, M. (2011). The expression pattern of TRPV1 in brain. *The Journal of Neuroscience, 31*(37), 13025–13027. http://doi.org/10.1523/JNEUROSCI.2589-11.2011.

Moore, F. L., & Miller, L. J. (1984). Stress-induced inhibition of sexual behavior: Corticosterone inhibits courtship behaviors of a male amphibian (Taricha granulosa). *Hormones and Behavior, 18*(4), 400–410.
Munro, S., Thomas, K. L., & Abu-Shaar, M. (1993). Molecular characterization of a peripheral receptor for cannabinoids. *Nature, 365*(6441), 61–65. http://doi.org/10.1038/365061a0.
Murphy, L. L., Steger, R. W., Smith, M. S., & Bartke, A. (1990). Effects of delta-9-tetrahydrocannabinol, cannabinol and cannabidiol, alone and in combinations, on luteinizing hormone and prolactin release and on hypothalamic neurotransmitters in the male rat. *Neuroendocrinology, 52*(4), 316–321.
Nahar, J., Haam, J., Chen, C., Jiang, Z., Glatzer, N. R., Muglia, L. J., et al. (2015). Rapid nongenomic glucocorticoid actions in male mouse hypothalamic neuroendocrine cells are dependent on the nuclear glucocorticoid receptor. *Endocrinology, 156*(8), 2831–2842. http://doi.org/10.1210/en.2015-1273.
Nakazato, M., Murakami, N., Date, Y., Kojima, M., Matsuo, H., Kangawa, K., et al. (2001). A role for ghrelin in the central regulation of feeding. *Nature, 409*(6817), 194–198. http://doi.org/10.1038/35051587.
Norsted, E., Gömüç, B., & Meister, B. (2008). Protein components of the blood–brain barrier (BBB) in the mediobasal hypothalamus. *Journal of Chemical Neuroanatomy, 36*(2), 107–121. http://doi.org/10.1016/j.jchemneu.2008.06.002.
Numan, M., & Insel, T. R. (2003). Hormones, brain and behavior. In G. F. Ball, J. Balthazart, & R. J. Nelson (Eds.), *The neurobiology of parental behavior*. Secaucus, NJ: Springer.
Ohno-Shosaku, T., Tsubokawa, H., Mizushima, I., Yoneda, N., Zimmer, A., & Kano, M. (2002). Presynaptic cannabinoid sensitivity is a major determinant of depolarization-induced retrograde suppression at hippocampal synapses. *The Journal of Neuroscience, 22*(10), 3864–3872. Retrieved from, http://www.ncbi.nlm.nih.gov/pubmed/12019305.
Olazabal, D. E., & Young, L. J. (2006). Oxytocin receptors in the nucleus accumbens facilitate "spontaneous" maternal behavior in adult female prairie voles. *Neuroscience, 141*(2), 559–568. http://doi.org/10.1016/j.neuroscience.2006.04.017.
Oliet, S. H. R., Baimoukhametova, D. V., Piet, R., & Bains, J. S. (2007). Retrograde regulation of GABA transmission by the tonic release of oxytocin and endocannabinoids governs postsynaptic firing. *The Journal of Neuroscience, 27*(6), 1325–1333. http://doi.org/10.1523/JNEUROSCI.2676-06.2007.
Orchinik, M., Murray, T. F., & Moore, F. L. (1991). A corticosteroid receptor in neuronal membranes. *Science, 252*(5014), 1848–1851.
Pagotto, U., Marsicano, G., Fezza, F., Theodoropoulou, M., Grübler, Y., Stalla, J., et al. (2001). Normal human pituitary gland and pituitary adenomas express cannabinoid receptor type 1 and synthesize endogenous cannabinoids: First evidence for a direct role of cannabinoids on hormone modulation at the human pituitary level. *The Journal of Clinical Endocrinology and Metabolism, 86*(6), 2687–2696. http://doi.org/10.1210/jcem.86.6.7565.
Paronis, C. A., Thakur, G. A., Bajaj, S., Nikas, S. P., Vemuri, V. K., Makriyannis, A. M., et al. (2013). Diuretic effects of cannabinoids. *The Journal of Pharmacology and Experimental Therapeutics, 344*(1), 8–14. http://doi.org/10.1124/jpet.112.199331.
Patel, S., Roelke, C. T., Rademacher, D. J., & Hillard, C. J. (2005). Inhibition of restraint stress-induced neural and behavioural activation by endogenous cannabinoid signalling. *The European Journal of Neuroscience, 21*(4), 1057–1069. http://doi.org/10.1111/j.1460-9568.2005.03916.x.
Pedersen, C. A., Ascher, J. A., Monroe, Y. L., & Prange, A. J., Jr. (1982). Oxytocin induces maternal behavior in virgin female rats. *Science, 216*(4546), 648–650.

Pedersen, C. A., & Boccia, M. L. (2002). Oxytocin links mothering received, mothering bestowed and adult stress responses. *Stress*, *5*(4), 259–267. http://doi.org/10.1080/1025389021000037586.

Pelleymounter, M. A., Joppa, M., Carmouche, M., Cullen, M. J., Brown, B., Murphy, B., et al. (2000). Role of corticotropin-releasing factor (CRF) receptors in the anorexic syndrome induced by CRF. *The Journal of Pharmacology and Experimental Therapeutics*, *293*(3), 799–806. Retrieved from, http://www.ncbi.nlm.nih.gov/pubmed/10869378.

Rademacher, D. J., Meier, S. E., Shi, L., Ho, W.-S. S. V., Jarrahian, A., & Hillard, C. J. (2008). Effects of acute and repeated restraint stress on endocannabinoid content in the amygdala, ventral striatum, and medial prefrontal cortex in mice. *Neuropharmacology*, *54*(1), 108–116. http://doi.org/10.1016/j.neuropharm.2007.06.012.

Radley, J. J., Gosselink, K. L., & Sawchenko, P. E. (2009). A discrete GABAergic relay mediates medial prefrontal cortical inhibition of the neuroendocrine stress response. *The Journal of Neuroscience*, *29*(22), 7330–7340. http://doi.org/10.1523/JNEUROSCI.5924-08.2009.

Repaci, A., Gambineri, A., Pagotto, U., & Pasquali, R. (2011). Ghrelin and reproductive disorders. *Molecular and Cellular Endocrinology*, *340*(1), 70–79. http://doi.org/10.1016/j.mce.2011.02.022.

Romero, E. M., Fernandez, B., Sagredo, O., Gomez, N., Uriguen, L., Guaza, C., et al. (2002). Antinociceptive, behavioural and neuroendocrine effects of CP 55,940 in young rats. *Brain Research. Developmental Brain Research*, *136*(2), 85–92. Retrieved from, http://www.ncbi.nlm.nih.gov/pubmed/12101025.

Rose, J. D., Marrs, G. S., & Moore, F. L. (1998). Rapid, corticosterone-induced disruption of medullary sensorimotor integration related to suppression of amplectic clasping in behaving roughskin newts (Taricha granulosa). *Hormones and Behavior*, *34*(3), 268–282. http://doi.org/10.1006/hbeh.1998.1483.

Rose, J. D., Moore, F. L., & Orchinik, M. (1993). Rapid neurophysiological effects of corticosterone on medullary neurons: Relationship to stress-induced suppression of courtship clasping in an amphibian. *Neuroendocrinology*, *57*(5), 815–824.

Rossi, S., De Chiara, V., Musella, A., Kusayanagi, H., Mataluni, G., Bernardi, G., et al. (2008). Chronic psychoemotional stress impairs cannabinoid-receptor-mediated control of GABA transmission in the striatum. *The Journal of Neuroscience*, *28*(29), 7284–7292. http://doi.org/10.1523/JNEUROSCI.5346-07.2008.

Ruginsk, S. G., Uchoa, E. T., Elias, L. L., & Antunes-Rodrigues, J. (2010). CB(1) modulation of hormone secretion, neuronal activation and mRNA expression following extracellular volume expansion. *Experimental Neurology*, *224*(1), 114–122. http://doi.org/10.1016/j.expneurol.2010.03.001.

Ruginsk, S. G., Uchoa, E. T., Elias, L. L., & Antunes-Rodrigues, J. (2012). Cannabinoid CB1 receptor mediates glucocorticoid effects on hormone secretion induced by volume and osmotic changes. *Clinical and Experimental Pharmacology & Physiology*, *39*(2), 151–154. http://doi.org/10.1111/j.1440-1681.2011.05658.x.

Ruginsk, S. G., Uchoa, E. T., Elias, L. L., & Antunes-Rodrigues, J. (2013). Anandamide modulates the neuroendocrine responses induced by extracellular volume expansion. *Clinical and Experimental Pharmacology & Physiology*, *40*(10), 698–705. http://doi.org/10.1111/1440-1681.12155.

Saario, S. M., Savinainen, J. R., Laitinen, J. T., Järvinen, T., & Niemi, R. (2004). Monoglyceride lipase-like enzymatic activity is responsible for hydrolysis of 2-arachidonoylglycerol in rat cerebellar membranes. *Biochemical Pharmacology*, *67*(7), 1381–1387. Retrieved from, http://www.sciencedirect.com/science/article/pii/S0006295203009420.

Sampaio, L. S., Taveira Da Silva, R., Lima, D., Sampaio, C. L. C., Iannotti, F. A., Mazzarella, E., et al. (2014). The endocannabinoid system in renal cells: Regulation

of Na(+) transport by CB1 receptors through distinct cell signalling pathways. *British Journal of Pharmacology*, *172*(19), 4615–4625. http://doi.org/10.1111/bph.13050.
Sapolsky, R. M., Armanini, M. P., Packan, D. R., Sutton, S. W., & Plotsky, P. M. (1990). Glucocorticoid feedback inhibition of adrenocorticotropic hormone secretagogue release. Relationship to corticosteroid receptor occupancy in various limbic sites. *Neuroendocrinology*, *51*(3), 328–336. Retrieved from, http://www.ncbi.nlm.nih.gov/pubmed/2157997.
Schwartz, M. W., Woods, S. C., Porte, D., Jr., Seeley, R. J., & Baskin, D. G. (2000). Control of food intake. *Nature*, *404*, 661–671. http://doi.org/10.1038/35007534.
Shahrokh, D. K., Zhang, T. Y., Diorio, J., Gratton, A., & Meaney, M. J. (2010). Oxytocin-dopamine interactions mediate variations in maternal behavior in the rat. *Endocrinology*, *151*(5), 2276–2286. http://doi.org/10.1210/en.2009-1271.
Soderstrom, K., Leid, M., Moore, F. L., & Murray, T. F. (2000). Behaviroal, pharmacological, and molecular characterization of an amphibian cannabinoid receptor. *Journal of Neurochemistry*, *75*(1), 413–423.
Sofia, R. D., Dixit, B. N., & Barry, H. (1977). The effect of repeated administration of delta9-tetrahydrocannabinol on serotonin metabolism in the rat brain. *Archives Internationales de Pharmacodynamie et de Thérapie*, *229*(1), 52–58.
Soria-Gómez, E., Bellocchio, L., Reguero, L., Lepousez, G., Martin, C., Bendahmane, M., et al. (2014). The endocannabinoid system controls food intake via olfactory processes. *Nature Neuroscience*, *17*(3), 407–415. http://doi.org/10.1038/nn.3647.
Soria-Gómez, E., Massa, F., Bellocchio, L., Rueda-Orozco, P. E., Ciofi, P., Cota, D., et al. (2014). Cannabinoid type-1 receptors in the paraventricular nucleus of the hypothalamus inhibit stimulated food intake. *Neuroscience*, *263*, 46–53. http://doi.org/10.1016/j.neuroscience.2014.01.005.
Sudbury, J. R., Ciura, S., Sharif-Naeini, R., & Bourque, C. W. (2010). Osmotic and thermal control of magnocellular neurosecretory neurons—Role of an N-terminal variant of trpv1. *The European Journal of Neuroscience*, *32*(12), 2022–2030. http://doi.org/10.1111/j.1460-9568.2010.07512.x.
Sumislawski, J. J., Ramikie, T. S., & Patel, S. (2011). Reversible gating of endocannabinoid plasticity in the amygdala by chronic stress: A potential role for monoacylglycerol lipase inhibition in the prevention of stress-induced behavioral adaptation. *Neuropsychopharmacology*, *36*(13), 2750–2761. http://doi.org/10.1038/npp.2011.166.
Sun, Y.-X., Tsuboi, K., Okamoto, Y., Tonai, T., Murakami, M., Kudo, I., et al. (2004). Biosynthesis of anandamide and N-palmitoylethanolamine by sequential actions of phospholipase A2 and lysophospholipase D. *The Biochemical Journal*, *380*(Pt. 3), 749–756. http://doi.org/10.1042/BJ20040031.
Tasker, J. G. (2006). Rapid glucocorticoid actions in the hypothalamus as a mechanism of homeostatic integration. *Obesity (Silver Spring, Md)*, *14*(Suppl. 5 (August)), 259S–265S. http://doi.org/10.1038/oby.2006.320.
Tasker, J. G., Di, S., & Malcher-Lopes, R. (2005). Rapid central corticosteroid effects: Evidence for membrane glucocorticoid receptors in the brain. *Integrative and Comparative Biology*, *45*(4), 665–671. http://doi.org/10.1093/icb/45.4.665.
Tasker, J. G., & Herman, J. P. (2011). Mechanisms of rapid glucocorticoid feedback inhibition of the hypothalamic-pituitary-adrenal axis. *Stress (Amsterdam, Netherlands)*, *14*(4), 398–406. http://doi.org/10.3109/10253890.2011.586446.
Ting, C.-H., Chi, C.-W., Li, C.-P., & Chen, C.-Y. (2015). Differential modulation of endogenous cannabinoid CB1 and CB2 receptors in spontaneous and splice variants of ghrelin-induced food intake in conscious rats. *Nutrition*, *31*(1), 230–235. http://doi.org/10.1016/j.nut.2014.06.008.
Travers, J. B., Travers, S. P., & Norgren, R. (1987). Gustatory neural processing in the hindbrain. *Annual Review of Neuroscience*, *10*, 595–632.

Tucci, S., Rogers, E. K., Korbonits, M., & Kirkham, T. C. (2004). The cannabinoid CB1 receptor antagonist SR141716 blocks the orexigenic effects of intrahypothalamic ghrelin. *British Journal of Pharmacology*, *143*(5), 520–523. http://doi.org/10.1038/sj.bjp.0705968.

Tyrey, L. (1978). delta-9-Tetrahydrocannabinol suppression of episodic luteinizing hormone secretion in the ovariectomized rat. *Endocrinology*, *102*(6), 1808–1814. http://doi.org/10.1210/endo-102-6-1808.

Van Sickle, M. D., Duncan, M., Kingsley, P. J., Mouihate, A., Urbani, P., Mackie, K., et al. (2005). Identification and functional characterization of brainstem cannabinoid CB2 receptors. *Science (New York)*, *310*(5746), 329–332. Retrieved from, http://www.ncbi.nlm.nih.gov/pubmed/16224028.

Varga, K., Lake, K., Martin, B. R., & Kunos, G. (1995). Novel antagonist implicates the CB1 cannabinoid receptor in the hypotensive action of anandamide. *European Journal of Pharmacology*, *278*(3), 279–283.

Varma, N., Carlson, G. C., Ledent, C., & Alger, B. E. (2001). Metabotropic glutamate receptors drive the endocannabinoid system in hippocampus. *The Journal of Neuroscience*, *21*(24), RC188. Retrieved from, http://www.ncbi.nlm.nih.gov/pubmed/11734603.

Vettor, R., Fabris, R., Pagano, C., & Federspil, G. (2002). Neuroendocrine regulation of eating behavior. *Journal of Endocrinological Investigation*, *25*(10), 836–854. http://doi.org/10.1007/BF03344047.

Vilela, F. C., & Giusti-Paiva, A. (2011). Glucocorticoids disrupt neuroendocrine and behavioral responses during lactation. *Endocrinology*, *152*(12), 4838–4845. http://doi.org/10.1210/en.2011-1096.

Vilela, F. C., Ruginsk, S. G., de Melo, C. M., & Giusti-Paiva, A. (2013). The CB1 cannabinoid receptor mediates glucocorticoid-induced effects on behavioural and neuronal responses during lactation. *Pflugers Arch*, *465*(8), 1197–1207. http://doi.org/10.1007/s00424-013-1238-8.

Wade, M. R., Degroot, A., & Nomikos, G. G. (2006). Cannabinoid CB1 receptor antagonism modulates plasma corticosterone in rodents. *European Journal of Pharmacology*, *551*(1–3), 162–167. http://doi.org/10.1016/j.ejphar.2006.08.083.

Wamsteeker, J. I., Kuzmiski, J. B., & Bains, J. S. (2010). Repeated stress impairs endocannabinoid signaling in the paraventricular nucleus of the hypothalamus. *The Journal of Neuroscience*, *30*(33), 11188–11196. http://doi.org/10.1523/JNEUROSCI.1046-10.2010.

Wang, M., Hill, M. N., Zhang, L., Gorzalka, B. B., Hillard, C. J., & Alger, B. E. (2012). Acute restraint stress enhances hippocampal endocannabinoid function via glucocorticoid receptor activation. *J Psychopharmacol*, *26*(1), 56–70. http://doi.org/10.1177/0269881111409606.

Wang, Y., Kaminski, N. E., & Wang, D. H. (2005). VR1-mediated depressor effects during high-salt intake: Role of anandamide. *Hypertension*, *46*(4), 986–991. http://doi.org/10.1161/01.HYP.0000174596.95607.fd.

Wang, W., Sun, D., Pan, B., Roberts, C. J., Sun, X., Hillard, C. J., et al. (2010). Deficiency in endocannabinoid signaling in the nucleus accumbens induced by chronic unpredictable stress. *Neuropsychopharmacology*, *35*(11), 2249–2261. http://doi.org/10.1038/npp.2010.99.

Watanabe, M., Fukuda, A., & Nabekura, J. (2014). The role of GABA in the regulation of GnRH neurons. *Frontiers in Neuroscience*, *8*, 387. Retrieved from, http://www.pubmedcentral.nih.gov/articlerender.fcgi?artid=4246667&tool=pmcentrez&rendertype=abstract.

Weidenfeld, J., Feldman, S., & Mechoulam, R. (1994). Effect of the brain constituent anandamide, a cannabinoid receptor agonist, on the hypothalamo-pituitary-adrenal axis in the rat. *Neuroendocrinology*, *59*(2), 110–112. Retrieved from, http://www.ncbi.nlm.nih.gov/pubmed/8127398.

Wenger, T., Croix, D., & Tramu, G. (1988). The effect of chronic prepubertal administration of marihuana (delta-9-tetrahydrocannabinol) on the onset of puberty and the post-pubertal reproductive functions in female rats. *Biology of Reproduction*, *39*(3), 540–545.

Wenger, T., Jamali, K. A., Juanéda, C., Léonardelli, J., & Tramu, G. (1997). Arachidonyl ethanolamide (anandamide) activates the parvocellular part of hypothalamic paraventricular nucleus. *Biochemical and Biophysical Research Communications*, *237*(3), 724–728. http://doi.org/10.1006/bbrc.1997.7222.

Wenger, T., Ledent, C., Csernus, V., & Gerendai, I. (2001). The central cannabinoid receptor inactivation suppresses endocrine reproductive functions. *Biochem Biophys Res Commun*, *284*(2), 363–368. http://doi.org/10.1006/bbrc.2001.4977.

Wenger, T., Rettori, V., Snyder, G. D., Dalterio, S., & McCann, S. M. (1987). Effects of delta-9-tetrahydrocannabinol on the hypothalamic-pituitary control of luteinizing hormone and follicle-stimulating hormone secretion in adult male rats. *Neuroendocrinology*, *46*(6), 488–493.

Williams, C. M., & Kirkham, T. C. (1999). Anandamide induces overeating: Mediation by central cannabinoid (CB1) receptors. *Psychopharmacology*, *143*(3), 315–317. http://doi.org/10.1007/s002130050953.

Williams, C. M., Rogers, P. J., & Kirkham, T. C. (1998). Hyperphagia in pre-fed rats following oral delta9-THC. *Physiology & Behavior*, *65*(2), 343–346. http://doi.org/10.1016/s0031-9384(98)00170-x.

Wilson, R. I., Kunos, G., & Nicoll, R. A. (2001). Presynaptic specificity of endocannabinoid signaling in the hippocampus. *Neuron*, *31*(3), 453–462. Retrieved from, http://www.ncbi.nlm.nih.gov/pubmed/11516401.

Wilson, R. I., & Nicoll, R. A. (2001). Endogenous cannabinoids mediate retrograde signalling at hippocampal synapses. *Nature*, *410*(6828), 588–592. Retrieved from, http://www.ncbi.nlm.nih.gov/pubmed/11279497.

Xu, A. W., Kaelin, C. B., Takeda, K., Akira, S., Schwartz, M. W., & Barsh, G. S. (2005). PI3K integrates the action of insulin and leptin on hypothalamic neurons. *Journal of Clinical Investigation*, *115*(4), 951–958. http://doi.org/10.1172/JCI200524301.

Yang, J., Brown, M. S., Liang, G., Grishin, N. V., & Goldstein, J. L. (2008). Identification of the acyltransferase that octanoylates ghrelin, an appetite-stimulating peptide hormone. *Cell*, *132*(3), 387–396. http://doi.org/10.1016/j.cell.2008.01.017.

Yoshida, R., Ohkuri, T., Jyotaki, M., Yasuo, T., Horio, N., Yasumatsu, K., et al. (2009). Endocannabinoids selectively enhance sweet taste. *Proceedings of the National Academy of Sciences*, *107*(2), 935–939. http://doi.org/10.1073/pnas.0912048107.

Ziegler, C. G., Mohn, C., Lamounier-Zepter, V., Rettori, V., Bornstein, S. R., Krug, A. W., et al. (2010). Expression and function of endocannabinoid receptors in the human adrenal cortex. *Hormone and Metabolic Research*, *42*(2), 88–92. http://doi.org/10.1055/s-0029-1241860.

CHAPTER SIX

The Role of the Brain's Endocannabinoid System in Pain and Its Modulation by Stress

Louise Corcoran*,†, Michelle Roche†,‡, David P. Finn*,†,1
*Pharmacology and Therapeutics, School of Medicine, National University of Ireland, Galway, Ireland
†Galway Neuroscience Centre and Centre for Pain Research, NCBES, National University of Ireland, Galway, Ireland
‡Physiology, School of Medicine, National University of Ireland, Galway, Ireland
[1]Corresponding author: e-mail address: david.finn@nuigalway.ie

Contents

1. Introduction 205
2. The Endocannabinoid System 207
3. The Endocannabinoid System in the Brain Regulates Pain 209
4. The Modulation of Pain by Stress: Role for the Brain's Endocannabinoid System 211
5. The Role of the Endocannabinoid System in the Rostral Ventromedial Medulla in Pain, Stress-Induced Analgesia, and Stress-Induced Hyperalgesia 213
   5.1 Pain 213
   5.2 SIA 227
   5.3 SIH 227
6. The Role of the Endocannabinoid System in the Periaquaductal Gray in Pain, Stress-Induced Analgesia, and Stress-Induced Hyperalgesia 228
   6.1 Pain 228
   6.2 SIA 231
   6.3 SIH 232
7. The Role of the Endocannabinoid System in the Amygdala in Pain, Stress-Induced Analgesia, and Stress-Induced Hyperalgesia 232
   7.1 Pain 232
   7.2 SIA 234
   7.3 SIH 235
8. The Role of the Endocannabinoid System in the Prefrontal Cortex in Pain, Stress-Induced Analgesia, and Stress-Induced Hyperalgesia 235
   8.1 Pain 235
   8.2 SIA 237
   8.3 SIH 237
9. Less Characterized Supraspinal Endocannabinoid Mechanisms in Pain Modulation 238
10. Concluding Remarks 241

Acknowledgments 242
References 242

International Review of Neurobiology, Volume 125
ISSN 0074-7742
http://dx.doi.org/10.1016/bs.irn.2015.10.003

Abstract

Stress has a complex, bidirectional modulatory influence on pain. Stress may either reduce (stress-induced analgesia) or exacerbate (stress-induced hyperalgesia) pain depending on the nature, duration, and intensity of the stressor. The endogenous cannabinoid (endocannabinoid) system is present throughout the neuroanatomical pathways that mediate and modulate responses to painful stimuli. The specific role of the endocannabinoid system in the brain in pain and the modulation of pain by stress is reviewed herein. We first provide a brief overview of the endocannabinoid system, followed by a review of the evidence that the brain's endocannabinoid system modulates pain. We provide a comprehensive evaluation of the role of the endocannabinoid system supraspinally, and particularly in the rostral ventromedial medulla, periaqueductal gray, amygdala, and prefrontal cortex, in pain, stress-induced analgesia, and stress-induced hyperalgesia. Increased understanding of endocannabinoid-mediated regulation of pain and its modulation by stress will inform the development of novel therapeutic approaches for pain and its comorbidity with stress-related disorders.

ABBREVIATIONS

2-AG 2-arachidonoyl glycerol
ABA accessory basal nucleus
AC adenylate cyclase
ACC anterior cingulate cortex
AEA anandamide
BLA basolateral nucleus
CB cannabinoid
CB_1 cannabinoid type 1
CB_2 cannabinoid type 2
CCI chronic constriction injury
CCK cholecystokinin
CeA central nucleus of the amygdala
CNS central nervous system
CUS chronic unpredictable stress
dlPAG dorsolateral periaqueductal gray
dmPAG dorsomedial periaqueductal gray
dPAG dorsal periaqueductal gray
eCB endocannabinoid
FAAH fatty acid amide hydrolase
FABP fatty acid-binding protein
FCA fear-conditioned analgesia
GiA gigantocellular reticular nucleus
HMBA 4-hydroxy-3-methoxybenzylamine
i.c.v. intracerebroventricular
IL infralimbic cortex
LA lateral nucleus
lPAG lateral periaqueductal gray

MAGL monoacylglycerol lipase
MAPK mitogen-activated protein kinase
MeA medial nucleus
mGlu metabotropic glutamate receptors
mPFC medial prefrontal cortex
NAPE *N*-arachidonoyl phosphatidylethanolamine
NGF nerve growth factor
NSAIDs nonsteroidal anti-inflammatory drugs
OEA *N*-oleoylethanolamide
OX orexin
PAG periaqueductal gray
PEA *N*-palmitoylethanolamide
PFC prefrontal cortex
PLD phospholipase D
PPARs peroxisome proliferator-activated receptors
PrL prelimbic cortex
RVM rostral ventromedial medulla
SD Sprague Dawley
SIA stress-induced analgesia
SIH stress-induced hyperalgesia
SNI spared nerve injury
SNL spinal nerve ligation
THC Δ^9-tetrahydrocannabinol
TRPV1 transient receptor potential vanilloid 1
vlPAG ventrolateral periaqueductal gray
VPL ventral posterolateral nucleus of the thalamus
WKY Wistar Kyoto

1. INTRODUCTION

Pain can be defined as "an unpleasant sensory and emotional experience associated with actual or potential tissue damage or described in terms of such damage" (International Association for the Study of Pain [IASP] Task Force on Taxonomy, 1994). Recent data indicate that approximately 20% of the population suffer from chronic pain, the majority of whom also suffer from some other disability or mood disturbance (Blyth et al., 2001; Demyttenaere et al., 2007; Vos et al., 2012). Chronic pain is usually defined as pain persisting for over 3 months. It may be neuropathic, inflammatory, or idiopathic in nature (Aguggia, 2003). Epidemiological studies of 289 diseases and injuries concluded that chronic pain conditions were among the 10 conditions resulting in the longest number of years lived with disability (Vos et al., 2012). Current pharmacotherapies for pain management lack efficacy in many patients, with ~40% of patients with chronic pain unsatisfied with their treatment (Breivik, Collett, Ventafridda, Cohen, & Gallacher, 2006). Furthermore, the annual economic cost of pain in the United States has been estimated at a staggering $560–$625 billion annually, including direct and indirect costs (for review, see Gaskin & Richard, 2012; McCarberg & Billington, 2006; Turk, 2002). Despite the efforts of the research community and the pharmaceutical industry to invest in and develop new drugs to manage pain, chronic pain in particular continues to represent a major unmet clinical need. Thus, further research is needed to understand fully the neurobiological mechanisms of pain, and its modulation, with a view to identifying novel targets and developing new, superior analgesics.

Increasing evidence over the past two decades has demonstrated that the endogenous cannabinoid (endocannabinoid; eCB) system has a regulatory role in pain processing and perception (Woodhams, Sagar, Burston, & Chapman, 2015). This regulatory function is facilitated by the expression of the eCB signaling machinery at neuronal synapses within all components of the pain pathway. Activation of cannabinoid (CB) receptors on presynaptic nerve terminals generally functions to reduce neurotransmission, resulting primarily in antinociception/analgesia. However, depending on physiological and pathological state, the tissue concentration of eCBs and expression levels of eCB-sensitive receptors can vary (Alexander & Kendall, 2007; Woodhams et al., 2015), and with it the regulatory potential of this system on nociceptive processing.

The intensity and severity of perceived pain does not necessarily correlate with the degree of tissue damage, injury, or inflammation occurring. The importance of context and modulation of pain by emotion is now widely recognized. Stress, fear, and anxiety exert important modulatory influences on pain (Asmundson & Katz, 2009; Burke, Finn, & Roche, 2015; Butler & Finn, 2009; Fitzgibbon, Finn, & Roche, 2015; Ford & Finn, 2008; Jennings, Okine, Roche, & Finn, 2014; Okine et al., 2014; Rhudy & Meagher, 2000, 2001; Wiech & Tracey, 2009). Regardless of arousal level, positive emotions generally act to inhibit pain, while negative emotions with low to moderate arousal tend to enhance pain, and negative emotions with high arousal inhibit pain (de Wied & Verbaten, 2001; Dougher, 1979; Meagher, Arnau, & Rhudy, 2001; Rhudy & Meagher, 2000, 2001, 2003a, 2003b). Thus, a complex relationship exists between emotion and pain processing. CB receptors are localized in brain regions involved in the modulation of pain including the rostral ventromedial medulla (RVM), the periaqueductal gray (PAG), amygdala, and prefrontal cortex (PFC) (Herkenham et al., 1991; Tsou, Brown, Sanudo-Pena, Mackie, & Walker, 1998) with these brain regions also key components of stress, fear, and anxiety circuitry. Stress and fear have been shown to alter levels of eCBs in these brain regions (Hill et al., 2013, 2005; Hohmann et al., 2005; Jennings et al., 2014; Olango, Roche, Ford, Harhen, & Finn, 2012; Patel, Cravatt, & Hillard, 2005; Rademacher et al., 2008) (for review, see Carrier, Patel, & Hillard, 2005; Morena, Patel, Bains, & Hill, 2015). Thus, the eCB system is an important common denominator in pain, stress, and fear and its role in the aforementioned brain regions in pain and the modulation of pain by stress is the main focus of this chapter.

We will consider the role of the supraspinal eCB system in acute and chronic pain, as well as its role in both stress-induced analgesia (SIA) and stress-induced hyperalgesia (SIH). The role of the spinal and peripheral eCB system in pain or stress–pain interactions is beyond the scope of this review but has been reviewed previously by ourselves and others (Butler & Finn, 2009; Finn, 2010; Hohmann & Suplita, 2006; Jennings et al., 2014; Maccarrone et al., 2015; Olango & Finn, 2014; Walker & Hohmann, 2005).

2. THE ENDOCANNABINOID SYSTEM

The medicinal properties of the *Cannabis sativa* plant have been known for millennia but it was not until the mid to late nineteenth century that its therapeutic potential was examined scientifically. The discovery of Δ^9-tetrahydrocannabinol (THC), the major psychoactive component of the plant *C. sativa* in the 1960s (Mechoulam & Gaoni, 1967) led to extensive studies that have revealed the mechanisms underlying the physiological and pharmacological effects of the eCB system.

The eCB system as we know it today consists of CB type 1 (CB_1) receptors (Devane, Dysarz, Johnson, Melvin, & Howlett, 1988; Matsuda, Lolait, Brownstein, Young, & Bonner, 1990) and CB type 2 (CB_2) receptors (Munro, Thomas, & Abu-Shaar, 1993), their endogenous ligands *N*-arachidonylethanolamide (anandamide, AEA) and 2-arachidonoyl glycerol (2-AG) (Devane et al., 1992; Mechoulam et al., 1995; Sugiura et al., 1995), and the enzymes responsible for their synthesis and degradation. AEA and 2-AG are the best characterized eCBs; however, there are a number of other endogenous ligands with affinity and activity at CB_1 and CB_2 receptors including 2-AG ether (noladin ether), virodhamine, *N*-arachidonyl dopamine, and others (for review, see Battista, Di

Tommaso, Bari, & Maccarrone, 2012; Di Marzo, 2008; Di Marzo, Stella, & Zimmer, 2015; Henry, Kerr, Finn, & Roche, 2015; Pertwee, 1997, 2001).

The CB receptors in the adult human brain and spinal cord are distributed in a heterogeneous fashion (Glass, Dragunow, & Faull, 1997). CB_1 receptors are the most abundant CB receptor subtype in the CNS (Glass et al., 1997; Herkenham et al., 1991; Pertwee, 1997), with particularly high density in brain regions that are key components of the descending inhibitory/facilitatory pain pathways and the stress/fear/anxiety circuitry. CB_2 receptors, although expressed in the CNS (Baek, Zheng, Darlington, & Smith, 2008; Concannon, Okine, Finn, & Dowd, 2015; Onaivi et al., 2006; Van Sickle et al., 2005; Zhang et al., 2014) are mainly distributed in the periphery with particularly high density on cells and tissues of the immune system (Berdyshev, 2000; Munro et al., 1993; Sugiura et al., 1995). CB_1 and CB_2 receptors are Gi/o protein-coupled receptors negatively coupled to adenylate cyclase (AC) (Howlett, 1985; Howlett, Mukhopadhyay, Shim, & Welsh, 1999) and positively coupled to mitogen-activated protein kinase (MAPK) (Bouaboula et al., 1995). Upon binding to CB_1 receptors, eCBs also inhibit N- and P/Q-type voltage-activated Ca^{2+} channels and induce inwardly rectifying K^+ currents, resulting in inhibition of neurotransmitter release (Demuth & Molleman, 2006).

The biosynthetic pathways for AEA are not fully characterized but the best described mechanism involves the formation of AEA from the precursor *N*-arachidonoyl phosphatidylethanolamine (NAPE), due to the hydrolytic activity of the phospholipase D enzyme known as NAPE-PLD (Bisogno, Ligresti, & Di Marzo, 2005). 2-AG is synthesized almost exclusively by phospholipase C (PLC) hydrolysis producing 1,2-diacylglycerol which is then converted to 2-AG by diacylglycerol lipases (DAGL) (Di Marzo, 2008; Howlett & Mukhopadhyay, 2000). For a more complete discussion of the biosynthetic routes for AEA and 2-AG, please refer to chapter "The endocannabinoid signaling system in the CNS: A primer" by Hillard. AEA is primarily degraded to arachidonic acid and ethanolamine by the enzyme fatty acid amide hydrolase (FAAH), located in the endoplasmic reticulum of the postsynaptic neuron (Cravatt et al., 1996; Giang & Cravatt, 1997) (for review, see Otrubova, Ezzili, & Boger, 2011). FAAH also catabolizes additional *N*-acylethanolamines including *N*-palmitoylethanolamide (PEA) and *N*-oleoylethanolamide (OEA) which themselves do not have appreciable activity at CB_1 or CB_2 receptors but which can elevate levels of AEA through substrate competition at FAAH (Di Marzo et al., 1994; Sugiura et al., 1995). In contrast, 2-AG is primarily metabolized to arachidonic acid and glycerol by the enzyme monoacylglycerol lipase (MAGL) (Ueda, Tsuboi, Uyama, & Ohnishi, 2011), with other enzymes including FAAH, ABHD6, and ABHD12 accounting for a modest degree of 2-AG catabolism (Blankman, Simon, & Cravatt, 2007; Goparaju, Ueda, Yamaguchi, & Yamamoto, 1998). FAAH is primarily a postsynaptic enzyme, whereas MAGL is presynaptic (Egertova, Cravatt, & Elphick, 2003; Gulyas et al., 2004; Tsou, Nogueron, et al., 1998) (for review, see Blankman & Cravatt, 2013; Di Marzo, 2008; Lichtman, Blankman, & Cravatt, 2010).

The mechanisms underlying eCB biosynthesis, signaling, and degradation are relatively well understood although controversy remains surrounding the mechanisms by which eCBs are transported across cell membranes. It has been proposed that due to their lipophilic nature, eCBs are readily transported via a simple diffusion mechanism (Glaser et al., 2003; Kaczocha, Hermann, Glaser, Bojesen, & Deutsch, 2006) while others suggest the existence of a protein-facilitated transport process (Beltramo & Piomelli, 2000; Hillard, Edgemond, Jarrahian, & Campbell, 1997). Most recently, a FAAH-like anandamide transporter has been described as the main mediator for AEA transport (Fu et al., 2012). Furthermore, fatty acid-binding proteins (FABPs) are small cytoplasmic lipid transport proteins (Furuhashi & Hotamisligil, 2008) located both peripherally (De Leon et al., 1996) and in the CNS (Yamamoto et al., 2009). FABP5 and FABP7 are capable of binding eCBs and regulating their signaling and catabolism by FAAH (Cravatt et al., 2001; Kaczocha, Glaser, & Deutsch, 2009; Kaczocha, Vivieca, Sun, Glaser, & Deutsch, 2012).

In addition to the two classical CB receptors (CB_1 and CB_2), several lines of evidence suggest that eCBs act at numerous other non-CB_1/non-CB_2 including the transient receptor potential vanilloid 1 (TRPV1), members of the nuclear receptor family of peroxisome proliferator-activated receptors (PPARs), and the G-protein-coupled receptors GPR55 and GPR119 (Alexander & Kendall, 2007; Brown, 2007; O'Sullivan, 2007).

3. THE ENDOCANNABINOID SYSTEM IN THE BRAIN REGULATES PAIN

Considerable effort has been invested in investigating the brain regions involved in mediating the antinociceptive effects of eCBs and CB receptor agonists. Later sections will discuss in more detail the role of the eCB system in individual brain regions in pain and its modulation by stress. Presented in this section is an overview of the studies that have identified a role of the eCB system, supraspinally, in the modulation of pain (and summarized in Fig. 1).

Strong evidence of a role for the supraspinal eCB system in the modulation of pain was provided by Hohmann, Tsou, and Walker (1999). Here, systemic administration of the CB receptor agonist WIN55,212-2 resulted in an antinociceptive effect in the tail flick test in rats. Transection of the spinal cord and thus blockade of descending pain processes inhibited the CB-induced suppression of noxious heat-evoked activity in the tail flick test, thus indicating that WIN55,212-2 acted supraspinally to mediate its antinociceptive efficacy. This study paved the way for the investigation of the role of supraspinal sites in CB-induced antinociception.

Antinociceptive activity of CB receptor agonists had been demonstrated in the mouse and rat tail flick tests following intracerebroventricular (i.c.v.)

Brain region	Receptor event	Functional consequence
ACC	CB_1 receptor activation	Decreased pain-related behavior/nociception
rACC	CB_1 receptor inhibition	Increased pain-related behavior/nociception
IL/PrL	CB_1 receptor activation coupled with TRPV1 inhibition	Decreased pain-related behavior/nociception
ACC	PPARα inhibition	Decreased pain-related behavior/nociception
BLA	CB_1 receptor inhibition	Decreased pain-related behavior/nociception
CeA	CB_1 receptor inhibition	Decreased pain-related behavior/nociception
CeA/BLA	CB_1 receptor activation	Decreased pain-related behavior/nociception
PAG	CB_1 receptor activation	Decreased pain-related behavior/nociception
PAG	TRPV1 activation, opposite at higher dose	Decreased pain-related behavior/nociception, opposite at higher dose
PAG	CB_1 receptor activation coupled with TRPV1 inhibition	Increased pain-related behavior/nociception
RVM	CB_1 receptor activation	Decreased pain-related behavior/nociception
GiA	CB_1 receptor activation	Decreased pain-related behavior/nociception

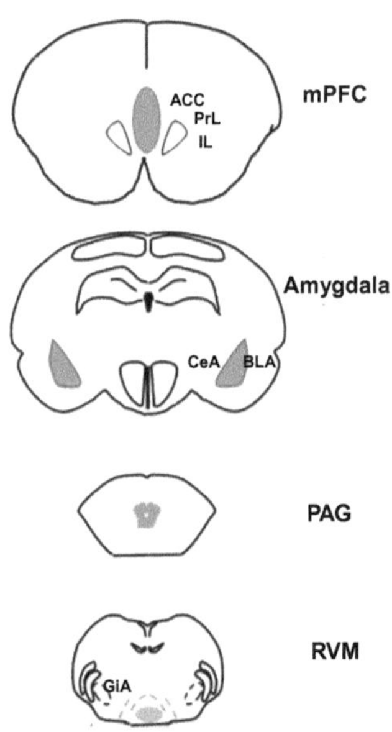

Figure 1 A synthesis of the literature reviewed herein on the role of the supraspinal endocannabinoid system in discrete brain regions in pain. mPFC, medial prefrontal cortex; ACC, anterior cingulate cortex; PrL, prelimbic cortex; IL, infralimbic cortex; BLA, basolateral amygdala; CeA, central nucleus of the amygdala; PAG, periaqueductal gray; RVM, rostral ventromedial medulla; GiA, gigantocellular reticular nucleus; TRPV1, transient receptor potential vanilloid 1; PPARs, peroxisome proliferator-activated receptors.

administration. Specifically, i.c.v. injection of the CB agonist WIN55, 212-2, Δ^9-THC, and CP-55,940 produced antinociception in the rat tail flick test (Lichtman, Cook, & Martin, 1996; Martin, Lai, Patrick, Tsou, & Walker, 1993). In spinally transected rats, i.c.v. administration of the CB_1 receptor antagonist/inverse agonist rimonabant completely blocked the antinociceptive effects of Δ^9-THC and CP-55,940 in the rat tail flick test, indicating that these effects are mediated through CB_1 receptors in the brain. The antagonist failed to block the effects of morphine, indicating its selectivity for CB receptors (Lichtman & Martin, 1997). However, when administered via the same route, i.c.v. administration of Δ^9-THC enhances the antinociceptive potency of morphine (Welch, Thomas, & Patrick, 1995), suggesting a synergistic interaction between the opioid and CB systems. Similar to Martin et al. (1993), i.c.v. injection of WIN55,212-2 and THC produced dose-related antinociceptive effects in the mouse tail flick test (Raffa, Stone, & Hipp, 1999). In addition, i.c.v. administration of WIN55,212-2 induces antinociception in the mouse tail flick and paw withdrawal test (Fang et al., 2012). It was also shown that rimonabant has greater efficacy in the mouse tail flick test at a supraspinal rather than spinal level when blocking the action of THC, HU210, CP-55,940, and AEA (Welch, Huffman, & Lowe, 1998). Thus, taken together, supraspinal CB_1 receptors are important in modulating pain processes. Although i.c.v. administration of pharmacological agents is a useful means of investigating the contribution of the brain in general, alternative approaches are required to study the role of the eCB system in specific brain regions in pain and its modulation by stress. These approaches and the results obtained are discussed later in the review for each of the key brain regions that comprise the descending pain pathways (RVM, PAG, amygdala, and PFC). Additionally, less characterized mechanisms and targets will also be discussed toward the end of the review. Lines of evidence implicating the supraspinal eCB system in pain, stress, and their interaction will be considered briefly now.

4. THE MODULATION OF PAIN BY STRESS: ROLE FOR THE BRAIN'S ENDOCANNABINOID SYSTEM

Both painful (Alexander & Kendall, 2007; Kwilasz, Abdullah, Poklis, Lichtman, & Negus, 2014; Walker, Huang, Strangman, Tsou, & Sanudo-Pena, 1999; Woodhams et al., 2015) and aversive (stress/fear) (Hill et al., 2013, 2005; Hohmann et al., 2005; Jennings et al., 2014; Olango et al.,

2012; Patel et al., 2005; Rademacher et al., 2008) stimuli have been shown to alter eCB levels and expression of key components of the eCB system in supraspinal regions (for review, see Morena et al., 2015). As highlighted earlier, emotion and stress can profoundly impact on nociceptive processing, with chronic stress paradigms shown to enhance pain perception under a variety of experimental conditions. Chronic unpredictable stress (CUS), a widely used model for inducing anxiety and depressive-like behavior in mice, has been shown to enhance thermal (hot plate test) and mechanical (von Frey) hyperalgesia. It has also been shown to induce long lasting widespread hyperalgesia following intramuscular injection of nerve growth factor (NGF) (Lomazzo et al., 2015). The FAAH and MAGL inhibitors URB597 and JZL184 attenuated the CUS-induced anxiety-related behavior in the light–dark box and thermal hyperalgesia in the hot plate test. URB597 significantly reduced the widespread hyperalgesia induced by combining CUS and NGF in this study, while JZL184 had no significant effect. Both drugs enhanced the levels of AEA and 2-AG, respectively, in the midbrain and cingulate cortex (Lomazzo et al., 2015). These data highlight the strong potential for pharmacological inventions aimed at increasing eCB levels supraspinally in both anxiety- and pain-related disorders.

I.c.v. administration of rimonabant increases levels of the stress hormones, adrenocorticotropic hormone, and corticosterone, in rats, suggesting a role for supraspinal CB_1 receptors in the neuroendocrine response to stress (Manzanares, Corchero, & Fuentes, 1999). $CB_1(-/-)$ knockout mice develop normal mechanical hypersensitivity but more pronounced anxiety-related behavior following partial sciatic nerve ligation, indicating a potential role for the EC system in chronic comorbid pain/anxiety disorders (Racz, Nent, Erxlebe, & Zimmer, 2015). Indeed, the acquisition, expression, and extinction of fear-related behavior have all been shown to involve eCB signaling (see Chhatwal & Ressler, 2007). Our group has shown an interaction between the eCB and opioid systems in fear-conditioned analgesia (FCA). FCA was modeled by assessing formalin-evoked nociceptive behavior in an arena previously paired with footshock. Systemic administration of the FAAH inhibitor URB597 enhanced FCA in rats, an effect blocked by the CB_1 and CB_2 receptor antagonists rimonabant and SR144528, respectively (Butler, Rea, Lang, Gavin, & Finn, 2008). These findings corroborated and extended our earlier work demonstrating that CB_1 receptors play a key role in mediating FCA (Finn et al., 2004). The use of transgenic mice lacking components of the eCB system further implicates a role for eCBs in SIA (Valverde, Ledent,

Beslot, Parmentier, & Roques, 2000) and studies that have investigated the role of the EC system in specific brain regions in FCA/SIA will be discussed in detail below.

It has been shown that the neuropeptide cholecystokinin (CCK) plays a role in pain sensitivity via its regulation of opioid tone in the CNS. A recent study has demonstrated an interaction between CCK and ECs in the regulation of SIA (Kurrikoff, Inno, Matsui, & Vasar, 2008). Intraperitoneal rimonabant prevented SIA, in the tail flick test, in response to footshock in wild-type mice. SIA was present in CCK type 2 receptor-deficient mice regardless of rimonabant treatment while naloxone weakened SIA in both wild-type and CCK type 2 receptor-deficient mice. The CCK_2 receptor gene, along with genes implicated in eCB-mediated neurotransmission, were upregulated in the mesolimbic area of the brain. CCK_2 receptors may therefore modulate the action of eCBs. This study demonstrates a clear involvement of the CCK_2 receptor in eCB-mediated SIA.

See Table 1 for a summary of studies (excluding those focused on the RVM, PAG, amygdala, and PFC) investigating the role of the brain's eCB system in pain and its modulation by stress. Tables 2–5, Fig. 2, and the sections that follow below then deal with the role of the eCB system within the RVM (Table 2), PAG (Table 3), amygdala (Table 4), and PFC (Table 5) in pain and its modulation by stress.

5. THE ROLE OF THE ENDOCANNABINOID SYSTEM IN THE ROSTRAL VENTROMEDIAL MEDULLA IN PAIN, STRESS-INDUCED ANALGESIA, AND STRESS-INDUCED HYPERALGESIA

5.1 Pain

The RVM is made up of the nucleus raphe magnus, the nucleus gigantocellularis pars alpha (GiA), and the adjacent reticular formation; and is a major component of the descending inhibitory pain pathway (Meng, Manning, Martin, & Fields, 1998). CB_1 receptors have been shown to be expressed in the RVM using receptor autoradiography and immunohistochemistry (Glass et al., 1997; Herkenham et al., 1991; Herzberg, Eliav, Bennett, & Kopin, 1997; Mailleux, Parmentier, & Vanderhaeghen, 1992; Thomas, Wei, & Martin, 1992; Tsou, Brown, et al., 1998). The RVM contains ON and OFF cells which are involved in descending facilitation and inhibition of nociception, respectively (Vanegas, Barbaro, & Fields, 1984), and it projects to the dorsal horn of the spinal cord and the trigeminal

Table 1 Summary of Studies Investigating the Role of the Brain's Endocannabinoid System in Pain and Its Modulation by Stress (Excluding Studies on RVM, PAG, Amygdala, and PFC Which Are Summarized in Tables 2–5)

Class of Pharmacological Compound	Pharmacological or Genetic Intervention	Route of Administration	Species	Model	Effect	References
CB_1/CB_2 agonist	WIN55,212-2	i.v.	Rat	Spinal transection; tail flick	Inhibits noxious activity; blocked by spinal transection	Hohmann et al. (1999)
CB_1/CB_2 agonist	THC; CP-55,940	i.c.v.	Rat	Spinal transection; tail flick	Antinociception; blocked by rimonabant	Lichtman and Martin (1991, 1997)
CB_1/CB_2 agonist	THC	i.c.v.	Mouse	Tail flick	Enhances the antinociceptive effects of morphine	Welch et al. (1995)
CB_1/CB_2 agonist	WIN55,212-2; CP-55,940	i.c.v.	Rat	Tail flick	Antinociception	Martin et al. (1993)
CB_1/CB_2 agonist	WIN55,212-2; THC	i.c.v.	Mouse	Tail flick	Dose-related antinociception	Raffa et al. (1999)
CB_1/CB_2 agonist	WIN55,212-2	i.v.	Rats	Pressure stimulus to hindpaw; electrophysiological recording of nociceptive neurons in thalamus	Decrease in nociceptive transmission in the thalamus	Martin, Hohmann, and Walker (1996)

CB_1/CB_2 agonist	WIN55,212-2	Microinjection; GiA, thalamus, noradrenergic A5 region	Rats	Tail flick	Antinociception when administered to each region	Martin et al. (1999)
CB_1/CB_2 agonist	WIN55,212-2	i.c.v.	Mice	Tail flick; paw withdrawal in the lamp-foot-flick assay	Antinociception	Fang et al. (2012)
CB_1antagonist/ inverse agonist	Rimonabant	i.c.v.; i.t.; intraperitoneal	Mouse	Tail flick	Rimonabant exhibits greater efficacy supraspinally rather than spinally	Welch et al. (1998)
CB_2 agonist	JWH-133	Intra-VPL	Rats	Spinal nerve ligation	Reduced noxious activity in SNL rats; blocked by SR144528 (CB_2 antagonist)	Jhaveri et al. (2008)
FAAH inhibitor	FAAH (−/−); URB597 PF3945		Mice	Nitroglycerin-induced migraine-like pain	FAAH (−/−), URB597, and PF3945 reduce nociceptive behavior; blocked by rimonabant	Nozaki, Markert, and Zimmer (2015)
	FAAH (−/−)		Mice	Tail immersion, hot plate, formalin tests, CCI and carrageenan	FAAH (−/−) increased response latency in tail immersion and hot plate	Lichtman, Shelton,

Continued

Table 1 Summary of Studies Investigating the Role of the Brain's Endocannabinoid System in Pain and Its Modulation by Stress (Excluding Studies on RVM, PAG, Amygdala, and PFC Which Are Summarized in Tables 2–5)—cont'd

Class of Pharmacological Compound	Pharmacological or Genetic Intervention	Route of Administration	Species	Model	Effect	References
					test; reduced formalin-evoked nociceptive behavior	Advani, and Cravatt (2004)
FAAH inhibitor; MAGL inhibitor	URB597; JZL184	i.p.	Mice	CUS; NGF hyperalgesia; tail flick; hot plate	Anti-hyperalgesic in carrageenan model	Lomazzo et al. (2015)
TRPV1 antagonist	A-784168	i.c.v.	Rats	Model of osteoarthritis (sodium monoiodoacetate); complete Freund's adjuvant (chronic inflammatory pain)	Effects in CUS mice—enhanced levels of AEA and 2-AG; decreased thermal hyperalgesia; URB597 decreased NGF hyperalgesia	Cui et al. (2006)
TRPV1 antagonist	Capsazepine	i.c.v.	Mice	Formalin	Decreased weight bearing; decreased chronic inflammatory thermal hyperalgesia	Santos and Calixto (1997)
	GPR55(−/−)		Mice	Inflammatory mechanical hyperalgesia (von Frey; complete Freund's adjuvant)	Attenuation of nociceptive behavior	Castane et al. (2006) and Staton et al. (2008)

	GPR55(−/−)		Mice	Nerve ligation; mechanical hyperalgesia	Hyperalgesia absent in GPR55(−/−)	Staton et al. (2008)
PPARγ agonist	Rosiglitazone; 15d-PGJ(2)	i.c.v.	Rats	Plantar carrageenan model of inflammatory pain	Mechanical hyperalgesia absent following nerve ligation in GPR55(−/−)	Morgenweck et al. (2010)
				SIA—footshock; tail flick	Anti-inflammatory and antihyperalgesia effects	Kurrikoff et al. (2008)
CB_1 antagonist/ inverse agonist; opioid antagonist	Rimonabant; naloxone; CCK_2 knockout mice	i.p.	Mice		Rimonabant prevented SIA, an effect not seen in CCK_2 knockout mice; naloxone weakened SIA in wild-type and CCK_2 knockout mice	
FAAH inhibitor	URB597	i.p.	Rats	FCA—conditioned fear (footshock) and formalin test	URB597 enhances FCA; attenuated by rimonabant, SR144528 and naloxone	Butler et al. (2008)
FAAH inhibitor	URB597	intra-ventral hippocampus	Rats	FCA–conditioned fear (footshock) and formalin test	URB597 enhanced FCA, an effect blocked by rimonabant.	Ford et al. (2011)

$CB_{1/2}$, Cannabinoid receptor type 1/2; i.v., intravenous; i.c.v., intracerebroventricular; GiA, nucleus reticularis gigantocellularis pars alpha; i.t., intra-thecal; intra-VPL, ventral posterolateral nucleus; FAAH, fatty acid amide hydrolase; SNL, spinal nerve ligation; (−/−), knock out; CCI, chronic constriction injury; MAGL, monoacylglycerol lipase; PPARα, peroxisome proliferator-activated receptor alpha; TRPV1, transient receptor potential cation channel subfamily V member 1; CUS, chronic unpredictable stress; NGF, nerve growth factor; AEA, anandamide; 2-AG, 2-arachidonoylglycerol; GPR55, G protein-coupled receptor 55; CCK2, cholecystokinin 2; SIA, stress-induced analgesia; FCA, fear-conditioned analgesia.

Table 2 Summary of Studies Investigating the Role of the Endocannabinoid System in the RVM in Pain and Its Modulation by Stress

Class of Pharmacological Compound	Pharmacological or Genetic Intervention	Route of Administration	Species	Model	Effect	References
CB_1/CB_2 agonists	WIN55,212-2; HU210	Intra-RVM	Rat	Tail flick	Nociceptive behavior suppression	Martin, Tsou, and Walker (1998)
CB_1/CB_2 agonist	WIN55,212-2	Intra-RVM	Rat	Tail flick	Increased tail flick latency; inhibition of ON-cell activity; increase in OFF-cell activity; effects blocked by rimonabant	Meng and Johansen (2004)
CB_1/CB_2 agonist	WIN55,212-2	Intra-GiA	Rat	Tail flick	Increased antinociception; blocked by rimonabant	Monhemius, Azami, Green, and Roberts (2001)
CB_1/CB_2 agonist	WIN55,212-2	Intra-GiA	Rat	Partial nerve ligation; formalin test	Decrease in formalin-evoked nociceptive behavior following nerve ligation; reversed by rimonabant	Monhemius et al. (2001)
CB_1 antagonist/ inverse agonist; dual FAAH/ TRPV1 inhibitor	Rimonabant; AA-5-HT	Intra-RVM;	Rats	SIA—Footshock, Formalin	Suppression of SIA by rimonabant; enhancement of SIA by AA-5-HT	Suplita, Farthing, Gutierrez, and Hohmann (2005)

CB_1 antagonist/ inverse agonist; FAAH inhibitor	AM251; URB597	Intraperitoneal (AM251 and URB597); intra-RVM AM251	Wistar Kyoto and Sprague Dawley rats	Formalin	Systemic AM251 potentiates hyperalgesia in WKY, URB597 attenuates hyperalgesia in WKY associated with impaired pain-related mobilization of ECs in RVM of WKY rats as seen from intra-RVM AM251	Rea et al. (2014)

$CB_{1/2}$, cannabinoid receptor type 1/2; RVM, rostral ventromedial medulla; GiA, nucleus reticularis gigantocellularis pars alpha; SIA, stress-induced analgesia; WKY, Wistar Kyoto; TRPV1, transient receptor potential cation channel subfamily V member 1; FAAH, fatty acid amide hydrolase.

Table 3 Summary of Studies Investigating the Role of the Endocannabinoid System in the PAG in Pain and Its Modulation by Stress

Class of Pharmacological Compound	Pharmacological or Genetic Intervention	Route of Administration	Species	Model	Effect	References
CB_1/CB_2 agonist	HU-210	vlPAG; systemic	Rat	Hot plate test	HU210 enhanced antinociceptive effect of morphine and morphine enhanced the antinociceptive effect of HU210	Wilson-Poe, Pocius, Herschbach, and Morgan (2013)
CB_1/CB_2 agonist	HU210	Intra-dPAG	Rat	Formalin	Reduced formalin-evoked nociceptive behavior; blocked following rimonabant administration	Finn et al. (2003)
CB_1/CB_2 agonist	CP-55,940	Intra-vlPAG; intra-dPAG;	Rat	Tail flick	Intra-vlPAG microinjection produced antinociception; intra-dPAG had no effect	Lichtman et al. (1996)
CB_1/CB_2 agonist	WIN55,212-2	Intra-dPAG	Rats	Tail flick	Increased tail flick latency	Martin, Patrick, Coffin, Tsou, and Walker (1995)
CB_1 antagonist/inverse agonist	AM251	Intra-PAG; intra-RVM	Rat	Metazinol-induced antinociception in a carrageenan model of inflammation	Reverses metazinol-induced analgesia	Escobar et al. (2012)

TRPV1 agonist	Capsaicin (low dose); capsaicin (high dose)	Intra-dlPAG	Rat	Plantar test	Low dose—antinociception; high dose—blocked antinociception	Palazzo et al. (2002)
TRPV1 agonist; TRPV1 antagonist	Capsaicin; capsazepine	Intra-dlPAG	Rat	Tail flick	Capsaicin—hyperalgesia followed by antinociception; capsazepine—blocked hyperalgesic effect of capsaicin	McGaraughty et al. (2003)
FAAH inhibitor	URB597	Intra-vlPAG	Rat	Plantar test	Low dose—hyperalgesia—coadministration with AM251 converted hyperalgesia to analgesia—coadministration with capsazepine and AM251 abolished any effect of URB597; high dose—antinociception; intermediate dose—biphasic response – blocked by AM251, hyperalgesic following coadministration of URB597 with capsazepine	Maione et al. (2006)

Continued

Table 3 Summary of Studies Investigating the Role of the Endocannabinoid System in the PAG in Pain and Its Modulation by Stress—cont'd

Class of Pharmacological Compound	Pharmacological or Genetic Intervention	Route of Administration	Species	Model	Effect	References
Dual FAAH/TRPV1 inhibitor; CB_1 antagonist/inverse agonist; TRPV1 antagonist; FAAH inhibitor	AA-5-HT; AM251; I-RTX; URB597	Intra-PAG	Rat	Tail flick	AA-5-HT-induced antinociception—blocked by AM251 and I-RTX; URB597 and I-RTX-induced analgesia	de Novellis et al. (2008)
CB_1 antagonist/inverse agonist; MAGL inhibitor	Rimonabant; URB602	Intra-dlPAG	Rat	SIA—footshock and tail flick	Rimonabant-attenuation of SIA; URB602—enhances SIA	Hohmann et al. (2005)
CB_1 antagonist/inverse agonist; dual FAAH/TRPV1 inhibitor	Rimonabant; AA-5-HT	Intra-dlPAG	Rat	SIA—footshock and tail flick	Rimonabant-suppression of SIA; AA-5-HT—enhances SIA	Suplita et al. (2005)
CB_1 antagonist/inverse agonist	Rimonabant	Intra-dlPAG	Rat	FCA—formalin and footshock	Rimonabant attenuated FCA	Olango et al. (2012)

$CB_{1/2}$, cannabinoid receptor type 1/2; TRPV1, transient receptor potential cation channel subfamily V member 1; FAAH, fatty acid amide hydrolase; MAGL, monoacylglycerol lipase; PAG, periaqueductal gray; vlPAG, ventrolateral periaqueductal gray; dPAG, dorsal periaqueductal gray; RVM, rostral ventromedial medulla; dlPAG, dorsolateral periaqueductal gray; SIA, stress-induced analgesia; FCA, fear-conditioned analgesia.

Table 4 Summary of Studies Investigating the Role of the Endocannabinoid System in the Amygdala in Pain and Its Modulation by Stress

Class of Pharmacological Compound	Pharmacological or Genetic Intervention	Route of Administration	Species	Model	Effect	References
CB_1/CB_2 agonist	WIN55,212-2	Intra-BLA; intra-CeA	Rat	Tail flick	Increased tail flick latency	Martin et al. (1999)
CB_1/CB_2 agonist	WIN55,212-2	Intra-BLA	Rat	Formalin; tail flick	Increased tail flick latency; decreased formalin-evoked nociceptive behavior—attenuated via AM251	Hasanein, Parviz, Keshavarz, and Javanmardi (2007)
CB_1/CB_2 agonist	WIN55,212-2;	Intra-CeA, intra-BLA	Rats	Tail flick	Antinociception	Manning, Martin, and Meng (2003)
CB_1/CB_2 agonist	WIN55,212-2	Intramuscularly	Rhesus monkey	Warm-water tail-withdrawal assay	Dose-dependent antinociception; attenuated via bilateral amygdala lesions	Manning, Merin, Meng, and Amaral (2001)
CB_1 antagonist/ inverse agonist	Rimonabant	Intra-BLA	Rat	SIA—tail flick, footshock	Suppression of SIA	Connell, Bolton, Olsen, Piomelli, and Hohmann (2006)
CB_1 antagonist/ inverse agonist	Rimonabant	Intra-BLA	Rat	FCA—formalin, footshock	Reduced formalin-evoked nociceptive behavior; no effect on FCA	Roche, O'Connor, Diskin, and Finn (2007) and Roche et al. (2010)

Continued

Table 4 Summary of Studies Investigating the Role of the Endocannabinoid System in the Amygdala in Pain and Its Modulation by Stress—cont'd

Class of Pharmacological Compound	Pharmacological or Genetic Intervention	Route of Administration	Species	Model	Effect	References
FAAH inhibitor	URB597	Intraperitoneal	Rat	FCA—formalin, footshock	URB597 enhances FCA; weakened by rimonabant and SR144528; FCA associated with enhanced phospho-ERK in the amygdala	Butler et al. (2008)
CB_1 antagonist/inverse agonist	AM251	Intraperitoneal; intra-BLA; intra-CeA	Rat	FCA—formalin, footshock	AM251 prevents expression of FCA following intraperitoneal and intra-BLA but not intra-CeA injection	Rea et al. (2013)

CB_1, cannabinoid receptor type 1; FAAH, fatty acid amide hydrolase; BLA, basolateral amygdala; CeA, central nucleus of the amygdala; SIA, stress-induced analgesia; FCA, fear-conditioned analgesia.

Table 5 Summary of Studies Investigating the Role of the Endocannabinoid System in the PFC in Pain and Its Modulation by Stress

Class of Pharmacological Compound	Pharmacological or Genetic Intervention	Route of Administration	Species	Model	Effect	References
Dual FAAH/TRPV1 inhibitor	AA-5-HT	Intra-PrL; intra-IL	Rats	Noxious mechanical stimuli (Von Frey); SNI rats	Reduced mechanical allodynia in rats following SNI	Giordano et al. (2012)
Dual FAAH/TRPV1 inhibitor; FAAH inhibitor; TRPV1 antagonist	AA-5-HT; URB597; I-RTX	Intra-PrL; intra-IL	Rats	SNI rats	All drugs produced antinociception; AA-5-HT produced antinociception more efficiently than URB597 or I-RTX	de Novellis et al. (2011)
PPARα antagonist; PPARα agonist	GW6471; GW7647	Intra-mPFC	Rats	Formalin	GW6471, but not GW7647, delayed onset of the second phase of formalin-evoked nociceptive behavior	Okine et al. (2014)
CB_1 antagonist/ inverse agonist	AM251	Intra-PrL	Rat	FCA—tail flick, footshock	Attenuated FCA	Freitas, Salgado-Rohner, Hallak, Crippa, and Coimbra (2013)

CB_1, Cannabinoid receptor type 1; TRPV1, transient receptor potential cation channel subfamily V member 1; FAAH, fatty acid amide hydrolase; PPARα, peroxisome proliferator-activated receptor alpha; SNI, spared nerve injury; mPFC, medial prefrontal cortex; IL, infralimbic cortex; PrL, prelimbic cortex; FCA, fear-conditioned analgesia.

Brain region	Receptor event	Functional consequence
BLA	CB_1 receptor inhibition	Decreased SIA
Amygdala	eCB dysfunction in WKY rats	Decreased SIH
dlPAG	CB_1 receptor inhibition	Decreased SIA
dlPAG	CB_1 receptor activation	Increased SIA
dlPAG	CB_1 receptor activation coupled with TRPV1 inhibition	Increased SIA
RVM	CB_1 receptor inhibition	Decreased SIA
RVM	CB_1 receptor activation coupled with TRPV1 inhibition	Increased SIA
RVM	eCB dysfunction in WKY rats	Decreased SIH

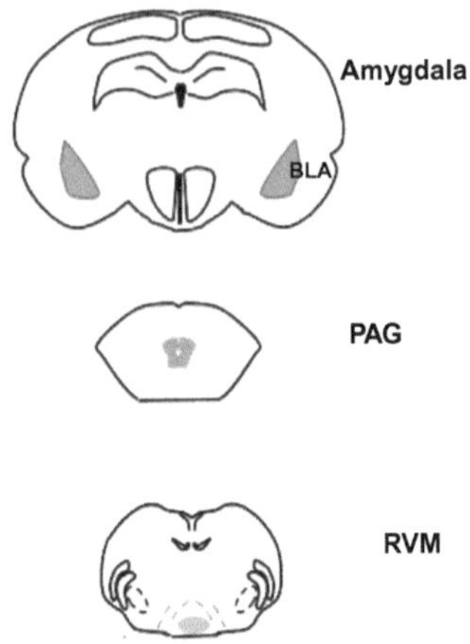

Figure 2 A synthesis of the literature reviewed herein on the role of the supraspinal endocannabinoid system in discrete brain regions in stress-induced analgesia (SIA) and stress-induced hyperalgesia (SIH). eCB, endocannabinoid; BLA, basolateral amygdala; PAG, periaqueductal gray; dlPAG, dorsolateral PAG; RVM, rostral ventromedial medulla; TRPV1, transient receptor potential vanilloid 1; WKY, Wistar Kyoto rat.

nucleus to exert bidirectional control over nociception (Aicher, Hermes, Whittier, & Hegarty, 2012; Basbaum & Fields, 1984).

The RVM shares connections with the PAG, forming the PAG-RVM pathway (Basbaum & Fields, 1984). CBs activate descending analgesia via this pathway through a process of "GABA disinhibition." According to the GABA disinhibition hypothesis of analgesia, CB_1 receptor-mediated inhibition of GABAergic interneurons in the PAG and RVM results in disinhibition of projection neurons within the descending inhibitory pain pathway, resulting in analgesia (Basbaum & Fields, 1984; Lau & Vaughan, 2014; Szabo & Schlicker, 2005).

Microinjection of the CB receptor agonists WIN55,212-2 and HU210 into the RVM suppressed nociceptive behaviors in the tail flick test, an effect attenuated by coadministration with the CB_1 receptor antagonist rimonabant (Martin et al., 1998). Nociceptive behavior remained unchanged upon CB receptor agonist injection outside of the RVM. CBs induce antinociception by modulating neuronal activity in the RVM and inactivation of the RVM prevents CB-induced analgesia (Meng et al., 1998). As previously mentioned, ON cells in the RVM increase firing in response to painful stimuli, whereas OFF cells decrease firing, facilitating, and inhibiting pain, respectively. Intra-RVM microinjection of WIN55,212-2 increases tail flick latencies while inhibiting ON-cell activity and increasing OFF-cell activity, thus decreasing nociception. Coinfusion with rimonabant blocked these effects, indicating a role for the eCB-CB_1 receptor system in the RVM in nociception (Meng & Johansen, 2004).

Microinjection of WIN55,212-2 into the GiA resulted in behavioral analgesia in the rat tail flick test, an effect blocked by coadministration of rimonabant (Monhemius et al., 2001). In the same study, animals with partial sciatic nerve ligation were given intra-GiA WIN55,212-2 and rimonabant and intraplantar formalin, contralaterally to the site of nerve ligation. Formalin-evoked nociceptive behavior was significantly reduced in partial nerve ligated rats, an effect reversed by microinjection of rimonabant into the GiA. This study demonstrated a role for the CB_1 receptor in GiA-mediated antinociception and modulation of nociceptive transmission in both acute pain and chronic neuropathic pain (Monhemius et al., 2001).

5.2 SIA

To our knowledge, only one study to date has investigated the role of the eCB system in the RVM in the modulation of pain by acute stress (SIA). Intra-RVM administration of rimonabant attenuated SIA in a rat model that combined footshock and a tail flick test. The FAAH inhibitor and TRPV1 antagonist AA-5-HT, administered systemically or intra-RVM, enhanced SIA in rats in a CB_1 receptor-dependent manner (Suplita et al., 2005). This study provides evidence for an important role of CB_1 receptors in the RVM in mediating and modulating SIA.

5.3 SIH

While there is evidence for a role of the RVM in SIH (for review, see Jennings et al., 2014), few studies have specifically investigated the role of the eCB system in the RVM in SIH. Genetic background plays a key role in determining the effect of stress on pain. The Wistar Kyoto (WKY) rat displays increased sensitivity to noxious stimuli and exhibits a depressive/anxiety-like phenotype and hypersensitivity to stress, compared with other rat strains including Sprague Dawley (SD) rats (Burke et al., 2010; O'Mahony et al., 2010). We have recently reported an impairment in pain-related mobilization of the eCBs AEA and 2-AG, along with their synthesizing enzymes, NAPE-PLD and DAGL, respectively, in the RVM of WKY rats compared with SD rats, following intraplantar injection of formalin (Rea et al., 2014). Systemic administration of AM251 potentiated while systemic administration of the FAAH inhibitor URB597 attenuated hyperalgesia to formalin injection in WKY rats, but not SD rats, an effect mediated by CB_1 receptors in the RVM. These data suggest eCB dysfunction in the RVM underlies the hypersensitivity to noxious stimuli in WKY rat model of negative affective state (Rea et al., 2014).

See Table 2 for a summary of studies investigating the role of the eCB system in the RVM in pain and its modulation by stress.

6. THE ROLE OF THE ENDOCANNABINOID SYSTEM IN THE PERIAQUADUCTAL GRAY IN PAIN, STRESS-INDUCED ANALGESIA, AND STRESS-INDUCED HYPERALGESIA

6.1 Pain

The periaquaductal gray (PAG) is a midbrain/brainstem structure that can be divided into four columns along its rostro-caudal axis: dorsomedial (dmPAG), dorsolateral (dlPAG), lateral (lPAG), and ventrolateral (vlPAG) columns (Bandler & Keay, 1996). Exposure to an aversive stimulus activates the descending inhibitory pain pathway, of which the PAG is a key component. The PAG, via the RVM, modulates nociceptive transmission at the level of the spinal cord (Fields, Heinricher, & Mason, 1991). The PAG possesses a larger density of CB receptors than other brainstem structures (Herkenham et al., 1991). CBs act in the PAG to inhibit GABAergic and glutamatergic synaptic transmission and to produce analgesia by a disinhibitory mechanism (Vaughan, Connor, Bagley, & Christie, 2000).

CB_1 receptor-mediated antinociception and increased levels of AEA were reported following electrical stimulation of the dorsal PAG (dPAG) and lPAG (Walker et al., 1999). These authors also showed that subcutaneous injection of formalin elicited a pain response in rats and substantially increased AEA levels in the PAG, measured by *in vivo* microdialysis. Increased levels of the eCBs, AEA, and 2-AG were also seen in the PAG and RVM of rats 7 days post chronic constriction injury (CCI) of the sciatic nerve, when hyperalgesia and mechanical allodynia were observed to be maximal (Petrosino et al., 2007).

Intra-vlPAG administration of morphine in rats enhanced the antinociceptive effect of the CB_1 receptor agonist HU-210 in the hot plate test (Wilson-Poe et al., 2013). Likewise, intra-vlPAG and systemic administration of HU-210 enhanced the antinociceptive effect of morphine (Wilson,

Maher, & Morgan, 2008). This study provides evidence for a dual role of morphine and CBs in pain and antinociception. Formalin-evoked nociceptive behavior was reduced following microinjection of HU210 into the dorsal PAG (dPAG) of rats, an effect blocked by coadministration with the CB_1 receptor antagonist rimonabant (Finn et al., 2003). Microinjection of CP-55,940 into the vlPAG, but not the posterior dlPAG or the anterior vlPAG, areas produces antinociception in the rat tail flick test (Lichtman et al., 1996). WIN55,212-2 increased tail flick latencies following microinjection into the rat dlPAG (Martin et al., 1995). Microinjection of WIN55,212-2 into the PAG increased the latency of the nociceptive response in the plantar test in rats, an effect blocked by coadministration with rimonabant. MPEP, a metabotropic glutamate receptor mGlu5 antagonist, also completely blocked the antinociceptive effect of WIN55,212-2 (Palazzo et al., 2001), indicating a CB_1-glutamatergic interaction in the PAG in mediating CB-induced analgesia.

Studies investigating the analgesic effect of the nonsteroidal anti-inflammatory drugs (NSAIDs) in supraspinal structures indicate a role for eCBs and CB_1 receptors in the PAG and RVM. Inflammation-induced hyperalgesia can be attenuated by microinjection of the NSAID metazinol into the PAG (Vazquez, Escobar, Ramirez, & Vanegas, 2007). Injection of the CB_1 receptor antagonist AM251 into the PAG or RVM reverses metazinol-induced analgesia, suggesting a role for the eCB system in these brain regions in NSAID-induced analgesia (Escobar et al., 2012).

TRPV1, a target of AEA, is expressed in the PAG (Palazzo, Rossi, & Maione, 2008) and a role for TRPV1 in pain modulation in the PAG has also been demonstrated. Intra-dlPAG injection of the TRPV1 agonist capsaicin increased the latency of nociceptive responses in the rat plantar test (Palazzo et al., 2002). A higher dose administered to the same region produced opposite effects, decreasing the latency of nociceptive responses and inducing hyperalgesia followed by analgesia (McGaraughty et al., 2003). Similar to Palazzo et al. (2002), intra-vlPAG administration of capsaicin also increased the latency of nociceptive responses in the hot-plate responses in rats (Liao, Lee, Ho, & Chiou, 2011) (for review, see Starowicz, Nigam, & Di Marzo, 2007). Thus, TRPV1 agonism in the PAG elicits antinociceptive effects in several pain models. For a recent review, see Madasu, Roche, and Finn (2015).

Intra-vlPAG injection of the FAAH inhibitor URB597 produced a robust hyperalgesic response at low doses, an analgesic response at high doses, and a biphasic effect on nociception at intermediate doses, in the rat plantar test (Maione et al., 2006). AEA and 2-AG levels were increased in a dose-dependent manner following URB597 administration into the vlPAG. Coadministration of a low dose of URB597 with the CB_1 receptor antagonist AM251 converted the hyperalgesic effect to an analgesic one, while coadministration of URB597 with both the TRPV1 antagonist capsazepine and AM251 abolished all effects. In comparison, the early hyperalgesic effect of the intermediate dose of URB597 was blocked by AM251, while the later URB597-induced analgesic effect became hyperalgesic following TRPV1 antagonism. CB_1 receptor-dependent analgesia was seen at the highest dose of intra-vlPAG URB597 administration (Maione et al., 2006). The URB597-induced antinociceptive effects (TRPV1-mediated) and pronociceptive effects (CB_1 receptor mediated) were associated with enhanced or reduced RVM OFF cell activity, respectively, suggesting URB597-induced alteration in the activity of excitatory PAG output neurons. This study indicates a role for both CB_1 and TRPV1 receptors in the eCB-mediated control of the descending pain pathway.

Diabetes is frequently associated with neuropathy, with many patients suffering from hyperalgesia or allodynia. A role for TRPV1 and CB_1 receptors in the PAG has been proposed in diabetic thermal hyperalgesia (Mohammadi-Farani, Sahebgharani, Sepehrizadeh, Jaberi, & Ghazi-Khansari, 2010). Intra-vlPAG administration of capsaicin and WIN produced antinociception in the hot plate test of nondiabetic mice (Mohammadi-Farani et al., 2010). In contrast, the antinociceptive effects of intra-vlPAG capsaicin and WIN were reduced in hyperalgesic diabetic mice, an effect associated with CB_1 upregulation and TRPV1 downregulation in the vlPAG (Mohammadi-Farani et al., 2010). Taken together, the data demonstrate that diabetic neuropathy is associated with altered eCB signaling in the PAG, effects which may underlie the associated hyperalgesia and allodynia.

Systemic administration of the FAAH inhibitor and TRPV1 antagonist AA-5-HT produced antinociceptive effects in both rats and mice treated with formalin and in rats with CCI of the sciatic nerve (Maione et al., 2007), effects associated with increased levels of AEA in both the PAG and RVM. These antinociceptive effects were blocked by both CB_1 receptor and TRPV1 antagonists. Intra-vlPAG injection of AA-5-HT increased eCB levels and induced a pronociceptive effect at low doses and an antinociceptive effect at higher doses in the rat tail flick test (de Novellis et al., 2008). These effects were blocked by antagonism of vlPAG CB_1 receptors (AM251) or TRPV1 (I-RTX). Furthermore, administration of the FAAH inhibitor URB597 with the TRPV1 antagonist I-RTX into the vlPAG also induced antinociceptive effects in the rat tail flick test and inhibited RVM ON- and OFF-cell activity (de Novellis et al., 2008), thus indicating that the antinociceptive effects of FAAH substrates in the vlPAG may be mediated by CB_1 receptors. In the formalin test of inflammatory pain, intra-PAG AA-5-HT prevented the changes in the ON and OFF cell firing activity induced by intraplantar injection of formalin. Since CB_1 and TRPV1 antagonists blocked the effects of AA-5-HT (de Novellis et al., 2008), it suggests that these two eCB receptors in the PAG may be responsible for AA-5-HT-induced analgesia. Furthermore, intra-PAG administration of the GPR55 agonist LPI reduced the nociceptive threshold in the rat hot plate test, an effect blocked upon pretreatment with the GPR55 antagonist ML-193 (Deliu et al., 2015). This study suggests that altering GPR55 activity in the PAG may affect nociceptive behaviour. Taken together these studies suggest that CB_1 receptors, TRPV1 and GPR55 in the PAG all play important roles in modulating pain behavior.

Orexin (OX) A and B are peptides and endogenous agonists for the OX1 and OX2 receptors which are localized in the lateral and perifornical area of the hypothalamus (de Lecea et al., 1998; Sakurai et al., 1998; Tsujino & Sakurai, 2009). They exert antinociceptive effects (Chiou et al., 2010) including following direct administration into the PAG (Azhdari Zarmehri et al., 2012). Orexin A decreases GABA release in an eCB-dependant manner in the vlPAG. Activation of OX receptors in the vlPAG leads to antinociception, measured electrophysiologically in brain slices. Intra-vlPAG microinjection of orexin-A reduced hot-plate nociceptive responses in rats in a manner blocked AM 251 (Ho et al., 2011).

6.2 SIA

A number of studies have demonstrated an important role for the eCB system in the PAG in SIA/FCA. Intra-dPAG administration of the CB_1 receptor antagonist rimonabant attenuated SIA, observed as an increase in the tail flick latency following exposure of rats to footshock stress (Hohmann et al., 2005). The same dose of this drug administered i.c.v., intra-vlPAG, and intra-lPAG had no effect on SIA in this study, highlighting a particular role of CB_1 receptors in the dlPAG in mediating SIA. Increased levels of 2-AG were seen in the dlPAG directly after footshock stress, implicating this eCB in the dlPAG in SIA. Moreover, inhibition of the 2-AG-degrading enzyme MAGL in the dlPAG using URB602 enhanced SIA (Hohmann et al., 2005).

A subsequent study by the same group confirmed the CB_1 receptor-dependant attenuation of SIA following intra-dlPAG administration of rimonabant and the CB_1-dependant enhancement of SIA following AA-5-HT administration (Suplita et al., 2005). These studies provide evidence that the PAG is a key neural substrate for eCB-mediated SIA. Another follow-up study from this group showed that mGlu5 receptor activation mobilizes 2-AG in the dlPAG to produce SIA in rats (Gregg et al., 2012). Thus, unconditioned SIA mediated by CB_1 receptor stimulation in the PAG is under the control of glutamatergic neurotransmission via mGlu5 receptors.

Our group has reported a role for the eCB system in the PAG in a model of SIA associated with conditioned learned fear (FCA) (Olango et al., 2012). FCA in these studies was measured as the reduction of formalin-evoked nociceptive behavior upon re-exposure of rats to a conditioning arena previously paired with footshock. Systemic administration of the FAAH inhibitor URB597 enhanced FCA, an effect associated with reduced phospho-ERK1/2 expression in the PAG (Butler et al., 2008). FCA was attenuated by intra-dlPAG administration of rimonabant (Olango et al., 2012), confirming a role for CB_1 receptors in the dlPAG in mediating both conditioned and unconditioned forms of SIA.

6.3 SIH

While there is evidence for a role of the PAG in SIH (for review, see Jennings et al., 2014), there is currently a paucity of studies addressing the role of the eCB system in the PAG in SIH and this is an area that warrants investigation.

See Table 3 for a summary of studies investigating the role of the eCB system in the PAG in pain and its modulation by stress.

7. THE ROLE OF THE ENDOCANNABINOID SYSTEM IN THE AMYGDALA IN PAIN, STRESS-INDUCED ANALGESIA, AND STRESS-INDUCED HYPERALGESIA

7.1 Pain

The amygdala is a key region of the limbic system located in the medial temporal lobe. It contains a number of different nuclei including the lateral nucleus (LA), basolateral nucleus (BLA), the central nucleus (CeA), accessory basal nucleus (ABA), and the medial nucleus (MeA). The amygdala plays a key role in the interaction between pain and emotion. The CeA, in particular, is involved in the emotional-affective component of persistent pain (Neugebauer, Galhardo, Maione, & Mackey, 2009; Neugebauer, Li, Bird, & Han, 2004), while the BLA may be involved in the modulation of acute or tonic nociceptive processing (Oliveira & Prado, 1998). The amygdala is a key region of the ascending and descending pain pathways and shares connections with other key regions including the PFC and PAG. Pain-related changes have been identified in the amygdala in animals and humans using PET and fMRI neuroimaging studies (Neugebauer et al., 2004).

All components of the eCB system are expressed in the amygdala, although CB_1 receptors are expressed in highest density in the BLA (Herkenham et al., 1991; Tsou, Brown, et al., 1998). The amygdala contributes to the antinociceptive effects produced by systemically administered CBs. WIN55,212-2 produces dose-dependent antinociceptive effects in rats characterized as increased tail flick latencies (Manning et al., 2003). Intra-CeA, but not intra-BLA, administration of muscimol, significantly attenuated these antinociceptive effects of systemically administered WIN55, 212-2. Moreover, unilateral CeA inactivation via muscimol reduced the suppression of formalin-evoked c-Fos expression by WIN55,212-2 in the superficial dorsal horn of the spinal cord but not in the deeper "nociceptive" laminae (Manning et al., 2003). Another study from the same group found that the amygdala also plays a role in antinociception in non-human primates (Manning et al., 2001). WIN55,212-2 produced dose-dependent analgesia in rhesus monkeys. Bilateral lesions to the amygdala of the monkeys significantly reduced CB-induced analgesia. Both of these lesion studies indicate that the eCB system in the amygdala, in particular the CeA, can mediate antinociceptive effects.

Tail flick latencies have been shown to be increased upon microinjection of WIN55,212-2 into the CeA and BLA in rats (Hasanein et al., 2007; Martin et al., 1999). Furthermore, intra-BLA administration of WIN55,212-2 has also been shown to reduce formalin-evoked nociceptive behavior in rats, an effect attenuated by intra-BLA administration of the CB_1 receptor antagonist AM251 (Hasanein et al., 2007). Interestingly, intra-BLA administration of rimonabant has also been shown to attenuate formalin-evoked nociceptive behavior and associated increases in c-Fos immunoreactivity in the hippocampus and RVM in rats (Roche et al., 2007, 2010), although intra-BLA administration of a different CB_1 receptor antagonist, AM251, did not exert a similar effect (Rea et al., 2013).

Using fMRI, it has been shown that the amygdala may play a role in the modulation of pain perception by Δ^9-THC in humans (Lee et al., 2013).

Cutaneous ongoing pain and hyperalgesia induced by capsaicin were monitored in healthy cannabis-naive volunteers. Δ^9-THC reduced "painfulness" but not the intensity of pain and hyperalgesia, an effect positively correlated with amygdala activity. A Δ^9-THC-related reduction in sensory-limbic functional activity was also seen between the amygdala and primary sensorimotor areas (Lee et al., 2013).

While the evidence points to a clear role for the eCB system in the amygdala in antinociception, there is a paucity of studies investigating its impact on the emotional aspect of pain. As a region with a clear role for the interaction between pain and emotion, it is necessary to further investigate this area and the role of the eCB system therein.

7.2 SIA

The amygdala plays a role in both unconditioned and conditioned SIA (Helmstetter, 1992; Helmstetter & Bellgowan, 1993; Helmstetter, Bellgowan, & Poore, 1995; Werka, 1994, 1997; Werka & Marek, 1990). Intra-BLA microinjection of rimonabant has been shown to suppress unconditioned SIA in rats exposed to footshock stress and then tested in the tail flick test, whereas intra-CeA microinjection had no effect on this form of SIA (Connell et al., 2006). Intra-BLA administration of FAAH and MAGL inhibitors, however, had no effect on SIA (Connell et al., 2006), suggesting that CB_1 receptors in the BLA, but not CeA, mediate SIA, although the role of the individual eCBs requires further investigation. Roche et al. (2007, 2010) reported no effect of unilateral or bilateral intra-BLA administration of rimonabant on FCA in rats (Roche et al., 2010, 2007). However, a subsequent study showed that the expression of FCA in rats was reduced following systemic or intra-BLA, but not intra-CeA, administration of a different CB_1 receptor antagonist, AM251 (Rea et al., 2013).

URB597 enhances the expression of FCA when administered via the intraperitoneal route, an effect blocked by CB_1, CB_2, or μ-opioid receptor antagonists (Butler et al., 2008). Interestingly, FCA in this study was associated with increased expression of phospho-ERK2 in the amygdaloid complex. In contrast, the URB597-induced enhancement of FCA was associated with reduced phospho-ERK1 and phospho-ERK2 expression in the amygdala. This dichotomy is not consistent with a causal role of ERK signaling in FCA (Butler et al., 2008).

CB_1 receptors are expressed on GABAergic and glutamatergic neurons in the BLA (Herkenham et al., 1991; Katona et al., 2001). Expression of FCA in rats was reduced following systemic or intra-BLA, but not intra-

CeA, administration of the CB_1 receptor antagonist AM251 (Rea et al., 2013), an effect attenuated by intra-BLA administration of both the $GABA_A$ receptor antagonist, bicuculline, and the mGlu5 receptor antagonist, MPEP, suggesting that CB_1 receptors in the BLA facilitate the expression of FCA, through a mechanism which is likely to involve the modulation of GABAergic and glutamatergic signaling. FCA was associated with increased levels of AEA in the left BLA (side contralateral to intraplantar formalin injection). Fear-conditioned, formalin-treated rats displayed increased levels of 2-AG and PEA in the left and right BLA, respectively (Rea et al., 2013).

It is clear, therefore, that the eCB system in the amygdala, and specifically the BLA, plays an important role in mediating both unconditioned and conditioned SIA with likely interactions with GABAergic and glutamatergic signaling.

7.3 SIH

A recent study from our group investigated the effects of repeated exposure to forced swim stress on formalin-evoked nociceptive behavior in rats in stress normo-responsive (SD) and stress hyper-responsive (WKY) rat strains. Formalin-evoked nociceptive behavior was increased in SD rats following 10 days of forced swim stress (Jennings, Okine, Olango, Roche, & Finn, 2015). AEA levels were reduced in the contralateral amygdala (relative to formalin injection) of SD rats but not WKY rats. There were also strain differences in components of the eCB system within the amygdala. For example, decreased levels of AEA and 2-AG were observed in the ipsilateral amygdala of SD, but not WKY, rats. Lower levels of CB_1 receptor mRNA were seen in the ipsilateral, but not contralateral, amygdala of WKY rats. These data indicate a role for the eCB system in the amygdala in SIH as well as implicating it in the strain differences seen in WKY and SD rats (Jennings et al., 2015). Additional studies are warranted to fully understand the role of the eCB system in the amygdala in SIH.

See Table 4 for a summary of studies investigating the role of eCB system in the amygdala in pain and its modulation by stress.

8. THE ROLE OF THE ENDOCANNABINOID SYSTEM IN THE PREFRONTAL CORTEX IN PAIN, STRESS-INDUCED ANALGESIA, AND STRESS-INDUCED HYPERALGESIA

8.1 Pain

The PFC is involved in both the top-down descending modulation of pain and also in the affective dimension of the pain experience. The medial PFC (mPFC) is comprised of the prelimbic cortex (PrL), infralimbic cortex (IL), and anterior cingulate cortex (ACC). Imaging studies have shown that the PFC is consistently activated by noxious stimuli (Casey, Minoshima, Morrow, & Koeppe, 1996; Davis, Taylor, Crawley, Wood, & Mikulis, 1997; Derbyshire et al., 1997; May et al., 1998; Millan, 1999; Neal, Pearson, & Powell, 1990; Svensson, Minoshima, Beydoun, Morrow, & Casey, 1997). CB_1 receptors are expressed in the PFC (Herkenham et al., 1991; Sim-Selley, Vogt, Vogt, & Childers, 2002; Tsou, Brown, et al., 1998). This, along with its projections to the PAG and amygdala (Diorio, Viau, & Meaney, 1993; Little & Carter, 2013; Marchand & Hagino, 1983), suggests a role for the EC system in the PFC in pain.

CB_1 receptors in the rodent mPFC are expressed on GABAergic interneurons (Marsicano & Lutz, 1999; Wedzony & Chocyk, 2009). CB_1 receptors on presynaptic axon terminals face pyramidal neurons with postsynaptic mGluR5 (Lafourcade et al., 2007). A rat arthritis pain model, induced via intra-articular injections of kaolin and carrageenan through the patellar ligament, shows hyperactivity in amygdala output neurons and abnormal inhibition of mPFC pyramidal neurons (Ji et al., 2010). Another study investigated the effect of mGluR5 and CB_1 receptor activation on the activity of the mPFC cells in rats in the previously described arthritis pain model (Ji & Neugebauer, 2014). Coactivation of mGluR5 and CB_1 receptors increased mPFC activity and inhibited pain-related neuronal activity in the CeA in the arthritis pain model. Thus, there appears to be an inverse link between activation of mPFC neurons and amygdala output and a role for the eCB system in this top-down cortical control (Ji & Neugebauer, 2014). Further evidence for a role of the eCB system in the PFC in arthritic conditions comes from work demonstrating that osteoarthritis pain is associated with increased 2-AG levels in the PFC of mice in the monosodium iodoacetate model of arthritis (La Porta et al., 2015).

CB_1 receptor activity is decreased in the rostral ACC 10 days post CCI in mice, compared with sham controls (Hoot et al., 2010). CB_1 receptor levels in the rostral ACC of CCI and sham rats remained unchanged and there were no significant differences in the levels of 2-AG or AEA in the ACC between CCI and sham-operated mice. The ACC is associated with the affective component of pain (Kulkarni et al., 2005; Kuo, Chiou, Liang, & Yen, 2009; LaBuda & Fuchs, 2005; Treede, Kenshalo, Gracely, & Jones, 1999). It is possible therefore that reduced CB_1 receptor activity in the ACC is associated with the negative affective component of neuropathic pain.

TRPV1 expression is increased, in glutamatergic neurons, in the mPFC (namely the PrL and IL) following spared nerve injury (SNI) (Giordano et al., 2012). Intra-PL/IL administration of AA-5-HT reduced mechanical allodynia in rats following SNI to a greater extent than that seen with a FAAH inhibitor or TRPV1 antagonist alone (Giordano et al., 2012). SNI-induced neuropathic pain is also associated with increased levels of endovanilloids and eCBs in the mPFC. Intra-PrL/IL injection of AA-5-HT produced antinociceptive effects more efficiently (de Novellis et al., 2011). These studies suggest that both the eCB and endovanilloid systems in the mPFC may play a role in neuropathic pain. Therapies which target both of these systems may prove useful in the treatment of chronic neuropathic pain.

We have studied the role of PPARα in the mPFC in formalin-evoked nociceptive behavior in rats. The PPARα antagonist GW6471 delayed the onset of the second phase of formalin-evoked nociceptive behavior. This reduction in nociceptive behavior was associated with a reduction in the levels of *N*-palmitoylethanolamide and *N*-oleoylethanolamide (PPARα ligands) in the mPFC (Okine et al., 2014). Together these data suggest a facilitatory role for PPARα in the mPFC in formalin-evoked nociceptive behavior.

8.2 SIA

Lesion studies have indicated a role for the PFC in acquisition, consolidation, and extinction of conditioned fear in rodents (Sierra-Mercado, Corcoran, Lebron-Milad, & Quirk, 2006). This region has also been shown to project to other regions important in fear neurocircuitry, including the previously discussed amygdala and PAG (LeDoux, 2000). CB_1 receptors in the PrL cortex are involved in the amplification of panic-like aversive reactions and SIA. Thus, microinjection of bicuculline into the dorsomedial and ventromedial hypothalamus-induced aversive panic-like behavior and SIA, an effect attenuated by microinjection of AM251 into the PrL (Freitas et al., 2013). This work suggests that CB_1 receptor signaling in the PrL may facilitate or augment SIA induced by stimulation of the hypothalamus. Further investigation of the roles of the eCB system in the PrL, IL, and ACC in SIA and FCA is warranted.

8.3 SIH

To our knowledge, there have been no published studies to date investigating the role of the eCB system in the PFC in SIH.

See Table 5 for a summary of studies investigating the role of the eCB system in the PFC in pain and its modulation by stress.

9. LESS CHARACTERIZED SUPRASPINAL ENDOCANNABINOID MECHANISMS IN PAIN MODULATION

Systemic administration of the CB receptor agonist WIN55,212-2 dose-dependently inhibited stimulus-evoked activity, in the form of graded pressure stimuli to the paw, of nociceptive neurons in the ventroposterolateral thalamus (VPL) of anesthetized rats (Martin et al., 1996). Further evidence for a role of CB_1 receptors in the thalamus in mediating and modulating nociceptive responding was observed following microinjection of WIN55,212-2 into the thalamus which resulted in antinociceptive effects in the tail flick test in rats (Martin et al., 1999). Similar effects were observed following microinjection into the alpha part of the gigantocellular reticular nucleus (GiA) and the noradrenergic A5 region. Furthermore, intralocus coeruleus microinjection of the hypothalamic peptide orexin-A decreased formalin-evoked nociceptive behavior in rats (Mohammad-Pour Kargar, Azizi, Mirnajafi-Zadeh, Reza Mani, & Semnanian, 2015), an effect reversed following pretreatment with either the OX1 receptor antagonist SB-334867 or the CB_1 receptor antagonist AM251. Intra-locus coeruleus microinjection of SB-334867 and AM251 alone induced hyperalgesia (Mohammad-Pour Kargar et al., 2015). The results from this study suggest a new mechanism by which orexin-A modulates nociceptive information in the locus coeruleus via interaction with CB_1 receptors.

There is now increasing evidence supporting the role of CB_2 receptors in the supraspinal modulation of pain (for review, see Guindon & Hohmann, 2008). For example, microinjection of the CB_2 receptor agonist JWH-133 into the ventral posterolateral nucleus of the thalamus (VPL) has been shown to reduce noxious activity, recorded with a multichannel electrode array in VPL neurons, in a rat model of neuropathic pain (spinal nerve ligation; SNL) (Jhaveri et al., 2008). No significant differences in the levels of eCBs in the thalamus of SNL rats compared to sham rats were observed (Jhaveri et al., 2008). The results from this study suggest that CB_2 receptors in the thalamus may contribute to the modulation of neuropathic pain responses.

The eCB system has also been proposed to play a role migraine-related pain (for review, see Greco, Gasperi, Maccarrone, & Tassorelli, 2010;

Russo, 2004; Smith & Wagner, 2014). FAAH-deficient mice (FAAH (−/−)) express less nitroglycerin-induced migraine-like pain, with similar effects observed following pharmacological inhibition of FAAH inhibitors using URB597 and PF3945. Administration of the CB_1 receptor antagonist rimonabant blocked these antinociceptive effects in this migraine model, demonstrating a key role for CB_1 receptors in mediating the effects of the FAAH substrates (i.e., AEA) (Nozaki et al., 2015). Similarly, several other studies have demonstrated that genetic and/or pharmacological inhibition of FAAH is associated with increased AEA levels in the brain, and associated with antinociceptive effects in several pain models (Kwilasz et al., 2014; Lichtman, Shelton, et al., 2004). For example, URB597 (intraperitoneal), a selective FAAH inhibitor, produced antinociception in the form of CB_1 dependant decreases in acid-stimulated stretching in a lactic acid model of pain, an effect associated with increased AEA levels in the brain (Kwilasz et al., 2014). Increased FAAH activity and an increased density of CB-binding sites have also been found in the hypothalamus in animal models of migraine (nitroglycerin-induced hyperalgesia) (Greco, Gasperi, Sandrini, et al., 2010). It is clear that elevation of brain eCB levels produces robust modulatory effects in mouse models of pain antinociception (Cravatt et al., 2001; Holt, Comelli, Costa, & Fowler, 2005; Jayamanne et al., 2006; Lichtman, Leung, et al., 2004; Lichtman, Shelton, et al., 2004), suggesting supraspinal CB_1 receptor-dependant antinociception.

PET imaging with a CB_1 receptor radioligand demonstrated that patients with in functional dyspepsia have higher CB_1 receptor availability in the hypothalamus and ACC (Ly et al., 2015). It is possible that eCB dysfunction and abnormal brain activity in these areas may be related to the pain felt in patients with functional dyspepsia; however, further work is warranted (Ly et al., 2015).

As previously mentioned, eCBs act on other non-CB_1/non-CB_2 receptors, such as the ligand-gated ion channel, TRPV1. TRPV1 on primary afferent neurons plays a key role in the sensation of pain and thermal hyperalgesia (Caterina et al., 2000). However, increasing evidence suggests a role for TRPV1 in pain modulation in supraspinal regions (Madasu et al., 2015). Central administration of the TRPV1 antagonist A-784168 induced potent analgesia in the rat sodium monoiodoacetate model of osteoarthritic pain and reduced thermal hyperalgesia and mechanical allodynia in the complete Freund's adjuvant model of inflammatory pain (Cui et al., 2006). Moreover, i.c.v. administration of TRPV1 antagonists reduced nociceptive behavior in the rat formalin test (Santos & Calixto, 1997). Following spinal cord injury, CB_1 and TRPV1 receptors interact and play a role in the plastic changes that occur in the rat brain (Knerlich-Lukoschus et al., 2011). In the same study, 7 days following spinal cord lesion, CB_1 receptor immunoreactivity was increased in the thalamus and hippocampus and downregulated in the ACC, amygdala and PAG, brain regions related to pain, emotion, learning, and memory in rats. Double labeling studies revealed that TRPV1 was coexpressed with CB_1 (Knerlich-Lukoschus et al., 2011). Thus, alterations in CB_1-TRPV1 expression/activity may underlie the emotional-affective and somatosensory pain responses following spinal cord lesion.

Paracetamol (acetaminophen) is a well-recognized and potent analgesic drug (Toms, McQuay, Derry, & Moore, 2008) and a number of recent studies have demonstrated that paracetamol is metabolized to the TRPV1 agonist and AEA transport blocker AM404, which contributes to the antinociceptive activity of paracetamol (Hogestatt et al., 2005; Mallet et al., 2010; Zygmunt, Chuang, Movahed, Julius, & Hogestatt, 2000). The breakdown of paracetamol to AM404 occurs in the brain and is dependent on FAAH (Hogestatt et al., 2005). The deacetylated paracetamol metabolite 4-aminophenol and 4-hydroxy-3-methoxybenzylamine (HMBA) produces antinociception in a variety of rodent (both mice and rats) pain tests (Barriere et al., 2013) and is metabolized in the brain to form AM404 plus HPODA or arvanil plus olvanil. The antinociceptive effects of arvanil were dependent on FAAH, TRPV1, and CB_1 receptors (Barriere et al., 2013). FAAH-dependent generation of TRPV1-active analgesic drug metabolites may be useful in the production of novel pain therapeutics (Barriere et al., 2013).

GPR55, a putative novel CB receptor, has recently been shown to be involved in the development of hyperalgesia in models of inflammatory and neuropathic pain. Inflammatory mechanical hyperalgesia was absent in GPR55(−/−) knockout mice (Castane et al., 2006; Staton et al., 2008). Furthermore, following partial sciatic nerve ligation, GPR55 (−/−) mice failed to express mechanical hyperalgesia (Staton et al., 2008). Together, these results suggest a pro-nociceptive role for GPR55. However, as discussed below, only one study to date has investigated the role of this novel target supraspinally in the modulation of pain or stress (Deliu et al., 2015).

The PPARs are also targets for eCBs and may play a role in eCB-induced analgesia. PPARγ agonists produced anti-inflammatory and antihyperalgesic effects in carrageenan-treated rats, effects which were supraspinally mediated (Morgenweck et al., 2010). Similarly, i.c.v. administration of PPARα

ligands produced anti-inflammatory and antihyperalgesic effects in mice and rats in the carrageenan model of inflammation (D'Agostino et al., 2007, 2009; Taylor, Kriedt, Nagalingam, Dadia, & Badr, 2005). Thus, central PPARs play an important role in inflammatory nociceptive processing and responding.

See Table 1 for a synthesis of the studies described above.

10. CONCLUDING REMARKS

This review has provided a detailed overview of the role of the supraspinal eCB system in pain, SIA, and SIH. Work in animal models has provided clear evidence that activation of supraspinal CB receptors (particularly CB_1) or elevation of supraspinal eCB levels can induce antinociception. Although our understanding of the physiological, biochemical, and molecular mechanisms mediating these processes has become much clearer in recent years, there is still a need for further work to provide full details on the neurobiological mechanism underlying these effects.

We have highlighted a few areas where further investigation is warranted. In particular, studies investigating the role of the supraspinal eCB system in SIA and SIH are still relatively sparse, and particularly for SIH. Further studies in this area would enhance our understanding of pain-stress interactions. Pain is processed via many interconnected receptor-mediated pathways utilizing different neurotransmitter systems including GABA, glutamate, monoamines, opioids, and the eCB system, among others. In order to fully understand the role of the supraspinal eCB system in pain, and its modulation by stress, it is imperative that we look at the interactions between the supraspinal eCB system and other neurotransmitter systems. To date there has been a relative paucity of studies investigating these interactions.

The majority of studies discussed in this review and the bulk of our understanding on this topic has come from laboratory animal studies. While there are many clinical trials investigating the effects of CBs in pain and psychiatric disorders (International Clinical Trials Registry Platform), to our knowledge there are no studies that specifically target the supraspinal eCB system to evaluate its effect on pain or comorbid pain and stress disorders, likely owing to the technical and ethical challenges that would be involved. It will be necessary to develop therapeutic approaches relevant to the clinical setting without overt side effects and these may include sub-region specific targeting of CB_1 receptors, elevation on eCB levels rather than potent CB_1

receptor agonism and targeting of CB_2 receptors or other non-CB_1 receptor targets of relevance within the eCB system—all of which have been reviewed in this manuscript. Given the high incidence of pain disorders and their comorbidity with stress-related disorders, there is an urgent need to fully understand the neurobiological mechanisms underpinning supraspinal modulation of pain, SIA, and SIH and develop new, more effective treatments with more favorable adverse side effect profiles.

ACKNOWLEDGMENTS

This work was supported by a grant from Science Foundation Ireland (10/IN.1/B2976) and a PhD fellowship to L.C. from the National University of Ireland Galway and from the Irish Research Council.

Conflict of Interest: None.

REFERENCES

Aguggia, M. (2003). Neurophysiology of pain. *Neurological Science*, *24*(Suppl. 2), S57–S60.

Aicher, S. A., Hermes, S. M., Whittier, K. L., & Hegarty, D. M. (2012). Descending projections from the rostral ventromedial medulla (RVM) to trigeminal and spinal dorsal horns are morphologically and neurochemically distinct. *Journal of Chemical Neuroanatomy*, *43*(2), 103–111.

Alexander, S. P., & Kendall, D. A. (2007). The complications of promiscuity: Endocannabinoid action and metabolism. *British Journal of Pharmacology*, *152*(5), 602–623.

Asmundson, G. J., & Katz, J. (2009). Understanding the co-occurrence of anxiety disorders and chronic pain: State-of-the-art. *Depression and Anxiety*, *26*(10), 888–901.

Azhdari Zarmehri, H., Semnanian, S., Fathollahi, Y., Erami, E., Khakpay, R., Azizi, H., et al. (2012). Intra-periaqueductal gray matter microinjection of orexin-A decreases formalin-induced nociceptive behaviors in adult male rats. *The Journal of Pain*, *12*(2), 280–287.

Baek, J. H., Zheng, Y., Darlington, C. L., & Smith, P. F. (2008). Cannabinoid CB2 receptor expression in the rat brainstem cochlear and vestibular nuclei. *Acta Oto-Laryngologica*, *128*(9), 961–967.

Bandler, R., & Keay, K. A. (1996). Columnar organization in the midbrain periaqueductal gray and the integration of emotional expression. *Progress in Brain Research*, *107*, 285–300.

Barriere, D. A., Mallet, C., Blomgren, A., Simonsen, C., Daulhac, L., Libert, F., et al. (2013). Fatty acid amide hydrolase-dependent generation of antinociceptive drug metabolites acting on TRPV1 in the brain. *PLoS One*, *8*(8), e70690.

Basbaum, A. I., & Fields, H. L. (1984). Endogenous pain control systems: Brainstem spinal pathways and endorphin circuitry. *Annual Review of Neuroscience*, 7, 309–338.

Battista, N., Di Tommaso, M., Bari, M., & Maccarrone, M. (2012). The endocannabinoid system: An overview. *Frontiers in Behavioral Neuroscience*, *6*, 9.

Beltramo, M., & Piomelli, D. (2000). Carrier-mediated transport and enzymatic hydrolysis of the endogenous cannabinoid 2-arachidonylglycerol. *Neuroreport*, *11*(6), 1231–1235.

Berdyshev, E. V. (2000). Cannabinoid receptors and the regulation of immune response. *Chemistry and Physics of Lipids*, *108*(1–2), 169–190.

Bisogno, T., Ligresti, A., & Di Marzo, V. (2005). The endocannabinoid signalling system: Biochemical aspects. *Pharmacology Biochemistry and Behavior*, *81*(2), 224–238.

Blankman, J. L., & Cravatt, B. F. (2013). Chemical probes of endocannabinoid metabolism. *Pharmacological Reviews*, *65*(2), 849–871.

Blankman, J. L., Simon, G. M., & Cravatt, B. F. (2007). A comprehensive profile of brain enzymes that hydrolyze the endocannabinoid 2-arachidonoylglycerol. *Chemistry & Biology*, *14*(12), 1347–1356.

Blyth, F. M., March, L. M., Brnabic, A. J., Jorm, L. R., Williamson, M., & Cousins, M. J. (2001). Chronic pain in Australia: A prevalence study. *Pain*, *89*(2–3), 127–134.

Bouaboula, M., Poinot-Chazel, C., Bourrie, B., Canat, X., Calandra, B., Rinaldi-Carmona, M., et al. (1995). Activation of mitogen-activated protein kinases by stimulation of the central cannabinoid receptor CB1. *The Biochemical Journal*, *312*(Pt. 2), 637–641.

Breivik, H., Collett, B., Ventafridda, V., Cohen, R., & Gallacher, D. (2006). Survey of chronic pain in Europe: Prevalence, impact on daily life, and treatment. *European Journal of Pain*, *10*(4), 287–333.

Brown, A. J. (2007). Novel cannabinoid receptors. *British Journal of Pharmacology*, *152*(5), 567–575.

Burke, N. N., Finn, D. P., & Roche, M. (2015). Neuroinflammatory mechanisms linking pain and depression: Pain in psychiatric disorders. *Modern Trends in Pharmacopsychiatry*, *30*, 36–50.

Burke, N. N., Hayes, E., Calpin, P., Kerr, D. M., Moriarty, O., Finn, D. P., et al. (2010). Enhanced nociceptive responding in two rat models of depression is associated with alterations in monoamine levels in discrete brain regions. *Neuroscience*, *171*(4), 1300–1313.

Butler, R. K., & Finn, D. P. (2009). Stress-induced analgesia. *Progress in Neurobiology*, *88*(3), 184–202.

Butler, R. K., Rea, K., Lang, Y., Gavin, A. M., & Finn, D. P. (2008). Endocannabinoid-mediated enhancement of fear-conditioned analgesia in rats: Opioid receptor dependency and molecular correlates. *Pain*, *140*(3), 491–500.

Carrier, E. J., Patel, S., & Hillard, C. J. (2005). Endocannabinoids in neuroimmunology and stress. *Current Drug Targets. CNS and Neurological Disorders*, *4*(6), 657–665.

Casey, K. L., Minoshima, S., Morrow, T. J., & Koeppe, R. A. (1996). Comparison of human cerebral activation pattern during cutaneous warmth, heat pain, and deep cold pain. *Journal of Neurophysiology*, *76*(1), 571–581.

Castane, A., Celerier, E., Martin, M., Ledent, C., Parmentier, M., Maldonado, R., et al. (2006). Development and expression of neuropathic pain in CB1 knockout mice. *Neuropharmacology*, *50*(1), 111–122.

Caterina, M. J., Leffler, A., Malmberg, A. B., Martin, W. J., Trafton, J., Petersen-Zeitz, K. R., et al. (2000). Impaired nociception and pain sensation in mice lacking the capsaicin receptor. *Science*, *288*(5464), 306–313.

Chhatwal, J. P., & Ressler, K. J. (2007). Modulation of fear and anxiety by the endogenous cannabinoid system. *CNS Spectrums*, *12*(3), 211–220.

Chiou, L. C., Lee, H. J., Ho, Y. C., Chen, S. P., Liao, Y. Y., Ma, C. H., et al. (2010). Orexins/hypocretins: Pain regulation and cellular actions. *Current Pharmaceutical Design*, *16*(28), 3089–3100.

Concannon, R. M., Okine, B. N., Finn, D. P., & Dowd, E. (2015). Differential upregulation of the cannabinoid CB(2) receptor in neurotoxic and inflammation-driven rat models of Parkinson's disease. *Experimental Neurology*, *269*, 133–141.

Connell, K., Bolton, N., Olsen, D., Piomelli, D., & Hohmann, A. G. (2006). Role of the basolateral nucleus of the amygdala in endocannabinoid-mediated stress-induced analgesia. *Neuroscience Letters*, *397*(3), 180–184.

Cravatt, B. F., Demarest, K., Patricelli, M. P., Bracey, M. H., Giang, D. K., Martin, B. R., et al. (2001). Supersensitivity to anandamide and enhanced endogenous cannabinoid signaling in mice lacking fatty acid amide hydrolase. *Proceedings of the National Academy of Sciences of the United States of America*, *98*(16), 9371–9376.

Cravatt, B. F., Giang, D. K., Mayfield, S. P., Boger, D. L., Lerner, R. A., & Gilula, N. B. (1996). Molecular characterization of an enzyme that degrades neuromodulatory fatty-acid amides. *Nature, 384*(6604), 83–87.

Cui, M., Honore, P., Zhong, C., Gauvin, D., Mikusa, J., Hernandez, G., et al. (2006). TRPV1 receptors in the CNS play a key role in broad-spectrum analgesia of TRPV1 antagonists. *The Journal of Neuroscience, 26*(37), 9385–9393.

D'Agostino, G., La Rana, G., Russo, R., Sasso, O., Iacono, A., Esposito, E., et al. (2007). Acute intracerebroventricular administration of palmitoylethanolamide, an endogenous peroxisome proliferator-activated receptor-alpha agonist, modulates carrageenan-induced paw edema in mice. *The Journal of Pharmacology and Experimental Therapeutics, 322*(3), 1137–1143.

D'Agostino, G., La Rana, G., Russo, R., Sasso, O., Iacono, A., Esposito, E., et al. (2009). Central administration of palmitoylethanolamide reduces hyperalgesia in mice via inhibition of NF-kappaB nuclear signalling in dorsal root ganglia. *European Journal of Pharmacology, 613*(1–3), 54–59.

Davis, K. D., Taylor, S. J., Crawley, A. P., Wood, M. L., & Mikulis, D. J. (1997). Functional MRI of pain- and attention-related activations in the human cingulate cortex. *Journal of Neurophysiology, 77*(6), 3370–3380.

de Lecea, L., Kilduff, T. S., Peyron, C., Gao, X., Foye, P. E., Danielson, P. E., et al. (1998). The hypocretins: Hypothalamus-specific peptides with neuroexcitatory activity. *Proceedings of the National Academy of Sciences of the United States of America, 95*(1), 322–327.

De Leon, M., Welcher, A. A., Nahin, R. H., Liu, Y., Ruda, M. A., Shooter, E. M., et al. (1996). Fatty acid binding protein is induced in neurons of the dorsal root ganglia after peripheral nerve injury. *Journal of Neuroscience Research, 44*(3), 283–292.

de Novellis, V., Palazzo, E., Rossi, F., De Petrocellis, L., Petrosino, S., Guida, F., et al. (2008). The analgesic effect of N-arachidonoyl-serotonin, a FAAH inhibitor and TRPV1 receptor antagonist, associated with changes in rostral ventromedial medulla and locus coeruleus cell activity in rats. *Neuropharmacology, 55*(7), 1105–1113.

de Novellis, V., Vita, D., Gatta, L., Luongo, L., Bellini, G., De Chiaro, M., et al. (2011). The blockade of the transient receptor potential vanilloid type 1 and fatty acid amide hydrolase decreases symptoms and central sequelae in the medial prefrontal cortex of neuropathic rats. *Molecular Pain, 7*, 7.

de Wied, M., & Verbaten, M. N. (2001). Affective pictures processing, attention, and pain tolerance. *Pain, 90*(1–2), 163–172.

Deliu, E., Sperow, M., Console-Bram, L., Carter, R. L., Tilley, D. G., Kalamarides, D. J., et al. (2015). The lysophosphatidylinositol receptor GPR55 modulates pain perception in the periaqueductal gray. *Molecular Pharmacology, 88*(2), 265–272.

Demuth, D. G., & Molleman, A. (2006). Cannabinoid signalling. *Life Sciences, 78*(6), 549–563.

Demyttenaere, K., Bruffaerts, R., Lee, S., Posada-Villa, J., Kovess, V., Angermeyer, M. C., et al. (2007). Mental disorders among persons with chronic back or neck pain: Results from the World Mental Health Surveys. *Pain, 129*(3), 332–342.

Derbyshire, S. W., Jones, A. K., Gyulai, F., Clark, S., Townsend, D., & Firestone, L. L. (1997). Pain processing during three levels of noxious stimulation produces differential patterns of central activity. *Pain, 73*(3), 431–445.

Devane, W. A., Dysarz, F. A., 3rd, Johnson, M. R., Melvin, L. S., & Howlett, A. C. (1988). Determination and characterization of a cannabinoid receptor in rat brain. *Molecular Pharmacology, 34*(5), 605–613.

Devane, W. A., Hanus, L., Breuer, A., Pertwee, R. G., Stevenson, L. A., Griffin, G., et al. (1992). Isolation and structure of a brain constituent that binds to the cannabinoid receptor. *Science, 258*(5090), 1946–1949.

Di Marzo, V. (2008). Endocannabinoids: Synthesis and degradation. *Reviews of Physiology, Biochemistry and Pharmacology*, *160*, 1–24.

Di Marzo, V., Fontana, A., Cadas, H., Schinelli, S., Cimino, G., Schwartz, J. C., et al. (1994). Formation and inactivation of endogenous cannabinoid anandamide in central neurons. *Nature*, *372*(6507), 686–691.

Di Marzo, V., Stella, N., & Zimmer, A. (2015). Endocannabinoid signalling and the deteriorating brain. *Nature Reviews Neuroscience*, *16*(1), 30–42.

Diorio, D., Viau, V., & Meaney, M. J. (1993). The role of the medial prefrontal cortex (cingulate gyrus) in the regulation of hypothalamic-pituitary-adrenal responses to stress. *The Journal of Neuroscience*, *13*(9), 3839–3847.

Dougher, M. J. (1979). Sensory decision theory analysis of the effects of anxiety and experimental instructions on pain. *Journal of Abnormal Psychology*, *88*(2), 137–144.

Egertova, M., Cravatt, B. F., & Elphick, M. R. (2003). Comparative analysis of fatty acid amide hydrolase and cb(1) cannabinoid receptor expression in the mouse brain: Evidence of a widespread role for fatty acid amide hydrolase in regulation of endocannabinoid signaling. *Neuroscience*, *119*(2), 481–496.

Escobar, W., Ramirez, K., Avila, C., Limongi, R., Vanegas, H., & Vazquez, E. (2012). Metamizol, a non-opioid analgesic, acts via endocannabinoids in the PAG-RVM axis during inflammation in rats. *European Journal of Pain*, *16*(5), 676–689.

Fang, Q., Han, Z. L., Li, N., Wang, Z. L., He, N., & Wang, R. (2012). Effects of neuropeptide FF system on CB(1) and CB(2) receptors mediated antinociception in mice. *Neuropharmacology*, *62*(2), 855–864.

Fields, H. L., Heinricher, M. M., & Mason, P. (1991). Neurotransmitters in nociceptive modulatory circuits. *Annual Review of Neuroscience*, *14*, 219–245.

Finn, D. P. (2010). Endocannabinoid-mediated modulation of stress responses: Physiological and pathophysiological significance. *Immunobiology*, *215*(8), 629–646.

Finn, D. P., Beckett, S. R., Richardson, D., Kendall, D. A., Marsden, C. A., & Chapman, V. (2004). Evidence for differential modulation of conditioned aversion and fear-conditioned analgesia by CB1 receptors. *The European Journal of Neuroscience*, *20*(3), 848–852.

Finn, D. P., Jhaveri, M. D., Beckett, S. R., Roe, C. H., Kendall, D. A., Marsden, C. A., et al. (2003). Effects of direct periaqueductal grey administration of a cannabinoid receptor agonist on nociceptive and aversive responses in rats. *Neuropharmacology*, *45*(5), 594–604.

Fitzgibbon, M., Finn, D. P., & Roche, M. (2015). High times for painful blues: The endocannabinoid system in pain-depression comorbidity. *International Journal of Neuropsychopharmacology*. http://dx.doi.org/10.1093/ijnp/pyv095.

Ford, G. K., & Finn, D. P. (2008). Clinical correlates of stress-induced analgesia: Evidence from pharmacological studies. *Pain*, *140*(1), 3–7.

Ford, G. K., Kieran, S., Dolan, K., Harhen, B., & Finn, D. P. (2011). A role for the ventral hippocampal endocannabinoid system in fear-conditioned analgesia and fear responding in the presence of nociceptive tone in rats. *Pain*, *152*(11), 2495–2504.

Freitas, R. L., Salgado-Rohner, C. J., Hallak, J. E., Crippa, J. A., & Coimbra, N. C. (2013). Involvement of prelimbic medial prefrontal cortex in panic-like elaborated defensive behaviour and innate fear-induced antinociception elicited by GABAA receptor blockade in the dorsomedial and ventromedial hypothalamic nuclei: Role of the endocannabinoid CB1 receptor. *The International Journal of Neuropsychopharmacology*, *16*(8), 1781–1798.

Fu, J., Bottegoni, G., Sasso, O., Bertorelli, R., Rocchia, W., Masetti, M., et al. (2012). A catalytically silent FAAH-1 variant drives anandamide transport in neurons. *Nature Neuroscience*, *15*(1), 64–69.

Furuhashi, M., & Hotamisligil, G. S. (2008). Fatty acid-binding proteins: Role in metabolic diseases and potential as drug targets. *Nature Reviews Drug Discovery*, 7(6), 489–503.

Gaskin, D. J., & Richard, P. (2012). The economic costs of pain in the United States. *The Journal of Pain*, *13*(8), 715–724.

Giang, D. K., & Cravatt, B. F. (1997). Molecular characterization of human and mouse fatty acid amide hydrolases. *Proceedings of the National Academy of Sciences of the United States of America*, *94*(6), 2238–2242.

Giordano, C., Cristino, L., Luongo, L., Siniscalco, D., Petrosino, S., Piscitelli, F., et al. (2012). TRPV1-dependent and -independent alterations in the limbic cortex of neuropathic mice: Impact on glial caspases and pain perception. *Cerebral Cortex*, *22*(11), 2495–2518.

Glaser, S. T., Abumrad, N. A., Fatade, F., Kaczocha, M., Studholme, K. M., & Deutsch, D. G. (2003). Evidence against the presence of an anandamide transporter. *Proceedings of the National Academy of Sciences of the United States of America*, *100*(7), 4269–4274.

Glass, M., Dragunow, M., & Faull, R. L. (1997). Cannabinoid receptors in the human brain: A detailed anatomical and quantitative autoradiographic study in the fetal, neonatal and adult human brain. *Neuroscience*, 77(2), 299–318.

Goparaju, S. K., Ueda, N., Yamaguchi, H., & Yamamoto, S. (1998). Anandamide amidohydrolase reacting with 2-arachidonoylglycerol, another cannabinoid receptor ligand. *FEBS Letters*, *422*(1), 69–73.

Greco, R., Gasperi, V., Maccarrone, M., & Tassorelli, C. (2010a). The endocannabinoid system and migraine. *Experimental Neurology*, *224*(1), 85–91.

Greco, R., Gasperi, V., Sandrini, G., Bagetta, G., Nappi, G., Maccarrone, M., et al. (2010b). Alterations of the endocannabinoid system in an animal model of migraine: Evaluation in cerebral areas of rat. *Cephalalgia*, *30*(3), 296–302.

Gregg, L. C., Jung, K. M., Spradley, J. M., Nyilas, R., Suplita, R. L., 2nd, Zimmer, A., et al. (2012). Activation of type 5 metabotropic glutamate receptors and diacylglycerol lipase-alpha initiates 2-arachidonoylglycerol formation and endocannabinoid-mediated analgesia. *The Journal of Neuroscience*, *32*(28), 9457–9468.

Guindon, J., & Hohmann, A. G. (2008). Cannabinoid CB2 receptors: A therapeutic target for the treatment of inflammatory and neuropathic pain. *British Journal of Pharmacology*, *153*(2), 319–334.

Gulyas, A. I., Cravatt, B. F., Bracey, M. H., Dinh, T. P., Piomelli, D., Boscia, F., et al. (2004). Segregation of two endocannabinoid-hydrolyzing enzymes into pre- and postsynaptic compartments in the rat hippocampus, cerebellum and amygdala. *The European Journal of Neuroscience*, *20*(2), 441–458.

Hasanein, P., Parviz, M., Keshavarz, M., & Javanmardi, K. (2007). CB1 receptor activation in the basolateral amygdala produces antinociception in animal models of acute and tonic nociception. *Clinical and Experimental Pharmacology & Physiology*, *34*(5–6), 439–449.

Helmstetter, F. J. (1992). The amygdala is essential for the expression of conditional hypoalgesia. *Behavioral Neuroscience*, *106*(3), 518–528.

Helmstetter, F. J., & Bellgowan, P. S. (1993). Lesions of the amygdala block conditional hypoalgesia on the tail flick test. *Brain Research*, *612*(1–2), 253–257.

Helmstetter, F. J., Bellgowan, P. S., & Poore, L. H. (1995). Microinfusion of mu but not delta or kappa opioid agonists into the basolateral amygdala results in inhibition of the tail flick reflex in pentobarbital-anesthetized rats. *The Journal of Pharmacology and Experimental Therapeutics*, *275*(1), 381–388.

Henry, R. J., Kerr, D. M., Finn, D. P., & Roche, M. (2015). For whom the endocannabinoid tolls: Modulation of innate immune function and implications for psychiatric disorders. *Progress in Neuropsychopharmacology and Biological Psychiatry*, *64*, 167–180.

Herkenham, M., Lynn, A. B., Johnson, M. R., Melvin, L. S., de Costa, B. R., & Rice, K. C. (1991). Characterization and localization of cannabinoid receptors in rat

brain: A quantitative in vitro autoradiographic study. *The Journal of Neuroscience*, *11*(2), 563–583.
Herzberg, U., Eliav, E., Bennett, G. J., & Kopin, I. J. (1997). The analgesic effects of R(+)-WIN 55,212-2 mesylate, a high affinity cannabinoid agonist, in a rat model of neuropathic pain. *Neuroscience Letters*, *221*(2–3), 157–160.
Hill, M. N., Kumar, S. A., Filipski, S. B., Iverson, M., Stuhr, K. L., Keith, J. M., et al. (2013). Disruption of fatty acid amide hydrolase activity prevents the effects of chronic stress on anxiety and amygdalar microstructure. *Molecular Psychiatry*, *18*(10), 1125–1135.
Hill, M. N., Patel, S., Carrier, E. J., Rademacher, D. J., Ormerod, B. K., Hillard, C. J., et al. (2005). Downregulation of endocannabinoid signaling in the hippocampus following chronic unpredictable stress. *Neuropsychopharmacology*, *30*(3), 508–515.
Hillard, C. J., Edgemond, W. S., Jarrahian, A., & Campbell, W. B. (1997). Accumulation of N-arachidonoylethanolamine (anandamide) into cerebellar granule cells occurs via facilitated diffusion. *Journal of Neurochemistry*, *69*(2), 631–638.
Ho, Y. C., Lee, H. J., Tung, L. W., Liao, Y. Y., Fu, S. Y., Teng, S. F., et al. (2011). Activation of orexin 1 receptors in the periaqueductal gray of male rats leads to antinociception via retrograde endocannabinoid (2-arachidonoylglycerol)-induced disinhibition. *The Journal of Neuroscience*, *31*(41), 14600–14610.
Hogestatt, E. D., Jonsson, B. A., Ermund, A., Andersson, D. A., Bjork, H., Alexander, J. P., et al. (2005). Conversion of acetaminophen to the bioactive N-acylphenolamine AM404 via fatty acid amide hydrolase-dependent arachidonic acid conjugation in the nervous system. *The Journal of Biological Chemistry*, *280*(36), 31405–31412.
Hohmann, A. G., & Suplita, R. L., 2nd. (2006). Endocannabinoid mechanisms of pain modulation. *The AAPS Journal*, *8*(4), E693–E708.
Hohmann, A. G., Suplita, R. L., Bolton, N. M., Neely, M. H., Fegley, D., Mangieri, R., et al. (2005). An endocannabinoid mechanism for stress-induced analgesia. *Nature*, *435*(7045), 1108–1112.
Hohmann, A. G., Tsou, K., & Walker, J. M. (1999). Cannabinoid suppression of noxious heat-evoked activity in wide dynamic range neurons in the lumbar dorsal horn of the rat. *Journal of Neurophysiology*, *81*(2), 575–583.
Holt, S., Comelli, F., Costa, B., & Fowler, C. J. (2005). Inhibitors of fatty acid amide hydrolase reduce carrageenan-induced hind paw inflammation in pentobarbital-treated mice: Comparison with indomethacin and possible involvement of cannabinoid receptors. *British Journal of Pharmacology*, *146*(3), 467–476.
Hoot, M. R., Sim-Selley, L. J., Poklis, J. L., Abdullah, R. A., Scoggins, K. L., Selley, D. E., et al. (2010). Chronic constriction injury reduces cannabinoid receptor 1 activity in the rostral anterior cingulate cortex of mice. *Brain Research*, *1339*, 18–25.
Howlett, A. C. (1985). Cannabinoid inhibition of adenylate cyclase. Biochemistry of the response in neuroblastoma cell membranes. *Molecular Pharmacology*, *27*(4), 429–436.
Howlett, A. C., & Mukhopadhyay, S. (2000). Cellular signal transduction by anandamide and 2-arachidonoylglycerol. *Chemistry and Physics of Lipids*, *108*(1–2), 53–70.
Howlett, A. C., Mukhopadhyay, S., Shim, J. Y., & Welsh, W. J. (1999). Signal transduction of eicosanoid CB1 receptor ligands. *Life Sciences*, *65*(6–7), 617–625.
Jayamanne, A., Greenwood, R., Mitchell, V. A., Aslan, S., Piomelli, D., & Vaughan, C. W. (2006). Actions of the FAAH inhibitor URB597 in neuropathic and inflammatory chronic pain models. *British Journal of Pharmacology*, *147*(3), 281–288.
Jennings, E. M., Okine, B. N., Olango, W. M., Roche, M., & Finn, D. P. (2015). Repeated forced swim stress differentially affects formalin-evoked nociceptive behaviour and the endocannabinoid system in stress normo-responsive and stress hyper-responsive rat strains. *Progress in Neuro-Psychopharmacology and Biological Psychiatry*, *64*, 181–189.
Jennings, E. M., Okine, B. N., Roche, M., & Finn, D. P. (2014). Stress-induced hyperalgesia. *Progress in Neurobiology*, *121*, 1–18.

Jhaveri, M. D., Elmes, S. J., Richardson, D., Barrett, D. A., Kendall, D. A., Mason, R., et al. (2008). Evidence for a novel functional role of cannabinoid CB(2) receptors in the thalamus of neuropathic rats. *The European Journal of Neuroscience*, *27*(7), 1722–1730.

Ji, G., & Neugebauer, V. (2014). CB1 augments mGluR5 function in medial prefrontal cortical neurons to inhibit amygdala hyperactivity in an arthritis pain model. *The European Journal of Neuroscience*, *39*(3), 455–466.

Ji, G., Sun, H., Fu, Y., Li, Z., Pais-Vieira, M., Galhardo, V., et al. (2010). Cognitive impairment in pain through amygdala-driven prefrontal cortical deactivation. *The Journal of Neuroscience*, *30*(15), 5451–5464.

Kaczocha, M., Glaser, S. T., & Deutsch, D. G. (2009). Identification of intracellular carriers for the endocannabinoid anandamide. *Proceedings of the National Academy of Sciences of the United States of America*, *106*(15), 6375–6380.

Kaczocha, M., Hermann, A., Glaser, S. T., Bojesen, I. N., & Deutsch, D. G. (2006). Anandamide uptake is consistent with rate-limited diffusion and is regulated by the degree of its hydrolysis by fatty acid amide hydrolase. *The Journal of Biological Chemistry*, *281*(14), 9066–9075.

Kaczocha, M., Vivieca, S., Sun, J., Glaser, S. T., & Deutsch, D. G. (2012). Fatty acid-binding proteins transport N-acylethanolamines to nuclear receptors and are targets of endocannabinoid transport inhibitors. *The Journal of Biological Chemistry*, *287*(5), 3415–3424.

Katona, I., Rancz, E. A., Acsady, L., Ledent, C., Mackie, K., Hajos, N., et al. (2001). Distribution of CB1 cannabinoid receptors in the amygdala and their role in the control of GABAergic transmission. *The Journal of Neuroscience*, *21*(23), 9506–9518.

Knerlich-Lukoschus, F., Noack, M., von der Ropp-Brenner, B., Lucius, R., Mehdorn, H. M., & Held-Feindt, J. (2011). Spinal cord injuries induce changes in CB1 cannabinoid receptor and C-C chemokine expression in brain areas underlying circuitry of chronic pain conditions. *Journal of Neurotrauma*, *28*(4), 619–634.

Kulkarni, B., Bentley, D. E., Elliott, R., Youell, P., Watson, A., Derbyshire, S. W., et al. (2005). Attention to pain localization and unpleasantness discriminates the functions of the medial and lateral pain systems. *The European Journal of Neuroscience*, *21*(11), 3133–3142.

Kuo, C. C., Chiou, R. J., Liang, K. C., & Yen, C. T. (2009). Differential involvement of the anterior cingulate and primary sensorimotor cortices in sensory and affective functions of pain. *Journal of Neurophysiology*, *101*(3), 1201–1210.

Kurrikoff, K., Inno, J., Matsui, T., & Vasar, E. (2008). Stress-induced analgesia in mice: Evidence for interaction between endocannabinoids and cholecystokinin. *The European Journal of Neuroscience*, *27*(8), 2147–2155.

Kwilasz, A. J., Abdullah, R. A., Poklis, J. L., Lichtman, A. H., & Negus, S. S. (2014). Effects of the fatty acid amide hydrolase inhibitor URB597 on pain-stimulated and pain-depressed behavior in rats. *Behavioural Pharmacology*, *25*(2), 119–129.

La Porta, C., Bura, A. S., Llorente-Onaindia, J., Pastor, A., Navarrete, F., Garcia-Gutierrez, M. S., et al. (2015). Role of the endocannabinoid system in the emotional manifestations of osteoarthritis pain. *Pain*, *156*(10), 2001–2012.

LaBuda, C. J., & Fuchs, P. N. (2005). Attenuation of negative pain affect produced by unilateral spinal nerve injury in the rat following anterior cingulate cortex activation. *Neuroscience*, *136*(1), 311–322.

Lafourcade, M., Elezgarai, I., Mato, S., Bakiri, Y., Grandes, P., & Manzoni, O. J. (2007). Molecular components and functions of the endocannabinoid system in mouse prefrontal cortex. *PLoS One*, *2*(8), e709.

Lau, B. K., & Vaughan, C. W. (2014). Descending modulation of pain: The GABA disinhibition hypothesis of analgesia. *Current Opinion in Neurobiology*, *29*, 159–164.

LeDoux, J. E. (2000). Emotion circuits in the brain. *Annual Review of Neuroscience*, *23*, 155–184.

Lee, M. C., Ploner, M., Wiech, K., Bingel, U., Wanigasekera, V., Brooks, J., et al. (2013). Amygdala activity contributes to the dissociative effect of cannabis on pain perception. *Pain, 154*(1), 124–134.

Liao, H. T., Lee, H. J., Ho, Y. C., & Chiou, L. C. (2011). Capsaicin in the periaqueductal gray induces analgesia via metabotropic glutamate receptor-mediated endocannabinoid retrograde disinhibition. *British Journal of Pharmacology, 163*(2), 330–345.

Lichtman, A. H., Blankman, J. L., & Cravatt, B. F. (2010). Endocannabinoid overload. *Molecular Pharmacology, 78*(6), 993–995.

Lichtman, A. H., Cook, S. A., & Martin, B. R. (1996). Investigation of brain sites mediating cannabinoid-induced antinociception in rats: Evidence supporting periaqueductal gray involvement. *The Journal of Pharmacology and Experimental Therapeutics, 276*(2), 585–593.

Lichtman, A. H., Leung, D., Shelton, C. C., Saghatelian, A., Hardouin, C., Boger, D. L., et al. (2004). Reversible inhibitors of fatty acid amide hydrolase that promote analgesia: Evidence for an unprecedented combination of potency and selectivity. *The Journal of Pharmacology and Experimental Therapeutics, 311*(2), 441–448.

Lichtman, A. H., & Martin, B. R. (1991). Cannabinoid-induced antinociception is mediated by a spinal alpha 2-noradrenergic mechanism. *Brain Research, 559*(2), 309–314.

Lichtman, A. H., & Martin, B. R. (1997). The selective cannabinoid antagonist SR 141716A blocks cannabinoid-induced antinociception in rats. *Pharmacology Biochemistry and Behavior, 57*(1–2), 7–12.

Lichtman, A. H., Shelton, C. C., Advani, T., & Cravatt, B. F. (2004). Mice lacking fatty acid amide hydrolase exhibit a cannabinoid receptor-mediated phenotypic hypoalgesia. *Pain, 109*(3), 319–327.

Little, J. P., & Carter, A. G. (2013). Synaptic mechanisms underlying strong reciprocal connectivity between the medial prefrontal cortex and basolateral amygdala. *The Journal of Neuroscience, 33*(39), 15333–15342.

Lomazzo, E., Bindila, L., Remmers, F., Lerner, R., Schwitter, C., Hoheisel, U., et al. (2015). Therapeutic potential of inhibitors of endocannabinoid degradation for the treatment of stress-related hyperalgesia in an animal model of chronic pain. *Neuropsychopharmacology, 40*(2), 488–501.

Ly, H. G., Ceccarini, J., Weltens, N., Bormans, G., Van Laere, K., Tack, J., et al. (2015). Increased cerebral cannabinoid-1 receptor availability is a stable feature of functional dyspepsia: A [F]MK-9470 PET study. *Psychotherapy and Psychosomatics, 84*(3), 149–158.

Maccarrone, M., Bab, I., Biro, T., Cabral, G. A., Dey, S. K., Di Marzo, V., et al. (2015). Endocannabinoid signaling at the periphery: 50 years after THC. *Trends in Pharmacological Sciences, 36*(5), 277–296.

Madasu, M. K., Roche, M., & Finn, D. P. (2015). Supraspinal TRPV1 in pain and psychiatric disorders. *Modern Trends in Pharmacopsychiatry, 30*, 36–50.

Mailleux, P., Parmentier, M., & Vanderhaeghen, J. J. (1992). Distribution of cannabinoid receptor messenger RNA in the human brain: An in situ hybridization histochemistry with oligonucleotides. *Neuroscience Letters, 143*(1–2), 200–204.

Maione, S., Bisogno, T., de Novellis, V., Palazzo, E., Cristino, L., Valenti, M., et al. (2006). Elevation of endocannabinoid levels in the ventrolateral periaqueductal grey through inhibition of fatty acid amide hydrolase affects descending nociceptive pathways via both cannabinoid receptor type 1 and transient receptor potential vanilloid type-1 receptors. *The Journal of Pharmacology and Experimental Therapeutics, 316*(3), 969–982.

Maione, S., De Petrocellis, L., de Novellis, V., Moriello, A. S., Petrosino, S., Palazzo, E., et al. (2007). Analgesic actions of N-arachidonoyl-serotonin, a fatty acid amide hydrolase inhibitor with antagonistic activity at vanilloid TRPV1 receptors. *British Journal of Pharmacology, 150*(6), 766–781.

Mallet, C., Barriere, D. A., Ermund, A., Jonsson, B. A., Eschalier, A., Zygmunt, P. M., et al. (2010). TRPV1 in brain is involved in acetaminophen-induced antinociception. *PLoS One*, *5*(9), e12748.

Manning, B. H., Martin, W. J., & Meng, I. D. (2003). The rodent amygdala contributes to the production of cannabinoid-induced antinociception. *Neuroscience*, *120*(4), 1157–1170.

Manning, B. H., Merin, N. M., Meng, I. D., & Amaral, D. G. (2001). Reduction in opioid- and cannabinoid-induced antinociception in rhesus monkeys after bilateral lesions of the amygdaloid complex. *The Journal of Neuroscience*, *21*(20), 8238–8246.

Manzanares, J., Corchero, J., & Fuentes, J. A. (1999). Opioid and cannabinoid receptor-mediated regulation of the increase in adrenocorticotropin hormone and corticosterone plasma concentrations induced by central administration of delta(9)-tetrahydrocannabinol in rats. *Brain Research*, *839*(1), 173–179.

Marchand, J. E., & Hagino, N. (1983). Afferents to the periaqueductal gray in the rat. A horseradish peroxidase study. *Neuroscience*, *9*(1), 95–106.

Marsicano, G., & Lutz, B. (1999). Expression of the cannabinoid receptor CB1 in distinct neuronal subpopulations in the adult mouse forebrain. *The European Journal of Neuroscience*, *11*(12), 4213–4225.

Martin, W. J., Coffin, P. O., Attias, E., Balinsky, M., Tsou, K., & Walker, J. M. (1999). Anatomical basis for cannabinoid-induced antinociception as revealed by intracerebral microinjections. *Brain Research*, *822*(1–2), 237–242.

Martin, W. J., Hohmann, A. G., & Walker, J. M. (1996). Suppression of noxious stimulus-evoked activity in the ventral posterolateral nucleus of the thalamus by a cannabinoid agonist: Correlation between electrophysiological and antinociceptive effects. *The Journal of Neuroscience*, *16*(20), 6601–6611.

Martin, W. J., Lai, N. K., Patrick, S. L., Tsou, K., & Walker, J. M. (1993). Antinociceptive actions of cannabinoids following intraventricular administration in rats. *Brain Research*, *629*(2), 300–304.

Martin, W. J., Patrick, S. L., Coffin, P. O., Tsou, K., & Walker, J. M. (1995). An examination of the central sites of action of cannabinoid-induced antinociception in the rat. *Life Sciences*, *56*(23–24), 2103–2109.

Martin, W. J., Tsou, K., & Walker, J. M. (1998). Cannabinoid receptor-mediated inhibition of the rat tail-flick reflex after microinjection into the rostral ventromedial medulla. *Neuroscience Letters*, *242*(1), 33–36.

Matsuda, L. A., Lolait, S. J., Brownstein, M. J., Young, A. C., & Bonner, T. I. (1990). Structure of a cannabinoid receptor and functional expression of the cloned cDNA. *Nature*, *346*(6284), 561–564.

May, A., Kaube, H., Buchel, C., Eichten, C., Rijntjes, M., Juptner, M., et al. (1998). Experimental cranial pain elicited by capsaicin: A PET study. *Pain*, *74*(1), 61–66.

McCarberg, B. H., & Billington, R. (2006). Consequences of neuropathic pain: Quality-of-life issues and associated costs. *The American Journal of Managed Care*, *12*(9 Suppl.), S263–S268.

McGaraughty, S., Chu, K. L., Bitner, R. S., Martino, B., El Kouhen, R., Han, P., et al. (2003). Capsaicin infused into the PAG affects rat tail flick responses to noxious heat and alters neuronal firing in the RVM. *Journal of Neurophysiology*, *90*(4), 2702–2710.

Meagher, M. W., Arnau, R. C., & Rhudy, J. L. (2001). Pain and emotion: Effects of affective picture modulation. *Psychosomatic Medicine*, *63*(1), 79–90.

Mechoulam, R., Ben-Shabat, S., Hanus, L., Ligumsky, M., Kaminski, N. E., Schatz, A. R., et al. (1995). Identification of an endogenous 2-monoglyceride, present in canine gut, that binds to cannabinoid receptors. *Biochemical Pharmacology*, *50*(1), 83–90.

Mechoulam, R., & Gaoni, Y. (1967). The absolute configuration of delta-1-tetrahydrocannabinol, the major active constituent of hashish. *Tetrahedron Letters*, *12*, 1109–1111.

Meng, I. D., & Johansen, J. P. (2004). Antinociception and modulation of rostral ventromedial medulla neuronal activity by local microinfusion of a cannabinoid receptor agonist. *Neuroscience*, *124*(3), 685–693.

Meng, I. D., Manning, B. H., Martin, W. J., & Fields, H. L. (1998). An analgesia circuit activated by cannabinoids. *Nature*, *395*(6700), 381–383.

Millan, M. J. (1999). The induction of pain: An integrative review. *Progress in Neurobiology*, *57*(1), 1–164.

Mohammadi-Farani, A., Sahebgharani, M., Sepehrizadeh, Z., Jaberi, E., & Ghazi-Khansari, M. (2010). Diabetic thermal hyperalgesia: Role of TRPV1 and CB1 receptors of periaqueductal gray. *Brain Research*, *1328*, 49–56.

Mohammad-Pour Kargar, H., Azizi, H., Mirnajafi-Zadeh, J., Reza Mani, A., & Semnanian, S. (2015). Microinjection of orexin-A into the rat locus coeruleus nucleus induces analgesia via cannabinoid type-1 receptors. *Brain Research*, *1624*, 424–432.

Monhemius, R., Azami, J., Green, D. L., & Roberts, M. H. (2001). CB1 receptor mediated analgesia from the Nucleus Reticularis Gigantocellularis pars alpha is activated in an animal model of neuropathic pain. *Brain Research*, *908*(1), 67–74.

Morena, M., Patel, S., Bains, J. S., & Hill, M. N. (2015). Neurobiological interactions between stress and the endocannabinoid system. *Neuropsychopharmacology*. http://dx.doi.org/10.1038/npp.2015.166.

Morgenweck, J., Abdel-Aleem, O. S., McNamara, K. C., Donahue, R. R., Badr, M. Z., & Taylor, B. K. (2010). Activation of peroxisome proliferator-activated receptor gamma in brain inhibits inflammatory pain, dorsal horn expression of Fos, and local edema. *Neuropharmacology*, *58*(2), 337–345.

Munro, S., Thomas, K. L., & Abu-Shaar, M. (1993). Molecular characterization of a peripheral receptor for cannabinoids. *Nature*, *365*(6441), 61–65.

Neal, J. W., Pearson, R. C., & Powell, T. P. (1990). The ipsilateral cortico-cortical connections of area 7b, PF, in the parietal and temporal lobes of the monkey. *Brain Research*, *524*(1), 119–132.

Neugebauer, V., Galhardo, V., Maione, S., & Mackey, S. C. (2009). Forebrain pain mechanisms. *Brain Research Reviews*, *60*(1), 226–242.

Neugebauer, V., Li, W., Bird, G. C., & Han, J. S. (2004). The amygdala and persistent pain. *The Neuroscientist*, *10*(3), 221–234.

Nozaki, C., Markert, A., & Zimmer, A. (2015). Inhibition of FAAH reduces nitroglycerin-induced migraine-like pain and trigeminal neuronal hyperactivity in mice. *European Neuropsychopharmacology*, *25*(8), 1388–1396.

O'Mahony, S. M., Bulmer, D. C., Coelho, A. M., Fitzgerald, P., Bongiovanni, C., Lee, K., et al. (2010). 5-HT(2B) receptors modulate visceral hypersensitivity in a stress-sensitive animal model of brain-gut axis dysfunction. *Neurogastroenterology and Motility*, *22*(5), 573–578. e124.

O'Sullivan, S. E. (2007). Cannabinoids go nuclear: Evidence for activation of peroxisome proliferator-activated receptors. *British Journal of Pharmacology*, *152*(5), 576–582.

Okine, B. N., Rea, K., Olango, W. M., Price, J., Herdman, S., Madasu, M. K., et al. (2014). A role for PPARalpha in the medial prefrontal cortex in formalin-evoked nociceptive responding in rats. *British Journal of Pharmacology*, *171*(6), 1462–1471.

Olango, W. M., & Finn, D. P. (2014). Neurobiology of stress-induced hyperalgesia. *Current Topics in Behavioral Neurosciences*, *20*, 251–280.

Olango, W. M., Roche, M., Ford, G. K., Harhen, B., & Finn, D. P. (2012). The endocannabinoid system in the rat dorsolateral periaqueductal grey mediates fear-conditioned analgesia and controls fear expression in the presence of nociceptive tone. *British Journal of Pharmacology*, *165*(8), 2549–2560.

Oliveira, M. A., & Prado, W. A. (1998). Antinociception induced by stimulating amygdaloid nuclei in rats: Changes produced by systemically administered antagonists. *Brazilian Journal of Medical and Biological Research*, *31*(5), 681–690.

Onaivi, E. S., Ishiguro, H., Gong, J. P., Patel, S., Perchuk, A., Meozzi, P. A., et al. (2006). Discovery of the presence and functional expression of cannabinoid CB2 receptors in brain. *Annals of the New York Academy of Sciences*, *1074*, 514–536.

Otrubova, K., Ezzili, C., & Boger, D. L. (2011). The discovery and development of inhibitors of fatty acid amide hydrolase (FAAH). *Bioorganic & Medicinal Chemistry Letters*, *21*(16), 4674–4685.

Palazzo, E., de Novellis, V., Marabese, I., Cuomo, D., Rossi, F., Berrino, L., et al. (2002). Interaction between vanilloid and glutamate receptors in the central modulation of nociception. *European Journal of Pharmacology*, *439*(1–3), 69–75.

Palazzo, E., Marabese, I., de Novellis, V., Oliva, P., Rossi, F., Berrino, L., et al. (2001). Metabotropic and NMDA glutamate receptors participate in the cannabinoid-induced antinociception. *Neuropharmacology*, *40*(3), 319–326.

Palazzo, E., Rossi, F., & Maione, S. (2008). Role of TRPV1 receptors in descending modulation of pain. *Molecular and Cellular Endocrinology*, *286*(1–2 Suppl. 1), S79–S83.

Patel, S., Cravatt, B. F., & Hillard, C. J. (2005). Synergistic interactions between cannabinoids and environmental stress in the activation of the central amygdala. *Neuropsychopharmacology*, *30*(3), 497–507.

Pertwee, R. G. (1997). Pharmacology of cannabinoid CB1 and CB2 receptors. *Pharmacology & Therapeutics*, *74*(2), 129–180.

Pertwee, R. G. (2001). Cannabinoid receptors and pain. *Progress in Neurobiology*, *63*(5), 569–611.

Petrosino, S., Palazzo, E., de Novellis, V., Bisogno, T., Rossi, F., Maione, S., et al. (2007). Changes in spinal and supraspinal endocannabinoid levels in neuropathic rats. *Neuropharmacology*, *52*(2), 415–422.

Racz, I., Nent, E., Erxlebe, E., & Zimmer, A. (2015). CB1 receptors modulate affective behaviour induced by neuropathic pain. *Brain Research Bulletin*, *114*, 42–48.

Rademacher, D. J., Meier, S. E., Shi, L., Ho, W. S., Jarrahian, A., & Hillard, C. J. (2008). Effects of acute and repeated restraint stress on endocannabinoid content in the amygdala, ventral striatum, and medial prefrontal cortex in mice. *Neuropharmacology*, *54*(1), 108–116.

Raffa, R. B., Stone, D. J., Jr., & Hipp, S. J. (1999). Differential cholera-toxin sensitivity of supraspinal antinociception induced by the cannabinoid agonists delta9-THC, WIN 55,212-2 and anandamide in mice. *Neuroscience Letters*, *263*(1), 29–32.

Rea, K., Olango, W. M., Harhen, B., Kerr, D. M., Galligan, R., Fitzgerald, S., et al. (2013). Evidence for a role of GABAergic and glutamatergic signalling in the basolateral amygdala in endocannabinoid-mediated fear-conditioned analgesia in rats. *Pain*, *154*(4), 576–585.

Rea, K., Olango, W. M., Okine, B. N., Madasu, M. K., McGuire, I. C., Coyle, K., et al. (2014). Impaired endocannabinoid signalling in the rostral ventromedial medulla underpins genotype-dependent hyper-responsivity to noxious stimuli. *Pain*, *155*(1), 69–79.

Rhudy, J. L., & Meagher, M. W. (2000). Fear and anxiety: Divergent effects on human pain thresholds. *Pain*, *84*(1), 65–75.

Rhudy, J. L., & Meagher, M. W. (2001). Noise stress and human pain thresholds: Divergent effects in men and women. *The Journal of Pain*, *2*(1), 57–64.

Rhudy, J. L., & Meagher, M. W. (2003a). Individual differences in the emotional reaction to shock determine whether hypoalgesia is observed. *Pain Medicine*, *4*(3), 244–256.

Rhudy, J. L., & Meagher, M. W. (2003b). Negative affect: Effects on an evaluative measure of human pain. *Pain*, *104*(3), 617–626.

Roche, M., Johnston, P., Mhuircheartaigh, O. N., Olango, W. M., Mackie, K., & Finn, D. P. (2010). Effects of intra-basolateral amygdala administration of rimonabant on nociceptive behaviour and neuronal activity in the presence or absence of contextual fear. *European Journal of Pain*, *14*(5), 487–495.

Roche, M., O'Connor, E., Diskin, C., & Finn, D. P. (2007). The effect of CB(1) receptor antagonism in the right basolateral amygdala on conditioned fear and associated analgesia in rats. *The European Journal of Neuroscience*, *26*(9), 2643–2653.

Russo, E. B. (2004). Clinical endocannabinoid deficiency (CECD): Can this concept explain therapeutic benefits of cannabis in migraine, fibromyalgia, irritable bowel syndrome and other treatment-resistant conditions? *Neuro Endocrinology Letters*, *25*(1–2), 31–39.

Sakurai, T., Amemiya, A., Ishii, M., Matsuzaki, I., Chemelli, R. M., Tanaka, H., et al. (1998). Orexins and orexin receptors: A family of hypothalamic neuropeptides and G protein-coupled receptors that regulate feeding behavior. *Cell*, *92*(5). 1 page following 696.

Santos, A. R., & Calixto, J. B. (1997). Ruthenium red and capsazepine antinociceptive effect in formalin and capsaicin models of pain in mice. *Neuroscience Letters*, *235*(1–2), 73–76.

Sierra-Mercado, D., Jr., Corcoran, K. A., Lebron-Milad, K., & Quirk, G. J. (2006). Inactivation of the ventromedial prefrontal cortex reduces expression of conditioned fear and impairs subsequent recall of extinction. *The European Journal of Neuroscience*, *24*(6), 1751–1758.

Sim-Selley, L. J., Vogt, L. J., Vogt, B. A., & Childers, S. R. (2002). Cellular localization of cannabinoid receptors and activated G-proteins in rat anterior cingulate cortex. *Life Sciences*, *71*(19), 2217–2226.

Smith, S. C., & Wagner, M. S. (2014). Clinical endocannabinoid deficiency (CECD) revisited: Can this concept explain the therapeutic benefits of cannabis in migraine, fibromyalgia, irritable bowel syndrome and other treatment-resistant conditions? *Neuro Endocrinology Letters*, *35*(3), 198–201.

Starowicz, K., Nigam, S., & Di Marzo, V. (2007). Biochemistry and pharmacology of endovanilloids. *Pharmacology & Therapeutics*, *114*(1), 13–33.

Staton, P. C., Hatcher, J. P., Walker, D. J., Morrison, A. D., Shapland, E. M., Hughes, J. P., et al. (2008). The putative cannabinoid receptor GPR55 plays a role in mechanical hyperalgesia associated with inflammatory and neuropathic pain. *Pain*, *139*(1), 225–236.

Sugiura, T., Kondo, S., Sukagawa, A., Nakane, S., Shinoda, A., Itoh, K., et al. (1995). 2-Arachidonoylglycerol: A possible endogenous cannabinoid receptor ligand in brain. *Biochemical and Biophysical Research Communications*, *215*(1), 89–97.

Suplita, R. L., 2nd, Farthing, J. N., Gutierrez, T., & Hohmann, A. G. (2005). Inhibition of fatty-acid amide hydrolase enhances cannabinoid stress-induced analgesia: Sites of action in the dorsolateral periaqueductal gray and rostral ventromedial medulla. *Neuropharmacology*, *49*(8), 1201–1209.

Svensson, P., Minoshima, S., Beydoun, A., Morrow, T. J., & Casey, K. L. (1997). Cerebral processing of acute skin and muscle pain in humans. *Journal of Neurophysiology*, *78*(1), 450–460.

Szabo, B., & Schlicker, E. (2005). Effects of cannabinoids on neurotransmission. *Handbook of Experimental Pharmacology*, *168*, 327–365.

Task Force on Taxonomy of the International Association for the Study of Pain. (1994). *Classification of chronic pain: Descriptions of chronic pain syndromes and definitions of pain terms*. (2nd ed.). Seattle: IASP Press (Part 3).

Taylor, B. K., Kriedt, C., Nagalingam, S., Dadia, N., & Badr, M. (2005). Central administration of perfluorooctanoic acid inhibits cutaneous inflammation. *Inflammation Research*, *54*(6), 235–242.

Thomas, B. F., Wei, X., & Martin, B. R. (1992). Characterization and autoradiographic localization of the cannabinoid binding site in rat brain using [3H]11-OH-delta 9-THC-DMH. *The Journal of Pharmacology and Experimental Therapeutics*, *263*(3), 1383–1390.

Toms, L., McQuay, H. J., Derry, S., & Moore, R. A. (2008). Single dose oral paracetamol (acetaminophen) for postoperative pain in adults. *Cochrane Database of Systematic Reviews*, *4*. CD004602.

Treede, R. D., Kenshalo, D. R., Gracely, R. H., & Jones, A. K. (1999). The cortical representation of pain. *Pain*, *79*(2–3), 105–111.
Tsou, K., Brown, S., Sanudo-Pena, M. C., Mackie, K., & Walker, J. M. (1998). Immunohistochemical distribution of cannabinoid CB1 receptors in the rat central nervous system. *Neuroscience*, *83*(2), 393–411.
Tsou, K., Nogueron, M. I., Muthian, S., Sanudo-Pena, M. C., Hillard, C. J., Deutsch, D. G., et al. (1998). Fatty acid amide hydrolase is located preferentially in large neurons in the rat central nervous system as revealed by immunohistochemistry. *Neuroscience Letters*, *254*(3), 137–140.
Tsujino, N., & Sakurai, T. (2009). Orexin/hypocretin: A neuropeptide at the interface of sleep, energy homeostasis, and reward system. *Pharmacological Reviews*, *61*(2), 162–176.
Turk, D. C. (2002). Clinical effectiveness and cost-effectiveness of treatments for patients with chronic pain. *The Clinical Journal of Pain*, *18*(6), 355–365.
Ueda, N., Tsuboi, K., Uyama, T., & Ohnishi, T. (2011). Biosynthesis and degradation of the endocannabinoid 2-arachidonoylglycerol. *Biofactors*, *37*(1), 1–7.
Valverde, O., Ledent, C., Beslot, F., Parmentier, M., & Roques, B. P. (2000). Reduction of stress-induced analgesia but not of exogenous opioid effects in mice lacking CB1 receptors. *The European Journal of Neuroscience*, *12*(2), 533–539.
Van Sickle, M. D., Duncan, M., Kingsley, P. J., Mouihate, A., Urbani, P., Mackie, K., et al. (2005). Identification and functional characterization of brainstem cannabinoid CB2 receptors. *Science*, *310*(5746), 329–332.
Vanegas, H., Barbaro, N. M., & Fields, H. L. (1984). Tail-flick related activity in medullospinal neurons. *Brain Research*, *321*(1), 135–141.
Vaughan, C. W., Connor, M., Bagley, E. E., & Christie, M. J. (2000). Actions of cannabinoids on membrane properties and synaptic transmission in rat periaqueductal gray neurons in vitro. *Molecular Pharmacology*, *57*(2), 288–295.
Vazquez, E., Escobar, W., Ramirez, K., & Vanegas, H. (2007). A nonopioid analgesic acts upon the PAG-RVM axis to reverse inflammatory hyperalgesia. *The European Journal of Neuroscience*, *25*(2), 471–479.
Vos, T., Flaxman, A. D., Naghavi, M., Lozano, R., Michaud, C., Ezzati, M., et al. (2012). Years lived with disability (YLDs) for 1160 sequelae of 289 diseases and injuries 1990-2010: A systematic analysis for the Global Burden of Disease Study 2010. *Lancet*, *380*(9859), 2163–2196.
Walker, J. M., & Hohmann, A. G. (2005). Cannabinoid mechanisms of pain suppression. *Handbook of Experimental Pharmacology*, *168*, 509–554.
Walker, J. M., Huang, S. M., Strangman, N. M., Tsou, K., & Sanudo-Pena, M. C. (1999). Pain modulation by release of the endogenous cannabinoid anandamide. *Proceedings of the National Academy of Sciences of the United States of America*, *96*(21), 12198–12203.
Wedzony, K., & Chocyk, A. (2009). Cannabinoid CB1 receptors in rat medial prefrontal cortex are colocalized with calbindin- but not parvalbumin- and calretinin-positive GABA-ergic neurons. *Pharmacological Reports*, *61*(6), 1000–1007.
Welch, S. P., Huffman, J. W., & Lowe, J. (1998). Differential blockade of the antinociceptive effects of centrally administered cannabinoids by SR141716A. *The Journal of Pharmacology and Experimental Therapeutics*, *286*(3), 1301–1308.
Welch, S. P., Thomas, C., & Patrick, G. S. (1995). Modulation of cannabinoid-induced antinociception after intracerebroventricular versus intrathecal administration to mice: Possible mechanisms for interaction with morphine. *The Journal of Pharmacology and Experimental Therapeutics*, *272*(1), 310–321.
Werka, T. (1994). Post-stress analgesia in rats with partial amygdala lesions. *Acta Neurobiologiae Experimentalis (Wars)*, *54*(2), 127–132.
Werka, T. (1997). The effects of the medial and cortical amygdala lesions on post-stress analgesia in rats. *Behavioural Brain Research*, *86*(1), 59–65.

Werka, T., & Marek, P. (1990). Post-stress analgesia after lesions to the central nucleus of the amygdala in rats. *Acta Neurobiologiae Experimentalis (Wars)*, *50*(1–2), 13–22.
Wiech, K., & Tracey, I. (2009). The influence of negative emotions on pain: Behavioral effects and neural mechanisms. *NeuroImage*, *47*(3), 987–994.
Wilson, A. R., Maher, L., & Morgan, M. M. (2008). Repeated cannabinoid injections into the rat periaqueductal gray enhance subsequent morphine antinociception. *Neuropharmacology*, *55*(7), 1219–1225.
Wilson-Poe, A. R., Pocius, E., Herschbach, M., & Morgan, M. M. (2013). The periaqueductal gray contributes to bidirectional enhancement of antinociception between morphine and cannabinoids. *Pharmacology Biochemistry and Behavior*, *103*(3), 444–449.
Woodhams, S. G., Sagar, D. R., Burston, J. J., & Chapman, V. (2015). The role of the endocannabinoid system in pain. *Handbook of Experimental Pharmacology*, *227*, 119–143.
Yamamoto, T., Yamamoto, A., Watanabe, M., Matsuo, T., Yamazaki, N., Kataoka, M., et al. (2009). Classification of FABP isoforms and tissues based on quantitative evaluation of transcript levels of these isoforms in various rat tissues. *Biotechnology Letters*, *31*(11), 1695–1701.
Zhang, H. Y., Gao, M., Liu, Q. R., Bi, G. H., Li, X., Yang, H. J., et al. (2014). Cannabinoid CB2 receptors modulate midbrain dopamine neuronal activity and dopamine-related behavior in mice. *Proceedings of the National Academy of Sciences of the United States of America*, *111*(46), E5007–E5015.
Zygmunt, P. M., Chuang, H., Movahed, P., Julius, D., & Hogestatt, E. D. (2000). The anandamide transport inhibitor AM404 activates vanilloid receptors. *European Journal of Pharmacology*, *396*(1), 39–42.

CHAPTER SEVEN

Endocannabinoid Signaling in Motivation, Reward, and Addiction: Influences on Mesocorticolimbic Dopamine Function

Claudia Sagheddu*, Anna Lisa Muntoni†, Marco Pistis*,†, Miriam Melis*,1

*Division of Neuroscience and Clinical Pharmacology, Department of Biomedical Sciences, Cittadella Universitaria di Monserrato, University of Cagliari, Monserrato (CA), Italy
†Neuroscience Institute, Section of Cagliari, National Research Council of Italy, Monserrato (CA), Italy
[1]Corresponding author: e-mail address: myriam@unica.it

Contents

1. Introduction 258
2. The Neurobiology of Reward 260
   2.1 Role of Dopamine 260
   2.2 Mechanisms Modulating Dopamine Signaling 262
3. The Influence of Endocannabinoid Signaling on Reward 263
   3.1 Endocannabinoid Influence on Food Reward 264
   3.2 Endocannabinoid Influence on Sexual Behavior 266
   3.3 Endocannabinoid Influence on Maternal Care 267
   3.4 Endocannabinoid Influence on Social Interaction 269
4. The Role of the Endocannabinoid System in Motivation 271
   4.1 Motivation—A Leitmotif 271
   4.2 Effects of Altered Cannabinoid Signaling on Motivation 273
5. Drug Addiction and the Endocannabinoid System: A Reciprocal Modulation of Synaptic Plasticity 277
   5.1 Etiology and Neural Mechanisms of Addiction 277
   5.2 Drug-Induced Alterations in Brain Endocannabinoid Signaling 279
6. Concluding Remarks 283
Acknowledgments 285
References 285

Abstract

Evidence suggests that the endocannabinoid system has been conserved in the animal kingdom for 500 million years, and this system influences many critical behavioral

International Review of Neurobiology, Volume 125
ISSN 0074-7742
http://dx.doi.org/10.1016/bs.irn.2015.10.004

processes including associative learning, reward signaling, goal-directed behavior, motor skill learning, and action-habit transformation. Additionally, the neurotransmitter dopamine has long been recognized to play a critical role in the processing of natural rewards, as well as of motivation that regulates approach and avoidance behavior. This motivational role of dopamine neurons is also based upon the evidence provided by several studies investigating disorders of dopamine pathways such as drug addiction and Parkinson's disease. From an evolutionary point of view, individuals engage in behaviors aimed at maximizing and minimizing positive and aversive consequences, respectively. Accordingly, those with the greatest *fitness* have a better potential to survival. Hence, deviations from *fitness* can be viewed as a part of the evolutionary process by means of natural selection. Given the long evolutionary history of both the endocannabinoid and dopaminergic systems, it is plausible that they must serve as fundamental and basic modulators of physiological functions and needs. Notably, endocannabinoids regulate dopamine neuronal activity and its influence on behavioral output. The goal of this chapter is to examine the endocannabinoid influence on dopamine signaling specifically related to (i) those behavioral processes that allow us to successfully adapt to ever-changing environments (i.e., reward signaling and motivational processes) and (ii) derangements from behavioral flexibility that underpin drug addiction.

1. INTRODUCTION

Evolution theory has guided contemporary biological research, and the most basic objective of evolution by natural selection is species self-preservation. Hence, phylogenetically conserved mechanisms drive basic and vital processes through the evaluation of sensory information to adjust behavioral output in a flexible manner in order to maximize reward, and to minimize aversive consequences. Effectively, our ancestors developed important behavioral strategies, which integrated basic information needed for survival, in order to satisfy both basic physiological and safety needs. Consequently, evolution by natural selection has provided human beings with opposable thumbs, upright posture, and a brain that made us the most adaptable creature on Earth. Indeed, dynamic interaction between genetic and environmental factors affects synaptic plasticity, shapes the brain, and the resulting cognitive processing and behavior. Derangements from the optimization of such adaptive mechanisms impinge on the plastic nature of the brain, and this might be viewed as a part of the evolutionary process. However, their cost in terms of health that impacts society at multiple levels is very high.

Motives, emotions, and self-control likely developed in response to selection pressures in ever-changing physical/social/cultural environments. Emotions, predictive of and/or following reward, guide purposeful behavior. Motivation drives an organism to initiate and to persist in certain behaviors over others. Self-control can delay behavior and grants flexibility in that it allows increased probability that behavioral responses are adaptive in the local environment.

Million years ago, our ancestors evolved a set of motivations aimed at increasing species preservation, including courting, raising offspring, and ensuring offspring success/survival. In this context, dopamine (DA), as an evolutionarily ancient neurotransmitter present in almost all multicellular organisms, is key to brain's processing of rewards, and drives several behaviors, ranging from a single motor action to very complicated series of actions. DA is also involved in encoding salience, value learning, decision-making processes, working memory, and behavioral flexibility. This slow-acting classic neurotransmitter operates as a neuromodulator that induces activity-dependent changes in synaptic strength and, therefore, influences emotional, motivational, cognitive, and motor processes. Thus, DA is key in experience-dependent plasticity, which can be in turn dynamically affected by both short- and long-term activity-dependent forms of plasticity. Short-term forms of synaptic plasticity are rapid means for a bidirectional and reversible modulation of synaptic strength, and serve as important mechanisms aimed at adjusting synaptic functions during computation. As such, they perform as dynamic filters and regulate proper scaling of synaptic inputs. On the other hand, long-term forms of synaptic plasticity are indicative of long-lasting adaptations of individual synapses, circuits, or neural networks, which underlie changes in behavior. In this scenario, an evolutionary and ancient neuromodulator, that is the endocannabinoid (eCB) system, has been shown to finely tune neuronal excitability and diverse forms of synaptic plasticity throughout the brain, including the DA circuits (Melis, Greco, & Tonini, 2014; Melis & Pistis, 2012; Wang & Lupica, 2014). The eCB system dates back to the unicellular common ancestor of animals and plants (Elphick, 2012), since the metabolic apparatus to build and degrade these lipids can be found throughout the entire animal kingdom. This system is unique when compared to other conventional neurotransmitter systems because it represents a crucial means by which postsynaptic neuronal activity adjusts synaptic gain at presynaptic compartment(s) and influences the transmission of information relevant to behavior (Melis, Greco, et al., 2014). In addition, eCBs allow a fine integration of chemical signals generated by different neurotransmitters and synaptic neuromodulators with changes in neuronal cell excitability. Since the relative level of these lipid-signaling molecules depends upon the release of neurotransmitters, such as DA itself, they might contribute to the several regulatory mechanisms underlying the processing of behavioral states at the synaptic level. As such, not only they are essential mediators of diverse forms of synaptic plasticity but also regulators of homosynaptic and heterosynaptic metaplasticity, which in turn serve as influential keys of emotional and cognitive behavior. In this chapter, we will focus on the influence exerted by the eCB system on ascending mesocorticolimbic DA pathway in the control of reward, motivation, and drug addiction. With regard to the modulation of corticolimbic and corticostriatal circuits by eCBs in the control of motivated behaviors, we refer the reader to other articles where it has been extensively reviewed (Fernandez-Ruiz, Hernandez, & Ramos, 2010; Gardner, 2011; Melis, Greco, et al., 2014).

2. THE NEUROBIOLOGY OF REWARD

2.1 Role of Dopamine

In 1943, Maslow posited that people are motivated to satisfy basic, as well as higher needs, to pursue sensory pleasure and happiness. According to his theory, we first seek to satisfy physiological/survival needs, such as satiety, warmth, and safety. Subsequently, we fulfill the next hierarchical levels within the pyramid toward self-esteem and self-actualization needs, until motivation drives us to the final stage: the realization of one's own potential (Maslow, 1943). Hence, the satisfaction of primary or superior needs, although they are various and subjective, elicits reward, which represents the experience of general wellness as an "award," resulting from well-performed tasks and/or fulfilled expectations.

Satisfaction of needs involves a cost-benefit decision process. Before we act, we assess whether or not the expected outcome is worth the effort. The *incentive-sensitization model* (Berridge & Aldridge, 2008; Robinson & Berridge, 1993, 2000, 2008) associates the anticipation of reward (wanting) with outcome evaluation (liking) (Berridge, Robinson, & Aldridge, 2009). According to this model, the incentive salience represents the motivational state (either conscious or not) behind the "wanting," while the hedonic impact experience (i.e., goal-succeeding pleasure or aversion) is the "liking" (Berridge et al., 2009). Under pathological conditions, such as mood disorders and drug addiction, the *incentive-sensitization* paradigm is aberrant. For instance, the incentive salience increases after repeated drug consumption, while hedonic impact decreases over time. Hence, the desire for drugs is excessive (pathological "wanting") despite reduced reward (Berridge et al., 2009).

In most circumstances, rewards are appetitive stimuli generally associated with pleasure, such as food, sex, exercise, music, and subjective hedonic experiences (e.g., viewing a certain landscape). The intense reward resulting from ingestion of some substances (e.g., drugs of abuse) or engagement in some behaviors (e.g., gambling, shopping) can lead to compulsive behaviors focused on obtaining further reward, eventually resulting in addiction (Olsen, 2011). Despite variability in the types of possible rewards, humans and other animals share common brain substrates for encoding pleasure and reward. Understanding of the neurobiology of reward requires a multidisciplinary approach encompassing the study of neurotransmitters, neuromodulators, and circuits that direct behavioral output (Volkow & Baler, 2014) and the mechanisms for their molecular control, such as genetics, and epigenetics (Befort, 2015; Butelman, Yuferov, & Kreek, 2012; Feng & Nestler, 2013; Nestler, 2014; Robledo, 2010; Solinas, Goldberg, & Piomelli, 2008). In this regard, a major focus has been on the role of DA in the mesolimbic projections from the ventral tegmental area (VTA) to the ventral striatum/nucleus accumbens (VStr/NAc) and in the mesocortical projections to the prefrontal cortex (PFC). Indeed, most substances abused by humans acutely increase extracellular DA levels in NAc (Di Chiara & Imperato, 1988) and PFC of laboratory animals (Maisonneuve, Keller, & Glick, 1990; Pistis et al., 2002). However, whether or not elevations of DA levels *per se* induce pleasure—while a lack of this effect results in anhedonia—is still a matter of debate (Berridge & Kringelbach, 2015; Ikemoto, 2010). Nonetheless, DA is involved in those behaviors aimed at obtaining reward, namely, motivation. For example, DA receptor antagonists in the VStr disrupt instrumental responding for sucrose without reducing sucrose consumption (Ikemoto & Panksepp, 1996). In addition, optogenetic phasic stimulation of VTA DA neurons produces conditioned place preference in freely moving mice (Tsai et al., 2009), and selective optical intracranial self-stimulation of VTA DA neurons has been correlated to the strength of behavioral responding in freely moving rats (Witten et al., 2011).

2.2 Mechanisms Modulating Dopamine Signaling

Spontaneously active DA neurons exhibit either a tonic and/or a phasic activity that robustly influences DA release (Grace & Bunney, 1984a; Grace, Floresco, Goto, & Lodge, 2007; Hyland, Reynolds, Hay, Perk, & Miller, 2002). Hence, single spikes are thought to control basal extrasynaptic (i.e., tonic) DA levels on a second timescale, while bursts of action potentials induce transient (i.e., phasic) DA release at the synapse, which is temporally and spatially controlled on a hundreds of milliseconds timescale (Grace & Bunney, 1984b; Grace et al., 2007). In turn, the switch to burst-firing mode produces a sudden elevation of extracellular DA levels (Garris, Ciolkowski, Pastore, & Wightman, 1994; Gonon, 1988; Grace, 1991), which has been associated with the prediction for expected reward in both rodents (Gronier & Rasmussen, 1998) and primates (Schultz, 1998; Schultz & Dickinson, 2000).

DA cell spontaneous activity is tuned by synaptic inputs (Floresco, West, Ash, Moore, & Grace, 2003; Kitai, Shepard, Callaway, & Scroggs, 1999; Pignatelli & Bonci, 2015). Indeed, GABA released by local interneurons (Erhardt, Mathe, Chergui, Engberg, & Svensson, 2002), and cells within NAc (Walaas & Fonnum, 1980), ventral pallidum (Ribak, Vaughn, & Roberts, 1980), and the rostromedial tegmental nucleus (RMTg) (Jhou, Fields, Baxter, Saper, & Holland, 2009) accounts for inactivation of DA neurons (Floresco et al., 2003; Hong & Hikosaka, 2011; Lecca, Melis, Luchicchi, Muntoni, & Pistis, 2012). Furthermore, GABA receptor activation has been shown to modulate DA cell-firing pattern, and GABA receptor antagonism can regularize DA cell activity and shift-firing patterns to burst mode (Erhardt et al., 2002; Lobb, Wilson, & Paladini, 2010; Paladini & Tepper, 1999). Glutamate release from terminals mainly arising from the PFC, the pedunculopontine tegmentum, and the bed nucleus of the stria terminalis (Floresco et al., 2003; Georges & Aston-Jones, 2001, 2002; Jalabert, Aston-Jones, Herzog, Manzoni, & Georges, 2009; Lokwan, Overton, Berry, & Clark, 1999, 2000), is essential for burst firing of spontaneously active midbrain DA cells (Floresco et al., 2003; Grace & Bunney, 1984b; Lokwan et al., 1999). In addition, acetylcholine, whose terminals mainly arise from the laterodorsal tegmentum (LDTg) (Lodge & Grace, 2006), is thought to be necessary for both tonic modulation of DA cell activity and glutamate-induced switching to burst-firing mode. In particular, activation of postsynaptic nicotinic acetylcholine receptors has been found to regulate spontaneous DA cell activity and to increase their firing frequency, whereas stimulation of presynaptic nicotinic receptors induces bursting activity *in vivo* (Mameli-Engvall et al., 2006; Schilstrom, Rawal, Mameli-Engvall, Nomikos, & Svensson, 2003). Activation of muscarinic acetylcholine receptors also increases firing rate *in vivo* (Gronier & Rasmussen, 1998) and produces burst-firing *in vitro* (Zhang, Liu, & Chen, 2005). Muscarinic activation results in diverse and opposing effects on DA cells, which depend upon the pattern of acetylcholine released and activating postsynaptic receptors (Fiorillo & Williams, 2000). Moreover, DA itself finely modulates neural processing of reward at the downstream levels. In fact, medium spiny neurons (MSNs) of the NAc integrate afferents received from the PFC, ventral subiculum, and basolateral amygdala (BLA) (Grace et al., 2007), and send their projections to VTA GABA interneurons (Bocklisch et al., 2013), which in turn inhibit DA cell firing. Such a multilevel control of reward circuitry suggests the prominence of this process for suitable subsistence, so that disruption of the natural reward neural processing is thought to be involved in aberrant compulsive behaviors associated with psychiatric disorders such as drug addiction.

3. THE INFLUENCE OF ENDOCANNABINOID SIGNALING ON REWARD

The eCB system is involved in the neurobiological mechanisms underlying reward-seeking behaviors and motivation processing (Melis, Muntoni, & Pistis, 2012; Melis & Pistis, 2012), since it influences and interacts with other neurotransmitter systems implicated in the mediation of basic reward functions such as the endogenous opioid and DAergic systems (Befort, 2015; Fattore et al., 2005; Gardner, 2005; Solinas & Goldberg, 2005). In particular, the past decade has provided compelling evidence for the pivotal role played by the eCB system in fine-tuning DA cell activity and, therefore, its influence on reward. In fact, the key discoveries that midbrain DA neurons release eCBs (Melis, Perra, et al., 2004; Melis, Pistis, et al., 2004; Riegel & Lupica, 2004), possess the machinery for the synthesis and degradation of several lipid molecules with eCB-like activity (mainly 2-arachidonoylglycerol, 2-AG, and *N*-acylethanolamines, NAEs) (Matyas et al., 2008), and that type 1 cannabinoid (CB1) receptors are expressed on axon terminals impinging upon these neurons (Marinelli et al., 2007; Melis, Sagheddu, et al., 2014) provide the likely substrates upon which these unconventional neurotransmitters act within the VTA to affect reward processes. The presence of CB1 receptors in other structures related to motivation and reward, like the hippocampus, BLA, hypothalamus, and PFC, also supports eCB system function in the regulation of natural and nonnatural reward (Fattore, Melis, Fadda, Pistis, & Fratta, 2010; Maldonado, Valverde, & Berrendero, 2006). In this section, we will focus on the role played by eCBs in reward evoked by natural stimuli, such as food, sex, maternal care, and social interaction.

3.1 Endocannabinoid Influence on Food Reward

Pleasurable and social aspects of food consumption elicit a strong motivational drive. When tasty food is available, self-control is required in order to avoid overeating. The eCB system not only influences the motivation toward feeding but also its hedonic value, probably by modifying palatability (Friemel, Zimmer, & Schneider, 2014; Higgs, Williams, & Kirkham, 2003; for review, see Farrimond, Mercier, Whalley, & Williams, 2011; Jager & Witkamp, 2014). The role of CB1 receptors and their endogenous ligands in the modulation of both homeostatic and nonhomeostatic aspects of appetite and food intake was revealed through investigations into the effects of exogenous cannabinoids (Di Marzo, Ligresti, & Cristino, 2009; Friemel et al., 2014; Jager & Witkamp, 2014; Scherma, Fattore, Castelli, Fratta, & Fadda, 2014). In fact, the ability of *Cannabis* derivatives to stimulate desire for food ingestion in humans has long been recognized (Abel, 1971; Kirkham & Williams, 2001). Noteworthy, appetite-stimulating properties of plant cannabinoids have provided the basis for their clinical application to enhance eating and weight gain in patients with several debilitating diseases (Foltin, Brady, & Fischman, 1986; Foltin, Fischman, & Byrne, 1988; Mattes, Engelman, Shaw, & Elsohly, 1994; Plasse, 1991). Conversely, the appetite-suppressing effects of CB1 antagonism have been exploited to produce weight loss in obese and overweight subjects (Di Marzo & Despres, 2009; Tonstad, 2006), though the clinical use of CB1 antagonists such as Rimonabant for this and other purposes is no longer permitted due to the incidence of negative psychiatric side effects (Di Marzo & Despres, 2009). Consistently with human studies, Δ^9-THC, the psychoactive ingredient of Cannabis, and eCBs (namely, anandamide, AEA; 2-AG) also promote feeding in animals (Fattore et al., 2010; Jager & Witkamp, 2014; Scherma et al., 2014). In particular, eCBs act as orexigenic agents via dose-dependent activation of CB1 receptors in hypothalamus and limbic forebrain that stimulate appetite (Cota et al., 2003; Jamshidi & Taylor, 2001) and increase food intake (Kirkham, Williams, Fezza, & Di Marzo, 2002). For example, AEA promotes and decreases food ingestion at low and high doses, respectively (Jamshidi & Taylor, 2001; Williams & Kirkham, 1999). AEA levels are decreased by the "satiety hormone" leptin (Di Marzo et al., 2001), thus suggesting that it might take part in the mechanisms through which leptin reduces food intake. Conversely, under conditions of leptin deficiency (i.e., obesity), 2-AG synthesis is enhanced in the lateral hypothalamus, leading to enhanced CB1-mediated decreases in inhibitory signaling thereby leading to increased orexin-A release and hyperphagia (Cristino et al., 2013). (Endo) cannabinoid effects on appetitive behavior have also been demonstrated using operant models in which animals are required to progressively "work harder" to obtain food rewards (Barbano, Castane, Martin-Garcia, & Maldonado, 2009; Gallate, Saharov, Mallet, & McGregor, 1999; Guegan et al., 2013). Conversely, CB1 receptor antagonists dose dependently reduce the motivation for food in rats (Fattore et al., 2010; Solinas & Goldberg, 2005). In line with these findings, genetic deletion of CB1 receptors confers mice with a hypophagic phenotype including a decreased sensitivity to sucrose reward and a reduced motivation to work for food (Cota et al., 2003; Ravinet Trillou, Delgorge, Menet, Arnone, & Soubrie, 2004; Sanchis-Segura, Cline, Marsicano, Lutz, & Spanagel, 2004). Moreover, CB1 antagonists reduce appetitive and consummatory responses to rewarding (sweet) foods/drinks (Maccioni, Pes, Carai, Gessa, & Colombo, 2008; Salamone, McLaughlin, Sink, Makriyannis, & Parker, 2007; Thornton-Jones, Vickers, & Clifton, 2005). In particular, these drugs have been reported to reliably suppress feeding and to reduce conditioned place preference for food, preference for sucrose, and chocolate reinforcing properties (Fattore et al., 2010). Also, human studies indicate that CB1 receptor blockade reduces functional magnetic resonance (fMRI) response to pleasurable stimuli (i.e., palatable food) in key reward areas such as the VStr and orbitofrontal cortex (OFC) (Horder, Harmer, Cowen, & McCabe, 2010). The eCB regulation of food intake and palatability is also mediated through limbic forebrain areas (e.g., the NAc) that have been recognized as a "hot spot" for sensory pleasure (Berridge & Kringelbach, 2015; Friemel et al., 2014; Mahler, Smith, & Berridge, 2007). Hence, the NAc not only is a crucial structure of brain reward system, but also a fundamental center in nonhomeostatic control of eating (Salamone, Correa, Mingote, & Weber, 2003). The enhancement of the eCB signaling in this nucleus exerts potent CB1-dependent hyperphagic actions (Kirkham et al., 2002; Soria-Gomez et al., 2007) and amplifies the natural "liking" reactions of rats to sweet compounds (Mahler et al., 2007). Moreover, a dose of a CB1 receptor agonist capable to induce food intake in fully fed animals activates the corticostriatal-hypothalamic pathway, which includes the NAc, and is central to the motivational drive to eat (Dodd, Stark, McKie, Williams, & Luckman, 2009). Within this circuit, DA is key for both rewarding and addictive properties of food, and Rimonabant blocks increased DA levels in the NAc evoked by presentation of novel palatable foods (Melis et al., 2007) as well as DA output in the medial PFC associated to feeding in hungry rats (Dazzi et al., 2014).

3.2 Endocannabinoid Influence on Sexual Behavior

With regard to sexual behavior, preclinical evidence suggests that cannabinoids affect the consummatory "rewarding" phase by acting centrally, specifically in the hypothalamus (Fattore et al., 2010; Gorzalka, Hill, & Chang, 2010). That *Cannabis* influences human sexual behavior and pleasure is well established (Abel, 1981; Halikas, Weller, & Morse, 1982; Koff, 1974). In fact, the effects of *Cannabis* intake on sexual behavior and arousal appear to be dose dependent in both men and women (Fattore et al., 2010; Gorzalka et al., 2010). On male sexuality, Abel (1981) concluded that a small amount of *Cannabis* can enhance sexual activity, but larger quantities may inhibit sexual motivation. This supported the evidence provided by Koff (1974) from a large sample of Indian men who were chronic *Cannabis* users (Chopra & Jandu, 1976). Overall, studies on male sexuality and *Cannabis* use indicate that *Cannabis* facilitates sexual desire, while simultaneously hindering erectile functioning (Fattore et al., 2010; Gorzalka et al., 2010). This is in contrast with studies on female sexuality and *Cannabis* consumption, which suggest positive effects on both sexual desire and function (Gorzalka et al., 2010). In fact, *Cannabis* use has been reported to have, with only one exception (Johnson, Phelps, & Cottle, 2004, #985), a facilitatory effect on subjective indices of sexual function, including orgasmic function, satisfaction, enjoyment, and/or pleasure (Fattore et al., 2010; Gorzalka et al., 2010). However, this influence is also dose dependent, with beneficial effects only at low doses (Fattore et al., 2010; Gorzalka et al., 2010).

Given the large amount of self-report data described above, animal studies are important complements to understand human findings. In laboratory animals, eCBs also display a bell-shaped dose–response curve on sexual behavior (Martinez-Gonzalez et al., 2004). Remarkably, low doses of AEA seem to positively affect sexual motivation in adult female rats by altering approach behavior (Memos, Vela, Tabone, & Guarraci, 2014). However, there is considerable evidence on the role of the eCB system in the inhibition of penile erection. For example, antagonism of CB1 receptors by AM251 promotes sexual behavior in male rats, whereas inhibition of AEA degradation (FAAH inhibition via URB597) has no effect on copulation (Gorzalka, Morrish, & Hill, 2008). In addition to facilitating ejaculation, it has been observed that Rimonabant induces penile erections (Melis, Succu, Mascia, & Argiolas, 2004; Melis et al., 2006; Succu et al., 2006), through CNS mechanisms (Castelli et al., 2007). Results obtained by pharmacologically modulating eCB levels (through FAAH inhibition by URB597 or inhibition of putative eCB transporters by AM404) are at variance with those observed after systemic administration of either AEA or direct CB1 receptor agonists, since the latter significantly impaired sexual behavior (Ferrari, Ottani, & Giuliani, 2000; Martinez-Gonzalez et al., 2004; Murphy, Gher, Steger, & Bartke, 1994; Shrenker & Bartke, 1985). It is likely that the effects of URB597 and AM404 are distinct from direct CB1 receptor agonists because they potentiate AEA signaling only at synapses, where neural signaling is increased by sexual activity rather than inducing widespread and indiscriminant CB1 receptor activation (Freund, Katona, & Piomelli, 2003). Importantly, a significant relationship between circulating eCB concentrations and sexual function has been observed in women (Klein, Hill, Chang, Hillard, & Gorzalka, 2012). In particular, these authors reported that increases in both physiological and subjective indices of sexual arousal were significantly associated with decreased AEA and 2-AG levels. Despite the nature of the association found in this research was contrary to author's predictions, the results are in line with preclinical and clinical literature supporting the idea that CB1 receptor activation might, indeed, be related to decreased sexual functioning in nonhuman and human females (Gorzalka et al., 2010, #548; Johnson et al., 2004, #985). Also, it should be mentioned that eCB serum levels not necessarily reflect brain concentrations (Klein et al., 2012, #549). A better knowledge of the influence of eCB system in physiological processes involved in sexual behavior not only will contribute to advance in developing drugs for treating sexual dysfunctions (Gordon, Bromley, Gorski, & Zimmermann, 1978; Salerian, 2004; Turley & Floody, 1981) but also to identify potential sexual side effects of pharmaceutical agents already marketed for nonsexual disorders.

3.3 Endocannabinoid Influence on Maternal Care

Maternal care is the first experience of social interaction. This natural, deeply emotional, behavior has a crucial impact on infant survival, proper development, and social competences throughout life. Indeed, either infants or pups have high inherent value to their respective mothers (for extensive reviews, see Manduca, Campolongo, & Trezza, 2012; Trezza, Campolongo, & Vanderschuren, 2011). Breastfeeding in humans (Kavanaugh, Meier, Zimmermann, & Mead, 1997) and pup suckling in animals (Ferris et al., 2005) represent rewarding and reinforcing stimuli that promote maternal behavior and bonding. The responsiveness of a mother toward her young depends on, and is influenced by, changes in the activity within multiple regions of the reward network (see Numan & Stolzenberg, 2009, for a comprehensive review). Indeed, lesion, depletion, and brain imaging studies have revealed that pup suckling activates the mesocorticolimbic DA system in lactating dams and is reinforcing (Febo, Numan, & Ferris, 2005; Ferris et al., 2005; Hansen, Harthon, Wallin, Lofberg, & Svensson, 1991; Numan & Smith, 1984). Consistently, extracellular NAc DA levels increase during anticipatory and consummatory phases of maternal behavior (Champagne et al., 2004; Hansen, Bergvall, & Nyiredi, 1993) and are positively associated with time spent nursing the offspring (Champagne et al., 2004). In addition to DA, oxytocin activity within the reward/motivational system serves as an important neurobiological substrate underlying maternal behavior (Love, 2014). Although less investigated in humans, mother–child relationships have also been linked with activity in the mesocorticolimbic circuit. Several fMRI studies have shown that mothers looking at pictures of their own children exhibited enhanced activation in areas such as OFC, medial PFC, insula, VStr, amygdala, hypothalamus, and VTA/substantia nigra (Bartels & Zeki, 2004; Leibenluft, Gobbini, Harrison, & Haxby, 2004; Strathearn, Fonagy, Amico, & Montague, 2009; Strathearn, Li, Fonagy, & Montague, 2008; but see Wittfoth-Schardt et al., 2012). Notably, CB1 receptors and their mRNA are highly expressed in the above-mentioned brain areas, which are implicated in forming long-lasting emotional bonds, including parenting (Hollis, Coddington, & Moore, 2006; Tsou, Brown, Sanudo-Pena, Mackie, & Walker, 1998). Thus, the presence of CB1 receptors in these circuits is suggestive of an eCB influence on maternal care, though further investigations are required in order to confirm it. Converging preclinical evidence indicates that the eCB system also affects the behavioral response of offspring to maternal care, including the initiation of suckling and the ontogeny of emotionality (Fride et al., 2009; Manduca et al., 2012; Schneider, 2009). Therefore, eCBs might provide a link between social and emotional responsiveness by also conferring rewarding value to suckling behavior (Chakrabarti, Kent, Suckling,

Bullmore, & Baron-Cohen, 2006). Nonetheless, this system appears to be crucial for establishment of social behavior during the first postpartum week by having a long-term impact on offspring socio-emotional development (Schechter, Pinhasov, Weller, & Fride, 2012; Schechter et al., 2013). Elucidating the influence of eCBs on maternal care and the responsiveness of offspring to this behavior, together with their underlying neural mechanisms, is of substantial interest, as early life social experiences epigenetically shape brain development and adult behavior, including social interaction (Champagne & Curley, 2009).

3.4 Endocannabinoid Influence on Social Interaction

In Maslow's pyramid of hierarchal levels of needs, the search for interpersonal relationships, which involve feelings of belongingness and social interaction, immediately follows safety needs. The importance of social behavior is demonstrated by a great deal of evidence showing that among all age groups supportive relationships are essential to the sense of physical and psychological well-being, and to happiness and satisfaction (Myers & Rocca, 2000; Reis, Aron, Clark, & Finkel, 2013; Walker & McGlone, 2013). Indeed, social interactions are rewarding. In fact, social activities play a crucial role for survival, and social activity early in life shapes brain development and adult behavior (Champagne & Curley, 2005). Among social activities, cooperative playing is highly rewarding not only for young (Calcagnetti & Schechter, 1992) and adult rats (Douglas, Varlinskaya, & Spear, 2004) but also for human (Alexander, Frohlich, & Fusco, 2012; Tabibnia & Lieberman, 2007) and nonhuman primates (for a comprehensive review on this topic, see Trezza, Baarendse, & Vanderschuren, 2010). Social play behavior is one of the earliest forms of nonmother-directed social interaction (Pellis & Pellis, 1998; Vanderschuren, Niesink, & Van Ree, 1997). Hence, social play activities during childhood and adolescence are determinant for the development of adult social competences, for the ability to deal with negative events, and for selection of appropriate behavioral patterns. In fact, rats isolated during adolescence exhibit aggressive behavior and severe social deficits at adulthood (Hol, Van den Berg, Van Ree, & Spruijt, 1999; Potegal & Einon, 1989). In this context, *Cannabis* intoxication has long been thought to increase sociability, though the consequences of *Cannabis* use might result in impaired sociability (see below). Most *Cannabis* users first experiment with the drug during adolescence, a critical phase not only for brain development but also for social competence and friendship formation. Accordingly, smoking marijuana promotes group cohesion of young people (Verkooijen, de Vries, & Nielsen, 2007). Animal studies focus on the consequences of prior cannabinoid exposure rather than acute intoxication and therefore cannot fully translate this observation and the anecdotal reports. In fact, chronic Δ^9-THC exposure reduces social interactions in mice (Frischknecht, Sieber, & Waser, 1982), and CB1 receptor activation during adolescence significantly decreases time spent in social interaction activity of adult rats (Genn, Tucci, Marco, Viveros, & File, 2004; Leweke & Schneider, 2011; O'Shea, McGregor, & Mallet, 2006; O'Shea, Singh, McGregor, & Mallet, 2004; Quinn et al., 2008; Realini et al., 2011). In particular, long-lasting behavioral disturbances are observed in mature rats, including disruption of social interactions and social play behavior that lead to social withdrawal and anhedonia (Schneider & Koch, 2003, 2005, 2007; Schneider, Schomig, & Leweke, 2008). Finally yet importantly, perinatal exposure to Δ^9-THC disrupts social interaction and play behavior in adult rats (Newsom & Kelly, 2008; Trezza et al., 2008), while prenatal exposure correlates with shorter duration of play during childhood, and with an increased risk for aggressive behavior and attention problems in girls, but not boys (Faden & Graubard, 2000). Hence, *Cannabis* exposure appears to reduce sociability. However, elegant studies have unveiled the differences between the effects of *on demand* activation of eCB system and direct CB1 receptor stimulation by exogenously administered drugs (Trezza et al., 2010, 2011; Trezza, Campolongo, et al., 2012; Trezza, Damsteegt, et al., 2012; Trezza & Vanderschuren, 2008a, 2008b, 2009). In particular, by boosting the eCB signaling (e.g., URB597 and VDM11, inhibitors of FAAH and of the putative eCB transporter, respectively), social play increases (Trezza, Campolongo, et al., 2012; Trezza & Vanderschuren, 2008a, 2008b, 2009), probably via AEA (or other NAEs) released in the amygdala and NAc of adolescent rats (Trezza, Damsteegt, et al., 2012). In adolescent rats, striatal AEA levels also participate in the emotional arousal resulting from exposure to a nonfamiliar social partner, and this may be important for coping responses to novel social contexts (Marco et al., 2011). On the other hand, agonists at CB1 receptors as well as CB1 receptor overexpression in medial PFC reduce sociability by disrupting cognitive functions (e.g., behavioral flexibility) and emotional reactivity required to perform adequate social interactions (Egerton, Allison, Brett, & Pratt, 2006; Klugmann, Goepfrich, Friemel, & Schneider, 2011). Notably, the positive effects of indirect CB1 receptor agonists on social play can be blocked by either DA or μ-opioid receptor antagonism

(Trezza & Vanderschuren, 2008a, 2008b, 2009). Conversely, play-enhancing actions of opiate agonists can be attenuated by either opiate or CB1 receptor antagonists (Trezza & Vanderschuren, 2008a). Endogenous opioid and cannabinoid systems have their own specific role in modulating play. Whereas the increased social play induced by eCBs is dependent upon the playful responsiveness of the test partner, morphine enhances social play independently from social activity level of the partner (Trezza & Vanderschuren, 2008b). Remarkably, while DA is involved in the facilitating effects of indirect cannabinoid agonists on social play, it is not implicated in play-stimulating effects of morphine (Trezza & Vanderschuren, 2008a). In this scenario, eCBs play a pivotal role by activating mesolimbic DA transmission (Trezza & Vanderschuren, 2008a, 2008b, 2009), which is deeply involved in modulating social play (Siviy & Panksepp, 2011; Trezza et al., 2010; Vanderschuren et al., 1997). An influence of eCB signaling in social behavior has also been observed in humans. In particular, CB1 receptor gene variations have been associated with a modulation of social reward responsiveness by reward-related forebrain areas (Chakrabarti et al., 2006). Collectively, these studies highlight the positive influence of the eCB system in regulating reward processing and behaviors, which could be pharmacologically exploited for the treatment of neuropsychiatric disorders characterized by aberrant emotions and social behaviors.

4. THE ROLE OF THE ENDOCANNABINOID SYSTEM IN MOTIVATION

4.1 Motivation—A Leitmotif

Motivation is a psychological construct used to explain behaviors that are energized and directed toward a goal (Bindra, 1968). These goals, such as food, sex, and maternal care, can be essential for individual and species survival. In humans, motivation has acquired a more complex and multifaceted connotation, as it is also directed toward objectives that are not normally essential to mere survival, such as school or sport scores, career, power, and money. Undoubtedly, these goals represent an extension of basic needs and indirectly contribute to the success of the individual within a society and among competitors. Hence, motivation guides us to carry out goal-directed behaviors for their complete satisfaction, and we expect effort-related rewards. The more the effort we perform to achieve the goal, the higher the motivational incentive behind it.

Incentive salience refers to the property of stimuli to raise attention and behavioral responses, on which motivation relies. We confer salience to otherwise neutral stimuli or goals by means of a value-related decision-making process, which is based on previous experiences. Reinforcement learning and expectancies of success also depend upon previously received reward (Bromberg-Martin, Matsumoto, & Hikosaka, 2010; Schultz, Dayan, & Montague, 1997). Thus, motivation refers to a learning- and memory-associative model, since learning is driven by expectations about future rewards and memory plays a major role in making decisions. Salience attribution and regulation of motivation are multistep processes that likely involve different "hard-wired" neuronal populations and brain circuits (Wise, 2004). Particularly, in primates, cognitive and behavioral studies demonstrate that DA neurons code unpredicted reward by changing tonic to phasic activity, while fully expected reward produces no response, and reward worse than prediction produces complete inhibition (Schultz, 2010, 2013; Schultz et al., 1997). These results provide convincing evidence implicating DA in the prediction error of saliency and reward for future events. OFC neurons exhibit anticipatory activity that is related to the nature of reward and are modulated by the motivational state of the animal (Hikosaka & Watanabe, 2000, 2004). Hence, different brain areas and neurons are involved in the expectancy of reward and value coding, suggesting the importance of motivational approach while behaving. Indeed, several psychiatric disorders, including drug addiction (according to DMS-V), involve a disruption of motivational processes. In fact, addiction is characterized by aberrant positive- and negative-reinforcement processes and deregulated reward system function (i.e., the pleasure experienced from drug consumption; Koob & Le Moal, 2001; Sjoerds, Luigjes, van den Brink, Denys, & Yucel, 2014; Wyvell & Berridge, 2000). As mentioned above, brain networks coding reward have a pivotal role in motivation and appetitive-directed behaviors (Grace et al., 2007). An understanding of the complex role of DA and non-DA neurons within the VTA has been progressively emerging (Bromberg-Martin et al., 2010; Grace et al., 2007; Haber & Knutson, 2010; Lammel et al., 2012). Nonetheless, optogenetic phasic activation of VTA DA cells triggers reward-associated behaviors (Lammel et al., 2012; Tsai et al., 2009), thus supporting that DA encodes for positive reinforcement and salience. Hence, spontaneous activity of DA neurons is activated by prediction error for reward (Paladini & Roeper, 2014; Schultz, 1998; Tobler, O'Doherty, Dolan, & Schultz, 2006), incentive motivation, and salience (Winton-Brown, Fusar-Poli,

Ungless, & Howes, 2014). Unexpected or cue-predicting rewards (Bromberg-Martin et al., 2010; Matsumoto & Hikosaka, 2009; Mirenowicz & Schultz, 1996; Schultz, 1998), aversive/stressful stimuli and/or reward omission differently modulate discrete subpopulations of midbrain DA cells (Lammel, Lim, & Malenka, 2014; Lammel et al., 2012; Schultz, 2013). Particularly, VTA DA projections to NAc or medial PFC differently contribute to both healthy and pathological states/behaviors, such as stress-induced depression and drug addiction (Lammel et al., 2014). Indeed, LDT-VTA(DA) → NAc and lateral habenula (LHb)-VTA (DA) → PFC circuits are mainly involved in positive and negative reinforcement, respectively (Lammel, Ion, Roeper, & Malenka, 2011; Lammel et al., 2012).

Acute aversive stimuli, which are a form of negative motivational cue, transiently suppress the majority of VTA DA cell firing (Cohen, Haesler, Vong, Lowell, & Uchida, 2012; Matsumoto & Hikosaka, 2009; Ungless, 2004). This is consistent with the activation of LHb (Lammel et al., 2012; Matsumoto & Hikosaka, 2007) and RMTg (Hong, Jhou, Smith, Saleem, & Hikosaka, 2011; Jhou et al., 2009) projections, as well as with the activation of local GABA interneurons impinging upon DA cells (Tan et al., 2012). Remarkably, this latter process is associated with place aversion (Tan et al., 2012), and passive avoidance (Danjo, Yoshimi, Funabiki, Yawata, & Nakanishi, 2014), two behavioral paradigms aimed at assessing negative motivational effects of stimuli. On the other hand, aversive stimuli induce phasic activity in a subset of DA cells in both rats (Brischoux, Chakraborty, Brierley, & Ungless, 2009) and monkeys (Matsumoto & Hikosaka, 2009; Schultz & Romo, 1987). Therefore, the encoding of choice between reward- or aversion-related behaviors depends on the integration of several inputs to VTA DA cells and varies according to the projection areas (e.g., VStr, OFC; Grace et al., 2007; Hollerman, Tremblay, & Schultz, 2000).

4.2 Effects of Altered Cannabinoid Signaling on Motivation

Information on how the eCB system regulates motivation derives mainly from the analysis of the effects of exogenous cannabinoid agonists, such as Δ^9-THC in *Cannabis* derivatives.

Several decades ago, one of the major concerns expressed over the possible long-term effects of marijuana use was the frequently described "amotivational syndrome" (McGlothlin & West, 1968). This was considered a sequel to chronic marijuana use that contributed "to the development of more passive, inward turning, amotivational personality characteristics" (Kolansky & Moore, 1971; Millman & Sbriglio, 1986). This syndrome consists in a wide range of symptoms including loss of energy and the drive for work, avolition, apathy, dullness, lethargy with mild-to-severe impairment of judgment, concentration, and memory (Tennant & Groesbeck, 1972). However, the observation of an amotivational syndrome was mainly based on case reports, observational reports, or small clinical studies. Hence, the existence and specificity of such a syndrome was questioned by a number of subsequent studies that found no compelling evidence that is a separate nosographic entity (Hall, Solowij, & Lemon, 1994). A parsimonious explanation for impaired motivation is to include this syndrome among symptoms of either chronic *Cannabis* intoxication (Negrete, Knapp, Douglas, & Smith, 1986) or other psychiatric conditions. In fact, the amotivational syndrome has rather aspecific symptoms, which are certainly not exclusive to *Cannabis*, as they can also be precipitated by several drugs of abuse, including alcohol, opioids, and psychostimulants. However, the notion that heavy cannabinoid consumption disrupts motivation has resurfaced, being supported by neuroimaging studies, which suggest that the hypodopaminergia associated with regular *Cannabis* use may underlie the reduced reward sensitivity and amotivation associated with chronic *Cannabis* use (van Hell et al., 2010). Particularly, van Hell et al. (2010) found in long-term *Cannabis* users a reduced reward anticipation in a monetary reward task coupled with attenuated brain activity during reward anticipation in the NAc when compared to control subjects. Notably, this effect was also present in a tobacco-smoking group, indicating that chronic nicotine abuse is also responsible. Noteworthy, *Cannabis* users specifically showed a decreased reward anticipation activity in the caudate nucleus compared to both control- and tobacco-smoking subjects (van Hell et al., 2010). Other studies investigated on the effect of chronic *Cannabis* use on DA synthesis measured with 3,4-dihydroxy-6-[^{18}F]-fluoro-l-phenylalanine ([^{18}F]-DOPA) positron emission tomography (PET), and its relationship between index of apathy in *Cannabis* user (Bloomfield, Morgan, Egerton, et al., 2014; Bloomfield, Morgan, Kapur, Curran, & Howes, 2014). Chronic *Cannabis* use is associated with a reduced DA synthesis capacity (Bloomfield, Morgan, Egerton, et al., 2014), and an inverse relationship between striatal DA synthesis and apathy was revealed (Bloomfield, Morgan, Kapur, et al., 2014). Altogether, these studies highlight the link among mesolimbic DA system, *Cannabis* use, and motivation. In fact, cannabinoids might reduce motivation and reward anticipation by disrupting DA transmission.

The involvement of the eCB system in neuronal mechanisms of motivation and reward is more directly supported by the effect of cannabinoid antagonists. In fact, clinical trials with Rimonabant in nonaddicted individuals indicate that this drug induces a marked reduction of motivation and reward, leading to anhedonia and frank depression in a number of individuals with one case of completed suicide (Christensen, Kristensen, Bartels, Bliddal, & Astrup, 2007). Thus, the antagonism of CB1 receptors induces a decreased motivation to search for natural rewards (Horder et al., 2010), and even precipitate suicidal ideation in vulnerable individuals. Consequently, it was withdrawn from clinical trials in humans. In this regard, animal studies demonstrate that cannabinoid antagonists decrease cue-evoked DA elevations and reward seeking in a number of experimental conditions (see these reviews and references therein Melis et al., 2012; Melis & Pistis, 2012; Oleson et al., 2012). Collectively, these findings suggest that eCBs in the VTA regulate reward seeking and motivation by controlling DA neuron firing rate and pattern, and DA release during reward-directed behavior (Melis & Pistis, 2012). This is line with the observations that Δ^9-THC acutely increases DA release in the human striatum (Bossong et al., 2009) and enhanced firing rate of rat mesolimbic DA neurons (French, Dillon, & Wu, 1997; Gessa, Melis, Muntoni, & Diana, 1998) as well as striatal DA levels (Chen et al., 1990; Fadda et al., 2006; Malone & Taylor, 1999; Tanda, Pontieri, & Di Chiara, 1997). On the other hand, chronic *Cannabis* use and withdrawal are associated with reduced DA synthesis in humans (Bloomfield, Morgan, Egerton, et al., 2014) and decreased DA levels in rat NAc (Tanda, Loddo, & Di Chiara, 1999) accompanied by a reduced firing activity of mesolimbic DA neurons (Diana, Melis, Muntoni, & Gessa, 1998). Noteworthy, both cannabinoid agonists and antagonists decrease motivation in humans. This apparent contradiction can be readily explained. In fact, chronic, but not acute, cannabinoid agonist administration induces a marked desensitization of CB1 receptors and decreases 2-AG content and signaling in relevant brain regions, such as the striatum (Di Marzo et al., 2000; Dudok et al., 2015; Gonzalez et al., 2004). On the contrary, either the increased eCB levels or the activation of CB1 receptors by exogenous agonists should enhance motivation to engage in rewarding activities. This notion has been difficult to substantiate in humans, with the exception of food and sex. As abovementioned, in fact,

Cannabis promotes strong craving for food and an intensification of its sensory and hedonic properties (Fattore et al., 2010; Kirkham, 2009a, 2009b). Conversely, Rimonabant was initially approved for obesity and metabolic syndrome treatment (Tonstad, 2006). Regarding sexual activity, as stated above, the effect of *Cannabis* depends on the sex and the dose (see Gorzalka et al., 2010 or Fattore et al., 2010 and references therein). The inverted U-shape effect of cannabinoids on sexual activity and motivation is also confirmed by animal studies, which indicate that an optimal cannabinoid tone promotes these behaviors by enhancing motivation and reward, whereas higher doses blunt motivation.

The link between DA transmission and the eCB system is evident when considering the placebo effect. Placebo is not only the inert treatment, but rather the full ritual of administration, which include a set of sensory and social stimuli informing the patient that a beneficial therapy is being given (Benedetti, Amanzio, Rosato, & Blanchard, 2011). The placebo effect is a psychobiological phenomenon composed of different mechanisms, particularly expectation of clinical improvement and pavlovian conditioning. Indeed, patient motivation to feel better determines the magnitude of placebo effect. Notwithstanding, DA transmission together with endogenous opioids and eCBs play a major role in this phenomenon (Benedetti et al., 2011). Hence, DA release enhances the placebo responses in Parkinson's disease and mediates placebo analgesia (de la Fuente-Fernandez et al., 2001; Scott et al., 2007). In an experimental pain model, enhanced DA release in the NAc correlated to placebo responsiveness and monetary reward (de la Fuente-Fernandez et al., 2001; Scott et al., 2007). These findings suggest that motivation and reward mechanisms mediated by DArgic transmission may in part mediate placebo responsiveness (Scott et al., 2007). Alongside the involvement of endogenous opioid system, the eCB system contributes (Benedetti et al., 2011; Pecina & Zubieta, 2015) to analgesic placebo response (Levine, Gordon, & Fields, 1978). This is supported by the influence of a functional polymorphism (Pro129Thr) of the gene encoding FAAH, the major hydrolyzing enzyme for AEA, in placebo-induced opioid and DA neurotransmission (Pecina et al., 2014). In this study, homozygous carriers of the FAAH Pro129Thr polymorphism showed significantly greater placebo responses, a more positive internal affective state during the placebo condition and a more positive recall of the placebo experience 24 h after the pain challenge compared with *Thr129* carriers. However, *Pro129/Pro129* homozygotes, who constitute nearly half of the population, have increased activity of the FAAH and, therefore, lower AEA levels

(Chiang, Gerber, Sipe, & Cravatt, 2004). These results are in contrast with animal models, where transgenic mice lacking FAAH display increased CB1-dependent analgesia (Cravatt et al., 1996). One possible explanation might be the development of tolerance to chronic CB1 receptor activation by eCBs, similarly to that produced by chronic exogenous cannabinoid administration.

5. DRUG ADDICTION AND THE ENDOCANNABINOID SYSTEM: A RECIPROCAL MODULATION OF SYNAPTIC PLASTICITY

5.1 Etiology and Neural Mechanisms of Addiction

As previously discussed, natural rewards activate brain circuits, and in particular, the mesocorticolimbic DA system that codes for motivated behaviors and processes the relevance of environmental stimuli for survival and life pleasantness. Beyond natural stimuli, animals find some psychoactive substances as voluptuary because they act on the same brain reward pathway (Volkow, Fowler, Wang, Baler, & Telang, 2009). Substances such as alcohol, nicotine, cocaine, morphine, and Δ^9-THC, but also repetition of some behaviors such as shopping and gambling, elicit pleasure by interfering with the typical activity of reward neural circuits.

Drugs of abuse act as positive reinforces, and many subjects seeking for wellness repeat their consumption thereby becoming abusers. Importantly, only a small percentage of them loses control over their behavior and becomes addicted to the substance(s). Remarkably, personality trait(s), individual vulnerability, and life events (Belcher, Volkow, Moeller, & Ferre, 2014; Sagheddu & Melis, 2015) contribute to the transition from acute drug use to dependence and/or addiction. In fact, drug addiction is recognized as a chronic psychiatric disorder, which is characterized by cyclical phases of drug use escalation, withdrawal and relapses despite negative consequences (Wise & Koob, 2014). Drug consumption initially arouses pleasure, which is most likely followed by feelings of general dysphoria. Doses and frequency of drug consumption usually increase progressively, thus inducing neuronal circuits to switch from homeostatic to allostatic dysregulation (see excellent reviews where the phenomenon is well described Edwards & Koob, 2010; Koob & Le Moal, 1997, 2001; Piazza & Deroche-Gamonet, 2013). Hence, positive- and negative-reinforcement mechanisms, which are the need for pleasure and avoidance of psychophysical discomfort, respectively, drive the compulsive seeking for the substance(s) (Kalivas, Volkow, & Seamans, 2005).

Drugs of abuse typically induce reward-related behaviors through neurobiological adaptations of mesocorticolimbic DA system (Baik, 2013a; Ikemoto & Bonci, 2014), including changes in DA neuron morphology (Mazei-Robison et al., 2014; Sklair-Tavron et al., 1996; Spiga, Lintas, Migliore, & Diana, 2010; Spiga, Puddu, Pisano, & Diana, 2005; Spiga et al., 2014), DA receptor function (Baik, 2013a, 2013b; Park, Volkow, Pan, & Du, 2013), and changes related to intracellular pathways, genes and/or epigenetic marks (Joffe, Grueter, & Grueter, 2014; Maze & Nestler, 2011; Mazei-Robison et al., 2011; Sadakierska-Chudy, Frankowska, & Filip, 2014; Zamora-Martinez & Edwards, 2014). The reward experienced after acute drug exposure is followed by aversive effects, which induce negative reinforcement encoded either directly or indirectly by VTA neurons (Matsui, Jarvie, Robinson, Hentges, & Williams, 2014; Volman et al., 2013). Particularly, increased AMPA receptor-mediated synaptic responses (Bellone & Luscher, 2006; Good & Lupica, 2010; Mameli, Balland, Lujan, & Luscher, 2007) have been associated with long-term potentiation (LTP) of glutamatergic synapses onto VTA DA neurons after both acute and chronic passive drug exposure (Argilli, Sibley, Malenka, England, & Bonci, 2008; Saal, Dong, Bonci, & Malenka, 2003; Ungless, Whistler, Malenka, & Bonci, 2001). Several neuronal populations within the VTA, and their afferents, are involved in aversive effects of drug consumption (Matsui et al., 2014; Volman et al., 2013), which represent negative motivation underlying the occurrence of relapses. Remarkable modifications of LHb synapses onto RMTg, but VTA, neurons have also been associated to drug-evoked hyperexcitability (Meye et al., 2015), suggestive of aversive signaling associated with negative symptoms of drug dependence. With respect to neuronal pathway modifications during different stages of drug addiction, evidence for the role of the eCB system, among the others, has emerged (Cachope & Cheer, 2014; Mark, Shabani, Dobbs, & Hansen, 2011; Melis & Pistis, 2012; Parolaro et al., 2010; Pistillo, Clementi, Zoli, & Gotti, 2015). In fact, given that DA neurons integrate intrinsic and extrinsic inputs to encode their own signal (Morikawa & Paladini, 2011), the eCB 2-AG participates in the regulation of DA cell excitability by modulating the strength of impinging synapses (Melis & Pistis, 2012; Wang & Lupica, 2014). Noteworthy, when the other brain structures involved in reward signaling and addiction are examined, eCBs other than 2-AG are found to exert a role in fine-tuning the whole-network activity by modulating both short- and long-term forms of synaptic plasticity (Fattore et al., 2010; Melis, Greco, et al., 2014).

The observation that both acute and chronic exposure to discrete and structurally unrelated drugs of abuse impacts on eCBs, and their synaptic functions, lends support to the dynamic and bidirectional interaction between drug-evoked plasticity of the mesocorticolimbic circuit and the eCB system. However, whether the eCB system represents the determinant or the consequence is still largely unknown. Nonetheless, the role of 2-AG released by the DA cell that shapes rewarding signals at target regions following/before and/or during exposure to substances of abuse has been substantiated. As a result, this lipid signal is involved in several and diverse phases of the drug addiction cycle (Cheer et al., 2007; Oleson et al., 2012; Oliere, Joliette-Riopel, Potvin, & Jutras-Aswad, 2014; Sagheddu & Melis, 2015). Hence, targeting the eCB system has appeared as promising avenue for the development of new pharmacological therapies to relieve symptoms (e.g., craving, relapse) associated with both drug and behavioral addiction. In particular, drugs that modulate the eCB system are currently under investigation (e.g., indirect agonists that enhance endogenous levels) especially since, as abovementioned, Rimonabant was withdrawn from the market.

5.2 Drug-Induced Alterations in Brain Endocannabinoid Signaling

Even though the eCB system is influenced during, and in turn affects, all of the different stages of addiction cycle, the investigation of drug-induced modifications of eCB function is still in its infancy, as the effects of only few drugs of abuse have been studied, and yet not fully elucidated. For instance, information on the acute actions of abused substances on eCB system function are available only for cocaine and Δ^9-THC (Fourgeaud et al., 2004; Mato et al., 2004; Mereu et al., 2015). Particularly, the first exposure to cocaine is able to abolish an eCB-mediated form of long-term depression (LTD) at corticolimbic synapses in the NAc (Fourgeaud et al., 2004). This impairment in synaptic plasticity appears to be due to a reduction in the ability of postsynaptic metabotropic glutamate receptors (mGluRs) to translate anterograde glutamate transmission into a retrograde eCB signaling acting at presynaptic CB1 receptors (Fourgeaud et al., 2004). Similarly, the first exposure to Δ^9-THC reduces eCB-mediated LTD at the same synapses, although through a functional tolerance of CB1 receptors (Mato et al., 2004). This aberrant plasticity and the corresponding behavioral output have been suggested to be the early underpinning for those changes in brain processing that allow the transition from recreational use to abuse, and ultimately to dependence. In fact, a context-independent behavioral and neurochemical sensitization following the first exposure to cocaine can be abolished by blocking CB1 receptors (Mereu et al., 2015). Hence, eCBs appear as key players from the earliest stage of cocaine use as well as of other drugs (Cheer et al., 2007), which contribute to long-term brain-adaptive responses, and the shaping of neural networks. As a result, they might indeed increase the vulnerability to subsequent stimulant abuse disorders (Mereu et al., 2015; Vezina & Leyton, 2009).

Often following acute intoxication is the phase where the drug is chronically consumed. This stage induces maladaptive brain changes and the development of tolerance. Notably, the eCB system also takes part in these phenomena through changes occurring in the mesocorticolimbic pathway (DePoy et al., 2013; Hoffman, Oz, Caulder, & Lupica, 2003; Mato, Robbe, Puente, Grandes, & Manzoni, 2005; Pan, Hillard, & Liu, 2008). Hence, escalation to drug intake is produced by a decrease in self-control that can be reversed and blocked by interfering with the eCB signaling (Hernandez et al., 2014; Kleijn et al., 2012; Pattij et al., 2007; Wiskerke, Stoop, Schetters, Schoffelmeer, & Pattij, 2011). In particular, eCB signaling appears to play a pivotal role in the facilitation and preservation of long-term changes produced by cocaine exposure in both reward evaluation and decision-making processes (Hernandez et al., 2014). This could be explained by the observation that CB1 receptor blockade prevents drug-induced changes in DA transients within the NAc (Cheer et al., 2007), an effect due to 2-AG released in the VTA and regulating reward seeking by shaping relevant patterns of DA release during reward-directed behavior (Oleson et al., 2012). In fact, repeated exposure to cocaine facilitates LTP of VTA DA neurons by reducing the strength of GABAergic inhibition in a long-term manner via activation of CB1 receptors by 2-AG (Pan et al., 2008). On the other hand, repeated Δ^9-THC administration reduces the coupling efficiency of CB1 receptors to $G_{(i/o)}$ transduction proteins at corticolimbic synapses. However, these synapses express normal LTD due to a reversible switch in its underlying mechanisms (Mato et al., 2005). In fact, compensatory mechanisms occur at corticolimbic synapses on MSNs to allow the recovery of activity-dependent LTD when CB1 receptor function is compromised. Thus, in the NAc, a form of presynaptic homeostasis mediated by presynaptic mGluRs (i.e., mGluR2/3) can actually rescue synaptic plasticity from Δ^9-THC-induced deficits. Notably, both receptor types are expressed on the same axons and use the same pathway to induce LTD, which are mutually occlusive (Mato et al., 2005). Finally, this form of homeostatic synaptic plasticity is engaged only upon conditions of sustained cortical activities inducing glutamate spillover able to activate presynaptic mGluR2/3, an effect not described by Hoffman et al. (2003). In particular, these authors found that chronic treatment with Δ^9-THC completely blocked LTD at the same synapses by reducing CB1 receptor sensitivity to both exogenous and endogenous cannabinoids. This tolerance appears to prevent eCBs from establishing LTD at excitatory synapses on MSNs of NAc. Remarkably, a cross-tolerance between CB1 and μ-opioid—but not adenosine type A1—receptors was also observed, suggestive of an altered G-protein coupling (or other effectors) shared by both CB1 and μ-opioid receptors (Manzanares et al., 1999). Similarly, CB1 receptor downregulation and desensitization (i.e., functional tolerance) has been reported in the dorsolateral striatum (DLS) following chronic Δ^9-THC exposure (Nazzaro et al., 2012). In addition, a loss of both LTD at cortico-striato-pallidal synapses and synaptic depotentiation associated to a switch from goal-directed actions to inflexible habitual strategies were also observed (Nazzaro et al., 2012). Similarly, a chronic intermittent alcohol treatment can enhance 2-AG levels, induce a downregulation of CB1 receptor signaling, and impair LTD at DLS corticostriatal synapses, which enables a form of NMDAR-mediated LTP (DePoy et al., 2013). In addition, the authors suggested that these profound remodeling processes most likely drive changes in learning and reward-directed behavior. Accordingly, human postmortem and PET studies have revealed a decreased CB1 receptor density in the VStr of patients with alcohol dependence (Hirvonen et al., 2013; Vinod et al., 2010), which persists several weeks into abstinence. In addition, fMRI studies have shown that long-term *Cannabis* users exhibit disrupted neural activity in the caudate nucleus, putamen, thalamus, inferior and superior frontal gyrus, and the parahippocampal gyrus (Jager et al., 2007; van Hell et al., 2010), which are brain areas involved in reward processing, motor control, and associative learning. It appears, therefore, that such deficits in eCB signaling might influence the control of goal-directed behavior, thus promoting the occurrence of habit-forming effects of not only Δ^9-THC and alcohol but more generally of drugs of abuse (Everitt & Robbins, 2005). In fact, by limiting synaptic rescaling, this drug-evoked loss of LTD in NAc has been posited to impair the responsiveness to new environmental information, which might contribute to behavioral inflexibility typical of the addicted state (Kasanetz et al., 2010). This latter is characterized by an inadequate response inhibition and salience attribution, drug craving, and compulsive drug-seeking and drug-taking behaviors despite negative consequences. Since 2004, following on from the development of an animal model of drug addiction (Deroche-Gamonet, Belin, & Piazza, 2004) meeting the diagnostic criteria for addiction in humans (DMS IV-TR), advances in the synaptic and molecular underpinnings of the addicted state have started (Chen et al., 2013; Kasanetz et al., 2010, 2013). In particular, using this paradigm, Kasanetz et al. (2010) observed a persistent loss of NMDA-dependent LTD at corticolimbic synapses onto NAc MSNs in cocaine-addicted animals. Notably, this deficit already occurs in all of the subjects chronically self-administering the drug, once that learning has been consolidated. However, while the majority of these individuals that maintain a controlled drug intake exhibit a recovery toward a normal NMDA-dependent LTD, those that manifest signs of addiction according to the *Deroche-Gamonet model* do not exhibit this plasticity. This corroborates the notion that dependence and addiction are different stages of the addiction cycle, and that specific changes in synaptic plasticity selectively occur in addicts (Piazza & Deroche-Gamonet, 2013). Remarkably, this latter state is not associated to profound changes in eCB system functions within the PFC (Kasanetz et al., 2013). On the contrary, a form of eCB-dependent LTD at excitatory synapses of prelimbic PFC (Lafourcade et al., 2007) is abolished in both dependent and addicted subjects. However, since another form of LTD (i.e., mGluR2/3 dependent) is selectively lost at these synapses of addicted rats, one might speculate that the unbalanced eCB signal might prevent the transition to addiction by granting behavioral flexibility that would prevent the loss of control over drug seeking and taking (Kasanetz et al., 2013). Conversely, CB1—but CB2—receptor content was found to be reduced in the cerebral cortex of long-term cocaine addicts (Alvaro-Bartolome & Garcia-Sevilla, 2013). In particular, since they were reduced and increased in membranes and cytosol, respectively, a CB1 receptor downregulation with redistribution and/or internalization is the most likely scenario occurring in the whole cortex.

The eCB system also influences the transition from aberrant positive reinforcement to negative reinforcement, which is often associated with acute withdrawal and the emergence of compulsivity. In fact, during short-term phases (3 days) of opiate withdrawal, enhanced levels of 2-AG mediate a robust short-term form of synaptic plasticity at inhibitory synapses (i.e., depolarization-induced suppression of inhibition, DSI) impinging upon NAc MSNs (Wang et al., 2014). This effect is ascribed to an increased 2-AG synthesis (via diacylglycerol lipase-α) without changes in its key degrading enzyme (i.e., monoacylglycerol lipase). Notably, upon the same conditions, DSI in the DStr is unaffected. Altogether, these findings might explain the reduction of opiate withdrawal syndrome observed following CB1 receptor antagonism (Rubino, Massi, Vigano, Fuzio, & Parolaro, 2000; Wills et al., 2014), which could prove therapeutically useful. Hence, NAc has an important role in regulating circuits that drive both somatic and aversive responses to opiate withdrawal (Harris & Aston-Jones, 1994, #544). Nevertheless, brain regions other than the NAc and the mesocorticolimbic system, play crucial roles in initiating somatic signs of opiate withdrawal as well as in mediating drug-seeking behaviors (e.g., relapse), and the specific analysis of the influence of eCB system on relapse mechanisms has been already reviewed in-depth elsewhere (De Vries & Schoffelmeer, 2005, #545; Melis & Pistis, 2012, #333; Scavone, Sterling, & Van Bockstaele, 2013, #546; Serrano & Parsons, 2011, #331). Finally, yet importantly, the eCB system is also involved in the predisposition (either inborn or not) to drug addiction. However, the examination of such aspect is beyond the goal of the present section, and we refer the reader to other articles where this subject matter has been extensively reviewed (Oliere et al., 2014; Sagheddu & Melis, 2015; Serrano & Parsons, 2011).

6. CONCLUDING REMARKS

The past three decades have provided us with a wealth of evidence—both direct and indirect—for the role played by the eCB system in the regulation of those fundamental processes (e.g., reward signaling, motivation) that allow diverse living organisms, including human beings, to evolve. Hence, eCBs as modulators of intra- and inter- neuron functions have most likely contributed to, being impacted by, adaptations resulting from learning experiences caused by environmental stimuli, which might have engendered changes in neural, cognitive, and behavioral outcomes. In fact, this system appears to be in a unique position to contribute to the activity-dependent modulation of synaptic plasticity, learning mechanisms, and cognitive processing (Alger, 2009; Iremonger, Wamsteeker Cusulin, & Bains, 2013; Melis, Greco, et al., 2014). Learning processes improve adaptability, and participate in guiding evolution by incorporating the evaluation of *fitness* consequences of one's actions that ultimately drive motivation. This likely shapes behaviors that not only allow the acquisition of sensory rewards and avoidance of aversive states but also that enhance skills and competency that drive motivation uniquely in humans. DA signaling plays a pivotal role in both ends of motivation spectrum, a phenomenon that is phylogenetically preserved. The role of eCBs in shaping DA neuronal activity and synaptic transmission within the mesocorticolimbic pathway is partly elucidated, and it is under intensive investigation.

Aside from compelling evidence for striking spatial segregation as well as functional division of task(s) of eCBs on specific inputs and cellular compartments, the canonical interpretation of their mobilization upon request should be revised, at least with regard to those brain areas such as the VTA, the Str, and hippocampus, where an endogenous tone has been unveiled (DePoy et al., 2013; Edwards, Zhang, & Alger, 2008; Huang & Woolley, 2012; Lee et al., 2015; Melis, De Felice, Lecca, Fattore, & Pistis, 2013). In addition, different states of engagement of eCB signaling (i.e., prior, tonic, and persistent) might represent adaptive or maladaptive modifications secondary to CB1 receptor downregulation and homeostatic changes in plasticity mediated by eCBs as well as other transmitters/modulators (Melis, Greco, et al., 2014). Notably, these states involve distinct molecular components of eCB signaling and can either trigger metaplasticity or represent a means for metaplastic control of the eCB system. As such, these interconnected functional states can dynamically affect neuronal circuits and behavior, and can be modified by the balance between interoceptive stimuli and sensory information (Melis, Greco, et al., 2014; Wamsteeker Cusulin, Senst, Teskey, & Bains, 2014). Although the salience of sensory stimuli might differ with species, sex, age, and context, the eCB system allows a critical modulation of the whole "motivation" network activity through a selective and narrow tuning of synapses impinging upon different neuronal populations—like a switch being turned on and off. In fact, eCBs can guarantee the dynamic accuracy of input information and modify the sensitivity of the postsynaptic cell to subtle changes in activity patterns of its own inputs. This may ultimately affect synaptic rescaling and change circuit functions. As a result, any homeostatic change in eCB system (e.g., CB1 receptors, metabolic enzymes) possibly influences any subsequent plasticity induction.

How eCBs have contributed to such a fine modulation of neural plasticity, which is bidirectionally affected by experience, and how changes—either structural and/or functional—of neural networks can be interpreted in an interactive model where genes, brain and behavior characterize and embody the human becoming, though remarkable, is still unknown. The understanding of how this system can be modified by experience (e.g., early life events, stress, drugs) via mechanisms involving diverse signaling cascades, in both presynaptic and postsynaptic compartments, is, in fact, only in its infancy. In addition, whether or not these changes can be (epi)genetically transmitted, and/or can be modified upon exposure to social/environmental variables is substantial, since not only might they determine changes in function of CB1 receptor and/or metabolic enzymes but also result in abnormal eCB-mediated plasticity. Hence, whether this ancient neuromodulator has represented a means, among the others, by which natural selection has used information to adaptively regulate physiology and behavior is unknown.

To date, fundamental questions remain unanswered. For example, it is unclear whether 2-AG is solely responsible for tuning plasticity within the VTA. Are other eCBs and/or NAEs involved in setting the threshold for subsequent plasticity? Is there a segregation within VTA cell subpopulations and/or a regional specificity with regard to eCB-mediated plasticity phenomena? Does tonic eCB signaling only occur at inhibitory synapses, and if so, what role does sex/gender play? Which regulatory mechanisms can finely tune the extent of such a tonic signaling, and do these depend on synaptic states? Is there a bias for a specific eCB to be mobilized upon prior neuronal activity?

Further investigation assessing whether inter- and intra- individual differences of eCB system in distinct molecular components of the mesocorticolimbic circuit do exist is crucial. In particular, it is fundamental to unravel the input-specific complexity of eCB metabolic apparatus and of the wide repertoire of intra- and inter- cellular processes mediated by these lipids. This knowledge not only might help to open new possibilities to treat diverse neuropsychiatric disorders but also be alternatively used, in addition to standard pharmaceutical therapy, to tailor pharmacological treatments and reduce side effects.

ACKNOWLEDGMENTS

We wish to thank the members of the Pistis Lab for helpful comments and suggestions regarding this chapter, and Raffaella Tonini for the many discussions and inputs on this topic.

REFERENCES

Abel, E. L. (1971). Changes in anxiety feelings following marihuana smoking. The alternation in feelings of anxiety resulting from the smoking of marihuana (Cannabis sativa L.). *The British Journal of Addiction to Alcohol and Other Drugs*, *66*, 185–187.

Abel, E. L. (1981). Marihuana and sex: A critical survey. *Drug and Alcohol Dependence*, *8*, 1–22.

Alexander, S. A., Frohlich, K. L., & Fusco, C. (2012). Playing for health? Revisiting health promotion to examine the emerging public health position on children's play. *Health Promotion International*, *29*, 155–164.

Alger, B. E. (2009). Endocannabinoid signaling in neural plasticity. *Current Topics in Behavioral Neurosciences*, *1*, 141–172.

Alvaro-Bartolome, M., & Garcia-Sevilla, J. A. (2013). Dysregulation of cannabinoid CB1 receptor and associated signaling networks in brains of cocaine addicts and cocaine-treated rodents. *Neuroscience, 247*, 294–308.
Argilli, E., Sibley, D. R., Malenka, R. C., England, P. M., & Bonci, A. (2008). Mechanism and time course of cocaine-induced long-term potentiation in the ventral tegmental area. *The Journal of Neuroscience, 28*, 9092–9100.
Baik, J. H. (2013a). Dopamine signaling in reward-related behaviors. *Frontiers in Neural Circuits*, 7, 152.
Baik, J. H. (2013b). Dopamine signaling in food addiction: Role of dopamine D2 receptors. *BMB Reports, 46*, 519–526.
Barbano, M. F., Castane, A., Martin-Garcia, E., & Maldonado, R. (2009). Delta-9-tetrahydrocannabinol enhances food reinforcement in a mouse operant conflict test. *Psychopharmacology, 205*, 475–487.
Bartels, A., & Zeki, S. (2004). The neural correlates of maternal and romantic love. *NeuroImage, 21*, 1155–1166.
Befort, K. (2015). Interactions of the opioid and cannabinoid systems in reward: Insights from knockout studies. *Frontiers in Pharmacology, 6*, 6.
Belcher, A. M., Volkow, N. D., Moeller, F. G., & Ferre, S. (2014). Personality traits and vulnerability or resilience to substance use disorders. *Trends in Cognitive Sciences, 18*, 211–217.
Bellone, C., & Luscher, C. (2006). Cocaine triggered AMPA receptor redistribution is reversed in vivo by mGluR-dependent long-term depression. *Nature Neuroscience, 9*, 636–641.
Benedetti, F., Amanzio, M., Rosato, R., & Blanchard, C. (2011). Nonopioid placebo analgesia is mediated by CB1 cannabinoid receptors. *Nature Medicine, 17*, 1228–1230.
Berridge, K. C., & Aldridge, J. W. (2008). Decision utility, the brain, and pursuit of hedonic goals. *Social Cognition, 26*, 621–646.
Berridge, K. C., & Kringelbach, M. L. (2015). Pleasure systems in the brain. *Neuron, 86*, 646–664.
Berridge, K. C., Robinson, T. E., & Aldridge, J. W. (2009). Dissecting components of reward: 'Liking', 'wanting', and learning. *Current Opinion in Pharmacology, 9*, 65–73.
Bindra, D. (1968). Neuropsychological interpretation of the effects of drive and incentive-motivation on general activity and instrumental behavior. *Psychological Review, 75*, 1–22.
Bloomfield, M. A., Morgan, C. J., Egerton, A., Kapur, S., Curran, H. V., & Howes, O. D. (2014). Dopaminergic function in cannabis users and its relationship to cannabis-induced psychotic symptoms. *Biological Psychiatry, 75*, 470–478.
Bloomfield, M. A., Morgan, C. J., Kapur, S., Curran, H. V., & Howes, O. D. (2014). The link between dopamine function and apathy in cannabis users: An [18F]-DOPA PET imaging study. *Psychopharmacology, 231*, 2251–2259.
Bocklisch, C., Pascoli, V., Wong, J. C., House, D. R., Yvon, C., de Roo, M., et al. (2013). Cocaine disinhibits dopamine neurons by potentiation of GABA transmission in the ventral tegmental area. *Science, 341*, 1521–1525.
Bossong, M. G., van Berckel, B. N., Boellaard, R., Zuurman, L., Schuit, R. C., Windhorst, A. D., et al. (2009). Delta 9-tetrahydrocannabinol induces dopamine release in the human striatum. *Neuropsychopharmacology, 34*, 759–766.
Brischoux, F., Chakraborty, S., Brierley, D. I., & Ungless, M. A. (2009). Phasic excitation of dopamine neurons in ventral VTA by noxious stimuli. *Proceedings of the National Academy of Sciences of the United States of America, 106*, 4894–4899.
Bromberg-Martin, E. S., Matsumoto, M., & Hikosaka, O. (2010). Dopamine in motivational control: Rewarding, aversive, and alerting. *Neuron, 68*, 815–834.
Butelman, E. R., Yuferov, V., & Kreek, M. J. (2012). kappa-opioid receptor/dynorphin system: Genetic and pharmacotherapeutic implications for addiction. *Trends in Neurosciences, 35*, 587–596.

Cachope, R., & Cheer, J. F. (2014). Local control of striatal dopamine release. *Frontiers in Behavioral Neuroscience, 8*, 188.

Calcagnetti, D. J., & Schechter, M. D. (1992). Place conditioning reveals the rewarding aspect of social interaction in juvenile rats. *Physiology & Behavior, 51*, 667–672.

Castelli, M. P., Piras, A. P., Melis, T., Succu, S., Sanna, F., Melis, M. R., et al. (2007). Cannabinoid CB1 receptors in the paraventricular nucleus and central control of penile erection: Immunocytochemistry, autoradiography and behavioral studies. *Neuroscience, 147*, 197–206.

Chakrabarti, B., Kent, L., Suckling, J., Bullmore, E., & Baron-Cohen, S. (2006). Variations in the human cannabinoid receptor (CNR1) gene modulate striatal responses to happy faces. *The European Journal of Neuroscience, 23*, 1944–1948.

Champagne, F. A., Chretien, P., Stevenson, C. W., Zhang, T. Y., Gratton, A., & Meaney, M. J. (2004). Variations in nucleus accumbens dopamine associated with individual differences in maternal behavior in the rat. *The Journal of Neuroscience, 24*, 4113–4123.

Champagne, F. A., & Curley, J. P. (2005). How social experiences influence the brain. *Current Opinion in Neurobiology, 15*, 704–709.

Champagne, F. A., & Curley, J. P. (2009). Epigenetic mechanisms mediating the long-term effects of maternal care on development. *Neuroscience and Biobehavioral Reviews, 33*, 593–600.

Cheer, J. F., Wassum, K. M., Sombers, L. A., Heien, M. L., Ariansen, J. L., Aragona, B. J., et al. (2007). Phasic dopamine release evoked by abused substances requires cannabinoid receptor activation. *The Journal of Neuroscience, 27*, 791–795.

Chen, J. P., Paredes, W., Li, J., Smith, D., Lowinson, J., & Gardner, E. L. (1990). Delta 9-tetrahydrocannabinol produces naloxone-blockable enhancement of presynaptic basal dopamine efflux in nucleus accumbens of conscious, freely-moving rats as measured by intracerebral microdialysis. *Psychopharmacology, 102*, 156–162.

Chen, B. T., Yau, H. J., Hatch, C., Kusumoto-Yoshida, I., Cho, S. L., Hopf, F. W., et al. (2013). Rescuing cocaine-induced prefrontal cortex hypoactivity prevents compulsive cocaine seeking. *Nature, 496*, 359–362.

Chiang, K. P., Gerber, A. L., Sipe, J. C., & Cravatt, B. F. (2004). Reduced cellular expression and activity of the P129T mutant of human fatty acid amide hydrolase: Evidence for a link between defects in the endocannabinoid system and problem drug use. *Human Molecular Genetics, 13*, 2113–2119.

Chopra, G. S., & Jandu, B. S. (1976). Psychoclinical effects of long-term marijuana use in 275 Indian chronic users. A comparative assessment of effects in Indian and USA users. *Annals of the New York Academy of Sciences, 282*, 95–108.

Christensen, R., Kristensen, P. K., Bartels, E. M., Bliddal, H., & Astrup, A. (2007). Efficacy and safety of the weight-loss drug rimonabant: A meta-analysis of randomised trials. *Lancet, 370*, 1706–1713.

Cohen, J. Y., Haesler, S., Vong, L., Lowell, B. B., & Uchida, N. (2012). Neuron-type-specific signals for reward and punishment in the ventral tegmental area. *Nature, 482*, 85–88.

Cota, D., Marsicano, G., Tschop, M., Grubler, Y., Flachskamm, C., Schubert, M., et al. (2003). The endogenous cannabinoid system affects energy balance via central orexigenic drive and peripheral lipogenesis. *The Journal of Clinical Investigation, 112*, 423–431.

Cravatt, B. F., Giang, D. K., Mayfield, S. P., Boger, D. L., Lerner, R. A., & Gilula, N. B. (1996). Molecular characterization of an enzyme that degrades neuromodulatory fatty-acid amides. *Nature, 384*, 83–87.

Cristino, L., Busetto, G., Imperatore, R., Ferrandino, I., Palomba, L., Silvestri, C., et al. (2013). Obesity-driven synaptic remodeling affects endocannabinoid control of orexinergic neurons. *Proceedings of the National Academy of Sciences of the United States of America, 110*, E2229–E2238.

Danjo, T., Yoshimi, K., Funabiki, K., Yawata, S., & Nakanishi, S. (2014). Aversive behavior induced by optogenetic inactivation of ventral tegmental area dopamine neurons is mediated by dopamine D2 receptors in the nucleus accumbens. *Proceedings of the National Academy of Sciences of the United States of America, 111*, 6455–6460.

Dazzi, L., Talani, G., Biggio, F., Utzeri, C., Lallai, V., Licheri, V., et al. (2014). Involvement of the cannabinoid CB1 receptor in modulation of dopamine output in the prefrontal cortex associated with food restriction in rats. *PLoS One, 9*, e92224.

de la Fuente-Fernandez, R., Ruth, T. J., Sossi, V., Schulzer, M., Calne, D. B., & Stoessl, A. J. (2001). Expectation and dopamine release: Mechanism of the placebo effect in Parkinson's disease. *Science, 293*, 1164–1166.

DePoy, L., Daut, R., Brigman, J. L., MacPherson, K., Crowley, N., Gunduz-Cinar, O., et al. (2013). Chronic alcohol produces neuroadaptations to prime dorsal striatal learning. *Proceedings of the National Academy of Sciences of the United States of America, 110*, 14783–14788.

Deroche-Gamonet, V., Belin, D., & Piazza, P. V. (2004). Evidence for addiction-like behavior in the rat. *Science, 305*, 1014–1017.

De Vries, T. J., & Schoffelmeer, A. N. (2005). Cannabinoid CB1 receptors control conditioned drug seeking. *Trends in Pharmacological Sciences, 26*(8), 420–426.

Di Chiara, G., & Imperato, A. (1988). Drugs abused by humans preferentially increase synaptic dopamine concentrations in the mesolimbic system of freely moving rats. *Proceedings of the National Academy of Sciences of the United States of America, 85*, 5274–5278.

Di Marzo, V., Berrendero, F., Bisogno, T., González, S., Cavaliere, P., Romero, J., et al. (2000). Enhancement of anandamide formation in the limbic forebrain and reduction of endocannabinoid contents in the striatum of Δ^9-tetrahydrocannabinol-tolerant rats. *Journal of Neurochemistry, 74*, 1627–1635.

Di Marzo, V., & Despres, J. P. (2009). CB1 antagonists for obesity—What lessons have we learned from rimonabant? *Nature Reviews Endocrinology, 5*, 633–638.

Di Marzo, V., Goparaju, S. K., Wang, L., Liu, J., Batkai, S., Jarai, Z., et al. (2001). Leptin-regulated endocannabinoids are involved in maintaining food intake. *Nature, 410*, 822–825.

Di Marzo, V., Ligresti, A., & Cristino, L. (2009). The endocannabinoid system as a link between homoeostatic and hedonic pathways involved in energy balance regulation. *International Journal of Obesity, 33*(Suppl. 2), S18–S24.

Diana, M., Melis, M., Muntoni, A. L., & Gessa, G. L. (1998). Mesolimbic dopaminergic decline after cannabinoid withdrawal. *Proceedings of the National Academy of Sciences of the United States of America, 95*, 10269–10273.

Dodd, G. T., Stark, J. A., McKie, S., Williams, S. R., & Luckman, S. M. (2009). Central cannabinoid signaling mediating food intake: A pharmacological-challenge magnetic resonance imaging and functional histology study in rat. *Neuroscience, 163*, 1192–1200.

Douglas, L. A., Varlinskaya, E. I., & Spear, L. P. (2004). Rewarding properties of social interactions in adolescent and adult male and female rats: Impact of social versus isolate housing of subjects and partners. *Developmental Psychobiology, 45*, 153–162.

Dudok, B., et al. (2015). Cell-specific STORM super-resolution imaging reveals nanoscale organization of cannabinoid signaling. *Nature Neuroscience, 18*, 75–86.

Edwards, S., & Koob, G. F. (2010). Neurobiology of dysregulated motivational systems in drug addiction. *Future Neurology, 5*, 393–401.

Edwards, D. A., Zhang, L., & Alger, B. E. (2008). Metaplastic control of the endocannabinoid system at inhibitory synapses in hippocampus. *Proceedings of the National Academy of Sciences of the United States of America, 105*, 8142–8147.

Egerton, A., Allison, C., Brett, R. R., & Pratt, J. A. (2006). Cannabinoids and prefrontal cortical function: Insights from preclinical studies. *Neuroscience and Biobehavioral Reviews, 30*, 680–695.

Elphick, M. R. (2012). The evolution and comparative neurobiology of endocannabinoid signalling. *Philosophical Transactions of the Royal Society of London, Series B: Biological Sciences, 367*, 3201–3215.

Erhardt, S., Mathe, J. M., Chergui, K., Engberg, G., & Svensson, T. H. (2002). GABA (B) receptor-mediated modulation of the firing pattern of ventral tegmental area dopamine neurons in vivo. *Naunyn-Schmiedeberg's Archives of Pharmacology, 365*, 173–180.

Everitt, B. J., & Robbins, T. W. (2005). Neural systems of reinforcement for drug addiction: From actions to habits to compulsion. *Nature Neuroscience, 8*, 1481–1489.

Fadda, P., Scherma, M., Spano, M. S., Salis, P., Melis, V., Fattore, L., et al. (2006). Cannabinoid self-administration increases dopamine release in the nucleus accumbens. *Neuroreport, 17*, 1629–1632.

Faden, V. B., & Graubard, B. I. (2000). Maternal substance use during pregnancy and developmental outcome at age three. *Journal of Substance Abuse, 12*, 329–340.

Farrimond, J. A., Mercier, M. S., Whalley, B. J., & Williams, C. M. (2011). Cannabis sativa and the endogenous cannabinoid system: Therapeutic potential for appetite regulation. *Phytotherapy Research, 25*, 170–188.

Fattore, L., Deiana, S., Spano, S. M., Cossu, G., Fadda, P., Scherma, M., et al. (2005). Endocannabinoid system and opioid addiction: Behavioural aspects. *Pharmacology, Biochemistry and Behavior, 81*, 343–359.

Fattore, L., Melis, M., Fadda, P., Pistis, M., & Fratta, W. (2010). The endocannabinoid system and nondrug rewarding behaviours. *Experimental Neurology, 224*, 23–36.

Febo, M., Numan, M., & Ferris, C. F. (2005). Functional magnetic resonance imaging shows oxytocin activates brain regions associated with mother-pup bonding during suckling. *The Journal of Neuroscience, 25*, 11637–11644.

Feng, J., & Nestler, E. J. (2013). Epigenetic mechanisms of drug addiction. *Current Opinion in Neurobiology, 23*, 521–528.

Fernandez-Ruiz, J., Hernandez, M., & Ramos, J. A. (2010). Cannabinoid-dopamine interaction in the pathophysiology and treatment of CNS disorders. *CNS Neuroscience & Therapeutics, 16*, e72–e91.

Ferrari, F., Ottani, A., & Giuliani, D. (2000). Inhibitory effects of the cannabinoid agonist HU 210 on rat sexual behaviour. *Physiology & Behavior, 69*, 547–554.

Ferris, C. F., Kulkarni, P., Sullivan, J. M., Jr., Harder, J. A., Messenger, T. L., & Febo, M. (2005). Pup suckling is more rewarding than cocaine: Evidence from functional magnetic resonance imaging and three-dimensional computational analysis. *The Journal of Neuroscience, 25*, 149–156.

Fiorillo, C. D., & Williams, J. T. (2000). Cholinergic inhibition of ventral midbrain dopamine neurons. *The Journal of Neuroscience: The Official Journal of the Society for Neuroscience, 20*, 7855–7860.

Floresco, S. B., West, A. R., Ash, B., Moore, H., & Grace, A. A. (2003). Afferent modulation of dopamine neuron firing differentially regulates tonic and phasic dopamine transmission. *Nature Neuroscience, 6*, 968–973.

Foltin, R. W., Brady, J. V., & Fischman, M. W. (1986). Behavioral analysis of marijuana effects on food intake in humans. *Pharmacology, Biochemistry and Behavior, 25*, 577–582.

Foltin, R. W., Fischman, M. W., & Byrne, M. F. (1988). Effects of smoked marijuana on food intake and body weight of humans living in a residential laboratory. *Appetite, 11*, 1–14.

Fourgeaud, L., Mato, S., Bouchet, D., Hemar, A., Worley, P. F., & Manzoni, O. J. (2004). A single in vivo exposure to cocaine abolishes endocannabinoid-mediated long-term depression in the nucleus accumbens. *The Journal of Neuroscience, 24*, 6939–6945.

French, E. D., Dillon, K., & Wu, X. (1997). Cannabinoids excite dopamine neurons in the ventral tegmentum and substantia nigra. *Neuroreport, 8*, 649–652.

Freund, T. F., Katona, I., & Piomelli, D. (2003). Role of endogenous cannabinoids in synaptic signaling. *Physiological Reviews, 83*, 1017–1066.

Fride, E., Gobshtis, N., Dahan, H., Weller, A., Giuffrida, A., & Ben-Shabat, S. (2009). The endocannabinoid system during development: Emphasis on perinatal events and delayed effects. *Vitamins and Hormones*, *81*, 139–158.

Friemel, C. M., Zimmer, A., & Schneider, M. (2014). The CB1 receptor as an important mediator of hedonic reward processing. *Neuropsychopharmacology*, *39*, 2387–2396.

Frischknecht, H. R., Sieber, B., & Waser, P. G. (1982). Effects of multiple, chronic and early hashish exposure on mating behavior, nest-building and gestation in mice. *Comparative Biochemistry and Physiology Part C: Comparative Pharmacology*, *72*, 363–368.

Gallate, J. E., Saharov, T., Mallet, P. E., & McGregor, I. S. (1999). Increased motivation for beer in rats following administration of a cannabinoid CB1 receptor agonist. *European Journal of Pharmacology*, *370*, 233–240.

Gardner, E. L. (2005). Endocannabinoid signaling system and brain reward: Emphasis on dopamine. *Pharmacology, Biochemistry and Behavior*, *81*, 263–284.

Gardner, E. L. (2011). Addiction and brain reward and antireward pathways. *Advances in Psychosomatic Medicine*, *30*, 22–60.

Garris, P. A., Ciolkowski, E. L., Pastore, P., & Wightman, R. M. (1994). Efflux of dopamine from the synaptic cleft in the nucleus accumbens of the rat brain. *The Journal of Neuroscience*, *14*, 6084–6093.

Genn, R. F., Tucci, S., Marco, E. M., Viveros, M. P., & File, S. E. (2004). Unconditioned and conditioned anxiogenic effects of the cannabinoid receptor agonist CP 55,940 in the social interaction test. *Pharmacology, Biochemistry and Behavior*, *77*, 567–573.

Georges, F., & Aston-Jones, G. (2001). Potent regulation of midbrain dopamine neurons by the bed nucleus of the stria terminalis. *The Journal of Neuroscience*, *21*, RC160.

Georges, F., & Aston-Jones, G. (2002). Activation of ventral tegmental area cells by the bed nucleus of the stria terminalis: A novel excitatory amino acid input to midbrain dopamine neurons. *The Journal of Neuroscience*, *22*, 5173–5187.

Gessa, G. L., Melis, M., Muntoni, A. L., & Diana, M. (1998). Cannabinoids activate mesolimbic dopamine neurons by an action on cannabinoid CB1 receptors. *European Journal of Pharmacology*, *341*, 39–44.

Gonon, F. G. (1988). Nonlinear relationship between impulse flow and dopamine released by rat midbrain dopaminergic neurons as studied by in vivo electrochemistry. *Neuroscience*, *24*, 19–28.

Gonzalez, S., Fernandez-Ruiz, J., Di Marzo, V., Hernandez, M., Arevalo, C., Nicanor, C., et al. (2004). Behavioral and molecular changes elicited by acute administration of SR141716 to delta9-tetrahydrocannabinol-tolerant rats: An experimental model of cannabinoid abstinence. *Drug and Alcohol Dependence*, *74*, 159–170.

Good, C. H., & Lupica, C. R. (2010). Afferent-specific AMPA receptor subunit composition and regulation of synaptic plasticity in midbrain dopamine neurons by abused drugs. *The Journal of Neuroscience: The Official Journal of the Society for Neuroscience*, *30*, 7900–7909.

Gordon, J. H., Bromley, B. L., Gorski, R. A., & Zimmermann, E. (1978). Delta9-tetrahydrocannabinol enhancement of lordosis behavior in estrogen treated female rats. *Pharmacology, Biochemistry and Behavior*, *8*, 603–608.

Gorzalka, B. B., Hill, M. N., & Chang, S. C. (2010). Male-female differences in the effects of cannabinoids on sexual behavior and gonadal hormone function. *Hormones and Behavior*, *58*, 91–99.

Gorzalka, B. B., Morrish, A. C., & Hill, M. N. (2008). Endocannabinoid modulation of male rat sexual behavior. *Psychopharmacology*, *198*, 479–486.

Grace, A. A. (1991). Phasic versus tonic dopamine release and the modulation of dopamine system responsivity: A hypothesis for the etiology of schizophrenia. *Neuroscience*, *41*, 1–24.

Grace, A. A., & Bunney, B. S. (1984a). The control of firing pattern in nigral dopamine neurons: Single spike firing. *The Journal of Neuroscience*, *4*, 2866–2876.

Grace, A. A., & Bunney, B. S. (1984b). The control of firing pattern in nigral dopamine neurons: Burst firing. *The Journal of Neuroscience*, *4*, 2877–2890.
Grace, A. A., Floresco, S. B., Goto, Y., & Lodge, D. J. (2007). Regulation of firing of dopaminergic neurons and control of goal-directed behaviors. *Trends in Neurosciences*, *30*, 220–227.
Gronier, B., & Rasmussen, K. (1998). Activation of midbrain presumed dopaminergic neurones by muscarinic cholinergic receptors: An in vivo electrophysiological study in the rat. *British Journal of Pharmacology*, *124*, 455–464.
Guegan, T., Cutando, L., Ayuso, E., Santini, E., Fisone, G., Bosch, F., et al. (2013). Operant behavior to obtain palatable food modifies neuronal plasticity in the brain reward circuit. *European Neuropsychopharmacology*, *23*, 146–159.
Haber, S. N., & Knutson, B. (2010). The reward circuit: Linking primate anatomy and human imaging. *Neuropsychopharmacology*, *35*, 4–26.
Halikas, J., Weller, R., & Morse, C. (1982). Effects of regular marijuana use on sexual performance. *Journal of Psychoactive Drugs*, *14*, 59–70.
Hall, W., Solowij, N., & Lemon, J. (1994). *The health and social consequences of cannabis use. Monograph series*. Sidney: National Drug and Alcohol research Centre.
Hansen, S., Bergvall, A. H., & Nyiredi, S. (1993). Interaction with pups enhances dopamine release in the ventral striatum of maternal rats: A microdialysis study. *Pharmacology, Biochemistry and Behavior*, *45*, 673–676.
Hansen, S., Harthon, C., Wallin, E., Lofberg, L., & Svensson, K. (1991). Mesotelencephalic dopamine system and reproductive behavior in the female rat: Effects of ventral tegmental 6-hydroxydopamine lesions on maternal and sexual responsiveness. *Behavioral Neuroscience*, *105*, 588–598.
Harris, G. C., & Aston-Jones, G. (1994). Involvement of D2 dopamine receptors in the nucleus accumbens in the opiate withdrawal syndrome. *Nature*, *371*(6493), 155–157.
Hernandez, G., Oleson, E. B., Gentry, R. N., Abbas, Z., Bernstein, D. L., Arvanitogiannis, A., et al. (2014). Endocannabinoids promote cocaine-induced impulsivity and its rapid dopaminergic correlates. *Biological Psychiatry*, *75*, 487–498.
Higgs, S., Williams, C. M., & Kirkham, T. C. (2003). Cannabinoid influences on palatability: Microstructural analysis of sucrose drinking after delta(9)-tetrahydrocannabinol, anandamide, 2-arachidonoyl glycerol and SR141716. *Psychopharmacology*, *165*, 370–377.
Hikosaka, K., & Watanabe, M. (2000). Delay activity of orbital and lateral prefrontal neurons of the monkey varying with different rewards. *Cerebral Cortex*, *10*, 263–271.
Hikosaka, K., & Watanabe, M. (2004). Long- and short-range reward expectancy in the primate orbitofrontal cortex. *The European Journal of Neuroscience*, *19*, 1046–1054.
Hirvonen, J., Zanotti-Fregonara, P., Umhau, J. C., George, D. T., Rallis-Frutos, D., Lyoo, C. H., et al. (2013). Reduced cannabinoid CB1 receptor binding in alcohol dependence measured with positron emission tomography. *Molecular Psychiatry*, *18*, 916–921.
Hoffman, A. F., Oz, M., Caulder, T., & Lupica, C. R. (2003). Functional tolerance and blockade of long-term depression at synapses in the nucleus accumbens after chronic cannabinoid exposure. *The Journal of Neuroscience*, *23*, 4815–4820.
Hol, T., Van den Berg, C. L., Van Ree, J. M., & Spruijt, B. M. (1999). Isolation during the play period in infancy decreases adult social interactions in rats. *Behavioural Brain Research*, *100*, 91–97.
Hollerman, J. R., Tremblay, L., & Schultz, W. (2000). Involvement of basal ganglia and orbitofrontal cortex in goal-directed behavior. *Progress in Brain Research*, *126*, 193–215.
Hollis, D. M., Coddington, E. J., & Moore, F. L. (2006). Neuroanatomical distribution of cannabinoid receptor gene expression in the brain of the rough-skinned newt, Taricha granulosa. *Brain, Behavior and Evolution*, *67*, 135–149.

Hong, S., & Hikosaka, O. (2011). Dopamine-mediated learning and switching in cortico-striatal circuit explain behavioral changes in reinforcement learning. *Frontiers in Behavioral Neuroscience*, *5*, 15.

Hong, S., Jhou, T. C., Smith, M., Saleem, K. S., & Hikosaka, O. (2011). Negative reward signals from the lateral habenula to dopamine neurons are mediated by rostromedial tegmental nucleus in primates. *The Journal of Neuroscience*, *31*, 11457–11471.

Horder, J., Harmer, C. J., Cowen, P. J., & McCabe, C. (2010). Reduced neural response to reward following 7 days treatment with the cannabinoid CB1 antagonist rimonabant in healthy volunteers. *International Journal of Neuropsychopharmacology*, *13*, 1103–1113.

Huang, G. Z., & Woolley, C. S. (2012). Estradiol acutely suppresses inhibition in the hippocampus through a sex-specific endocannabinoid and mGluR-dependent mechanism. *Neuron*, *74*, 801–808.

Hyland, B. I., Reynolds, J. N., Hay, J., Perk, C. G., & Miller, R. (2002). Firing modes of midbrain dopamine cells in the freely moving rat. *Neuroscience*, *114*, 475–492.

Ikemoto, S. (2010). Brain reward circuitry beyond the mesolimbic dopamine system: A neurobiological theory. *Neuroscience and Biobehavioral Reviews*, *35*, 129–150.

Ikemoto, S., & Bonci, A. (2014). Neurocircuitry of drug reward. *Neuropharmacology*, *76*(Pt. B), 329–341.

Ikemoto, S., & Panksepp, J. (1996). Dissociations between appetitive and consummatory responses by pharmacological manipulations of reward-relevant brain regions. *Behavioral Neuroscience*, *110*, 331–345.

Iremonger, K. J., Wamsteeker Cusulin, J. I., & Bains, J. S. (2013). Changing the tune: Plasticity and adaptation of retrograde signals. *Trends in Neurosciences*, *36*, 471–479.

Jager, G., Van Hell, H. H., De Win, M. M., Kahn, R. S., Van Den Brink, W., Van Ree, J. M., et al. (2007). Effects of frequent cannabis use on hippocampal activity during an associative memory task. *European Neuropsychopharmacology*, *17*, 289–297.

Jager, G., & Witkamp, R. F. (2014). The endocannabinoid system and appetite: Relevance for food reward. *Nutrition Research Reviews*, *27*, 172–185.

Jalabert, M., Aston-Jones, G., Herzog, E., Manzoni, O., & Georges, F. (2009). Role of the bed nucleus of the stria terminalis in the control of ventral tegmental area dopamine neurons. *Progress in Neuropsychopharmacology Biological Psychiatry*, *33*, 1336–1346.

Jamshidi, N., & Taylor, D. A. (2001). Anandamide administration into the ventromedial hypothalamus stimulates appetite in rats. *British Journal of Pharmacology*, *134*, 1151–1154.

Jhou, T. C., Fields, H. L., Baxter, M. G., Saper, C. B., & Holland, P. C. (2009). The rostromedial tegmental nucleus (RMTg), a GABAergic afferent to midbrain dopamine neurons, encodes aversive stimuli and inhibits motor responses. *Neuron*, *61*, 786–800.

Joffe, M. E., Grueter, C. A., & Grueter, B. A. (2014). Biological substrates of addiction. *Wiley Interdisciplinary Reviews: Cognitive Science*, *5*, 151–171.

Johnson, S. D., Phelps, D. L., & Cottler, L. B. (2004). The association of sexual dysfunction and substance use among a community epidemiological sample. *Archives of Sexual Behavior*, *33*(1), 55–63.

Kalivas, P. W., Volkow, N., & Seamans, J. (2005). Unmanageable motivation in addiction: a pathology in prefrontal-accumbens glutamate transmission. *Neuron*, *45*(5), 647–650.

Kasanetz, F., Deroche-Gamonet, V., Berson, N., Balado, E., Lafourcade, M., Manzoni, O., et al. (2010). Transition to addiction is associated with a persistent impairment in synaptic plasticity. *Science*, *328*, 1709–1712.

Kasanetz, F., Lafourcade, M., Deroche-Gamonet, V., Revest, J. M., Berson, N., Balado, E., et al. (2013). Prefrontal synaptic markers of cocaine addiction-like behavior in rats. *Molecular Psychiatry*, *18*, 729–737.

Kavanaugh, K., Meier, P., Zimmermann, B., & Mead, L. (1997). The rewards outweigh the efforts: Breastfeeding outcomes for mothers of preterm infants. *Journal of Human Lactation*, *13*, 15–21.

Kirkham, T. C. (2009a). Cannabinoids and appetite: Food craving and food pleasure. *International Review of Psychiatry (Abingdon, England)*, *21*, 163–171.
Kirkham, T. C. (2009b). Endocannabinoids and the non-homeostatic control of appetite. *Current Topics in Behavioral Neurosciences*, *1*, 231–253.
Kirkham, T. C., & Williams, C. M. (2001). Endogenous cannabinoids and appetite. *Nutrition Research Reviews*, *14*, 65–86.
Kirkham, T. C., Williams, C. M., Fezza, F., & Di Marzo, V. (2002). Endocannabinoid levels in rat limbic forebrain and hypothalamus in relation to fasting, feeding and satiation: Stimulation of eating by 2-arachidonoyl glycerol. *British Journal of Pharmacology*, *136*, 550–557.
Kitai, S. T., Shepard, P. D., Callaway, J. C., & Scroggs, R. (1999). Afferent modulation of dopamine neuron firing patterns. *Current Opinion in Neurobiology*, *9*, 690–697.
Kleijn, J., Wiskerke, J., Cremers, T. I., Schoffelmeer, A. N., Westerink, B. H., & Pattij, T. (2012). Effects of amphetamine on dopamine release in the rat nucleus accumbens shell region depend on cannabinoid CB1 receptor activation. *Neurochemistry International*, *60*, 791–798.
Klein, C., Hill, M. N., Chang, S. C., Hillard, C. J., & Gorzalka, B. B. (2012). Circulating endocannabinoid concentrations and sexual arousal in women. *The Journal of Sexual Medicine*, *9*, 1588–1601.
Klugmann, M., Goepfrich, A., Friemel, C. M., & Schneider, M. (2011). AAV-mediated overexpression of the CB1 receptor in the mPFC of adult rats alters cognitive flexibility, social behavior, and emotional reactivity. *Frontiers in Behavioral Neuroscience*, *5*, 37.
Koff, W. C. (1974). Marijuana and sexual activity. *Journal of Sex Research*, *10*, 194–204.
Kolansky, H., & Moore, W. T. (1971). Effects of marihuana on adolescents and young adults. *JAMA*, *216*, 486–492.
Koob, G. F., & Le Moal, M. (1997). Drug abuse: Hedonic homeostatic dysregulation. *Science*, *278*, 52–58.
Koob, G. F., & Le Moal, M. (2001). Drug addiction, dysregulation of reward, and allostasis. *Neuropsychopharmacology*, *24*, 97–129.
Lafourcade, M., Elezgarai, I., Mato, S., Bakiri, Y., Grandes, P., & Manzoni, O. J. (2007). Molecular components and functions of the endocannabinoid system in mouse prefrontal cortex. *PLoS One*, *2*, e709.
Lammel, S., Ion, D. I., Roeper, J., & Malenka, R. C. (2011). Projection-specific modulation of dopamine neuron synapses by aversive and rewarding stimuli. *Neuron*, *70*, 855–862.
Lammel, S., Lim, B. K., & Malenka, R. C. (2014). Reward and aversion in a heterogeneous midbrain dopamine system. *Neuropharmacology*, *76*(Pt. B), 351–359.
Lammel, S., Lim, B. K., Ran, C., Huang, K. W., Betley, M. J., Tye, K. M., et al. (2012). Input-specific control of reward and aversion in the ventral tegmental area. *Nature*, *491*, 212–217.
Lecca, S., Melis, M., Luchicchi, A., Muntoni, A. L., & Pistis, M. (2012). Inhibitory inputs from rostromedial tegmental neurons regulate spontaneous activity of midbrain dopamine cells and their responses to drugs of abuse. *Neuropsychopharmacology*, *37*, 1164–1176.
Lee, S. H., Ledri, M., Toth, B., Marchionni, I., Henstridge, C. M., Dudok, B., et al. (2015). Multiple forms of endocannabinoid and endovanilloid signaling regulate the tonic control of GABA release. *The Journal of Neuroscience: The Official Journal of the Society for Neuroscience*, *35*, 10039–10057.
Leibenluft, E., Gobbini, M. I., Harrison, T., & Haxby, J. V. (2004). Mothers' neural activation in response to pictures of their children and other children. *Biological Psychiatry*, *56*, 225–232.
Levine, J. D., Gordon, N. C., & Fields, H. L. (1978). The mechanism of placebo analgesia. *Lancet*, *2*, 654–657.
Leweke, F. M., & Schneider, M. (2011). Chronic pubertal cannabinoid treatment as a behavioural model for aspects of schizophrenia: Effects of the atypical antipsychotic quetiapine. *International Journal of Neuropsychopharmacology*, *14*, 43–51.

Lobb, C. J., Wilson, C. J., & Paladini, C. A. (2010). A dynamic role for GABA receptors on the firing pattern of midbrain dopaminergic neurons. *Journal of Neurophysiology*, *104*, 403–413.

Lodge, D. J., & Grace, A. A. (2006). The laterodorsal tegmentum is essential for burst firing of ventral tegmental area dopamine neurons. *Proceedings of the National Academy of Sciences of the United States of America*, *103*, 5167–5172.

Lokwan, S. J., Overton, P. G., Berry, M. S., & Clark, D. (1999). Stimulation of the pedunculopontine tegmental nucleus in the rat produces burst firing in A9 dopaminergic neurons. *Neuroscience*, *92*, 245–254.

Lokwan, S. J., Overton, P. G., Berry, M. S., & Clark, D. (2000). The medial prefrontal cortex plays an important role in the excitation of A10 dopaminergic neurons following intravenous muscimol administration. *Neuroscience*, *95*, 647–656.

Love, T. M. (2014). Oxytocin, motivation and the role of dopamine. *Pharmacology, Biochemistry and Behavior*, *119*, 49–60.

Maccioni, P., Pes, D., Carai, M. A., Gessa, G. L., & Colombo, G. (2008). Suppression by the cannabinoid CB1 receptor antagonist, rimonabant, of the reinforcing and motivational properties of a chocolate-flavoured beverage in rats. *Behavioural Pharmacology*, *19*, 197–209.

Mahler, S. V., Smith, K. S., & Berridge, K. C. (2007). Endocannabinoid hedonic hotspot for sensory pleasure: Anandamide in nucleus accumbens shell enhances 'liking' of a sweet reward. *Neuropsychopharmacology*, *32*, 2267–2278.

Maisonneuve, I. M., Keller, R. W., & Glick, S. D. (1990). Similar effects of D-amphetamine and cocaine on extracellular dopamine levels in medial prefrontal cortex of rats. *Brain Research*, *535*, 221–226.

Maldonado, R., Valverde, O., & Berrendero, F. (2006). Involvement of the endocannabinoid system in drug addiction. *Trends in Neurosciences*, *29*, 225–232.

Malone, D. T., & Taylor, D. A. (1999). Modulation by fluoxetine of striatal dopamine release following delta9-tetrahydrocannabinol: A microdialysis study in conscious rats. *British Journal of Pharmacology*, *128*, 21–26.

Mameli, M., Balland, B., Lujan, R., & Luscher, C. (2007). Rapid synthesis and synaptic insertion of GluR2 for mGluR-LTD in the ventral tegmental area. *Science*, *317*, 530–533.

Mameli-Engvall, M., Evrard, A., Pons, S., Maskos, U., Svensson, T. H., Changeux, J. P., et al. (2006). Hierarchical control of dopamine neuron-firing patterns by nicotinic receptors. *Neuron*, *50*, 911–921.

Manduca, A., Campolongo, P., & Trezza, V. (2012). Cannabinoid modulation of mother-infant interaction: Is it just about milk? *Reviews in the Neurosciences*, *23*, 707–722.

Manzanares, J., Corchero, J., Romero, J., Fernandez-Ruiz, J. J., Ramos, J. A., & Fuentes, J. A. (1999). Pharmacological and biochemical interactions between opioids and cannabinoids. *Trends in Pharmacological Sciences*, *20*, 287–294.

Marco, E. M., Rapino, C., Caprioli, A., Borsini, F., Maccarrone, M., & Laviola, G. (2011). Social encounter with a novel partner in adolescent rats: Activation of the central endocannabinoid system. *Behavioural Brain Research*, *220*, 140–145.

Marinelli, S., Di Marzo, V., Florenzano, F., Fezza, F., Viscomi, M. T., van der Stelt, M., et al. (2007). N-arachidonoyl-dopamine tunes synaptic transmission onto dopaminergic neurons by activating both cannabinoid and vanilloid receptors. *Neuropsychopharmacology*, *32*, 298–308.

Mark, G. P., Shabani, S., Dobbs, L. K., & Hansen, S. T. (2011). Cholinergic modulation of mesolimbic dopamine function and reward. *Physiology & Behavior*, *104*, 76–81.

Martinez-Gonzalez, D., Bonilla-Jaime, H., Morales-Otal, A., Henriksen, S. J., Velazquez-Moctezuma, J., & Prospero-Garcia, O. (2004). Oleamide and anandamide effects on food intake and sexual behavior of rats. *Neuroscience Letters*, *364*, 1–6.

Maslow, A. H. (1943). A theory of human motivation. *Psychological Review*, *50*, 370–396.
Mato, S., Chevaleyre, V., Robbe, D., Pazos, A., Castillo, P. E., & Manzoni, O. J. (2004). A single in-vivo exposure to delta 9THC blocks endocannabinoid-mediated synaptic plasticity. *Nature Neuroscience*, *7*, 585–586.
Mato, S., Robbe, D., Puente, N., Grandes, P., & Manzoni, O. J. (2005). Presynaptic homeostatic plasticity rescues long-term depression after chronic delta 9-tetrahydrocannabinol exposure. *The Journal of Neuroscience*, *25*, 11619–11627.
Matsui, A., Jarvie, B. C., Robinson, B. G., Hentges, S. T., & Williams, J. T. (2014). Separate GABA afferents to dopamine neurons mediate acute action of opioids, development of tolerance, and expression of withdrawal. *Neuron*, *82*, 1346–1356.
Matsumoto, M., & Hikosaka, O. (2007). Lateral habenula as a source of negative reward signals in dopamine neurons. *Nature*, *447*, 1111–1115.
Matsumoto, M., & Hikosaka, O. (2009). Two types of dopamine neuron distinctly convey positive and negative motivational signals. *Nature*, *459*, 837–841.
Mattes, R. D., Engelman, K., Shaw, L. M., & Elsohly, M. A. (1994). Cannabinoids and appetite stimulation. *Pharmacology, Biochemistry and Behavior*, *49*, 187–195.
Matyas, F., Urban, G. M., Watanabe, M., Mackie, K., Zimmer, A., Freund, T. F., et al. (2008). Identification of the sites of 2-arachidonoylglycerol synthesis and action imply retrograde endocannabinoid signaling at both GABAergic and glutamatergic synapses in the ventral tegmental area. *Neuropharmacology*, *54*, 95–107.
Maze, I., & Nestler, E. J. (2011). The epigenetic landscape of addiction. *Annals of the New York Academy of Sciences*, *1216*, 99–113.
Mazei-Robison, M. S., Appasani, R., Edwards, S., Wee, S., Taylor, S. R., Picciotto, M. R., et al. (2014). Self-administration of ethanol, cocaine, or nicotine does not decrease the soma size of ventral tegmental area dopamine neurons. *PLoS One*, *9*, e95962.
Mazei-Robison, M. S., et al. (2011). Role for mTOR signaling and neuronal activity in morphine-induced adaptations in ventral tegmental area dopamine neurons. *Neuron*, *72*, 977–990.
McGlothlin, W. H., & West, L. J. (1968). The marihuana problem: An overview. *The American Journal of Psychiatry*, *125*, 126–134.
Melis, M., De Felice, M., Lecca, S., Fattore, L., & Pistis, M. (2013). Sex-specific tonic 2-arachidonoylglycerol signaling at inhibitory inputs onto dopamine neurons of Lister Hooded rats. *Frontiers in Integrative Neuroscience*, *7*, 93.
Melis, M., Greco, B., & Tonini, R. (2014). Interplay between synaptic endocannabinoid signaling and metaplasticity in neuronal circuit function and dysfunction. *The European Journal of Neuroscience*, *39*, 1189–1201.
Melis, M., Muntoni, A. L., & Pistis, M. (2012). Endocannabinoids and the processing of value-related signals. *Frontiers in Pharmacology*, *3*, 7.
Melis, M., Perra, S., Muntoni, A. L., Pillolla, G., Lutz, B., Marsicano, G., et al. (2004). Prefrontal cortex stimulation induces 2-arachidonoyl-glycerol-mediated suppression of excitation in dopamine neurons. *The Journal of Neuroscience*, *24*, 10707–10715.
Melis, M., & Pistis, M. (2012). Hub and switches: Endocannabinoid signalling in midbrain dopamine neurons. *Philosophical Transactions of the Royal Society of London, Series B: Biological Sciences*, *367*, 3276–3285.
Melis, M., Pistis, M., Perra, S., Muntoni, A. L., Pillolla, G., & Gessa, G. L. (2004). Endocannabinoids mediate presynaptic inhibition of glutamatergic transmission in rat ventral tegmental area dopamine neurons through activation of CB1 receptors. *The Journal of Neuroscience*, *24*, 53–62.
Melis, M., Sagheddu, C., De Felice, M., Casti, A., Madeddu, C., Spiga, S., et al. (2014). Enhanced endocannabinoid-mediated modulation of rostromedial tegmental nucleus drive onto dopamine neurons in Sardinian alcohol-preferring rats. *The Journal of Neuroscience*, *34*, 12716–12724.

Melis, M. R., Succu, S., Mascia, M. S., & Argiolas, A. (2004). Antagonism of cannabinoid CB1 receptors in the paraventricular nucleus of male rats induces penile erection. *Neuroscience Letters, 359*, 17–20.

Melis, M. R., Succu, S., Mascia, M. S., Sanna, F., Melis, T., Castelli, M. P., et al. (2006). The cannabinoid receptor antagonist SR-141716A induces penile erection in male rats: Involvement of paraventricular glutamic acid and nitric oxide. *Neuropharmacology, 50*, 219–228.

Melis, T., Succu, S., Sanna, F., Boi, A., Argiolas, A., & Melis, M. R. (2007). The cannabinoid antagonist SR 141716A (Rimonabant) reduces the increase of extra-cellular dopamine release in the rat nucleus accumbens induced by a novel high palatable food. *Neuroscience Letters, 419*, 231–235.

Memos, N. K., Vela, R., Tabone, C., & Guarraci, F. A. (2014). Endocannabinoid influence on partner preference in female rats. *Pharmacology, Biochemistry and Behavior, 124*, 380–388.

Mereu, M., Tronci, V., Chun, L. E., Thomas, A. M., Green, J. L., Katz, J. L., et al. (2015). Cocaine-induced endocannabinoid release modulates behavioral and neurochemical sensitization in mice. *Addiction Biology, 20*, 91–103.

Meye, F. J., Valentinova, K., Lecca, S., Marion-Poll, L., Maroteaux, M. J., Musardo, S., et al. (2015). Cocaine-evoked negative symptoms require AMPA receptor trafficking in the lateral habenula. *Nature Neuroscience, 18*, 376–378.

Millman, R. B., & Sbriglio, R. (1986). Patterns of use and psychopathology in chronic marijuana users. *The Psychiatric Clinics of North America, 9*, 533–545.

Mirenowicz, J., & Schultz, W. (1996). Preferential activation of midbrain dopamine neurons by appetitive rather than aversive stimuli. *Nature, 379*, 449–451.

Morikawa, H., & Paladini, C. A. (2011). Dynamic regulation of midbrain dopamine neuron activity: Intrinsic, synaptic, and plasticity mechanisms. *Neuroscience, 198*, 95–111.

Murphy, L. L., Gher, J., Steger, R. W., & Bartke, A. (1994). Effects of delta 9-tetrahydrocannabinol on copulatory behavior and neuroendocrine responses of male rats to female conspecifics. *Pharmacology, Biochemistry and Behavior, 48*, 1011–1017.

Myers, S. A., & Rocca, K. A. (2000). Students' state motivation and instructors' use of verbally aggressive messages. *Psychological Reports, 87*, 291–294.

Nazzaro, C., Greco, B., Cerovic, M., Baxter, P., Rubino, T., Trusel, M., et al. (2012). SK channel modulation rescues striatal plasticity and control over habit in cannabinoid tolerance. *Nature Neuroscience, 15*, 284–293.

Negrete, J. C., Knapp, W. P., Douglas, D. E., & Smith, W. B. (1986). Cannabis affects the severity of schizophrenic symptoms: Results of a clinical survey. *Psychological Medicine, 16*, 515–520.

Nestler, E. J. (2014). Epigenetic mechanisms of drug addiction. *Neuropharmacology, 76*(Pt. B), 259–268.

Newsom, R. J., & Kelly, S. J. (2008). Perinatal delta-9-tetrahydrocannabinol exposure disrupts social and open field behavior in adult male rats. *Neurotoxicology and Teratology, 30*, 213–219.

Numan, M., & Smith, H. G. (1984). Maternal behavior in rats: Evidence for the involvement of preoptic projections to the ventral tegmental area. *Behavioral Neuroscience, 98*, 712–727.

Numan, M., & Stolzenberg, D. S. (2009). Medial preoptic area interactions with dopamine neural systems in the control of the onset and maintenance of maternal behavior in rats. *Frontiers in Neuroendocrinology, 30*, 46–64.

Oleson, E. B., Beckert, M. V., Morra, J. T., Lansink, C. S., Cachope, R., Abdullah, R. A., et al. (2012). Endocannabinoids shape accumbal encoding of cue-motivated behavior via CB1 receptor activation in the ventral tegmentum. *Neuron, 73*, 360–373.

Oliere, S., Joliette-Riopel, A., Potvin, S., & Jutras-Aswad, D. (2014). Modulation of the endocannabinoid system: Vulnerability factor and new treatment target for stimulant addiction. *Frontiers in Psychiatry*, *4*, 109.

Olsen, C. M. (2011). Natural rewards, neuroplasticity, and non-drug addictions. *Neuropharmacology*, *61*, 1109–1122.

O'Shea, M., McGregor, I. S., & Mallet, P. E. (2006). Repeated cannabinoid exposure during perinatal, adolescent or early adult ages produces similar longlasting deficits in object recognition and reduced social interaction in rats. *Journal of Psychopharmacology*, *20*, 611–621.

O'Shea, M., Singh, M. E., McGregor, I. S., & Mallet, P. E. (2004). Chronic cannabinoid exposure produces lasting memory impairment and increased anxiety in adolescent but not adult rats. *Journal of Psychopharmacology*, *18*, 502–508.

Paladini, C. A., & Roeper, J. (2014). Generating bursts (and pauses) in the dopamine midbrain neurons. *Neuroscience*, *282C*, 109–121.

Paladini, C. A., & Tepper, J. M. (1999). GABA(A) and GABA(B) antagonists differentially affect the firing pattern of substantia nigra dopaminergic neurons in vivo. *Synapse*, *32*, 165–176.

Pan, B., Hillard, C. J., & Liu, Q. S. (2008). Endocannabinoid signaling mediates cocaine-induced inhibitory synaptic plasticity in midbrain dopamine neurons. *The Journal of Neuroscience*, *28*, 1385–1397.

Park, K., Volkow, N. D., Pan, Y., & Du, C. (2013). Chronic cocaine dampens dopamine signaling during cocaine intoxication and unbalances D1 over D2 receptor signaling. *The Journal of Neuroscience*, *33*, 15827–15836.

Parolaro, D., Rubino, T., Vigano, D., Massi, P., Guidali, C., & Realini, N. (2010). Cellular mechanisms underlying the interaction between cannabinoid and opioid system. *Current Drug Targets*, *11*, 393–405.

Pattij, T., Janssen, M. C., Schepers, I., Gonzalez-Cuevas, G., de Vries, T. J., & Schoffelmeer, A. N. (2007). Effects of the cannabinoid CB1 receptor antagonist rimonabant on distinct measures of impulsive behavior in rats. *Psychopharmacology*, *193*, 85–96.

Pecina, M., Martinez-Jauand, M., Hodgkinson, C., Stohler, C. S., Goldman, D., & Zubieta, J. K. (2014). FAAH selectively influences placebo effects. *Molecular Psychiatry*, *19*, 385–391.

Pecina, M., & Zubieta, J. K. (2015). Molecular mechanisms of placebo responses in humans. *Molecular Psychiatry*, *20*, 416–423.

Pellis, S. M., & Pellis, V. C. (1998). Play fighting of rats in comparative perspective: A schema for neurobehavioral analyses. *Neuroscience and Biobehavioral Reviews*, *23*, 87–101.

Piazza, P. V., & Deroche-Gamonet, V. (2013). A multistep general theory of transition to addiction. *Psychopharmacology*, *229*, 387–413.

Pignatelli, M., & Bonci, A. (2015). Role of dopamine neurons in reward and aversion: A synaptic plasticity perspective. *Neuron*, *86*, 1145–1157.

Pistillo, F., Clementi, F., Zoli, M., & Gotti, C. (2015). Nicotinic, glutamatergic and dopaminergic synaptic transmission and plasticity in the mesocorticolimbic system: Focus on nicotine effects. *Progress in Neurobiology*, *124*, 1–27.

Pistis, M., Ferraro, L., Pira, L., Flore, G., Tanganelli, S., Gessa, G. L., et al. (2002). Delta(9)-tetrahydrocannabinol decreases extracellular GABA and increases extracellular glutamate and dopamine levels in the rat prefrontal cortex: An in vivo microdialysis study. *Brain Research*, *948*, 155–158.

Plasse, T. F. (1991). Clinical use of dronabinol. *Journal of Clinical Oncology*, *9*, 2079–2080.

Potegal, M., & Einon, D. (1989). Aggressive behaviors in adult rats deprived of playfighting experience as juveniles. *Developmental Psychobiology*, *22*, 159–172.

Quinn, H. R., Matsumoto, I., Callaghan, P. D., Long, L. E., Arnold, J. C., Gunasekaran, N., et al. (2008). Adolescent rats find repeated delta(9)-THC less aversive than adult rats but display greater residual cognitive deficits and changes in hippocampal protein expression following exposure. *Neuropsychopharmacology*, *33*, 1113–1126.
Ravinet Trillou, C., Delgorge, C., Menet, C., Arnone, M., & Soubrie, P. (2004). CB1 cannabinoid receptor knockout in mice leads to leanness, resistance to diet-induced obesity and enhanced leptin sensitivity. *International Journal of Obesity and Related Metabolic Disorders*, *28*, 640–648.
Realini, N., Vigano, D., Guidali, C., Zamberletti, E., Rubino, T., & Parolaro, D. (2011). Chronic URB597 treatment at adulthood reverted most depressive-like symptoms induced by adolescent exposure to THC in female rats. *Neuropharmacology*, *60*, 235–243.
Reis, H. T., Aron, A., Clark, M. S., & Finkel, E. J. (2013). Ellen Berscheid, Elaine Hatfield, and the emergence of relationship science. *Perspectives on Psychological Science*, *8*, 558–572.
Ribak, C. E., Vaughn, J. E., & Roberts, E. (1980). GABAergic nerve terminals decrease in the substantia nigra following hemitransections of the striatonigral and pallidonigral pathways. *Brain Research*, *192*, 413–420.
Riegel, A. C., & Lupica, C. R. (2004). Independent presynaptic and postsynaptic mechanisms regulate endocannabinoid signaling at multiple synapses in the ventral tegmental area. *The Journal of Neuroscience*, *24*, 11070–11078.
Robinson, T. E., & Berridge, K. C. (1993). The neural basis of drug craving: An incentive-sensitization theory of addiction. *Brain Research Brain Research Reviews*, *18*, 247–291.
Robinson, T. E., & Berridge, K. C. (2000). The psychology and neurobiology of addiction: An incentive-sensitization view. *Addiction*, *95*(Suppl. 2), S91–S117.
Robinson, T. E., & Berridge, K. C. (2008). Review. The incentive sensitization theory of addiction: Some current issues. *Philosophical Transactions of the Royal Society of London, Series B: Biological Sciences*, *363*, 3137–3146.
Robledo, P. (2010). Cannabinoids, opioids and MDMA: Neuropsychological interactions related to addiction. *Current Drug Targets*, *11*, 429–439.
Rubino, T., Massi, P., Vigano, D., Fuzio, D., & Parolaro, D. (2000). Long-term treatment with SR141716A, the CB1 receptor antagonist, influences morphine withdrawal syndrome. *Life Sciences*, *66*, 2213–2219.
Saal, D., Dong, Y., Bonci, A., & Malenka, R. C. (2003). Drugs of abuse and stress trigger a common synaptic adaptation in dopamine neurons. *Neuron*, *37*, 577–582.
Sadakierska-Chudy, A., Frankowska, M., & Filip, M. (2014). Mitoepigenetics and drug addiction. *Pharmacology & Therapeutics*, *144*, 226–233.
Sagheddu, C., & Melis, M. (2015). Individual differences and vulnerability to drug addiction: A focus on the endocannabinoid system. *CNS & Neurological Disorders: Drug Targets*, *14*, 502–517.
Salamone, J. D., Correa, M., Mingote, S., & Weber, S. M. (2003). Nucleus accumbens dopamine and the regulation of effort in food-seeking behavior: Implications for studies of natural motivation, psychiatry, and drug abuse. *The Journal of Pharmacology and Experimental Therapeutics*, *305*, 1–8.
Salamone, J. D., McLaughlin, P. J., Sink, K., Makriyannis, A., & Parker, L. A. (2007). Cannabinoid CB1 receptor inverse agonists and neutral antagonists: Effects on food intake, food-reinforced behavior and food aversions. *Physiology & Behavior*, *91*, 383–388.
Salerian, A. J. (2004). Successful treatment of sexual dysfunction with dronabinol: A case report. *The Journal of Clinical Psychiatry*, *65*, 1146–1147.
Sanchis-Segura, C., Cline, B. H., Marsicano, G., Lutz, B., & Spanagel, R. (2004). Reduced sensitivity to reward in CB1 knockout mice. *Psychopharmacology*, *176*, 223–232.
Scavone, J. L., Sterling, R. C., & Van Bockstaele, E. J. (2013). Cannabinoid and opioid interactions: implications for opiate dependence and withdrawal. *Neuroscience*, *248*, 637–654.

Schechter, M., Pinhasov, A., Weller, A., & Fride, E. (2012). Blocking the postpartum mouse dam's CB1 receptors impairs maternal behavior as well as offspring development and their adult social-emotional behavior. *Behavioural Brain Research, 226*, 481–492.
Schechter, M., Weller, A., Pittel, Z., Gross, M., Zimmer, A., & Pinhasov, A. (2013). Endocannabinoid receptor deficiency affects maternal care and alters the dam's hippocampal oxytocin receptor and brain-derived neurotrophic factor expression. *Journal of Neuroendocrinology, 25*, 898–909.
Scherma, M., Fattore, L., Castelli, M. P., Fratta, W., & Fadda, P. (2014). The role of the endocannabinoid system in eating disorders: Neurochemical and behavioural preclinical evidence. *Current Pharmaceutical Design, 20*, 2089–2099.
Schilstrom, B., Rawal, N., Mameli-Engvall, M., Nomikos, G. G., & Svensson, T. H. (2003). Dual effects of nicotine on dopamine neurons mediated by different nicotinic receptor subtypes. *International Journal of Neuropsychopharmacology, 6*, 1–11.
Schneider, M. (2009). Cannabis use in pregnancy and early life and its consequences: Animal models. *European Archives of Psychiatry and Clinical Neuroscience, 259*, 383–393.
Schneider, M., & Koch, M. (2003). Chronic pubertal, but not adult chronic cannabinoid treatment impairs sensorimotor gating, recognition memory, and the performance in a progressive ratio task in adult rats. *Neuropsychopharmacology, 28*, 1760–1769.
Schneider, M., & Koch, M. (2005). Deficient social and play behavior in juvenile and adult rats after neonatal cortical lesion: Effects of chronic pubertal cannabinoid treatment. *Neuropsychopharmacology, 30*, 944–957.
Schneider, M., & Koch, M. (2007). The effect of chronic peripubertal cannabinoid treatment on deficient object recognition memory in rats after neonatal mPFC lesion. *European Neuropsychopharmacology, 17*, 180–186.
Schneider, M., Schomig, E., & Leweke, F. M. (2008). Acute and chronic cannabinoid treatment differentially affects recognition memory and social behavior in pubertal and adult rats. *Addiction Biology, 13*, 345–357.
Schultz, W. (1998). Predictive reward signal of dopamine neurons. *Journal of Neurophysiology, 80*, 1–27.
Schultz, W. (2010). Dopamine signals for reward value and risk: Basic and recent data. *Behavioral and Brain Functions, 6*, 24.
Schultz, W. (2013). Updating dopamine reward signals. *Current Opinion in Neurobiology, 23*, 229–238.
Schultz, W., Dayan, P., & Montague, P. R. (1997). A neural substrate of prediction and reward. *Science, 275*, 1593–1599.
Schultz, W., & Dickinson, A. (2000). Neuronal coding of prediction errors. *Annual Review of Neuroscience, 23*, 473–500.
Schultz, W., & Romo, R. (1987). Responses of nigrostriatal dopamine neurons to high-intensity somatosensory stimulation in the anesthetized monkey. *Journal of Neurophysiology, 57*, 201–217.
Scott, D. J., Stohler, C. S., Egnatuk, C. M., Wang, H., Koeppe, R. A., & Zubieta, J. K. (2007). Individual differences in reward responding explain placebo-induced expectations and effects. *Neuron, 55*, 325–336.
Serrano, A., & Parsons, L. H. (2011). Endocannabinoid influence in drug reinforcement, dependence and addiction-related behaviors. *Pharmacology & Therapeutics, 132*, 215–241.
Shrenker, P., & Bartke, A. (1985). Suppression of male copulatory behavior by delta 9-THC is not dependent on changes in plasma testosterone or hypothalamic dopamine or serotonin content. *Pharmacology, Biochemistry and Behavior, 22*, 415–420.
Siviy, S. M., & Panksepp, J. (2011). In search of the neurobiological substrates for social playfulness in mammalian brains. *Neuroscience and Biobehavioral Reviews, 35*, 1821–1830.

Sjoerds, Z., Luigjes, J., van den Brink, W., Denys, D., & Yucel, M. (2014). The role of habits and motivation in human drug addiction: A reflection. *Frontiers in Psychiatry*, *5*, 8.

Sklair-Tavron, L., Shi, W. X., Lane, S. B., Harris, H. W., Bunney, B. S., & Nestler, E. J. (1996). Chronic morphine induces visible changes in the morphology of mesolimbic dopamine neurons. *Proceedings of the National Academy of Sciences of the United States of America*, *93*, 11202–11207.

Solinas, M., & Goldberg, S. R. (2005). Motivational effects of cannabinoids and opioids on food reinforcement depend on simultaneous activation of cannabinoid and opioid systems. *Neuropsychopharmacology*, *30*, 2035–2045.

Solinas, M., Goldberg, S. R., & Piomelli, D. (2008). The endocannabinoid system in brain reward processes. *British Journal of Pharmacology*, *154*, 369–383.

Soria-Gomez, E., Matias, I., Rueda-Orozco, P. E., Cisneros, M., Petrosino, S., Navarro, L., et al. (2007). Pharmacological enhancement of the endocannabinoid system in the nucleus accumbens shell stimulates food intake and increases c-Fos expression in the hypothalamus. *British Journal of Pharmacology*, *151*, 1109–1116.

Spiga, S., Lintas, A., Migliore, M., & Diana, M. (2010). Altered architecture and functional consequences of the mesolimbic dopamine system in cannabis dependence. *Addiction Biology*, *15*, 266–276.

Spiga, S., Puddu, M. C., Pisano, M., & Diana, M. (2005). Morphine withdrawal-induced morphological changes in the nucleus accumbens. *The European Journal of Neuroscience*, *22*, 2332–2340.

Spiga, S., Talani, G., Mulas, G., Licheri, V., Fois, G. R., Muggironi, G., et al. (2014). Hampered long-term depression and thin spine loss in the nucleus accumbens of ethanol-dependent rats. *Proceedings of the National Academy of Sciences of the United States of America*, *111*, E3745–E3754.

Strathearn, L., Fonagy, P., Amico, J., & Montague, P. R. (2009). Adult attachment predicts maternal brain and oxytocin response to infant cues. *Neuropsychopharmacology*, *34*, 2655–2666.

Strathearn, L., Li, J., Fonagy, P., & Montague, P. R. (2008). What's in a smile? Maternal brain responses to infant facial cues. *Pediatrics*, *122*, 40–51.

Succu, S., Mascia, M. S., Sanna, F., Melis, T., Argiolas, A., & Melis, M. R. (2006). The cannabinoid CB1 receptor antagonist SR 141716A induces penile erection by increasing extra-cellular glutamic acid in the paraventricular nucleus of male rats. *Behavioural Brain Research*, *169*, 274–281.

Tabibnia, G., & Lieberman, M. D. (2007). Fairness and cooperation are rewarding: Evidence from social cognitive neuroscience. *Annals of the New York Academy of Sciences*, *1118*, 90–101.

Tan, K. R., Yvon, C., Turiault, M., Mirzabekov, J. J., Doehner, J., Labouebe, G., et al. (2012). GABA neurons of the VTA drive conditioned place aversion. *Neuron*, *73*, 1173–1183.

Tanda, G., Loddo, P., & Di Chiara, G. (1999). Dependence of mesolimbic dopamine transmission on delta9-tetrahydrocannabinol. *European Journal of Pharmacology*, *376*, 23–26.

Tanda, G., Pontieri, F. E., & Di Chiara, G. (1997). Cannabinoid and heroin activation of mesolimbic dopamine transmission by a common mu1 opioid receptor mechanism. *Science*, *276*, 2048–2050.

Tennant, F. S., Jr., & Groesbeck, C. J. (1972). Psychiatric effects of hashish. *Archives of General Psychiatry*, *27*, 133–136.

Thornton-Jones, Z. D., Vickers, S. P., & Clifton, P. G. (2005). The cannabinoid CB1 receptor antagonist SR141716A reduces appetitive and consummatory responses for food. *Psychopharmacology*, *179*, 452–460.

Tobler, P. N., O'Doherty, J. P., Dolan, R. J., & Schultz, W. (2006). Human neural learning depends on reward prediction errors in the blocking paradigm. *Journal of Neurophysiology*, *95*, 301–310.

Tonstad, S. (2006). Rimonabant: A cannabinoid receptor blocker for the treatment of metabolic and cardiovascular risk factors. *Nutrition, Metabolism, and Cardiovascular Diseases, 16*, 156–162.

Trezza, V., Baarendse, P. J., & Vanderschuren, L. J. (2010). The pleasures of play: Pharmacological insights into social reward mechanisms. *Trends in Pharmacological Sciences, 31*, 463–469.

Trezza, V., Campolongo, P., Cassano, T., Macheda, T., Dipasquale, P., Carratu, M. R., et al. (2008). Effects of perinatal exposure to delta-9-tetrahydrocannabinol on the emotional reactivity of the offspring: A longitudinal behavioral study in Wistar rats. *Psychopharmacology, 198*, 529–537.

Trezza, V., Campolongo, P., Manduca, A., Morena, M., Palmery, M., Vanderschuren, L. J., et al. (2012). Altering endocannabinoid neurotransmission at critical developmental ages: Impact on rodent emotionality and cognitive performance. *Frontiers in Behavioral Neuroscience, 6*, 2.

Trezza, V., Campolongo, P., & Vanderschuren, L. J. (2011). Evaluating the rewarding nature of social interactions in laboratory animals. *Developmental Cognitive Neuroscience, 1*, 444–458.

Trezza, V., Damsteegt, R., Manduca, A., Petrosino, S., Van Kerkhof, L. W., Pasterkamp, R. J., et al. (2012). Endocannabinoids in amygdala and nucleus accumbens mediate social play reward in adolescent rats. *The Journal of Neuroscience, 32*, 14899–14908.

Trezza, V., & Vanderschuren, L. J. (2008a). Cannabinoid and opioid modulation of social play behavior in adolescent rats: Differential behavioral mechanisms. *European Neuropsychopharmacology, 18*, 519–530.

Trezza, V., & Vanderschuren, L. J. (2008b). Bidirectional cannabinoid modulation of social behavior in adolescent rats. *Psychopharmacology, 197*, 217–227.

Trezza, V., & Vanderschuren, L. J. (2009). Divergent effects of anandamide transporter inhibitors with different target selectivity on social play behavior in adolescent rats. *The Journal of Pharmacology and Experimental Therapeutics, 328*, 343–350.

Tsai, H. C., Zhang, F., Adamantidis, A., Stuber, G. D., Bonci, A., de Lecea, L., et al. (2009). Phasic firing in dopaminergic neurons is sufficient for behavioral conditioning. *Science, 324*, 1080–1084.

Tsou, K., Brown, S., Sanudo-Pena, M. C., Mackie, K., & Walker, J. M. (1998). Immunohistochemical distribution of cannabinoid CB1 receptors in the rat central nervous system. *Neuroscience, 83*, 393–411.

Turley, W. A., Jr., & Floody, O. R. (1981). Delta-9-tetrahydrocannabinol stimulates receptive and proceptive sexual behaviors in female hamsters. *Pharmacology, Biochemistry and Behavior, 14*, 745–747.

Ungless, M. A. (2004). Dopamine: The salient issue. *Trends in Neurosciences, 27*, 702–706.

Ungless, M. A., Whistler, J. L., Malenka, R. C., & Bonci, A. (2001). Single cocaine exposure in vivo induces long-term potentiation in dopamine neurons. *Nature, 411*, 583–587.

van Hell, H. H., Vink, M., Ossewaarde, L., Jager, G., Kahn, R. S., & Ramsey, N. F. (2010). Chronic effects of cannabis use on the human reward system: An fMRI study. *European Neuropsychopharmacology, 20*, 153–163.

Vanderschuren, L. J., Niesink, R. J., & Van Ree, J. M. (1997). The neurobiology of social play behavior in rats. *Neuroscience and Biobehavioral Reviews, 21*, 309–326.

Verkooijen, K. T., de Vries, N. K., & Nielsen, G. A. (2007). Youth crowds and substance use: The impact of perceived group norm and multiple group identification. *Psychology of Addictive Behaviors, 21*, 55–61.

Vezina, P., & Leyton, M. (2009). Conditioned cues and the expression of stimulant sensitization in animals and humans. *Neuropharmacology, 56*(Suppl. 1), 160–168.

Vinod, K. Y., Kassir, S. A., Hungund, B. L., Cooper, T. B., Mann, J. J., & Arango, V. (2010). Selective alterations of the CB1 receptors and the fatty acid amide hydrolase in the ventral striatum of alcoholics and suicides. *Journal of Psychiatric Research, 44*, 591–597.

Volkow, N. D., & Baler, R. D. (2014). Addiction science: Uncovering neurobiological complexity. *Neuropharmacology*, *76*(Pt. B), 235–249.
Volkow, N. D., Fowler, J. S., Wang, G. J., Baler, R., & Telang, F. (2009). Imaging dopamine's role in drug abuse and addiction. *Neuropharmacology*, *56*(Suppl. 1), 3–8.
Volman, S. F., Lammel, S., Margolis, E. B., Kim, Y., Richard, J. M., Roitman, M. F., et al. (2013). New insights into the specificity and plasticity of reward and aversion encoding in the mesolimbic system. *The Journal of Neuroscience: The Official Journal of the Society for Neuroscience*, *33*, 17569–17576.
Walaas, I., & Fonnum, F. (1980). Biochemical evidence for gamma-aminobutyrate containing fibres from the nucleus accumbens to the substantia nigra and ventral tegmental area in the rat. *Neuroscience*, *5*, 63–72.
Walker, S. C., & McGlone, F. P. (2013). The social brain: Neurobiological basis of affiliative behaviours and psychological well-being. *Neuropeptides*, *47*, 379–393.
Wamsteeker Cusulin, J. I., Senst, L., Teskey, G. C., & Bains, J. S. (2014). Experience salience gates endocannabinoid signaling at hypothalamic synapses. *The Journal of Neuroscience*, *34*, 6177–6181.
Wang, H., & Lupica, C. R. (2014). Release of endogenous cannabinoids from ventral tegmental area dopamine neurons and the modulation of synaptic processes. *Progress in Neuro-Psychopharmacology & Biological Psychiatry*, *52*, 24–27.
Wang, X. Q., Ma, J., Cui, W., Yuan, W. X., Zhu, G., Yang, Q., et al. (2014). The endocannabinoid system regulates synaptic transmission in nucleus accumbens by increasing DAGL-alpha expression following short-term morphine withdrawal. *British Journal of Pharmacology*. (Ahead of print).
Williams, C. M., & Kirkham, T. C. (1999). Anandamide induces overeating: Mediation by central cannabinoid (CB1) receptors. *Psychopharmacology*, *143*, 315–317.
Wills, K. L., Vemuri, K., Kalmar, A., Lee, A., Limebeer, C. L., Makriyannis, A., et al. (2014). CB1 antagonism: Interference with affective properties of acute naloxone-precipitated morphine withdrawal in rats. *Psychopharmacology*, *231*, 4291–4300.
Winton-Brown, T. T., Fusar-Poli, P., Ungless, M. A., & Howes, O. D. (2014). Dopaminergic basis of salience dysregulation in psychosis. *Trends in Neurosciences*, *37*, 85–94.
Wise, R. A. (2004). Dopamine, learning and motivation. *Nature Reviews Neuroscience*, *5*, 483–494.
Wise, R. A., & Koob, G. F. (2014). The development and maintenance of drug addiction. *Neuropsychopharmacology*, *39*, 254–262.
Wiskerke, J., Stoop, N., Schetters, D., Schoffelmeer, A. N., & Pattij, T. (2011). Cannabinoid CB1 receptor activation mediates the opposing effects of amphetamine on impulsive action and impulsive choice. *PLoS One*, *6*, e25856.
Witten, I. B., Steinberg, E. E., Lee, S. Y., Davidson, T. J., Zalocusky, K. A., Brodsky, M., et al. (2011). Recombinase-driver rat lines: Tools, techniques, and optogenetic application to dopamine-mediated reinforcement. *Neuron*, *72*, 721–733.
Wittfoth-Schardt, D., Grunding, J., Wittfoth, M., Lanfermann, H., Heinrichs, M., Domes, G., et al. (2012). Oxytocin modulates neural reactivity to children's faces as a function of social salience. *Neuropsychopharmacology*, *37*, 1799–1807.
Wyvell, C. L., & Berridge, K. C. (2000). Intra-accumbens amphetamine increases the conditioned incentive salience of sucrose reward: Enhancement of reward "wanting" without enhanced "liking" or response reinforcement. *The Journal of Neuroscience*, *20*, 8122–8130.
Zamora-Martinez, E. R., & Edwards, S. (2014). Neuronal extracellular signal-regulated kinase (ERK) activity as marker and mediator of alcohol and opioid dependence. *Frontiers in Integrative Neuroscience*, *8*, 24.
Zhang, L., Liu, Y., & Chen, X. (2005). Carbachol induces burst firing of dopamine cells in the ventral tegmental area by promoting calcium entry through L-type channels in the rat. *The Journal of Physiology*, *568*, 469–481.

INDEX

Note: Page numbers followed by "*f*" indicate figures and "*t*" indicate tables.

A

Acute nausea
- brainstem circuitry, 142
- cannabinoid compounds, 132–133
- cannabinoid hyperemesis syndrome, 134
- conditioned gaping (*see* Nausea-induced conditioned gaping)
- conditioned taste avoidance, 138
- efficacy, 128, 129*t*
- endocannabinoid levels, 134–137
- insular cortex, 142–144
- motion sickness, 144–145
- neurocircuitry, brain, 128, 132*f*
- orally active and synthetic analog, 133
- Pica, 137–138
- Sativex, 133–134
- THC, 132–133

Adolescence
- behaviors, 50
- brain, 54–55
- cannabinoid exposure, 69–70
- chronic variable stress, 57
- CORT and ACTH, 55–57
- eCB system regulation, 61–64
- endocannabinoid system dysregulation, 71–74, 72*t*
- fear/anxiety-related disorder, 50
- long-term sex and region-dependent consequences, 57–58
- pubertal maturation, 52–53
- repeated restraint stress paradigm, 57
- stress exposure, 57–58, 68–69
- stress hyporesponsive period, 55–56
- transient developmental imbalance, 50–51

Amotivational syndrome, 273–275

Amygdala
- eCB system, 213, 226*f*
- pain and modulation, 223*t*, 232–234
- SIA, 234–235

Amygdalar eCB signaling, 63–64

Anandamide. *See* N-arachidonoylethanolamine (AEA)

Anticipatory nausea (AN)
- brain neurocircuitry, 128, 132*f*
- cannabinoid treatments, 146–148, 150
- conditioned gaping, 149–150
- conditioning theory, 145–146
- efficacy, 146, 147*t*
- FAAH inhibition, 151–152
- MAGL inhibition, 151–152

Antinociceptive activity, 210–211

2-Arachidonoylglycerol (2-AG), 60–62
- AA concentrations, 18–19
- Abhd6, 18
- Abhd12, 18
- bound receptors, 165
- cellular uptake and transport, 98
- DAGL, 12–14, 13*f*, 165–166
- DAG synthesis, 13*f*, 14–15
- degradation, 95–97, 165–166
- eCB actions, 165
- endocannabinoid levels, 134–137
- from lipid precursors, 165
- MAGL, 13*f*, 15–18
- metabolization, 208–209
- monoacylglycerol lipase, 165–166
- synthesis, 95–97, 165, 208–209

Atrial natriuretic peptide (ANP), 182–183

B

Brain
- AEA and 2-AG, 207–209
- amygdala (*see* Amygdala)
- antinociceptive activity, 210–211
- brain regions, 209–210, 210*f*
- CB_1 and CB_2 receptors, 207–208
- mechanisms, 209
- PAG (*see* Periaquaductal gray (PAG))
- in pain and modulation, 211–213, 214*t*
- PFC (*see* Prefrontal cortex (PFC))
- physiological and pharmacological effects, 207
- supraspinal sites, 210

Brainstem circuitry, 142

Burst like mechanism, 61–62

C

Cannabidiol (CBD), 101
Cannabinoid hyperemesis syndrome, 134
Cannabinoid (CB) receptors
  CB_1, 92, 164–165, 173
  CB_2, 92, 164–165, 173
  desensitization, 109
  dimerization, 108–109
  in ENS, 92–94
  GPCRs, selective coupling of, 107–108
  inhibition and activation, 91
  ligands, 91–92
  long-term depression, 27
  metaplasticity, 107
  pharmacology, 23–25
  short-term depression, 25–27
  signaling, 23
Cannabinoid regulation, stress response, 166–172, 169*f*
Central nervous system (CNS)
  AA, 9, 18–19
  Abhd6, 18
  Abhd12, 18
  DAGL, 12–14, 13*f*
  DAG synthesis, 13*f*, 14–15
  definition, 4
  FAAH, 9–11
  MAGL, 13*f*, 15–18
  multienzyme pathways, 6*f*, 7–8
  NAE-hydrolyzing acid amidase, 11
  NAPE-PLD, 5–7, 6*f*
  precursor synthesis, 4–5
Chemotherapy-induced nausea and vomiting (CINV), 128
Chronic pain, 205–206
Corticotropin-releasing hormone (CRH), 10, 53–55, 166–167, 169–170, 175–176, 179–180

D

DAGL. *See* Diacylglycerol lipases (DAGL)
Depolarization-induced suppression of inhibition (DSI), 166, 282–283
Deroche-Gamonet model, 280–282
Diacylglycerol lipases (DAGL), 12–14, 95–96, 208–209
Dopamine (DA)
  behavioral flexibility, 259–261
  decision-making processes, 259–261
  encoding salience, 259–260
  incentive-sensitization model, 260–261
  intense reward, 261
  mechanisms modulating signaling, 262–263
  multidisciplinary approach, 261
  synaptic plasticity, 259–260
  value learning, 259–260
  working memory, 259–260
Dorsolateral striatum (DLS), 280–282
Dronabinol, 132–133
Drug addiction
  aberrant plasticity and behavioral output, 279–280
  acute intoxication, 280–282
  cocaine use, 279–280
  cycle, 279–280
  Deroche-Gamonet model, 280–282
  DLS, 280–282
  DSI, 282–283
  eCB signaling, 280–282
  etiology and neural mechanisms of, 277–279
  functional tolerance, 280–282
  homeostatic synaptic plasticity, 280–282
  LTD, 279–280
  mesocorticolimbic pathway, 280–282
  opiate withdrawal syndrome, 282–283
  presynaptic homeostasis, 280–282

E

Electromyographic analysis, 139
Emotions, 259
Endocannabinoid signaling mechanisms
  addiction (*see* Drug addiction)
  adolescent (*see* Adolescence)
  activation of potassium channels, 99
  CB_1, 100
  cellular uptake and transport, 98
  in CNS (*see* Central nervous system (CNS))
  fatty acid binding proteins, 98–99
  inactivation of voltage-gated calcium channels, 99
  long-term depression and potentiation, 99
  motivation (*see* Motivation)

myenteric plexus, 99–100
negative coupling, 99
neuroendocrine (*see* Neuroendocrine regulation)
reward (*see* Reward)
spontaneous and fast excitatory postsynaptic potentials, 99–100
Endocannabinoid synthesis, 94–95
Endocannabinoid (eCB) system, 51–52
AEA, 60–61
2-AG, 60–61
brain (*see* Brain)
cannabis consumption, 60
CB1 and CB2 receptors, 60
G protein-coupled receptors, 60
HPA axis regulation, 61–64
nausea (*see* Acute nausea; Anticipatory nausea (AN))
ontogeny, 64–67, 65*t*
psychoactive component, 60
Endocannabinoid tone
altered intestinal motility, 106
clinical endocannabinoid deficiency, 106–107
clinical trials, 106–107
constitutive receptor activity, 105–106
with slow transit constipation, 106
Energy homeostasis
ANP, 182–183
CB receptors, 173, 178*f*, 179
CRH, 175–176
eCB receptors, 173–174
fluid, 177
ghrelin, 174–175
leptin, 176–177
oxytocin, 182–183
TRH, 176
TRPV1 receptors, 177–179, 178*f*
vasopressin, 180–182
Enteric nervous system (ENS)
brain-gut axis, 89, 114–115
cannabinoid receptors, 92–94
endocannabinoid tone, 105–107
enteric glia, 110
enteroendocrine cells, 114
GPR55, 103–104
intestinal inflammation, 115–116
metaplasticity, 107
neuromuscular transmission, 110–112
OEA, 104
PEA, 104–105
peptide endocannabinoids, 102
phytocannabinoids, 100–102
secretomotor transmission, 112–114
structure of, 89–91
TRPV1, 102–103
virodhamine, endogenous CB1 antagonist, 103
Evolution theory, 258

F

Fatty acid amide hydrolase (FAAH), 134–135, 208–209
enzymatic activity, 9
$FAAH^{-/-}$ mice, 11
function, 9
integral membrane protein, 9–10
in vitro treatment with endotoxin, 10–11
pharmacological inhibitors, 11
posttranslational processes, 10
promoter, 10
single-nucleotide polymorphism, 11
Fear/anxiety-related disorder, 50

G

Gastrointestinal (GI) tract, neurotransmission
enteroendocrine cells, 114
neuromuscular transmission, 110–112
secretomotor transmission, 112–114
Gatekeeper hypothesis, 61–63
Gonadotropin-releasing hormone (GnRH), 52–53, 184–185
G protein-coupled receptor 55 (GPR55), 103–104

H

Hypothalamic-pituitary-adrenal (HPA) axis
acute stress regulation, 168–170, 169*f*
adolescent (*see* Adolescence)
chronic stress regulation, 170–172
gonadal hormones development, 58–59
inhibitory endocannabinoid tone, 167–168
stress, 53–55

I

Incentive salience, 272–273
Incentive-sensitization model, 260–261
Insular cortex, 142–144

L

Learning and memory associative model, 272–273
Long-term depression (LTD), 279–280

M

Monoacylglycerol lipase (MAGL), 96–97, 134–135
Motion sickness, 144–145
Motivation
 acute aversive stimuli, 273
 amotivational syndrome, 273–275
 behaviors, 271
 chronic *cannabis* use, 273–276
 FAAH Pro129Thr polymorphism, 276–277
 incentive salience, 272–273
 initiate and persistent, 259
 inverted U-shape effect, 275–276
 learning and memory associative model, 272–273
 in neuronal mechanisms, 275–276
 placebo effect, 276–277
 psychological construct, 271
 and reward mechanisms, 276–277
Mouse and rat tail flick tests, 210–211

N

N-acyl phosphatidylethanolamine (NAPE), 60–61
N-arachidonoylethanolamine (AEA), 60–62
 AA, 9
 biosynthetic pathways, 208–209
 bound receptors, 165
 cellular uptake and transport, 98
 characterization, 207–208
 degradation, 97–98, 165–166
 eCB actions, 165
 endocannabinoid levels, 134–137
 FAAH, 9–11
 hydrolyze, 165–166
 from lipid precursors, 165
 multienzyme pathways, 6*f*, 7–8
 NAE-hydrolyzing acid amidase, 11
 NAPE-PLD, 5–7, 6*f*
 oxygenation, 21
 phospholipase A2, 165–166
 phospholipase C, 165–166
 phospholipase D, 165–166
 precursor synthesis, 4–5
 protein phosphatase activity, 165–166
 synthesis, 97–98, 165
 uptake, accumulation, and sequestration, 19–21
Nausea. *See* Acute nausea; Anticipatory nausea (AN)
Nausea-induced conditioned gaping
 AM-251, 139–140
 antiemetic treatments, 139
 electromyographic analysis, 139
 endocannabinoid system, 140–142
 LiCl, 139
 somatodendritic 5-HT_{1A} autoreceptors, 140
 taste-reactivity test, 139
Neurocircuitry, 128, 132*f*
Neuroendocrine regulation
 CB_1 receptor, 164
 CB_2 receptor, 164
 depolarization-induced suppression, 166
 energy (*see* Energy Homeostasis)
 reproduction (*see* Reproduction)
 steroids, 166
 stress response, 166–172, 169*f*
Neurotransmission
 enteroendocrine cells, 114
 neuromuscular transmission, 110–112
 secretomotor transmission, 112–114
Nonsteroidal antiinflammatory drugs (NSAIDs), 229

O

Oleoylethanolamide (OEA), 104
Ontogeny
 adolescent CB_1R binding, 66
 adolescent corticolimbic eCB system development, 64, 65*t*
 AEA and 2AG concentration, 66–67
 behavioral profiles, 67
 maturational trajectory, 66–67

structural and functional neurodevelopment, 64
Opiate withdrawal syndrome, 282–283

P

Pain
amygdala, 223*t*, 232–234
brain regions, 209–210, 210*f*
chronic, 205–206
context and modulation, 206–207
definition, 205–206
economic cost of, 205–206
epidemiological studies, 205–206
and modulation, 213, 214*t* (*see also* Supraspinal endocannabinoid mechanisms)
PAG, 220*t*
PFC, 225*t*, 235–236
processing and perception, 206
RVM, 213–227, 218*t*
Palmitoylethanolamide (PEA), 104–105
Peptide endocannabinoids, 102
Paracetamol (acetaminophen), 240
Periaqueductal gray (PAG)
diabetes, 230
disinhibitory mechanism, 228
dlPAG column, 228–229
dmPAG column, 228
eCB system, 213, 226*f*
intra-vlPAG and systemic administration, 228–231
lPAG column, 228–229
NSAID, 229
orexin (OX) A and B, 231
pain and modulation, 220*t*
SIA, 231–232
TRPV1, 229
vlPAG column, 228
Phospholipase C (PLC) hydrolysis, 208–209
Phytocannabinoids, 100–102
Pica, 137–138
Prefrontal cortex (PFC)
CB1 receptors, 236
eCB system, 213, 226*f*
pain and modulation, 225*t*, 235–236
PPARα, 237
SIA, 237
TRPV1 expression, 237
Pubertal maturation, 52–53

R

Reproduction
ghrelin, 185–186
glucocorticoid modulation, 187–189
GnRH neurons, 184–185
maternal behavior, 186–187
oxytocin neurons, 186–187
reproductive behavior and physiological function, 183–184
Reward
food, 264–266
maternal care, 267–269
sexual behavior, 266–267
social interaction, 269–271
Rostral ventromedial medulla (RVM)
eCB system, 213, 226*f*
pain and modulation, 213–227, 218*t*
SIA, 227

S

Sativex, 133–134
Secretomotor transmission, 112–114
Self-control, 259
SIA. *See* Stress-induced analgesia (SIA)
SIH. *See* Stress-induced hyperalgesia (SIH)
Steroids, 166
Stress, HPA axis
acute stress regulation, 168–170, 169*f*
amygdala, 54
brain, 54–55
chronic stress regulation, 55, 170–172
homeostatic functioning, 53–54
infralimbic region, 54
inhibitory endocannabinoid tone, 167–168
negative feedback regulation, 55
neuroendocrine, 53–54
prelimbic region, 54
promoting feedback inhibition, 55
Stress-induced analgesia (SIA)
amygdala, 234–235
PAG, 231–232
PFC, 237
RVM, 227

Stress-induced hyperalgesia (SIH)
amygdala, 235
PAG, 232
PFC, 237
RVM, 227–228
Supraspinal endocannabinoid mechanisms
CB_1 receptors, 238–239
CB_2 receptors, 238
FAAH, 238–239
GPR55, 240
paracetamol (acetaminophen), 240
PPARs, 240–241
TRPV1, 239–240
Synaptic activity
long-term depression, 27
short-term depression, 25–27

T

Δ^9-Tetrahydrocannabinol (THC)
acidic, 139, 150
animal models, 150
discovered and characterization, 60
diuretic agent, 177
high-affinity agonist, 132–133
in humans, 233–234
in managing acute nausea, 146–148
structural analog, 3
suppresses pulsatile LH secretion, 184
synthetic, 128
Thyrotropin-releasing hormone (TRH), 176
Tonic like mechanism, 61–62
Transient receptor potential vanilloid receptor 1 (TRPV1), 102–103

V

Virodhamine, endogenous CB1 antagonist, 103

CONTENTS OF RECENT VOLUMES

Volume 37

Section I: Selectionist Ideas and Neurobiology

Selectionist and Instructionist Ideas in Neuroscience
Olaf Sporns

Population Thinking and Neuronal Selection: Metaphors or Concepts?
Ernst Mayr

Selection and the Origin of Information
Manfred Eigen

Section II: Development and Neuronal Populations

Morphoregulatory Molecules and Selectional Dynamics during Development
Kathryn L. Crossin

Exploration and Selection in the Early Acquisition of Skill
Esther Thelen and Daniela Corbetta

Population Activity in the Control of Movement
Apostolos P. Georgopoulos

Section III: Functional Segregation and Integration in the Brain

Reentry and the Problem of Cortical Integration
Giulio Tononi

Coherence as an Organizing Principle of Cortical Functions
Wolf Singerl

Temporal Mechanisms in Perception
Ernst Pöppel

Section IV: Memory and Models

Selection versus Instruction: Use of Computer Models to Compare Brain Theories
George N. Reeke, Jr.

Memory and Forgetting: Long-Term and Gradual Changes in Memory Storage
Larry R. Squire

Implicit Knowledge: New Perspectives on Unconscious Processes
Daniel L. Schacter

Section V: Psychophysics, Psychoanalysis, and Neuropsychology

Phantom Limbs, Neglect Syndromes, Repressed Memories, and Freudian Psychology
V. S. Ramachandran

Neural Darwinism and a Conceptual Crisis in Psychoanalysis
Arnold H. Modell

A New Vision of the Mind
Oliver Sacks

INDEX

Volume 38

Regulation of $GABA_A$ Receptor Function and Gene Expression in the Central Nervous System
A. Leslie Morrow

Genetics and the Organization of the Basal Ganglia
Robert Hitzemann, Yeang Olan, Stephen Kanes, Katherine Dains, and Barbara Hitzemann

Structure and Pharmacology of Vertebrate $GABA_A$ Receptor Subtypes
Paul J. Whiting, Ruth M. McKernan, and Keith A. Wafford

Neurotransmitter Transporters: Molecular Biology, Function, and Regulation
Beth Borowsky and Beth J. Hoffman

Presynaptic Excitability
Meyer B. Jackson

Monoamine Neurotransmitters in Invertebrates and Vertebrates: An Examination of the Diverse Enzymatic Pathways Utilized to Synthesize and Inactivate Biogenic Amines
B. D. Sloley and A. V. Juorio

Neurotransmitter Systems in Schizophrenia
Gavin P. Reynolds

Physiology of Bergmann Glial Cells
Thomas Müller and Helmut Kettenmann

INDEX

Volume 39

Modulation of Amino Acid-Gated Ion Channels by Protein Phosphorylation
Stephen J. Moss and Trevor G. Smart

Use-Dependent Regulation of $GABA_A$ Receptors
Eugene M. Barnes, Jr.

Synaptic Transmission and Modulation in the Neostriatum
David M. Lovinger and Elizabeth Tyler

The Cytoskeleton and Neurotransmitter Receptors
Valerie J. Whatley and R. Adron Harris

Endogenous Opioid Regulation of Hippocampal Function
Michele L. Simmons and Charles Chavkin

Molecular Neurobiology of the Cannabinoid Receptor
Mary E. Abood and Billy R. Martin

Genetic Models in the Study of Anesthetic Drug Action
Victoria J. Simpson and Thomas E. Johnson

Neurochemical Bases of Locomotion and Ethanol Stimulant Effects
Tamara J. Phillips and Elaine H. Shen

Effects of Ethanol on Ion Channels
Fulton T. Crews, A. Leslie Morrow, Hugh Criswell, and George Breese

INDEX

Volume 40

Mechanisms of Nerve Cell Death: Apoptosis or Necrosis after Cerebral Ischemia
R. M. E. Chalmers-Redman, A. D. Fraser, W. Y. H. Ju, J. Wadia, N. A. Tatton, and W. G. Tatton

Changes in Ionic Fluxes during Cerebral Ischemia
Tibor Kristian and Bo K. Siesjo

Techniques for Examining Neuroprotective Drugs *in Vitro*
A. Richard Green and Alan J. Cross

Techniques for Examining Neuroprotective Drugs *in Vivo*
Mark P. Goldberg, Uta Strasser, and Laura L. Dugan

Calcium Antagonists: Their Role in Neuroprotection
A. Jacqueline Hunter

Sodium and Potassium Channel Modulators: Their Role in Neuroprotection
Tihomir P. Obrenovich

NMDA Antagonists: Their Role in Neuroprotection
Danial L. Small

Development of the NMDA Ion-Channel Blocker, Aptiganel Hydrochloride, as a Neuroprotective Agent for Acute CNS Injury
Robert N. McBurney

The Pharmacology of AMPA Antagonists and Their Role in Neuroprotection
Rammy Gill and David Lodge

GABA and Neuroprotection
Patrick D. Lyden

Adenosine and Neuroprotection
Bertil B. Fredholm

Interleukins and Cerebral Ischemia
Nancy J. Rothwell, Sarah A. Loddick, and Paul Stroemer

Nitrone-Based Free Radical Traps as Neuroprotective Agents in Cerebral Ischemia and Other Pathologies
Kenneth Hensley, John M. Carney, Charles A. Stewart, Tahera Tabatabaie, Quentin Pye, and Robert A. Floyd

Neurotoxic and Neuroprotective Roles of Nitric Oxide in Cerebral Ischemia
Turgay Dalkara and Michael A. Moskowitz

A Review of Earlier Clinical Studies on Neuroprotective Agents and Current Approaches
Nils-Gunnar Wahlgren

INDEX

Volume 41

Section I: Historical Overview

Rediscovery of an Early Concept
Jeremy D. Schmahmann

Section II: Anatomic Substrates

The Cerebrocerebellar System
Jeremy D. Schmahmann and Deepak N. Pandya

Cerebellar Output Channels
Frank A. Middleton and Peter L. Strick

Cerebellar-Hypothalamic Axis: Basic Circuits and Clinical Observations
Duane E. Haines, Espen Dietrichs, Gregory A. Mihailoff, and E. Frank McDonald

Section III. Physiological Observations

Amelioration of Aggression: Response to Selective Cerebellar Lesions in the Rhesus Monkey
Aaron J. Berman

Autonomic and Vasomotor Regulation
Donald J. Reis and Eugene V. Golanov

Associative Learning
Richard F. Thompson, Shaowen Bao, Lu Chen, Benjamin D. Cipriano, Jeffrey S. Grethe, Jeansok J. Kim, Judith K. Thompson, Jo Anne Tracy, Martha S. Weninger, and David J. Krupa

Visuospatial Abilities
Robert Lalonde

Spatial Event Processing
Marco Molinari, Laura Petrosini, and Liliana G. Grammaldo

Section IV: Functional Neuroimaging Studies

Linguistic Processing
Julie A. Fiez and Marcus E. Raichle

Sensory and Cognitive Functions
Lawrence M. Parsons and Peter T. Fox

Skill Learning
Julien Doyon

Section V: Clinical and Neuropsychological Observations

Executive Function and Motor Skill Learning
Mark Hallett and Jordon Grafman

Verbal Fluency and Agrammatism
Marco Molinari, Maria G. Leggio, and Maria C. Silveri

Classical Conditioning
Diana S. Woodruff-Pak

Early Infantile Autism
Margaret L. Bauman, Pauline A. Filipek, and Thomas L. Kemper

Olivopontocerebellar Atrophy and Friedreich's Ataxia: Neuropsychological Consequences of Bilateral versus Unilateral Cerebellar Lesions
Thérèse Botez-Marquard and Mihai I. Botez

Posterior Fossa Syndrome
Ian F. Pollack

Cerebellar Cognitive Affective Syndrome
Jeremy D. Schmahmann and Janet C. Sherman

Inherited Cerebellar Diseases
Claus W. Wallesch and Claudius Bartels

Neuropsychological Abnormalities in Cerebellar Syndromes—Fact or Fiction?
Irene Daum and Hermann Ackermann

Section VI: Theoretical Considerations

Cerebellar Microcomplexes
Masao Ito

Control of Sensory Data Acquisition
James M. Bower

Neural Representations of Moving Systems
Michael Paulin

How Fibers Subserve Computing Capabilities: Similarities between Brains and Machines
Henrietta C. Leiner and Alan L. Leiner

Cerebellar Timing Systems
Richard Ivry

Attention Coordination and Anticipatory Control
Natacha A. Akshoomoff, Eric Courchesne, and Jeanne Townsend

Context-Response Linkage
W. Thomas Thach

Duality of Cerebellar Motor and Cognitive Functions
James R. Bloedel and Vlastislav Bracha

Section VII: Future Directions

Therapeutic and Research Implications
Jeremy D. Schmahmann

Volume 42

Alzheimer Disease
Mark A. Smith

Neurobiology of Stroke
W. Dalton Dietrich

Free Radicals, Calcium, and the Synaptic Plasticity-Cell Death Continuum: Emerging Roles of the Trascription Factor NFκB
Mark P. Mattson

AP-I Transcription Factors: Short- and Long-Term Modulators of Gene Expression in the Brain
Keith Pennypacker

Ion Channels in Epilepsy
Istvan Mody

Posttranslational Regulation of Ionotropic Glutamate Receptors and Synaptic Plasticity
Xiaoning Bi, Steve Standley, and Michel Baudry

Heritable Mutations in the Glycine, $GABA_A$, and Nicotinic Acetylcholine Receptors Provide New Insights into the Ligand-Gated Ion Channel Receptor Superfamily
Behnaz Vafa and Peter R. Schofield

INDEX

Volume 43

Early Development of the *Drosophila* Neuromuscular Junction: A Model for Studying Neuronal Networks in Development
Akira Chiba

Development of Larval Body Wall Muscles
Michael Bate, Matthias Landgraf, and Mar Ruiz Gómez Bate

Development of Electrical Properties and Synaptic Transmission at the Embryonic Neuromuscular Junction
Kendal S. Broadie

Ultrastructural Correlates of Neuromuscular Junction Development
Mary B. Rheuben, Motojiro Yoshihara, and Yoshiaki Kidokoro

Assembly and Maturation of the *Drosophila* Larval Neuromuscular Junction
L. Sian Gramates and Vivian Budnik

Second Messenger Systems Underlying Plasticity at the Neuromuscular Junction
Frances Hannan and Yi Zhong

Mechanisms of Neurotransmitter Release
J. Troy Littleton, Leo Pallanck, and Barry Ganetzky

Vesicle Recycling at the *Drosophila* Neuromuscular Junction
Daniel T. Stimson and Mani Ramaswami

Ionic Currents in Larval Muscles of *Drosophila*
Satpal Singh and Chun-Fang Wu

Development of the Adult Neuromuscular System
Joyce J. Fernandes and Haig Keshishian

Controlling the Motor Neuron
James R. Trimarchi, Ping Jin, and Rodney K. Murphey

Volume 44

Human Ego-Motion Perception
A. V. van den Berg

Optic Flow and Eye Movements
M. Lappe and K.-P. Hoffman

The Role of MST Neurons during Ocular Tracking in 3D Space
K. Kawano, U. Inoue, A. Takemura, Y. Kodaka, and F. A. Miles

Visual Navigation in Flying Insects
M. V. Srinivasan and S.-W. Zhang

Neuronal Matched Filters for Optic Flow Processing in Flying Insects
H. G. Krapp

A Common Frame of Reference for the Analysis of Optic Flow and Vestibular Information
B. J. Frost and D. R. W. Wylie

Optic Flow and the Visual Guidance of Locomotion in the Cat
H. Sherk and G. A. Fowler

Stages of Self-Motion Processing in Primate Posterior Parietal Cortex
F. Bremmer, J.-R. Duhamel, S. B. Hamed, and W. Graf

Optic Flow Analysis for Self-Movement Perception
C. J. Duffy

Neural Mechanisms for Self-Motion Perception in Area MST
R. A. Andersen, K. V. Shenoy, J. A. Crowell, and D. C. Bradley

Computational Mechanisms for Optic Flow Analysis in Primate Cortex
M. Lappe

Human Cortical Areas Underlying the Perception of Optic Flow: Brain Imaging Studies
M. W. Greenlee

What Neurological Patients Tell Us about the Use of Optic Flow
L. M. Vaina and S. K. Rushton

INDEX

Volume 45

Mechanisms of Brain Plasticity: From Normal Brain Function to Pathology
Philip. A. Schwartzkroin

Brain Development and Generation of Brain Pathologies
Gregory L. Holmes and Bridget McCabe

Maturation of Channels and Receptors: Consequences for Excitability
David F. Owens and Arnold R. Kriegstein

Neuronal Activity and the Establishment of Normal and Epileptic Circuits during Brain Development
John W. Swann, Karen L. Smith, and Chong L. Lee

The Effects of Seizures of the Hippocampus of the Immature Brain
Ellen F. Sperber and Solomon L. Moshe

Abnormal Development and Catastrophic Epilepsies: The Clinical Picture and Relation to Neuroimaging
Harry T. Chugani and Diane C. Chugani

Cortical Reorganization and Seizure Generation in Dysplastic Cortex
G. Avanzini, R. Preafico, S. Franceschetti, G. Sancini, G. Battaglia, and V. Scaioli

Rasmussen's Syndrome with Particular Reference to Cerebral Plasticity: A Tribute to Frank Morrell
Fredrick Andermann and Yuonne Hart

Structural Reorganization of Hippocampal Networks Caused by Seizure Activity
Daniel H. Lowenstein

Epilepsy-Associated Plasticity in gamma-Amniobutyric Acid Receptor Expression, Function and Inhibitory Synaptic Properties
Douglas A. Coulter

Synaptic Plasticity and Secondary Epileptogenesis
Timothy J. Teyler, Steven L. Morgan, Rebecca N. Russell, and Brian L. Woodside

Synaptic Plasticity in Epileptogenesis: Cellular Mechanisms Underlying Long-Lasting Synaptic Modifications that Require New Gene Expression
Oswald Steward, Christopher S. Wallace, and Paul F. Worley

Cellular Correlates of Behavior
Emma R. Wood, Paul A. Dudchenko, and Howard Eichenbaum

Mechanisms of Neuronal Conditioning
David A. T. King, David J. Krupa, Michael R. Foy, and Richard F. Thompson

Plasticity in the Aging Central Nervous System
C. A. Barnes

Secondary Epileptogenesis, Kindling, and Intractable Epilepsy: A Reappraisal from the Perspective of Neuronal Plasticity
Thomas P. Sutula

Kindling and the Mirror Focus
Dan C. McIntyre and Michael O. Poulter

Partial Kindling and Behavioral Pathologies
Robert E. Adamec

The Mirror Focus and Secondary Epileptogenesis
B. J. Wilder

Hippocampal Lesions in Epilepsy: A Historical Review
Robert Naquet

Clinical Evidence for Secondary Epileptogensis
Hans O. Luders

Epilepsy as a Progressive (or Nonprogressive "Benign") Disorder
John A. Wada

Pathophysiological Aspects of Landau-Kleffner Syndrome: From the Active Epileptic Phase to Recovery
Marie-Noelle Metz-Lutz, Pierre Maquet, Annd De Saint Martin, Gabrielle Rudolf, Norma Wioland, Edouard Hirsch, and Chriatian Marescaux

Local Pathways of Seizure Propagation in Neocortex
Barry W. Connors, David J. Pinto, and Albert E. Telefeian

Multiple Subpial Transection: A Clinical Assessment
C. E. Polkey

The Legacy of Frank Morrell
Jerome Engel, Jr.

Volume 46

Neurosteroids: Beginning of the Story
Etienne E. Baulieu, P. Robel, and M. Schumacher

Biosynthesis of Neurosteroids and Regulation of Their Synthesis
Synthia H. Mellon and Hubert Vaudry

Neurosteroid 7-Hydroxylation Products in the Brain
Robert Morfin and Luboslav Stárka

Neurosteroid Analysis
Ahmed A. Alomary, Robert L. Fitzgerald, and Robert H. Purdy

Role of the Peripheral-Type Benzodiazepine Receptor in Adrenal and Brain Steroidogenesis
Rachel C. Brown and Vassilios Papadopoulos

Formation and Effects of Neuroactive Steroids in the Central and Peripheral Nervous System
Roberto Cosimo Melcangi, Valerio Magnaghi, Mariarita Galbiati, and Luciano Martini

Neurosteroid Modulation of Recombinant and Synaptic $GABA_A$ Receptors
Jeremy J. Lambert, Sarah C. Harney, Delia Belelli, and John A. Peters

$GABA_A$-Receptor Plasticity during Long-Term Exposure to and Withdrawal from Progesterone
Giovanni Biggio, Paolo Follesa, Enrico Sanna, Robert H. Purdy, and Alessandra Concas

Stress and Neuroactive Steroids
Maria Luisa Barbaccia, Mariangela Serra, Robert H. Purdy, and Giovanni Biggio

Neurosteroids in Learning and Memory Processes
Monique Vallée, Willy Mayo, George F. Koob, and Michel Le Moal

Neurosteroids and Behavior
Sharon R. Engel and Kathleen A. Grant

Ethanol and Neurosteroid Interactions in the Brain
A. Leslie Morrow, Margaret J. VanDoren, Rebekah Fleming, and Shannon Penland

Preclinical Development of Neurosteroids as Neuroprotective Agents for the Treatment of Neurodegenerative Diseases
Paul A. Lapchak and Dalia M. Araujo

Clinical Implications of Circulating Neurosteroids
Andrea R. Genazzani, Patrizia Monteleone, Massimo Stomati, Francesca Bernardi, Luigi Cobellis, Elena Casarosa, Michele Luisi, Stefano Luisi, and Felice Petraglia

Neuroactive Steroids and Central Nervous System Disorders
Mingde Wang, Torbjörn Bäckström, Inger Sundström, Göran Wahlström, Tommy Olsson, Di Zhu, Inga-Maj Johansson, Inger Björn, and Marie Bixo

Neuroactive Steroids in Neuropsychopharmacology
Rainer Rupprecht and Florian Holsboer

Current Perspectives on the Role of Neurosteroids in PMS and Depression
Lisa D. Griffin, Susan C. Conrad, and Synthia H. Mellon

Index

Volume 47

Introduction: Studying Gene Expression in Neural Tissues by *in Situ* Hybridization
W. Wisden and B. J. Morris

Part I: *In Situ* Hybridization with Radiolabelled Oligonucleotides
In Situ Hybridization with Oligonucleotide Probes
Wl. Wisden and B. J. Morris

Cryostat Sectioning of Brains
Victoria Revilla and Alison Jones

Processing Rodent Embryonic and Early Postnatal Tissue for *in Situ* Hybridization with Radiolabelled Oligonucleotides
David J. Laurie, Petra C. U. Schrotz, Hannah Monyer, and Ulla Amtmann

Processing of Retinal Tissue for *in Situ* Hybridization
Frank Müller

Processing the Spinal Cord for *in Situ* Hybridization with Radiolabelled Oligonucleotides
A. Berthele and T. R. Tölle

Processing Human Brain Tissue for *in Situ* Hybridization with Radiolabelled Oligonucleotides
Louise F. B. Nicholson

In Situ Hybridization of Astrocytes and Neurons Cultured *in Vitro*
L. A. Arizza-McNaughton, C. De Felipe, and S. P. Hunt

In Situ Hybridization on Organotypic Slice Cultures
A. Gerfin-Moser and H. Monyer

Quantitative Analysis of *in Situ* Hybridization Histochemistry
Andrew L. Gundlach and Ross D. O'Shea

Part II: Nonradioactive *in Situ* hybridization

Nonradioactive *in Situ* Hybridization Using Alkaline Phosphatase-Labelled Oligonucleotides
S. J. Augood, E. M. McGowan, B. R. Finsen, B. Heppelmann, and P. C. Emson

Combining Nonradioactive *in Situ* Hybridization with Immunohistological and Anatomical Techniques
Petra Wahle

Nonradioactive *in Situ* Hybridization: Simplified Procedures for Use in Whole Mounts of Mouse and Chick Embryos
Linda Ariza-McNaughton and Robb Krumlauf

INDEX

Volume 48

Assembly and Intracellular Trafficking of $GABA_A$ Receptors Eugene
Barnes

Subcellular Localization and Regulation of $GABA_A$ Receptors and Associated Proteins Bernhard Lüscher and Jean-Marc Fritschy D_1 Dopamine Receptors
Richard Mailman

Molecular Modeling of Ligand-Gated Ion Channels: Progress and Challenges
Ed Bertaccini and James R. Trudel

Alzheimer's Disease: Its Diagnosis and Pathogenesis
Jillian J. Kril and Glenda M. Halliday

DNA Arrays and Functional Genomics in Neurobiology
Christelle Thibault, Long Wang, Li Zhang, and Michael F. Miles

INDEX

Volume 49

What Is West Syndrome?
Olivier Dulac, Christine Soufflet, Catherine Chiron, and Anna Kaminski

The Relationship between encephalopathy and Abnormal Neuronal Activity in the Developing Brain
Frances E. Jensen

Hypotheses from Functional Neuroimaging Studies
Csaba Juhász, Harry T. Chugani, Ouo Muzik, and Diane C. Chugani

Infantile Spasms: Unique Sydrome or General Age-Dependent Manifestation of a Diffuse Encephalopathy?
M. A. Koehn and M. Duchowny

Histopathology of Brain Tissue from Patients with Infantile Spasms
Harry V. Vinters

Generators of Ictal and Interictal Electroencephalograms Associated with Infantile Spasms: Intracellular Studies of Cortical and Thalamic Neurons
M. Steriade and I. Timofeev

Cortical and Subcortical Generators of Normal and Abnormal Rhythmicity
David A. McCormick

Role of Subcortical Structures in the Pathogenesis of Infantile Spasms: What Are Possible Subcortical Mediators?
F. A. Lado and S. L. Moshé

What Must We Know to Develop Better Therapies?
Jean Aicardi

The Treatment of Infantile Spasms: An Evidence-Based Approach
Mark Mackay, Shelly Weiss, and O. Carter Snead III

ACTH Treatment of Infantile Spasms: Mechanisms of Its Effects in Modulation of Neuronal Excitability
K. L. Brunson, S. Avishai-Eliner, and T. Z. Baram

Neurosteroids and Infantile Spasms: The Deoxycorticosterone Hypothesis
Michael A. Rogawski and Doodipala S. Reddy

Are there Specific Anatomical and/or Transmitter Systems (Cortical or Subcortical) That Should Be Targeted?
Phillip C. Jobe

Medical versus Surgical Treatment: Which Treatment When
W. Donald Shields

Developmental Outcome with and without Successful Intervention
Rochelle Caplan, Prabha Siddarth, Gary Mathern, Harry Vinters, Susan Curtiss, Jennifer Levitt, Robert Asarnow, and W. Donald Shields

Infantile Spasms versus Myoclonus: Is There a Connection?
Michael R. Pranzatelli

Tuberous Sclerosis as an Underlying Basis for Infantile Spasm
Raymond S. Yeung

Brain Malformation, Epilepsy, and Infantile Spasms
M. Elizabeth Ross

Brain Maturational Aspects Relevant to Pathophysiology of Infantile Spasms
G. Auanzini, F. Panzica, and S. Franceschetti

Gene Expression Analysis as a Strategy to Understand the Molecular Pathogenesis of Infantile Spasms
Peter B. Crino

Infantile Spasms: Criteria for an Animal Model
Carl E. Stafstrom and Gregory L. Holmes

INDEX

Volume 50

Part I: Primary Mechanisms

How Does Glucose Generate Oxidative Stress In Peripheral Nerve?
Irina G. Obrosova

Glycation in Diabetic Neuropathy: Characteristics, Consequences, Causes, and Therapeutic Options
Paul J. Thornalley

Part II: Secondary Changes

Protein Kinase C Changes in Diabetes: Is the Concept Relevant to Neuropathy?
Joseph Eichberg

Are Mitogen-Activated Protein Kinases Glucose Transducers for Diabetic Neuropathies?
Tertia D. Purves and David R. Tomlinson

Neurofilaments in Diabetic Neuropathy
Paul Fernyhough and Robert E. Schmidt

Apoptosis in Diabetic Neuropathy
Aviva Tolkovsky

Nerve and Ganglion Blood Flow in Diabetes: An Appraisal
Douglas W. Zochodne

Part III: Manifestations

Potential Mechanisms of Neuropathic Pain in Diabetes
Nigel A. Calcutt

Electrophysiologic Measures of Diabetic Neuropathy: Mechanism and Meaning
Joseph C. Arezzo and Elena Zotova

Neuropathology and Pathogenesis of Diabetic Autonomic Neuropathy
Robert E. Schmidt

Role of the Schwann Cell in Diabetic Neuropathy
Luke Eckersley

Part IV: Potential Treatment

Polyol Pathway and Diabetic Peripheral Neuropathy
Peter J. Oates

Nerve Growth Factor for the Treatment of Diabetic Neuropathy: What Went Wrong, What Went Right, and What Does the Future Hold?
Stuart C. Apfel

Angiotensin-Converting Enzyme Inhibitors: Are there Credible Mechanisms for Beneficial Effects in Diabetic Neuropathy?
Rayaz A. Malik and David R. Tomlinson

Clinical Trials for Drugs Against Diabetic Neuropathy: Can We Combine Scientific Needs With Clinical Practicalities?
Dan Ziegler and Dieter Luft

INDEX

Volume 51

Energy Metabolism in the Brain
Leif Hertz and Gerald A. Dienel

The Cerebral Glucose-Fatty Acid Cycle: Evolutionary Roots, Regulation, and (Patho) physiological Importance
Kurt Heininger

Expression, Regulation, and Functional Role of Glucose Transporters (GLUTs) in Brain
Donard S. Dwyer, Susan J. Vannucci, and Ian A. Simpson

Insulin-Like Growth Factor-1 Promotes Neuronal Glucose Utilization During Brain Development and Repair Processes
Carolyn A. Bondy and Clara M. Cheng

CNS Sensing and Regulation of Peripheral Glucose Levels
Barry E. Levin, Ambrose A. Dunn-Meynell, and Vanessa H. Routh

Glucose Transporter Protein Syndromes
Darryl C. De Vivo, Dong Wang, Juan M. Pascual, and Yuan Yuan Ho

Glucose, Stress, and Hippocampal Neuronal Vulnerability
Lawrence P. Reagan

Glucose/Mitochondria in Neurological Conditions
John P. Blass

Energy Utilization in the Ischemic/Reperfused Brain
John W. Phillis and Michael H. O'Regan

Diabetes Mellitus and the Central Nervous System
Anthony L. McCall

Diabetes, the Brain, and Behavior: Is There a Biological Mechanism Underlying the Association between Diabetes and Depression?
A. M. Jacobson, J. A. Samson, K. Weinger, and C. M. Ryan

Schizophrenia and Diabetes
David C. Henderson and Elissa R. Ettinger

Psychoactive Drugs Affect Glucose Transport and the Regulation of Glucose Metabolism
Donard S. Dwyer, Timothy D. Ardizzone, and Ronald J. Bradley

INDEX

Volume 52

Neuroimmune Relationships in Perspective
Frank Hucklebridge and Angela Clow

Sympathetic Nervous System Interaction with the Immune System
Virginia M. Sanders and Adam P. Kohm

Mechanisms by Which Cytokines Signal the Brain
Adrian J. Dunn

Neuropeptides: Modulators of Immune Responses in Health and Disease
David S. Jessop

Brain–Immune Interactions in Sleep
Lisa Marshall and Jan Born

Neuroendocrinology of Autoimmunity
Michael Harbuz

Systemic Stress-Induced Th2 Shift and Its Clinical Implications
Ibia J. Elenkov

Neural Control of Salivary S-IgA Secretion
Gordon B. Proctor and Guy H. Carpenter

Stress and Secretory Immunity
Jos A. Bosch, Christopher Ring, Eco J. C. de Geus, Enno C. I. Veerman, and Arie V. Nieuw Amerongen

Cytokines and Depression
Angela Clow

Immunity and Schizophrenia: Autoimmunity, Cytokines, and Immune Responses
Fiona Gaughran

Cerebral Lateralization and the Immune System
Pierre J. Neveu

Behavioral Conditioning of the Immune System
Frank Hucklebridge

Psychological and Neuroendocrine Correlates of Disease Progression
Julie M. Turner-Cobb

The Role of Psychological Intervention in Modulating Aspects of Immune Function in Relation to Health and Well-Being
J. H. Gruzelier

INDEX

Volume 53

Section I: Mitochondrial Structure and Function

Mitochondrial DNA Structure and Function
Carlos T. Moraes, Sarika Srivastava, Ilias Kirkinezos, Jose Oca-Cossio, Corina van Waveren, Markus Woischnick, and Francisca Diaz

Oxidative Phosphorylation: Structure, Function, and Intermediary Metabolism
Simon J. R. Heales, Matthew E. Gegg, and John B. Clark

Import of Mitochondrial Proteins
Matthias F. Bauer, Sabine Hofmann, and Walter Neupert

Section II: Primary Respiratory Chain Disorders

Mitochondrial Disorders of the Nervous System: Clinical, Biochemical, and Molecular Genetic Features
Dominic Thyagarajan and Edward Byrne

Section III: Secondary Respiratory Chain Disorders

Friedreich's Ataxia
J. M. Cooper and J. L. Bradley

Wilson Disease
C. A. Davie and A. H. V. Schapira

Hereditary Spastic Paraplegia
Christopher J. McDermott and Pamela J. Shaw

Cytochrome c Oxidase Deficiency
Giacomo P. Comi, Sandra Strazzer, Sara Galbiati, and Nereo Bresolin

Section IV: Toxin Induced Mitochondrial Dysfunction

Toxin-Induced Mitochondrial Dysfunction
Susan E. Browne and M. Flint Beal

Section V: Neurodegenerative Disorders

Parkinson's Disease
L. V. P. Korlipara and A. H. V. Schapira

Huntington's Disease: The Mystery Unfolds?
Åsa Petersén and Patrik Brundin

Mitochondria in Alzheimer's Disease
Russell H. Swerdlow and Stephen J. Kish

Contributions of Mitochondrial Alterations, Resulting from Bad Genes and a Hostile Environment, to the Pathogenesis of Alzheimer's Disease
Mark P. Mattson

Mitochondria and Amyotrophic Lateral Sclerosis
Richard W. Orrell and Anthony H. V. Schapira

Section VI: Models of Mitochondrial Disease

Models of Mitochondrial Disease
Danae Liolitsa and Michael G. Hanna

Section VII: Defects of β Oxidation Including Carnitine Deficiency

Defects of β Oxidation Including Carnitine Deficiency
K. Bartlett and M. Pourfarzam

Section VIII: Mitochondrial Involvement in Aging

The Mitochondrial Theory of Aging: Involvement of Mitochondrial DNA Damage and Repair
Nadja C. de Souza-Pinto and Vilhelm A. Bohr

INDEX

Volume 54

Unique General Anesthetic Binding Sites Within Distinct Conformational States of the Nicotinic Acetylcholine Receptor
Hugo R. Ariaas, William, R. Kem, James R. Truddell, and Michael P. Blanton

Signaling Molecules and Receptor Transduction Cascades That Regulate NMDA Receptor-Mediated Synaptic Transmission
Suhas. A. Kotecha and John F. MacDonald

Behavioral Measures of Alcohol Self-Administration and Intake Control: Rodent Models
Herman H. Samson and Cristine L. Czachowski

Dopaminergic Mouse Mutants: Investigating the Roles of the Different Dopamine Receptor Subtypes and the Dopamine Transporter
Shirlee Tan, Bettina Hermann, and Emiliana Borrelli

Drosophila melanogaster, A Genetic Model System for Alcohol Research
Douglas J. Guarnieri and Ulrike Heberlein

INDEX

Volume 55

Section I: Virsu Vectors For Use in the Nervous System

Non-Neurotropic Adenovirus: a Vector for Gene Transfer to the Brain and Gene Therapy of Neurological Disorders
P. R. Lowenstein, D. Suwelack, J. Hu, X. Yuan, M. Jimenez-Dalmaroni, S. Goverdhama, and M.G. Castro

Adeno-Associated Virus Vectors
E. Lehtonen and L. Tenenbaum

Problems in the Use of Herpes Simplex Virus as a Vector
L. T. Feldman

Lentiviral Vectors
J. Jakobsson, C. Ericson, N. Rosenquist, and C. Lundberg

Retroviral Vectors for Gene Delivery to Neural Precursor Cells
K. Kageyama, H. Hirata, and J. Hatakeyama

Section II: Gene Therapy with Virus Vectors for Specific Disease of the Nervous System

The Principles of Molecular Therapies for Glioblastoma
G. Karpati and J. Nalbatonglu

Oncolytic Herpes Simplex Virus
J. C. C. Hu and R. S. Coffin

Recombinant Retrovirus Vectors for Treatment of Brain Tumors
N. G. Rainov and C. M. Kramm

Adeno-Associated Viral Vectors for Parkinson's Disease
I. Muramatsu, L. Wang, K. Ikeguchi, K-i Fujimoto, T. Okada, H. Mizukami, Y. Hanazono, A. Kume, I. Nakano, and K. Ozawa

HSV Vectors for Parkinson's Disease
D. S. Latchman

Gene Therapy for Stroke
K. Abe and W. R. Zhang

Gene Therapy for Mucopolysaccharidosis
A. Bosch and J. M. Heard

INDEX

Volume 56

Behavioral Mechanisms and the Neurobiology of Conditioned Sexual Responding
Mark Krause

NMDA Receptors in Alcoholism
Paula L. Hoffman

Processing and Representation of Species-Specific Communication Calls in the Auditory System of Bats
George D. Pollak, Achim Klug, and Eric E. Bauer

Central Nervous System Control of Micturition
Gert Holstege and Leonora J. Mouton

The Structure and Physiology of the Rat Auditory System: An Overview
Manuel Malmierca

Neurobiology of Cat and Human Sexual Behavior
Gert Holstege and J. R. Georgiadis

INDEX

Volume 57

Cumulative Subject Index of Volumes 1–25

Volume 58

Cumulative Subject Index of Volumes 26–50

Volume 59

Loss of Spines and Neuropil
Liesl B. Jones

Schizophrenia as a Disorder of Neuroplasticity
Robert E. McCullumsmith, Sarah M. Clinton, and James H. Meador-Woodruff

The Synaptic Pathology of Schizophrenia: Is Aberrant Neurodevelopment and Plasticity to Blame?
Sharon L. Eastwood

Neurochemical Basis for an Epigenetic Vision of Synaptic Organization
E. Costa, D. R. Grayson, M. Veldic, and A. Guidotti

Muscarinic Receptors in Schizophrenia: Is There a Role for Synaptic Plasticity?
Thomas J. Raedler

Serotonin and Brain Development
Monsheel S. K. Sodhi and Elaine Sanders-Bush

Presynaptic Proteins and Schizophrenia
William G. Honer and Clint E. Young

Mitogen-Activated Protein Kinase Signaling
Svetlana V. Kyosseva

Postsynaptic Density Scaffolding Proteins at Excitatory Synapse and Disorders of Synaptic Plasticity: Implications for Human Behavior Pathologies
Andrea de Bartolomeis and Germano Fiore

Prostaglandin-Mediated Signaling in Schizophrenia
S. Smesny

Mitochondria, Synaptic Plasticity, and Schizophrenia
Dorit Ben-Shachar and Daphna Laifenfeld

Membrane Phospholipids and Cytokine Interaction in Schizophrenia
Jeffrey K. Yao and Daniel P. van Kammen

Neurotensin, Schizophrenia, and Antipsychotic Drug Action
Becky Kinkead and Charles B. Nemeroff

Schizophrenia, Vitamin D, and Brain Development
Alan Mackay-Sim, François FÉron, Darryl Eyles, Thomas Burne, and John McGrath

Possible Contributions of Myelin and Oligodendrocyte Dysfunction to Schizophrenia
Daniel G. Stewart and Kenneth L. Davis

Brain-Derived Neurotrophic Factor and the Plasticity of the Mesolimbic Dopamine Pathway
Oliver Guillin, Nathalie Griffon, Jorge Diaz, Bernard Le Foll, Erwan Bezard, Christian Gross, Chris Lammers, Holger Stark, Patrick Carroll, Jean-Charles Schwartz, and Pierre Sokoloff

S100B in Schizophrenic Psychosis
Matthias Rothermundt, Gerald Ponath, and Volker Arolt

Oct-6 Transcription Factor
Maria Ilia

NMDA Receptor Function, Neuroplasticity, and the Pathophysiology of Schizophrenia
Joseph T. Coyle and Guochuan Tsai

INDEX

Volume 60

Microarray Platforms: Introduction and Application to Neurobiology
Stanislav L. Karsten, Lili C. Kudo, and Daniel H. Geschwind

Experimental Design and Low-Level Analysis of Microarray Data
B. M. Bolstad, F. Collin, K. M. Simpson, R. A. Irizarry, and T. P. Speed

Brain Gene Expression: Genomics and Genetics
Elissa J. Chesler and Robert W. Williams

DNA Microarrays and Animal Models of Learning and Memory
Sebastiano Cavallaro

Microarray Analysis of Human Nervous System Gene Expression in Neurological Disease
Steven A. Greenberg

DNA Microarray Analysis of Postmortem Brain Tissue
Károly Mirnics, Pat Levitt, and David A. Lewis

INDEX

Volume 61

Section I: High-Throughput Technologies

Biomarker Discovery Using Molecular Profiling Approaches
Stephen J. Walker and Arron Xu

Proteomic Analysis of Mitochondrial Proteins
Mary F. Lopez, Simon Melov, Felicity Johnson, Nicole Nagulko, Eva Golenko, Scott Kuzdzal, Suzanne Ackloo, and Alvydas Mikulskis

Section II: Proteomic Applications

NMDA Receptors, Neural Pathways, and Protein Interaction Databases
Holger Husi

Dopamine Transporter Network and Pathways
Rajani Maiya and R. Dayne Mayfield

Proteomic Approaches in Drug Discovery and Development
Holly D. Soares, Stephen A. Williams, Peter J. Snyder, Feng Gao, Tom Stiger, Christian Rohlff, Athula Herath, Trey Sunderland, Karen Putnam, and W. Frost White

Section III: Informatics

Proteomic Informatics
Steven Russell, William Old, Katheryn Resing, and Lawrence Hunter

Section IV: Changes in the Proteome by Disease

Proteomics Analysis in Alzheimer's Disease: New Insights into Mechanisms of Neurodegeneration
D. Allan Butterfield and Debra Boyd-Kimball

Proteomics and Alcoholism
Frank A. Witzmann and Wendy N. Strother

Proteomics Studies of Traumatic Brain Injury
Kevin K. W. Wang, Andrew Ottens, William Haskins, Ming Cheng Liu, Firas Kobeissy, Nancy Denslow, SuShing Chen, and Ronald L. Hayes

Influence of Huntington's Disease on the Human and Mouse Proteome
Claus Zabel and Joachim Klose

Section V: Overview of the Neuroproteome

Proteomics—Application to the Brain
Katrin Marcus, Oliver Schmidt, Heike Schaefer, Michael Hamacher, AndrÅ van Hall, and Helmut E. Meyer

INDEX

Volume 62

$GABA_A$ Receptor Structure–Function Studies: A Reexamination in Light of New Acetylcholine Receptor Structures
Myles H. Akabas

Dopamine Mechanisms and Cocaine Reward
Aiko Ikegami and Christine L. Duvauchelle

Proteolytic Dysfunction in Neurodegenerative Disorders
Kevin St. P. McNaught

Neuroimaging Studies in Bipolar Children and Adolescents
Rene L. Olvera, David C. Glahn, Sheila C. Caetano, Steven R. Pliszka, and Jair C. Soares

Chemosensory G-Protein-Coupled Receptor Signaling in the Brain
Geoffrey E. Woodard

Disturbances of Emotion Regulation after Focal Brain Lesions
Antoine Bechara

The Use of *Caenorhabditis elegans* in Molecular Neuropharmacology
Jill C. Bettinger, Lucinda Carnell, Andrew G. Davies, and Steven L. McIntire

INDEX

Volume 63

Mapping Neuroreceptors at work: On the Definition and Interpretation of Binding Potentials after 20 years of Progress
Albert Gjedde, Dean F. Wong, Pedro Rosa-Neto, and Paul Cumming

Mitochondrial Dysfunction in Bipolar Disorder: From ^{31}P-Magnetic Resonance Spectroscopic Findings to Their Molecular Mechanisms
Tadafumi Kato

Large-Scale Microarray Studies of Gene Expression in Multiple Regions of the Brain in Schizophrenia and Alzeimer's Disease
Pavel L. Katsel, Kenneth L. Davis, and Vahram Haroutunian

Regulation of Serotonin 2C Receptor PRE-mRNA Editing By Serotonin
Claudia Schmauss

The Dopamine Hypothesis of Drug Addiction: Hypodopaminergic State
Miriam Melis, Saturnino Spiga, and Marco Diana

Human and Animal Spongiform Encephalopathies are Autoimmune Diseases: A Novel Theory and Its supporting Evidence
Bao Ting Zhu

Adenosine and Brain Function
Bertil B. Fredholm, Jiang-Fan Chen, Rodrigo A. Cunha, Per Svenningsson, and Jean-Marie Vaugeois

INDEX

Volume 64

Section I. The Cholinergic System
John Smythies

Section II. The Dopamine System
John Symythies

Section III. The Norepinephrine System
John Smythies

Section IV. The Adrenaline System
John Smythies

Section V. Serotonin System
John Smythies

INDEX

Volume 65

Insulin Resistance: Causes and Consequences
Zachary T. Bloomgarden

Antidepressant-Induced Manic Conversion: A Developmentally Informed Synthesis of the Literature
Christine J. Lim, James F. Leckman, Christopher Young, and AndrÉs Martin

Sites of Alcohol and Volatile Anesthetic Action on Glycine Receptors
Ingrid A. Lobo and R. Adron Harris

Role of the Orbitofrontal Cortex in Reinforcement Processing and Inhibitory Control: Evidence from Functional Magnetic Resonance Imaging Studies in Healthy Human Subjects
Rebecca Elliott and Bill Deakin

Common Substrates of Dysphoria in Stimulant Drug Abuse and Primary Depression: Therapeutic Targets
Kate Baicy, Carrie E. Bearden, John Monterosso, Arthur L. Brody, Andrew J. Isaacson, and Edythe D. London

The Role of cAMP Response Element–Binding Proteins in Mediating Stress-Induced Vulnerability to Drug Abuse
Arati Sadalge Kreibich and Julie A. Blendy

G-Protein–Coupled Receptor Deorphanizations
Yumiko Saito and Olivier Civelli

Mechanistic Connections Between Glucose/ Lipid Disturbances and Weight Gain Induced by Antipsychotic Drugs
Donard S. Dwyer, Dallas Donohoe, Xiao-Hong Lu, and Eric J. Aamodt

Serotonin Firing Activity as a Marker for Mood Disorders: Lessons from Knockout Mice
Gabriella Gobbi

INDEX

Volume 66

Brain Atlases of Normal and Diseased Populations
Arthur W. Toga and Paul M. Thompson

Neuroimaging Databases as a Resource for Scientific Discovery
John Darrell Van Horn, John Wolfe, Autumn Agnoli, Jeffrey Woodward, Michael Schmitt, James Dobson, Sarene Schumacher, and Bennet Vance

Modeling Brain Responses
Karl J. Friston, William Penny, and Olivier David

Voxel-Based Morphometric Analysis Using Shape Transformations
Christos Davatzikos

The Cutting Edge of fMRI and High-Field fMRI
Dae-Shik Kim

Quantification of White Matter Using Diffusion-Tensor Imaging
Hae-Jeong Park

Perfusion fMRI for Functional Neuroimaging
Geoffrey K. Aguirre, John A. Detre, and Jiongjiong Wang

Functional Near-Infrared Spectroscopy: Potential and Limitations in Neuroimaging Studies
Yoko Hoshi

Neural Modeling and Functional Brain Imaging: The Interplay Between the Data-Fitting and Simulation Approaches
Barry Horwitz and Michael F. Glabus

Combined EEG and fMRI Studies of Human Brain Function
V. Menon and S. Crottaz-Herbette

INDEX

Volume 67

Distinguishing Neural Substrates of Heterogeneity Among Anxiety Disorders
Jack B. Nitschke and Wendy Heller

Neuroimaging in Dementia
K. P. Ebmeier, C. Donaghey, and N. J. Dougall

Prefrontal and Anterior Cingulate Contributions to Volition in Depression
Jack B. Nitschke and Kristen L. Mackiewicz

Functional Imaging Research in Schizophrenia
H. Tost, G. Ende, M. Ruf, F. A. Henn, and A. Meyer-Lindenberg

Neuroimaging in Functional Somatic Syndromes
Patrick B. Wood

Neuroimaging in Multiple Sclerosis
Alireza Minagar, Eduardo Gonzalez-Toledo, James Pinkston, and Stephen L. Jaffe

Stroke
Roger E. Kelley and Eduardo Gonzalez-Toledo

Functional MRI in Pediatric Neurobehavioral Disorders
Michael Seyffert and F. Xavier Castellanos

Structural MRI and Brain Development
Paul M. Thompson, Elizabeth R. Sowell, Nitin Gogtay, Jay N. Giedd, Christine N. Vidal, Kiralee M. Hayashi, Alex Leow, Rob Nicolson, Judith L. Rapoport, and Arthur W. Toga

Neuroimaging and Human Genetics
Georg Winterer, Ahmad R. Hariri, David Goldman, and Daniel R. Weinberger

Neuroreceptor Imaging in Psychiatry: Theory and Applications
W. Gordon Frankle, Mark Slifstein, Peter S. Talbot, and Marc Laruelle

INDEX

Volume 68

Fetal Magnetoencephalography: Viewing the Developing Brain *In Utero*
Hubert Preissl, Curtis L. Lowery, and Hari Eswaran

Magnetoencephalography in Studies of Infants and Children
Minna Huotilainen

Let's Talk Together: Memory Traces Revealed by Cooperative Activation in the Cerebral Cortex
Jochen Kaiser, Susanne Leiberg, and Werner Lutzenberger

Human Communication Investigated With Magnetoencephalography: Speech, Music, and Gestures
Thomas R. Knösche, Burkhard Maess, Akinori Nakamura, and Angela D. Friederici

Combining Magnetoencephalography and Functional Magnetic Resonance Imaging
Klaus Mathiak and Andreas J. Fallgatter

Beamformer Analysis of MEG Data
Arjan Hillebrand and Gareth R. Barnes

Functional Connectivity Analysis in Magnetoencephalography
Alfons Schnitzler and Joachim Gross

Human Visual Processing as Revealed by Magnetoencephalographys
Yoshiki Kaneoke, Shoko Watanabe, and Ryusuke Kakigi

A Review of Clinical Applications of Magnetoencephalography
Andrew C. Papanicolaou, Eduardo M. Castillo, Rebecca Billingsley-Marshall, Ekaterina Pataraia, and Panagiotis G. Simos

INDEX

Volume 69

Nematode Neurons: Anatomy and Anatomical Methods in Caenorhabditis elegans
David H. Hall, Robyn Lints, and Zeynep Altun

Investigations of Learning and Memory in Caenorhabditis elegans
Andrew C. Giles, Jacqueline K. Rose, and Catharine H. Rankin

Neural Specification and Differentiation
Eric Aamodt and Stephanie Aamodt

Sexual Behavior of the Caenorhabditis elegans Male
Scott W. Emmons

The Motor Circuit
Stephen E. Von Stetina, Millet Treinin, and David M. Miller III

Mechanosensation in Caenorhabditis elegans
Robert O'Hagan and Martin Chalfie

Volume 70

Spectral Processing by the Peripheral Auditory System Facts and Models
Enrique A. Lopez-Poveda

Basic Psychophysics of Human Spectral Processing
Brian C. J. Moore

Across-Channel Spectral Processing
John H. Grose, Joseph W. Hall III, and Emily Buss

Speech and Music Have Different Requirements for Spectral Resolution
Robert V. Shannon

Non-Linearities and the Representation of Auditory Spectra
Eric D. Young, Jane J. Yu, and Lina A. J. Reiss

Spectral Processing in the Inferior Colliculus
Kevin A. Davis

Neural Mechanisms for Spectral Analysis in the Auditory Midbrain, Thalamus, and Cortex
Monty A. Escabı´ and Heather L. Read

Spectral Processing in the Auditory Cortex
Mitchell L. Sutter

Processing of Dynamic Spectral Properties of Sounds
Adrian Rees and Manuel S. Malmierca

Representations of Spectral Coding in the Human Brain
Deborah A. Hall, PhD

Spectral Processing and Sound Source Determination
Donal G. Sinex

Spectral Information in Sound Localization
Simon Carlile, Russell Martin, and Ken McAnally

Plasticity of Spectral Processing
Dexter R. F. Irvine and Beverly A. Wright

Spectral Processing In Cochlear Implants
Colette M. McKay

Index

Volume 71

Autism: Neuropathology, Alterations of the GABAergic System, and Animal Models
Christoph Schmitz, Imke A. J. van Kooten, Patrick R. Hof, Herman van Engeland, Paul H. Patterson, and Harry W. M. Steinbusch

The Role of GABA in the Early Neuronal Development
Marta Jelitai and Emı´lia Madarasz

GABAergic Signaling in the Developing Cerebellum
Chitoshi Takayama

Insights into GABA Functions in the Developing Cerebellum
Mo´nica L. Fiszman

Role of GABA in the Mechanism of the Onset of Puberty in Non-Human Primates
Ei Terasawa

Rett Syndrome: A Rosetta Stone for Understanding the Molecular Pathogenesis of Autism
Janine M. LaSalle, Amber Hogart, and Karen N. Thatcher

GABAergic Cerebellar System in Autism: A Neuropathological and Developmental Perspective
Gene J. Blatt

Reelin Glycoprotein in Autism and Schizophrenia
S. Hossein Fatemi

Is There A Connection Between Autism, Prader-Willi Syndrome, Catatonia, and GABA?
Dirk M. Dhossche, Yaru Song, and Yiming Liu

Alcohol, GABA Receptors, and Neurodevelopmental Disorders
Ujjwal K. Rout

Effects of Secretin on Extracellular GABA and Other Amino Acid Concentrations in the Rat Hippocampus
Hans-Willi Clement, Alexander Pschibul, and Eberhard Schulz

Predicted Role of Secretin and Oxytocin in the Treatment of Behavioral and Developmental Disorders: Implications for Autism
Martha G. Welch and David A. Ruggiero

Immunological Findings in Autism
Hari Har Parshad Cohly and Asit Panja

Correlates of Psychomotor Symptoms in Autism
Laura Stoppelbein, Sara Sytsma-Jordan, and Leilani Greening

GABRB3 Gene Deficient Mice: A Potential Model of Autism Spectrum Disorder
Timothy M. DeLorey

The Reeler Mouse: Anatomy of a Mutant
Gabriella D'Arcangelo

Shared Chromosomal Susceptibility Regions Between Autism and Other Mental Disorders
Yvon C. Chagnon index

INDEX

Volume 72

Classification Matters for Catatonia and Autism in Children
Klaus-Jürgen Neumärker

A Systematic Examination of Catatonia-Like Clinical Pictures in Autism Spectrum Disorders
Lorna Wing and Amitta Shah

Catatonia in Individuals with Autism Spectrum Disorders in Adolescence and Early Adulthood: A Long-Term Prospective Study
Masataka Ohta, Yukiko Kano, and Yoko Nagai

Are Autistic and Catatonic Regression Related? A Few Working Hypotheses Involving GABA, Purkinje Cell Survival, Neurogenesis, and ECT
Dirk Marcel Dhossche and Ujjwal Rout

Psychomotor Development and Psychopathology in Childhood
Dirk M. J. De Raeymaecker

The Importance of Catatonia and Stereotypies in Autistic Spectrum Disorders
Laura Stoppelbein, Leilani Greening, and Angelina Kakooza

Prader–Willi Syndrome: Atypical Psychoses and Motor Dysfunctions
Willem M. A. Verhoeven and Siegfried Tuinier

Towards a Valid Nosography and Psychopathology of Catatonia in Children and Adolescents
David Cohen

Is There a Common Neuronal Basis for Autism and Catatonia?
Dirk Marcel Dhossche, Brendan T. Carroll, and Tressa D. Carroll

Shared Susceptibility Region on Chromosome 15 Between Autism and Catatonia
Yvon C. Chagnon

Current Trends in Behavioral Interventions for Children with Autism
Dorothy Scattone and Kimberly R. Knight

Case Reports with a Child Psychiatric Exploration of Catatonia, Autism, and Delirium
Jan N. M. Schieveld

ECT and the Youth: Catatonia in Context
Frank K. M. Zaw

Catatonia in Autistic Spectrum Disorders: A Medical Treatment Algorithm
Max Fink, Michael A. Taylor, and Neera Ghaziuddin

Psychological Approaches to Chronic Catatonia-Like Deterioration in Autism Spectrum Disorders
Amitta Shah and Lorna Wing

Section V: Blueprints

Blueprints for the Assessment, Treatment, and Future Study of Catatonia in Autism Spectrum Disorders
Dirk Marcel, Dhossche, Amitta Shah, and Lorna Wing

INDEX

Volume 73

Chromosome 22 Deletion Syndrome and Schizophrenia
Nigel M. Williams, Michael C. O'Donovan, and Michael J. Owen

Characterization of Proteome of Human Cerebrospinal Fluid
Jing Xu, Jinzhi Chen, Elaine R. Peskind, Jinghua Jin, Jimmy Eng, Catherine Pan, Thomas J. Montine, David R. Goodlett, and Jing Zhang

Hormonal Pathways Regulating Intermale and Interfemale Aggression
Neal G. Simon, Qianxing Mo, Shan Hu, Carrie Garippa, and Shi-Fang Lu

Neuronal GAP Junctions: Expression, Function, and Implications for Behavior
Clinton B. McCracken and David C. S. Roberts

Effects of Genes and Stress on the Neurobiology of Depression
J. John Mann and Dianne Currier

Quantitative Imaging with the Micropet Small-Animal Pet Tomograph
Paul Vaska, Daniel J. Rubins, David L. Alexoff, and Wynne K. Schiffer

Understanding Myelination through Studying its Evolution
Rüdiger Schweigreiter, Betty I. Roots, Christine Bandtlow, and Robert M. Gould

INDEX

Volume 74

Evolutionary Neurobiology and Art
C. U. M. Smith

Section I: Visual Aspects

Perceptual Portraits
Nicholas Wade

The Neuropsychology of Visual Art: Conferring Capacity
Anjan Chatterjee

Vision, Illusions, and Reality
Christopher Kennard

Localization in the Visual Brain
George K. York

Section II: Episodic Disorders

Neurology, Synaesthesia, and Painting
Amy Ione

Fainting in Classical Art
Philip Smith

Migraine Art in the Internet: A Study of 450 Contemporary Artists
Klaus Podoll

Sarah Raphael's Migraine with Aura as Inspiration for the Foray of Her Work into Abstraction
Klaus Podoll and Debbie Ayles

The Visual Art of Contemporary Artists with Epilepsy
Steven C. Schachter

Section III: Brain Damage

Creativity in Painting and Style in Brain-Damaged Artists
Julien Bogousslavsky

Artistic Changes in Alzheimer's Disease
Sebastian J. Crutch and Martin N. Rossor

Section IV: Cerebrovascular Disease

Stroke in Painters
H. Bäzner and M. Hennerici

Visuospatial Neglect in Lovis Corinth's Self-Portraits
Olaf Blanke

Art, Constructional Apraxia, and the Brain
Louis Caplan

Section V: Genetic Diseases

Neurogenetics in Art
Alan E. H. Emery

A Naïve Artist of St Ives
F. Clifford Rose

Van Gogh's Madness
F. Clifford Rose

Absinthe, The Nervous System and Painting
Tiina Rekand

Section VI: Neurologists as Artists

Sir Charles Bell, KGH, FRS, FRSE (1774–1842)
Christopher Gardner-Thorpe

Section VII: Miscellaneous

Peg Leg Frieda
Espen Dietrichs

The Deafness of Goya (1746–1828)
F. Clifford Rose

INDEX

Volume 75

Introduction on the Use of the *Drosophila* Embryonic/Larval Neuromuscular Junction as a Model System to Study Synapse Development and Function, and a Brief Summary of Pathfinding and Target Recognition
Catalina Ruiz-Cañada and Vivian Budnik

Development and Structure of Motoneurons
Matthias Landgraf and Stefan Thor

The Development of the *Drosophila* Larval Body Wall Muscles
Karen Beckett and Mary K. Baylies

Organization of the Efferent System and Structure of Neuromuscular Junctions in *Drosophila*
Andreas Prokop

Development of Motoneuron Electrical Properties and Motor Output
Richard A. Baines

Transmitter Release at the Neuromuscular Junction
Thomas L. Schwarz

Vesicle Trafficking and Recycling at the Neuromuscular Junction: Two Pathways for Endocytosis
Yoshiaki Kidokoro

Glutamate Receptors at the *Drosophila* Neuromuscular Junction
Aaron DiAntonio

Scaffolding Proteins at the *Drosophila* Neuromuscular Junction
Bulent Ataman, Vivian Budnik, and Ulrich Thomas

Synaptic Cytoskeleton at the Neuromuscular Junction
Catalina Ruiz-Cañada and Vivian Budnik

Plasticity and Second Messengers During Synapse Development
Leslie C. Griffith and Vivian Budnik

Retrograde Signaling that Regulates Synaptic Development and Function at the *Drosophila* Neuromuscular Junction
Guillermo Marqués and Bing Zhang

Activity-Dependent Regulation of Transcription During Development of Synapses
Subhabrata Sanyal and Mani Ramaswami

Experience-Dependent Potentiation of Larval Neuromuscular Synapses
Christoph M. Schuster

Selected Methods for the Anatomical Study of *Drosophila* Embryonic and Larval Neuromuscular Junctions
Vivian Budnik, Michael Gorczyca, and Andreas Prokop

INDEX

Volume 76

Section I: Physiological Correlates of Freud's Theories

The ID, the Ego, and the Temporal Lobe
Shirley M. Ferguson and Mark Rayport

ID, Ego, and Temporal Lobe Revisited
Shirley M. Ferguson and Mark Rayport

Section II: Stereotaxic Studies

Olfactory Gustatory Responses Evoked by Electrical Stimulation of Amygdalar Region in Man Are Qualitatively Modifiable by Interview Content: Case Report and Review
Mark Rayport, Sepehr Sani, and Shirley M. Ferguson

Section III: Controversy in Definition of Behavioral Disturbance

Pathogenesis of Psychosis in Epilepsy. The "Seesaw" Theory: Myth or Reality?
Shirley M. Ferguson and Mark Rayport

Section IV: Outcome of Temporal Lobectomy

Memory Function After Temporal Lobectomy for Seizure Control: A Comparative Neuropsy chiatric and Neuropsychological Study
Shirley M. Ferguson, A. John McSweeny, and Mark Rayport

Life After Surgery for Temporolimbic Seizures
Shirley M. Ferguson, Mark Rayport, and Carolyn A. Schell

Appendix I
Mark Rayport

Appendix II: Conceptual Foundations of Studies of Patients Undergoing Temporal Lobe Surgery for Seizure Control
Mark Rayport

INDEX

Volume 77

Regenerating the Brain
David A. Greenberg and Kunlin Jin

Serotonin and Brain: Evolution, Neuroplasticity, and Homeostasis
Efrain C. Azmitia

Therapeutic Approaches to Promoting Axonal Regeneration in the Adult Mammalian Spinal Cord
Sari S. Hannila, Mustafa M. Siddiq, and Marie T. Filbin

Evidence for Neuroprotective Effects of Antipsychotic Drugs: Implications for the Pathophysiology and Treatment of Schizophrenia
Xin-Min Li and Haiyun Xu

Neurogenesis and Neuroenhancement in the Pathophysiology and Treatment of Bipolar Disorder
Robert J. Schloesser, Guang Chen, and Husseini K. Manji

Neuroreplacement, Growth Factor, and Small Molecule Neurotrophic Approaches for Treating Parkinson's Disease
Michael J. O'Neill, Marcus J. Messenger, Viktor Lakics, Tracey K. Murray, Eric H. Karran, Philip G. Szekeres, Eric S. Nisenbaum, and Kalpana M. Merchant

Using *Caenorhabditis elegans* Models of Neurodegenerative Disease to Identify Neuroprotective Strategies
Brian Kraemer and Gerard D. Schellenberg

Neuroprotection and Enhancement of Neurite Outgrowth With Small Molecular Weight Compounds From Screens of Chemical Libraries
Donard S. Dwyer and Addie Dickson

INDEX

Volume 78

Neurobiology of Dopamine in Schizophrenia
Olivier Guillin, Anissa Abi-Dargham, and Marc Laruelle

The Dopamine System and the Pathophysiology of Schizophrenia: A Basic Science Perspective
Yukiori Goto and Anthony A. Grace

Glutamate and Schizophrenia: Phencyclidine, *N*-methyl-D-aspartate Receptors, and Dopamine–Glutamate Interactions
Daniel C. Javitt

Deciphering the Disease Process of Schizophrenia: The Contribution of Cortical GABA Neurons
David A. Lewis and Takanori Hashimoto

Alterations of Serotonin Transmission in Schizophrenia
Anissa Abi-Dargham

Serotonin and Dopamine Interactions in Rodents and Primates: Implications for Psychosis and Antipsychotic Drug Development
Gerard J. Marek

Cholinergic Circuits and Signaling in the Pathophysiology of Schizophrenia
Joshua A. Berman, David A. Talmage, and Lorna W. Role

Schizophrenia and the α7 Nicotinic Acetylcholine Receptor
Laura F. Martin and Robert Freedman

Histamine and Schizophrenia
Jean-Michel Arrang

Cannabinoids and Psychosis
Deepak Cyril D'Souza

Involvement of Neuropeptide Systems in Schizophrenia: Human Studies
Ricardo Cáceda, Becky Kinkead, and Charles B. Nemeroff

Brain-Derived Neurotrophic Factor in Schizophrenia and Its Relation with Dopamine
Olivier Guillin, Caroline Demily, and Florence Thibaut

Schizophrenia Susceptibility Genes: In Search of a Molecular Logic and Novel Drug Targets for a Devastating Disorder
Joseph A. Gogos

INDEX

Volume 79

The Destructive Alliance: Interactions of Leukocytes, Cerebral Endothelial Cells, and the Immune Cascade in Pathogenesis of Multiple Sclerosis
Alireza Minagar, April Carpenter, and J. Steven Alexander

Role of B Cells in Pathogenesis of Multiple Sclerosis
Behrouz Nikbin, Mandana Mohyeddin Bonab, Farideh Khosravi, and Fatemeh Talebian

The Role of CD4 T Cells in the Pathogenesis of Multiple Sclerosis
Tanuja Chitnis

The CD8 T Cell in Multiple Sclerosis: Suppressor Cell or Mediator of Neuropathology?
Aaron J. Johnson, Georgette L. Suidan, Jeremiah McDole, and Istvan Pirko

Immunopathogenesis of Multiple Sclerosis
Smriti M. Agrawal and V. Wee Yong

Molecular Mimicry in Multiple Sclerosis
Jane E. Libbey, Lori L. McCoy, and Robert S. Fujinami

Molecular "Negativity" May Underlie Multiple Sclerosis: Role of the Myelin Basic Protein Family in the Pathogenesis of MS
Abdiwahab A. Musse and George Harauz

Microchimerism and Stem Cell Transplantation in Multiple Sclerosis
Behrouz Nikbin, Mandana Mohyeddin Bonab, and Fatemeh Talebian

The Insulin-Like Growth Factor System in Multiple Sclerosis
Daniel Chesik, Nadine Wilczak, and Jacques De Keyser

Cell-Derived Microparticles and Exosomes in Neuroinflammatory Disorders
Lawrence L. Horstman, Wenche Jy, Alireza Minagar, Carlos J. Bidot, Joaquin J. Jimenez, J. Steven Alexander, and Yeon S. Ahn

Multiple Sclerosis in Children: Clinical, Diagnostic, and Therapeutic Aspects
Kevin Rostásy

Migraine in Multiple Sclerosis
Debra G. Elliott

Multiple Sclerosis as a Painful Disease
Meghan Kenner, Uma Menon, and Debra Elliott

Multiple Sclerosis and Behavior
James B. Pinkston, Anita Kablinger, and Nadejda Alekseeva

Cerebrospinal Fluid Analysis in Multiple Sclerosis
Francisco A. Luque and Stephen L. Jaffe

Multiple Sclerosis in Isfahan, Iran
Mohammad Saadatnia, Masoud Etemadifar, and Amir Hadi Maghzi

Gender Issues in Multiple Sclerosis
Robert N. Schwendimann and Nadejda Alekseeva

Differential Diagnosis of Multiple Sclerosis
Halim Fadil, Roger E. Kelley, and Eduardo Gonzalez-Toledo

Prognostic Factors in Multiple Sclerosis
Roberto Bergamaschi

Neuroimaging in Multiple Sclerosis
Robert Zivadinov and Jennifer L. Cox

Detection of Cortical Lesions Is Dependent on Choice of Slice Thickness in Patients with Multiple Sclerosis
Ondrej Dolezal, Michael G. Dwyer, Dana Horakova, Eva Havrdova, Alireza Minagar, Srivats Balachandran, Niels Bergsland, Zdenek Seidl, Manuela Vaneckova, David Fritz, Jan Krasensky, and Robert Zivadinov

The Role of Quantitative Neuroimaging Indices in the Differentiation of Ischemia from Demyelination: An Analytical Study with Case Presentation
Romy Hoque, Christina Ledbetter, Eduardo Gonzalez-Toledo, Vivek Misra, Uma Menon, Meghan Kenner, Alejandro A. Rabinstein, Roger E. Kelley, Robert Zivadinov, and Alireza Minagar

HLA-DRB1*1501, -DQB1*0301, -DQB1*0302, -DQB1*0602, and -DQB1*0603 Alleles Are Associated with More Severe Disease Outcome on MRI in Patients with Multiple Sclerosis
Robert Zivadinov, Laura Uxa, Alessio Bratina, Antonio Bosco, Bhooma Srinivasaraghavan, Alireza Minagar, Maja Ukmar, Su yen Benedetto, and Marino Zorzon

Glatiramer Acetate: Mechanisms of Action in Multiple Sclerosis
Tjalf Ziemssen and Wiebke Schrempf

Evolving Therapies for Multiple Sclerosis
Elena Korniychuk, John M. Dempster, Eileen O'Connor, J. Steven Alexander, Roger E. Kelley, Meghan Kenner, Uma Menon, Vivek Misra, Romy Hoque, Eduardo C. Gonzalez-Toledo, Robert N. Schwendimann, Stacy Smith, and Alireza Minagar

Remyelination in Multiple Sclerosis
Divya M. Chari

Trigeminal Neuralgia: A Modern-Day Review
Kelly Hunt and Ravish Patwardhan

Optic Neuritis and the Neuro-Ophthalmology of Multiple Sclerosis
Paramjit Kaur and Jeffrey L. Bennett

Neuromyelitis Optica: New Findings on Pathogenesis
Dean M. Wingerchuk

INDEX

Volume 80

Epilepsy in the Elderly: Scope of the Problem
Ilo E. Leppik

Animal Models in Gerontology Research
Nancy L. Nadon

Animal Models of Geriatric Epilepsy
Lauren J. Murphree, Lynn M. Rundhaugen, and Kevin M. Kelly

Life and Death of Neurons in the Aging Cerebral Cortex
John H. Morrison and Patrick R. Hof

An In Vitro Model of Stroke-Induced Epilepsy: Elucidation of the Roles of Glutamate and Calcium in the Induction and Maintenance of Stroke-Induced Epileptogenesis
Robert J. DeLorenzo, David A. Sun, Robert E. Blair, and Sompong Sambati

Mechanisms of Action of Antiepileptic Drugs
H. Steve White, Misty D. Smith, and Karen S. Wilcox

Epidemiology and Outcomes of Status Epilepticus in the Elderly
Alan R. Towne

Diagnosing Epilepsy in the Elderly
R. Eugene Ramsay, Flavia M. Macias, and A. James Rowan

Pharmacoepidemiology in Community-Dwelling Elderly Taking Antiepileptic Drugs
Dan R. Berlowitz and Mary Jo V. Pugh

Use of Antiepileptic Medications in Nursing Homes
Judith Garrard, Susan L. Harms, Lynn E. Eberly, and Ilo E. Leppik

Differential Diagnosis of Multiple Sclerosis
Halim Fadil, Roger E. Kelley, and Eduardo Gonzalez-Toledo

Prognostic Factors in Multiple Sclerosis
Roberto Bergamaschi

Neuroimaging in Multiple Sclerosis
Robert Zivadinov and Jennifer L. Cox

Detection of Cortical Lesions Is Dependent on Choice of Slice Thickness in Patients with Multiple Sclerosis
Ondrej Dolezal, Michael G. Dwyer, Dana Horakova, Eva Havrdova, Alireza Minagar, Srivats Balachandran, Niels Bergsland, Zdenek Seidl, Manuela Vaneckova, David Fritz, Jan Krasensky, and Robert Zivadinov

The Role of Quantitative Neuroimaging Indices in the Differentiation of Ischemia from Demyelination: An Analytical Study with Case Presentation
Romy Hoque, Christina Ledbetter, Eduardo Gonzalez-Toledo, Vivek Misra, Uma Menon, Meghan Kenner, Alejandro A. Rabinstein, Roger E. Kelley, Robert Zivadinov, and Alireza Minagar

HLA-DRB1*1501, -DQB1*0301,-DQB1 *0302,-DQB1*0602, and -DQB1*0603 Alleles Are Associated with More Severe Disease Outcome on MRI in Patients with Multiple Sclerosis
Robert Zivadinov, Laura Uxa, Alessio Bratina, Antonio Bosco, Bhooma Srinivasaraghavan, Alireza Minagar, Maja Ukmar, Su yen Benedetto, and Marino Zorzon

Glatiramer Acetate: Mechanisms of Action in Multiple Sclerosis
Tjalf Ziemssen and Wiebke Schrempf

Evolving Therapies for Multiple Sclerosis
Elena Korniychuk, John M. Dempster, Eileen O'Connor, J. Steven Alexander, Roger E. Kelley, Meghan Kenner, Uma Menon, Vivek Misra, Romy Hoque, Eduardo C. Gonzalez-Toledo, Robert N. Schwendimann, Stacy Smith, and Alireza Minagar

Remyelination in Multiple Sclerosis
Divya M. Chari

Trigeminal Neuralgia: A Modern-Day Review
Kelly Hunt and Ravish Patwardhan

Optic Neuritis and the Neuro-Ophthalmology of Multiple Sclerosis
Paramjit Kaur and Jeffrey L. Bennett

Neuromyelitis Optica: New Findings on Pathogenesis
Dean M. Wingerchuk

INDEX

Volume 81

Epilepsy in the Elderly: Scope of the Problem
Ilo E. Leppik

Animal Models in Gerontology Research
Nancy L. Nadon

Animal Models of Geriatric Epilepsy
Lauren J. Murphree, Lynn M. Rundhaugen, and Kevin M. Kelly

Life and Death of Neurons in the Aging Cerebral Cortex
John H. Morrison and Patrick R. Hof

An In Vitro Model of Stroke-Induced Epilepsy: Elucidation of the Roles of Glutamate and Calcium in the Induction and Maintenance of Stroke-Induced Epileptogenesis
Robert J. DeLorenzo, David A. Sun, Robert E. Blair, and Sompong Sambati

Mechanisms of Action of Antiepileptic Drugs
H. Steve White, Misty D. Smith, and Karen S. Wilcox

Epidemiology and Outcomes of Status Epilepticus in the Elderly
Alan R. Towne

Diagnosing Epilepsy in the Elderly
R. Eugene Ramsay, Flavia M. Macias, and A. James Rowan

Pharmacoepidemiology in Community-Dwelling Elderly Taking Antiepileptic Drugs
Dan R. Berlowitz and Mary Jo V. Pugh

Use of Antiepileptic Medications in Nursing Homes
Judith Garrard, Susan L. Harms, Lynn E. Eberly, and Ilo E. Leppik

Age-Related Changes in Pharmacokinetics: Predictability and Assessment Methods
Emilio Perucca

Factors Affecting Antiepileptic Drug Pharmacokinetics in Community-Dwelling Elderly
James C. Cloyd, Susan Marino, and Angela K. Birnbaum

Pharmacokinetics of Antiepileptic Drugs in Elderly Nursing Home Residents
Angela K. Birnbaum

The Impact of Epilepsy on Older Veterans
Mary Jo V. Pugh, Dan R. Berlowitz, and Lewis Kazis

Risk and Predictability of Drug Interactions in the Elderly
René H. Levy and Carol Collins

Outcomes in Elderly Patients With Newly Diagnosed and Treated Epilepsy
Martin J. Brodie and Linda J. Stephen

Recruitment and Retention in Clinical Trials of the Elderly
Flavia M. Macias, R. Eugene Ramsay, and A. James Rowan

Treatment of Convulsive Status Epilepticus
David M. Treiman

Treatment of Nonconvulsive Status Epilepticus
Matthew C. Walker

Antiepileptic Drug Formulation and Treatment in the Elderly: Biopharmaceutical Considerations
Barry E. Gidal

INDEX

Volume 82

Inflammatory Mediators Leading to Protein Misfolding and Uncompetitive/Fast Off-Rate Drug Therapy for Neurodegenerative Disorders
Stuart A. Lipton, Zezong Gu, and Tomohiro Nakamura

Innate Immunity and Protective Neuroinflammation: New Emphasis on the Role of Neuroimmune Regulatory Proteins
M. Griffiths, J. W. Neal, and P. Gasque

Glutamate Release from Astrocytes in Physiological Conditions and in Neurodegenerative Disorders Characterized by Neuroinflammation
Sabino Vesce, Daniela Rossi, Liliana Brambilla, and Andrea Volterra

The High-Mobility Group Box 1 Cytokine Induces Transporter-Mediated Release of Glutamate from Glial Subcellular Particles (Gliosomes) Prepared from *In Situ*-Matured Astrocytes
Giambattista Bonanno, Luca Raiteri, Marco Milanese, Simona Zappettini, Edon Melloni, Marco Pedrazzi, Mario Passalacqua, Carlo Tacchetti, Cesare Usai, and Bianca Sparatore

The Role of Astrocytes and Complement System in Neural Plasticity
Milos Pekny, Ulrika Wilhelmsson, Yalda Rahpeymai Bogestål, and Marcela Pekna

New Insights into the Roles of Metalloproteinases in Neurodegeneration and Neuroprotection
A. J. Turner and N. N. Nalivaeva

Relevance of High-Mobility Group Protein Box 1 to Neurodegeneration
Silvia Fossati and Alberto Chiarugi

Early Upregulation of Matrix Metalloproteinases Following Reperfusion Triggers Neuroinflammatory Mediators in Brain Ischemia in Rat
Diana Amantea, Rossella Russo, Micaela Gliozzi, Vincenza Fratto, Laura Berliocchi, G. Bagetta, G. Bernardi, and M. Tiziana Corasaniti

The (Endo)Cannabinoid System in Multiple Sclerosis and Amyotrophic Lateral Sclerosis
Diego Centonze, Silvia Rossi, Alessandro Finazzi-Agrò, Giorgio Bernardi, and Mauro Maccarrone

Chemokines and Chemokine Receptors: Multipurpose Players in Neuroinflammation
Richard M. Ransohoff, LiPing Liu, and Astrid E. Cardona

Systemic and Acquired Immune Responses in Alzheimer's Disease
Markus Britschgi and Tony Wyss-Coray

Neuroinflammation in Alzheimer's Disease and Parkinson's Disease: Are Microglia Pathogenic in Either Disorder?
Joseph Rogers, Diego Mastroeni, Brian Leonard, Jeffrey Joyce, and Andrew Grover

Cytokines and Neuronal Ion Channels in Health and Disease
Barbara Viviani, Fabrizio Gardoni, and Marina Marinovich

Cyclooxygenase-2, Prostaglandin E_2, and Microglial Activation in Prion Diseases
Luisa Minghetti and Maurizio Pocchiari

Glia Proinflammatory Cytokine Upregulation as a Therapeutic Target for Neurodegenerative Diseases: Function-Based and Target-Based Discovery Approaches
Linda J. Van Eldik, Wendy L. Thompson, Hantamalala Ralay Ranaivo, Heather A. Behanna, and D. Martin Watterson

Oxidative Stress and the Pathogenesis of Neurodegenerative Disorders
Ashley Reynolds, Chad Laurie, R. Lee Mosley, and Howard E. Gendelman

Differential Modulation of Type 1 and Type 2 Cannabinoid Receptors Along the Neuroimmune Axis
Sergio Oddi, Paola Spagnuolo, Monica Bari, Antonella D'Agostino, and Mauro Maccarrone

Effects of the HIV-1 Viral Protein Tat on Central Neurotransmission: Role of Group I Metabotropic Glutamate Receptors
Elisa Neri, Veronica Musante, and Anna Pittaluga

Evidence to Implicate Early Modulation of Interleukin-1β Expression in the Neuroprotection Afforded by 17β-Estradiol in Male Rats Undergone Transient Middle Cerebral Artery Occlusion
Olga Chiappetta, Micaela Gliozzi, Elisa Siviglia, Diana Amantea, Luigi A. Morrone, Laura Berliocchi, G. Bagetta, and M. Tiziana Corasaniti

A Role for Brain Cyclooxygenase-2 and Prostaglandin-E_2 in Migraine: Effects of Nitroglycerin
Cristina Tassorelli, Rosaria Greco, Marie Therèse Armentero, Fabio Blandini, Giorgio Sandrini, and Giuseppe Nappi

The Blockade of K+-ATP Channels has Neuroprotective Effects in an *In Vitro* Model of Brain Ischemia
Robert Nisticò, Silvia Piccirilli, L. Sebastianelli, Giuseppe Nisticò, G. Bernardi, and N. B. Mercuri

Retinal Damage Caused by High Intraocular Pressure-Induced Transient Ischemia is Prevented by Coenzyme Q10 in Rat
Carlo Nucci, Rosanna Tartaglione, Angelica Cerulli, R. Mancino, A. Spanò, Federica Cavaliere, Laura Rombolà, G. Bagetta, M. Tiziana Corasaniti, and Luigi A. Morrone

Evidence Implicating Matrix Metalloproteinases in the Mechanism Underlying Accumulation of IL-1β and Neuronal Apoptosis in the Neocortex of HIV/gp120-Exposed Rats
Rossella Russo, Elisa Siviglia, Micaela Gliozzi, Diana Amantea, Annamaria Paoletti, Laura Berliocchi, G. Bagetta, and M. Tiziana Corasaniti

Neuroprotective Effect of Nitroglycerin in a Rodent Model of Ischemic Stroke: Evaluation of Bcl-2 Expression
Rosaria Greco, Diana Amantea, Fabio Blandini, Giuseppe Nappi, Giacinto Bagetta, M. Tiziana Corasaniti, and Cristina Tassorelli

Index

Volume 83

Gender Differences in Pharmacological Response
Gail D. Anderson

Epidemiology and Classification of Epilepsy: Gender Comparisons
John C. McHugh and Norman Delanty

Hormonal Influences on Seizures: Basic Neurobiology
Cheryl A. Frye

Catamenial Epilepsy
Patricia E. Penovich and Sandra Helmers

Epilepsy in Women: Special Considerations for Adolescents
Mary L. Zupanc and Sheryl Haut

Contraception in Women with Epilepsy: Pharmacokinetic Interactions, Contraceptive Options, and Management
Caryn Dutton and Nancy Foldvary-Schaefer

Reproductive Dysfunction in Women with Epilepsy: Menstrual Cycle Abnormalities, Fertility, and Polycystic Ovary Syndrome
Jürgen Bauer and Déirdre Cooper-Mahkorn

Sexual Dysfunction in Women with Epilepsy: Role of Antiepileptic Drugs and Psychotropic Medications
Mary A. Gutierrez, Romila Mushtaq, and Glen Stimmel

Pregnancy in Epilepsy: Issues of Concern
John DeToledo

Teratogenicity and Antiepileptic Drugs: Potential Mechanisms
Mark S. Yerby

Antiepileptic Drug Teratogenesis: What are the Risks for Congenital Malformations and Adverse Cognitive Outcomes?
Cynthia L. Harden

Teratogenicity of Antiepileptic Drugs: Role of Pharmacogenomics
Raman Sankar and Jason T. Lerner

Antiepileptic Drug Therapy in Pregnancy I: Gestation-Induced Effects on AED Pharmacokinetics
Page B. Pennell and Collin A. Hovinga

Antiepileptic Drug Therapy in Pregnancy II: Fetal and Neonatal Exposure
Collin A. Hovinga and Page B. Pennell

Seizures in Pregnancy: Diagnosis and Management
Robert L. Beach and Peter W. Kaplan

Management of Epilepsy and Pregnancy: An Obstetrical Perspective
Julian N. Robinson and Jane Cleary-Goldman

Pregnancy Registries: Strengths, Weaknesses, and Bias Interpretation of Pregnancy Registry Data
Marianne Cunnington and John Messenheimer

Bone Health in Women With Epilepsy: Clinical Features and Potential Mechanisms
Alison M. Pack and Thaddeus S. Walczak

Metabolic Effects of AEDs: Impact on Body Weight, Lipids and Glucose Metabolism
Raj D. Sheth and Georgia Montouris

Psychiatric Comorbidities in Epilepsy
W. Curt Lafrance, Jr., Andres M. Kanner, and Bruce Hermann

Issues for Mature Women with Epilepsy
Cynthia L. Harden

Pharmacodynamic and Pharmacokinetic Interactions of Psychotropic Drugs with Antiepileptic Drugs
Andres M. Kanner and Barry E. Gidal

Health Disparities in Epilepsy: How Patient-Oriented Outcomes in Women Differ from Men
Frank Gilliam

Index

Volume 84

Normal Brain Aging: Clinical, Immunological, Neuropsychological, and Neuroimaging Features
Maria T. Caserta, Yvonne Bannon, Francisco Fernandez, Brian Giunta, Mike R. Schoenberg, and Jun Tan

Subcortical Ischemic Cerebrovascular Dementia
Uma Menon and Roger E. Kelley

Cerebrovascular and Cardiovascular Pathology in Alzheimer's Disease
Jack C. de la Torre

Neuroimaging of Cognitive Impairments in Vascular Disease
Carol Di Perri, Turi O. Dalaker, Mona K. Beyer, and Robert Zivadinov

Contributions of Neuropsychology and Neuroimaging to Understanding Clinical Subtypes of Mild Cognitive Impairment
Amy J. Jak, Katherine J. Bangen, Christina E. Wierenga, Lisa Delano-Wood, Jody Corey-Bloom, and Mark W. Bondi

Proton Magnetic Resonance Spectroscopy in Dementias and Mild Cognitive Impairment
H. Randall Griffith, Christopher C. Stewart, and Jan A. den Hollander

Application of PET Imaging to Diagnosis of Alzheimer's Disease and Mild Cognitive Impairment
James M. Noble and Nikolaos Scarmeas

The Molecular and Cellular Pathogenesis of Dementia of the Alzheimer's Type: An Overview
Francisco A. Luque and Stephen L. Jaffe

Alzheimer's Disease Genetics: Current Status and Future Perspectives
Lars Bertram

Frontotemporal Lobar Degeneration: Insights from Neuropsychology and Neuroimaging
Andrea C. Bozoki and Muhammad U. Farooq

Lewy Body Dementia
Jennifer C. Hanson and Carol F. Lippa

Dementia in Parkinson's Disease
Bradley J. Robottom and William J. Weiner

Early Onset Dementia
Halim Fadil, Aimee Borazanci, Elhachmia Ait Ben Haddou, Mohamed Yahyaoui, Elena Korniychuk, Stephen L. Jaffe, and Alireza Minagar

Normal Pressure Hydrocephalus
Glen R. Finney

Reversible Dementias
Anahid Kabasakalian and Glen R. Finney

INDEX

Volume 85

Involvement of the Prefrontal Cortex in Problem Solving
Hajime Mushiake, Kazuhiro Sakamoto, Naohiro Saito, Toshiro Inui, Kazuyuki Aihara, and Jun Tanji

GluK1 Receptor Antagonists and Hippocampal Mossy Fiber Function
Robert Nisticò, Sheila Dargan, Stephen M. Fitzjohn, David Lodge, David E. Jane, Graham L. Collingridge, and Zuner A. Bortolotto

Monoamine Transporter as a Target Molecule for Psychostimulants
Ichiro Sora, BingJin Li, Setsu Fumushima, Asami Fukui, Yosefu Arime, Yoshiyuki Kasahara, Hiroaki Tomita, and Kazutaka Ikeda

Targeted Lipidomics as a Tool to Investigate Endocannabinoid Function
Giuseppe Astarita, Jennifer Geaga, Faizy Ahmed, and Daniele Piomelli

The Endocannabinoid System as a Target for Novel Anxiolytic and Antidepressant Drugs
Silvana Gaetani, Pasqua Dipasquale, Adele Romano, Laura Righetti, Tommaso Cassano, Daniele Piomelli, and Vincenzo Cuomo

$GABA_A$ Receptor Function and Gene Expression During Pregnancy and Postpartum
Giovanni Biggio, Maria Cristina Mostallino, Paolo Follesa, Alessandra Concas, and Enrico Sanna

Early Postnatal Stress and Neural Circuit Underlying Emotional Regulation
Machiko Matsumoto, Mitsuhiro Yoshioka, and Hiroko Togashi

Roles of the Histaminergic Neurotransmission on Methamphetamine-Induced Locomotor Sensitization and Reward: A Study of Receptors Gene Knockout Mice
Naoko Takino, Eiko Sakurai, Atsuo Kuramasu, Nobuyuki Okamura, and Kazuhiko Yanai

Developmental Exposure to Cannabinoids Causes Subtle and Enduring Neurofunctional Alterations
Patrizia Campolongo, Viviana Trezza, Maura Palmery, Luigia Trabace, and Vincenzo Cuomo

Neuronal Mechanisms for Pain-Induced Aversion: Behavioral Studies Using a Conditioned Place Aversion Test
Masabumi Minami

Bv8/Prokineticins and their Receptors: A New Pronociceptive System
Lucia Negri, Roberta Lattanzi, Elisa Giannini, Michela Canestrelli, Annalisa Nicotra, and Pietro Melchiorri

$P2Y_6$-Evoked Microglial Phagocytosis
Kazuhide Inoue, Schuichi Koizumi, Ayako Kataoka, Hidetoshi Tozaki-Saitoh, and Makoto Tsuda

PPAR and Pain
Takehiko Maeda and Shiroh Kishioka

Involvement of Inflammatory Mediators in Neuropathic Pain Caused by Vincristine
Norikazu Kiguchi, Takehiko Maeda, Yuka Kobayashi, Fumihiro Saika, and Shiroh Kishioka

Nociceptive Behavior Induced by the Endogenous Opioid Peptides Dynorphins in Uninjured Mice: Evidence with Intrathecal *N*-ethylmaleimide Inhibiting Dynorphin Degradation
Koichi Tan-No, Hiroaki Takahashi, Osamu Nakagawasai, Fukie Niijima, Shinobu Sakurada, Georgy Bakalkin, Lars Terenius, and Takeshi Tadano

Mechanism of Allodynia Evoked by Intrathecal Morphine-3-Glucuronide in Mice
Takaaki Komatsu, Shinobu Sakurada, Sou Katsuyama, Kengo Sanai, and Tsukasa Sakurada

(−)-Linalool Attenuates Allodynia in Neuropathic Pain Induced by Spinal Nerve Ligation in C57/Bl6 Mice
Laura Berliocchi, Rossella Russo, Alessandra Levato, Vincenza Fratto, Giacinto Bagetta, Shinobu Sakurada, Tsukasa Sakurada, Nicola Biagio Mercuri, and Maria Tiziana Corasaniti

Intraplantar Injection of Bergamot Essential Oil into the Mouse Hindpaw: Effects on Capsaicin-Induced Nociceptive Behaviors
Tsukasa Sakurada, Hikari Kuwahata, Soh Katsuyama, Takaaki Komatsu, Luigi A. Morrone, M. Tiziana Corasaniti, Giacinto Bagetta, and Shinobu Sakurada

New Therapy for Neuropathic Pain
Hirokazu Mizoguchi, Chizuko Watanabe, Akihiko Yonezawa, and Shinobu Sakurada

Regulated Exocytosis from Astrocytes: Physiological and Pathological Related Aspects
Corrado Calì, Julie Marchaland, Paola Spagnuolo, Julien Gremion, and Paola Bezzi

Glutamate Release from Astrocytic Gliosomes Under Physiological and Pathological Conditions
Marco Milanese, Tiziana Bonifacino, Simona Zappettini, Cesare Usai, Carlo Tacchetti, Mario Nobile, and Giambattista Bonanno

Neurotrophic and Neuroprotective Actions of an Enhancer of Ganglioside Biosynthesis
Jin-ichi Inokuchi

Involvement of Endocannabinoid Signaling in the Neuroprotective Effects of Subtype 1 Metabotropic Glutamate Receptor Antagonists in Models of Cerebral Ischemia
Elisa Landucci, Francesca Boscia, Elisabetta Gerace, Tania Scartabelli, Andrea Cozzi, Flavio Moroni, Guido Mannaioni, and Domenico E. Pellegrini-Giampietro

NF-kappaB Dimers in the Regulation of Neuronal Survival
Ilenia Sarnico, Annamaria Lanzillotta, Marina Benarese, Manuela Alghisi, Cristina Baiguera, Leontino Battistin, PierFranco Spano, and Marina Pizzi

Oxidative Stress in Stroke Pathophysiology: Validation of Hydrogen Peroxide Metabolism as a Pharmacological Target to Afford Neuroprotection
Diana Amantea, Maria Cristina Marrone, Robert Nisticò, Mauro Federici, Giacinto Bagetta, Giorgio Bernardi, and Nicola Biagio Mercuri

Role of Akt and ERK Signaling in the Neurogenesis following Brain Ischemia
Norifumi Shioda, Feng Han, and Kohji Fukunaga

Prevention of Glutamate Accumulation and Upregulation of Phospho-Akt may Account for Neuroprotection Afforded by Bergamot Essential Oil against Brain Injury Induced by Focal Cerebral Ischemia in Rat
Diana Amantea, Vincenza Fratto, Simona Maida, Domenicantonio Rotiroti, Salvatore Ragusa, Giuseppe Nappi, Giacinto Bagetta, and Maria Tiziana Corasaniti

Identification of Novel Pharmacological Targets to Minimize Excitotoxic Retinal Damage
Rossella Russo, Domenicantonio Rotiroti, Cristina Tassorelli, Carlo Nucci, Giacinto Bagetta, Massimo Gilberto Bucci, Maria Tiziana Corasaniti, and Luigi Antonio Morrone

INDEX

Volume 86

Section One: Hybrid Bionic Systems

EMG-Based and Gaze-Tracking-Based Man–Machine Interfaces
Federico Carpi and Danilo De Rossi

Bidirectional Interfaces with the Peripheral Nervous System
Silvestro Micera and Xavier Navarro

Interfacing Insect Brain for Space Applications
Giovanni Di Pino, Tobias Seidl, Antonella Benvenuto, Fabrizio Sergi, Domenico Campolo, Dino Accoto, Paolo Maria Rossini, and Eugenio Guglielmelli

Section Two: Meet the Brain
Meet the Brain: Neurophysiology
John Rothwell

Fundamentals of Electroencefalography, Magnetoencefalography, and Functional Magnetic Resonance Imaging
Claudio Babiloni, Vittorio Pizzella, Cosimo Del Gratta, Antonio Ferretti, and Gian Luca Romani

Implications of Brain Plasticity to Brain–Machine Interfaces Operation: A Potential Paradox?
Paolo Maria Rossini

Section Three: Brain Machine Interfaces, A New Brain-to-Environment Communication Channel
An Overview of BMIs
Francisco Sepulveda

Neurofeedback and Brain–Computer Interface: Clinical Applications
Niels Birbaumer, Ander Ramos Murguialday, Cornelia Weber, and Pedro Montoya

Flexibility and Practicality: Graz Brain–Computer Interface Approach
Reinhold Scherer, Gernot R. Müller-Putz, and Gert Pfurtscheller

On the Use of Brain–Computer Interfaces Outside Scientific Laboratories: Toward an Application in Domotic Environments
F. Babiloni, F. Cincotti, M. Marciani, S. Salinari, L. Astolfi, F. Aloise, F. De Vico Fallani, and D. Mattia

Brain–Computer Interface Research at the Wadsworth Center: Developments in Noninvasive Communication and Control
Dean J. Krusienski and Jonathan R. Wolpaw

Watching Brain TV and Playing Brain Ball: Exploring Novel BCL Strategies Using Real–Time Analysis of Human Intercranial Data
Karim Jerbi, Samson Freyermuth, Lorella Minotti, Philippe Kahane, Alain Berthoz, and Jean-Philippe Lachaux

Section Four: Brain-Machine Interfaces and Space
Adaptive Changes of Rhythmic EEG Oscillations in Space: Implications for Brain–Machine Interface Applications
G. Cheron, A. M. Cebolla, M. Petieau, A. Bengoetxea, E. Palmero-Soler, A. Leroy, and B. Dan

Validation of Brain–Machine Interfaces During Parabolic Flight
José del R. Millán, Pierre W. Ferrez, and Tobias Seidl

Matching Brain–Machine Interface Performance to Space Applications
Luca Citi, Oliver Tonet, and Martina Marinelli

Brain–Machine Interfaces for Space Applications—Research, Technological Development, and Opportunities
Leopold Summerer, Dario Izzo, and Luca Rossini

INDEX

Volume 87

Peripheral Nerve Repair and Regeneration Research: A Historical Note
Bruno Battiston, Igor Papalia, Pierluigi Tos, and Stefano Geuna

Development of the Peripheral Nerve
Suleyman Kaplan, Ersan Odaci, Bunyami Unal, Bunyamin Sahin, and Michele Fornaro

Histology of the Peripheral Nerve and Changes Occurring During Nerve Regeneration
Stefano Geuna, Stefania Raimondo, Giulia Ronchi, Federica Di Scipio, Pierluigi Tos, Krzysztof Czaja, and Michele Fornaro

Methods and Protocols in Peripheral Nerve Regeneration Experimental Research: Part I—Experimental Models
Pierluigi Tos, Giulia Ronchi, Igor Papalia, Vera Sallen, Josette Legagneux, Stefano Geuna, and Maria G. Giacobini-Robecchi

Methods and Protocols in Peripheral Nerve Regeneration Experimental Research: Part II—Morphological Techniques
Stefania Raimondo, Michele Fornaro, Federica Di Scipio, Giulia Ronchi, Maria G. Giacobini-Robecchi, and Stefano Geuna

Methods and Protocols in Peripheral Nerve Regeneration Experimental Research: Part III—Electrophysiological Evaluation
Xavier Navarro and Esther Udina

Methods and Protocols in Peripheral Nerve Regeneration Experimental Research: Part IV—Kinematic Gait Analysis to Quantify Peripheral Nerve Regeneration in the Rat
Luís M. Costa, Maria J. Simões, Ana C. Maurício and Artur S.P. Varejão

Current Techniques and Concepts in Peripheral Nerve Repair
Maria Siemionow and Grzegorz Brzezicki

Artificial Scaffolds for Peripheral Nerve Reconstruction
Valeria Chiono, Chiara Tonda-Turo, and Gianluca Ciardelli

Conduit Luminal Additives for Peripheral Nerve Repair
Hede Yan, Feng Zhang, Michael B. Chen, and William C. Lineaweaver

Tissue Engineering of Peripheral Nerves
Bruno Battiston, Stefania Raimondo, Pierluigi Tos, Valentina Gaidano, Chiara Audisio, Anna Scevola, Isabelle Perroteau, and Stefano Geuna

Mechanisms Underlying The End-to-Side Nerve Regeneration
Eleana Bontioti and Lars B. Dahlin

Experimental Results in End-To-Side Neurorrhaphy
Alexandros E. Beris and Marios G. Lykissas

End-to-Side Nerve Regeneration: From the Laboratory Bench to Clinical Applications
Pierluigi Tos, Stefano Artiaco, Igor Papalia, Ignazio Marcoccio, Stefano Geuna, and Bruno Battiston

Novel Pharmacological Approaches to Schwann Cells as Neuroprotective Agents for Peripheral Nerve Regeneration
Valerio Magnaghi, Patrizia Procacci, and Ada Maria Tata

Melatonin and Nerve Regeneration
Ersan Odaci and Suleyman Kaplan

Transthyretin: An Enhancer of Nerve Regeneration
Carolina E. Fleming, Fernando Milhazes Mar, Filipa Franquinho, and Mónica M. Sousa

Enhancement of Nerve Regeneration and Recovery by Immunosuppressive Agents
Damien P. Kuffler

The Role of Collagen in Peripheral Nerve Repair
Guido Koopmans, Birgit Hasse, and Nektarios Sinis

Gene Therapy Perspectives for Nerve Repair
Serena Zacchigna and Mauro Giacca

Use of Stem Cells for Improving Nerve Regeneration
Giorgio Terenghi, Mikael Wiberg, and Paul J. Kingham

Transplantation of Olfactory Ensheathing Cells for Peripheral Nerve Regeneration
Christine Radtke, Jeffery D. Kocsis, and Peter M. Vogt

Manual Stimulation of Target Muscles has Different Impact on Functional Recovery after Injury of Pure Motor or Mixed Nerves
Nektarios Sinis, Thodora Manoli, Frank Werdin, Armin Kraus, Hans E. Schaller, Orlando Guntinas-Lichius, Maria Grosheva, Andrey Irintchev, Emanouil Skouras, Sarah Dunlop, and Doychin N. Angelov

Electrical Stimulation for Improving Nerve Regeneration: Where do we Stand?
Tessa Gordon, Olewale A. R. Sulaiman, and Adil Ladak

Phototherapy in Peripheral Nerve Injury: Effects on Muscle Preservation and Nerve Regeneration
Shimon Rochkind, Stefano Geuna, and Asher Shainberg

Age-Related Differences in the Reinnervation after Peripheral Nerve Injury
Uroš Kovačič, Janez Sketelj, and Fajko F. Bajrović

Neural Plasticity After Nerve Injury and Regeneration
Xavier Navarro

Future Perspective in Peripheral Nerve Reconstruction
Lars Dahlin, Fredrik Johansson, Charlotta Lindwall, and Martin Kanje

INDEX

Volume 88

Effects Of Psychostimulants On Neurotrophins: Implications For Psychostimulant-Induced Neurotoxicity
Francesco Angelucci, Valerio Ricci, Gianfranco Spalletta, Carlo Caltagirone, Aleksander A. Mathé, and Pietro Bria

Dosing Time-Dependent Actions of Psychostimulants
H. Manev and T. Uz

Dopamine-Induced Behavioral Changes and Oxidative Stress in Methamphetamine-Induced Neurotoxicity
Taizo Kita, Ikuko Miyazaki, Masato Asanuma, Mika Takeshima, and George C. Wagner

Acute Methamphetamine Intoxication: Brain Hyperthermia, Blood–Brain Barrier, Brain Edema, and morphological cell abnormalities
Eugene A. Kiyatkin and Hari S. Sharma

Molecular Bases of Methamphetamine-Induced Neurodegeneration
Jean Lud Cadet and Irina N. Krasnova

Involvement of Nicotinic Receptors in Methamphetamine- and MDMA-Induced Neurotoxicity: Pharmacological Implications
E. Escubedo, J. Camarasa, C. Chipana, S. García-Ratés, and D.Pubill

Ethanol Alters the Physiology of Neuron–Glia Communication
Antonio González and Ginés M. Salido

Therapeutic Targeting of "DARPP-32": A Key Signaling Molecule in the Dopiminergic Pathway for the Treatment of Opiate Addiction
Supriya D. Mahajan, Ravikumar Aalinkeel, Jessica L. Reynolds, Bindukumar B. Nair, Donald E. Sykes, Zihua Hu, Adela Bonoiu, Hong Ding, Paras N. Prasad, and Stanley A. Schwartz

Pharmacological and Neurotoxicological Actions Mediated By Bupropion and Diethylpropion
Hugo R. Arias, Abel Santamaría, and Syed F. Ali

Neural and Cardiac Toxicities Associated With 3,4-Methylenedioxymethamphetamine (MDMA)
Michael H. Baumann and Richard B. Rothman

Cocaine-Induced Breakdown of the Blood–Brain Barrier and Neurotoxicity
Hari S. Sharma, Dafin Muresanu, Aruna Sharma, and Ranjana Patnaik

Cannabinoid Receptors in Brain: Pharmacogenetics, Neuropharmacology, Neurotoxicology, and Potential Therapeutic Applications
Emmanuel S. Onaivi

Intermittent Dopaminergic Stimulation causes Behavioral Sensitization in the Addicted Brain and Parkinsonism
Francesco Fornai, Francesca Biagioni, Federica Fulceri, Luigi Murri, Stefano Ruggieri, Antonio Paparelli

The Role of the Somatotrophic Axis in Neuroprotection and Neuroregeneration of the Addictive Brain
Fred Nyberg

INDEX

Volume 89

Molecular Profiling of Striatonigral and Striatopallidal Medium Spiny Neurons: Past, Present, and Future
Mary Kay Lobo

BAC to Degeneration: Bacterial Artificial Chromosome (Bac)-Mediated Transgenesis for Modeling Basal Ganglia Neurodegenerative Disorders
Xiao-Hong Lu

Behavioral Outcome Measures for the Assessment of Sensorimotor Function in Animal Models of Movement Disorders
Sheila M. Fleming

The Role of DNA Methylation in the Central Nervous System and Neuropsychiatric Disorders
Jian Feng and Guoping Fan

Heritability of Structural Brain Traits: An Endo-phenotype Approach to Deconstruct Schizophrenia
Nil Kaymaz and J. Van Os

The Role of Striatal NMDA Receptors in Drug Addiction
Yao-Ying Ma, Carlos Cepeda, and Cai-Lian Cui

Deciphering Rett Syndrome With Mouse Genetics, Epigenomics, and Human Neurons
Jifang Tao, Hao Wu, and Yi Eve Sun

INDEX

Volume 90

Part I: Introduction

Introductory Remarks on the History and Current Applications of TCS
Matthew B. Stern

Method and Validity of Transcranial Sonography in Movement Disorders
David Školoudík and Uwe Walter

Transcranial Sonography—Anatomy
Heiko Huber

Part II: Transcranial Sonography in Parkinsons Disease

Transcranial Sonography in Relation to SPECT and MIBG
Yoshinori Kajimoto, Hideto Miwa and Tomoyoshi Kondo

Diagnosis of Parkinson's Disease—Transcranial Sonography in Relation to MRI
Ludwig Niehaus and Kai Boelmans

Early Diagnosis of Parkinson's Disease
Alexandra Gaenslen and Daniela Berg

Transcranial Sonography in the Premotor Diagnosis of Parkinson's Disease
Stefanie Behnke, Ute Schroder and Daniela Berg

Pathophysiology of Transcranial Sonography Signal Changes in the Human Substantia Nigra
K. L. Double, G. Todd and S. R. Duma

Transcranial Sonography for the Discrimination of Idiopathic Parkinson's Disease from the Atypical Parkinsonian Syndromes
A. E. P. Bouwmans, A. M. M. Vlaar, K. Srulijes, W. H. Mess AND W. E. J. Weber

Transcranial Sonography in the Discrimination of Parkinson's Disease Versus Vascular Parkinsonism
Pablo Venegas-Francke

TCS in Monogenic Forms of Parkinson's Disease
Kathrin Brockmann and Johann Hagenah

Part III—Transcranial Sonography in other Movement Disorders and Depression

Transcranial Sonography in Brain Disorders with Trace Metal Accumulation
Uwe Walter

Transcranial Sonography in Dystonia
Alexandra Gaenslen

Transcranial Sonography in Essential Tremor
Heike Stockner and Isabel Wurster

VII—Transcranial Sonography in Restless Legs Syndrome
Jana Godau and Martin Sojer

Transcranial Sonography in Ataxia
Christos Krogias, Thomas Postert and Jens Eyding

Transcranial Sonography in Huntington's Disease
Christos Krogias, Jens Eyding and Thomas Postert

Transcranial Sonography in Depression
Milija D. Mijajlovic

Part IV: Future Applications and Conclusion

Transcranial Sonography-Assisted Stereotaxy and Follow-Up of Deep Brain Implants in Patients with Movement Disorders
Uwe Walter

Conclusions
Daniela Berg

INDEX

Volume 91

The Role of microRNAs in Drug Addiction: A Big Lesson from Tiny Molecules
Andrzej Zbigniew Pietrzykowski

The Genetics of Behavioral Alcohol Responses in *Drosophila*
Aylin R. Rodan and Adrian Rothenfluh

Neural Plasticity, Human Genetics, and Risk for Alcohol Dependence
Shirley Y. Hill

Using Expression Genetics to Study the Neurobiology of Ethanol and Alcoholism
Sean P. Farris, Aaron R. Wolen and Michael F. Miles

Genetic Variation and Brain Gene Expression in Rodent Models of Alcoholism: Implications for Medication Development
Karl Björk, Anita C. Hansson and W. olfgang H. Sommer

Identifying Quantitative Trait Loci (QTLs) and Genes (QTGs) for Alcohol-Related Phenotypes in Mice
Lauren C. Milner and Kari J. Buck

Glutamate Plasticity in the Drunken Amygdala: The Making of an Anxious Synapse
Brian A. Mccool, Daniel T. Christian, Marvin R. Diaz and Anna K. Läck

Ethanol Action on Dopaminergic Neurons in the Ventral Tegmental Area: Interaction with Intrinsic Ion Channels and Neurotransmitter Inputs
Hitoshi Morikawa and Richard A. Morrisett

Alcohol and the Prefrontal Cortex
Kenneth Abernathy, L. Judson Chandler and John J. Woodward

BK Channel and Alcohol, A Complicated Affair
Gilles Erwan Martin

A Review of Synaptic Plasticity at Purkinje Neurons with a Focus on Ethanol-Induced Cerebellar Dysfunction
C. Fernando Valenzuela, Britta Lindquist and Paula A. Zflmudio-Bulcock

INDEX

Volume 92

The Development of the Science of Dreaming
Claude Gottesmann

Dreaming as Inspiration: Evidence from Religion, Philosophy, Literature, and Film
Kelly Bulkeley

Developmental Perspective: Dreaming Across the Lifespan and What This Tells Us
Melissa M. Burnham and Christian Conte

REM and NREM Sleep Mentation
Patrick Mcnamara, Patricia Johnson, Deirdre McLaren, Erica Harris,Catherine Beauharnais and Sanford Auerbach

Neuroimaging of Dreaming: State of the Art and Limitations
Caroline Kussé, Vincenzo Muto, Laura Mascetti, Luca Matarazzo, Ariane Foret, Anahita Shaffii-Le Bourdiec and Pierre Maquet

Memory Consolidation, The Diurnal Rhythm of Cortisol, and The Nature of Dreams: A New Hypothesis
Jessica D. Payne

Characteristics and Contents of Dreams
Michael Schredl

Trait and Neurobiological Correlates of Individual Differences in Dream Recall and Dream Content
Mark Blagrove and Edward F. Pace-Schott

Consciousness in Dreams
David Kahn and Tzivia Gover

The Underlying Emotion and the Dream: Relating Dream Imagery to the Dreamer's Underlying Emotion can Help Elucidate the Nature of Dreaming
Ernest Hartmann

Dreaming, Handedness, and Sleep Architecture: Interhemispheric Mechanisms
Stephen D. Christman and Ruth E. Propper

To What Extent Do Neurobiological Sleep-Waking Processes Support Psychoanalysis?
Claude Gottesmann

The Use of Dreams in Modern Psychotherapy
Clara E. Hill and Sarah Knox

INDEX

Volume 93

Underlying Brain Mechanisms that Regulate Sleep-Wakefulness Cycles
Irma Gvilia

What Keeps Us Awake?—the Role of Clocks and Hourglasses, Light, and Melatonin
Christian Cajochen, Sarah Chellappa and Christina Schmidt

Suprachiasmatic Nucleus and Autonomic Nervous System Influences on Awakening From Sleep
Andries Kalsbeek, Chun-xia Yi, Susanne E. la Fleur, Ruud m. Buijs, and Eric Fliers

Preparation for Awakening: Self-Awakening Vs. Forced Awakening: Preparatory Changes in the Pre-Awakening Period
Mitsuo Hayashi, Noriko Matsuura and Hiroki Ikeda

Circadian and Sleep Episode Duration Influences on Cognitive Performance Following the Process of Awakening
Robert L. Matchock

The Cortisol Awakening Response in Context
Angela Clow, Frank Hucklebridge and Lisa Thorn

Causes and Correlates of Frequent Night Awakenings in Early Childhood
Amy Jo Schwichtenberg and Beth Goodlin-Jones

Pathologies of Awakenings: The Clinical Problem of Insomnia Considered From Multiple Theory Levels
Douglas E. Moul

The Neurochemistry of Awakening: Findings from Sleep Disorder Narcolepsy
Seiji Nishino and Yohei Sagawa

INDEX

Volume 94

5-HT_6 Medicinal Chemistry
Kevin G. Liu and Albert J. Robichaud

Patents
Nicolas Vincent Ruiz and Gloria Oranias

5-HT_6 Receptor Charactertization
Teresa Riccioni

5-HT_6 Receptor Signal Transduction: Second Messenger Systems
Xavier Codony, Javier Burgueño, Maria Javier Ramírez and José Miguel Vela

Electrophysiology of 5-HT_6 Receptors
Annalisa Tassone, Graziella Madeo, Giuseppe Sciamanna, Antonio Pisani and Paola Bonsi

Genetic Variations and Association
Massimo Gennarelli and Annamaria Cattaneo

Pharmacokinetics of 5-HT_6 Receptor Ligands
Angelo Mancinelli

INDEX

Volume 95

Introductory Remarks: Catechol-O-Methyltransferase Inhibition–An Innovative Approach to Enhance L-dopa Therapy in Parkinson's Disease with Dual Enzyme Inhibition
Erkki Nissinen

The Catechol-O-Methyltransferase Gene: its Regulation and Polymorphisms
Elizabeth M. Tunbridge

Distribution and Functions of Catechol-O-Methyltransferase Proteins: Do Recent Findings Change the Picture?
Timo T. Myöhänen and Pekka T. Männistö

Catechol-O-Methyltransferase Enzyme: Cofactor S-Adenosyl-L-Methionine and Related Mechanisms
Thomas Müller

Biochemistry and Pharmacology of Catechol-O-Methyltransferase Inhibitors
Erkki nissinen and Pekka T. Männisto

The Chemistry of Catechol-O-Methyltransferase Inhibitors
David A. Learmonth, Lászlǒ E. Kiss, and Patrício Soares-da-Silva

Toxicology and Safety of COMT Inhibitors
Kristiina Haasio

Catechol-O-Methyltransferase Inhibitors in Preclinical Models as Adjuncts of L-dopa Treatment
Concepció Marin and J. A. Obeso

Problems with the Present Inhibitors and a Relevance of New and Improved COMT Inhibitors in Parkinson's Disease
Seppo Kaakkola

Catechol-O-Methyltransferase and Pain
Oleg Kambur and Pekka T. Männistö

INDEX

Volume 96

The Central Role of 5-HT_6 Receptors in Modulating Brain Neurochemistry
Lee A. Dawson

5-HT_6 Receptor Memory and Amnesia: Behavioral Pharmacology – Learning and Memory Processes
Alfredo Meneses, G. Pérez-García, R. Tellez, T. Ponce-Lopez and C. Castillo

Behavioral Pharmacology: Potential Antidepressant and Anxiolytic Properties
Anna Wesołowska and Magdalena Jastrzbska-Wisek

The 5-HT_6 Receptor as a Target for Developing Novel Antiobesity Drugs
David Heal, Jane Gosden and Sharon Smith

Behavioral and Neurochemical Pharmacology of 5-HT_6 Receptors Related to Reward and Reinforcement
Gaetano Di Chiara, Valentina Valentini and Sandro Fenu

5-HT_6 Receptor Ligands and their Antipsychotic Potential
Jørn Arnt and Christina Kurre Olsen

5-HT_6 Receptor Ligands as Antidementia Drugs
Ellen Siobhan Mitchell

Other 5-HT_6 Receptor-Mediated Effects
Franco Borsini

INDEX

Volume 97

Behavioral Pharmacology of Orofacial Movement Disorders
Noriaki Koshikawa, Satoshi Fujita and Kazunori Adachi

Regulation of Orofacial Movement: Dopamine Receptor Mechanisms and Mutant Models
John L. Waddington, Gerard J. O'Sullivan and Katsunori Tomiyama

Regulation of Orofacial Movement: Amino Acid Mechanisms and Mutant Models
Katsunori Tomiyama, Colm M.P. O'Tuathaigh, and John L. Waddington

The Trigeminal Circuits Responsible for Chewing
Karl-Gunnar Westberg and Arlette Kolta

Ultrastructural Basis for Craniofacial Sensory Processing in the Brainstem
Yong Chul Bae and Atsushi Yoshida

Mechanisms of Nociceptive Transduction and Transmission: A Machinery for Pain Sensation and Tools for Selective Analgesia
Alexander M. Binshtok

Peripheral and Central Mechanisms of Orofacial Inflammatory Pain
Barry J. Sessle

The Role of Trigeminal Interpolaris-Caudalis Transition Zone in Persistent Orofacial Pain
Ke Ren and Ronald Dubner

Physiological Mechanisms of Neuropathic Pain: The Orofacial Region
Koichi Iwata, Yoshiki Imamura, Kuniya Honda and Masamichi Shinoda

Neurobiology of Estrogen Status in Deep Craniofacial Pain
David A Bereiter and Keiichiro Okamoto

Macroscopic Connection of Rat Insular Cortex: Anatomical Bases Underlying its Physiological Functions
Masayuki Kobayashi

The Balance Between Excitation And Inhibition And Functional Sensory Processing in the Somatosensory Cortex
Zhi Zhang and Qian-Quan Sun

INDEX

Volume 98

An Introduction to Dyskinesia—the Clinical Spectrum
Ainhi Ha and Joseph Jankovic

L-dopa-induced Dyskinesia—Clinical Presentation, Genetics, And Treatment
L.K. Prashanth, Susan Fox and Wassilios G. Meissner

Experimental Models of L-DOPA-induced Dyskinesia
Tom H. Johnston and Emma L. Lane

Molecular Mechanisms of L-DOPA-induced Dyskinesia
Gilberto Fisone and Erwan Bezard

New Approaches to Therapy
Jonathan Brotchie and Peter Jenner

Surgical Approach to L-DOPA-induced Dyskinesias
Tejas Sankar and Andres M. Lozano

Clinical and Experimental Experiences of Graft-induced Dyskinesia
Emma L. Lane

Tardive Dyskinesia: Clinical Presentation and Treatment
P.N. van Harten and D.E. Tenback

Epidemiology and Risk Factors for (Tardive) Dyskinesia
D.E. Tenback and P.N. van Harten

Genetics of Tardive Dyskinesia
Heon-Jeong Lee and Seung-Gul Kang

Animal Models of Tardive Dyskinesia
S.K. Kulkarni and Ashish Dhir

Surgery for Tardive Dyskinesia
Stephane Thobois, Alice Poisson and Philippe Damier

Huntington's Disease: Clinical Presentation and Treatment
M.J.U. Novak and S.J. Tabrizi

Genetics and Neuropathology of Huntington's Disease: Huntington's Disease
Anton Reiner, Ioannis Dragatsis and Paula Dietrich

Pathogenic Mechanisms in Huntington's Disease
Lesley Jones and Alis Hughes

Experimental Models of HD And Reflection on Therapeutic Strategies
Olivia L. Bordiuk, Jinho Kim and Robert J. Ferrante

Cell-based Treatments for Huntington's Disease
Stephen B. Dunnett and Anne E. Rosser

Clinical Phenomenology of Dystonia
Carlo Colosimo and Alfredo Berardelli

Genetics and Pharmacological Treatment of Dystonia
Susan Bressman and Matthew James

Experimental Models of Dystonia
A. Tassone, G. Sciamanna, P. Bonsi, G. Martella and A. Pisani

Surgical Treatment of Dystonia
John Yianni, Alexander L. Green and Tipu Z. Aziz

INDEX

Volume 99

Seizure and Epilepsy: Studies of Seizure-disorders in *Drosophila*
Louise Parker, Iris C. Howlett, Zeid M. Rusan and Mark A. Tanouye

Homeostatic Control of Neural Activity: A *Drosophila* Model for Drug Tolerance and Dependence
Alfredo Ghezzi and Nigel S. Atkinson

Attention in *Drosophila*
Bruno van Swinderen

The roles of *Fruitless* and *Doublesex* in the Control of Male Courtship
Brigitte Dauwalder

Circadian Plasticity: from Structure to Behavior
Lia Frenkel and María Fernanda Ceriani

Learning and Memory in *Drosophila*: Behavior, Genetics, and Neural Systems
Lily Kahsai and Troy Zars

Studying Sensorimotor Processing with Physiology in Behaving *Drosophila*
Johannes D. Seelig and Vivek Jayaraman

Modeling Human Trinucleotide Repeat Diseases in *Drosophila*
Zhenming Yu and Nancy M. Bonini

From Genetics to Structure to Function: Exploring Sleep in *Drosophila*
Daniel Bushey and Chiara Cirelli

INDEX

Volume 100

Structural Properties of Human Monoamine Oxidases A and B
Claudia Binda, Andrea Mattevi and Dale E. Edmondson

Behavioral Outcomes of Monoamine Oxidase Deficiency: Preclinical and Clinical Evidence
Marco Bortolato and Jean C. Shih

Kinetic Behavior and Reversible Inhibition of Monoamine Oxidases—Enzymes that Many Want Dead
Keith F. Tipton, Gavin P. Davey and Andrew G. McDonald

The Pharmacology of Selegiline
Kálmán Magyar

Type A Monoamine Oxidase Regulates Life and Death of Neurons in Neurodegeneration and Neuroprotection
Makoto Naoi, Wakako Maruyama, Keiko Inaba-Hasegawa and Yukihiro Akao

Multimodal Drugs and their Future for Alzheimer's and Parkinson's Disease
Cornelis J. Van der Schyf and Werner J. Geldenhuys

Neuroprotective Profile of the Multitarget Drug Rasagiline in Parkinson's Disease
Orly Weinreb, Tamar Amit, Peter Riederer, Moussa B.H. Youdim and Silvia A. Mandel

Rasagiline in Parkinson's Disease
L.M. Chahine and M.B. Stern

Selective Inhibitors of Monoamine Oxidase Type B and the "Cheese Effect"
John P.M. Finberg and Ken Gillman

A Novel Anti-Alzheimer's Disease Drug, Ladostigil: Neuroprotective, Multimodal Brain-Selective Monoamine Oxidase and Cholinesterase Inhibitor
Orly Weinreb, Tamar Amit, Orit Bar-Am and Moussa B.H. Youdim

Novel MAO-B Inhibitors: Potential Therapeutic Use of the Selective MAO-B Inhibitor PF9601N in Parkinson's Disease
Mercedes Unzeta and Elisenda Sanz

Index

Volume 101

General Overview: Biomarkers in Neuroscience Research
Michaela D. Filiou and Christoph W. Turck

Imaging Brain Microglial Activation Using Positron Emission Tomography and Translocator Protein-Specific Radioligands
David R.J. Owen and Paul M. Matthews

The Utility of Gene Expression in Blood Cells for Diagnosing Neuropsychiatric Disorders
Christopher H. Woelk, Akul Singhania, Josué Pérez-Santiago, Stephen J. Glatt and Ming T. Tsuang

Proteomic Technologies for Biomarker Studies in Psychiatry: Advances and Needs
Daniel Martins-de-Souza, Paul C. Guest, Natacha Vanattou-Saifoudine, Laura W. Harris and Sabine Bahn

Converging Evidence of Blood-Based Biomarkers for Schizophrenia: An update
Man K. Chan, Paul C. Guest, Yishai Levin, Yagnesh Umrania, Emanuel Schwarz, Sabine Bahn and Hassan Rahmoune

Abnormalities in Metabolism and Hypothalamic–Pituitary–Adrenal Axis Function in Schizophrenia
Paul C. Guest, Daniel Martins-de-Souza, Natacha Vanattou-Saifoudine, Laura W. Harris and Sabine Bahn

Immune and Neuroimmune Alterations in Mood Disorders and Schizophrenia
Roosmarijn C. Drexhage, Karin Weigelt, Nico van Beveren, Dan Cohen, Marjan A. Versnel, Willem A. Nolen and Hemmo A. Drexhage

Behavioral and Molecular Biomarkers in Translational Animal Models for Neuropsychiatric Disorders
Zoltán Sarnyai, Murtada Alsaif, Sabine Bahn, Agnes Ernst, Paul C. Guest, Eva Hradetzky, Wolfgang Kluge, Viktoria Stelzhammer and Hendrik Wesseling

Stem Cell Models for Biomarker Discovery in Brain Disease
Alan Mackay-Sim, George Mellick and Stephen Wood

The Application of Multiplexed Assay Systems for Molecular Diagnostics
Emanuel Schwarz, Nico J.M. VanBeveren, Paul C. Guest, Rauf Izmailov and Sabine Bahn

Algorithm Development for Diagnostic Biomarker Assays
Rauf Izmailov, Paul C. Guest, Sabine Bahn and Emanuel Schwarz

Challenges of Introducing New Biomarker Products for Neuropsychiatric Disorders into the Market
Sabine Bahn, Richard Noll, Anthony Barnes, Emanuel Schwarz and Paul C. Guest

Toward Personalized Medicine in the Neuropsychiatric Field
Erik H.F. Wong, Jayne C. Fox, Mandy Y.M. Ng and Chi-Ming Lee

Clinical Utility of Serum Biomarkers for Major Psychiatric Disorders
Nico J.M. van Beveren and Witte J.G. Hoogendijk

The Future: Biomarkers, Biosensors, Neuroinformatics, and E-Neuropsychiatry
Christopher R. Lowe

Subject Index

Volume 102

The Function and Mechanisms of Nurr1 Action in Midbrain Dopaminergic Neurons, from Development and Maintenance to Survival
Yu Luo

Monoclonal Antibodies as Novel Neurotherapeutic Agents in CNS Injury and Repair
Aruna Sharma and Hari Shanker Sharma

The Blood–Brain Barrier in Alzheimer's Disease: Novel Therapeutic Targets and Nanodrug delivery
Hari Shanker Sharma, Rudy J. Castellani, Mark A. Smith and Aruna Sharma

Neurovascular Aspects of Amyotrophic Lateral Sclerosis
Maria Carolina O. Rodrigues, Diana G. Hernandez-Ontiveros, Michael K. Louis, Alison E. Willing, Cesario V. Borlongan, Paul R. Sanberg, Júlio C. Voltarelli and Svitlana Garbuzova-Davis

Quercetin in Hypoxia-Induced Oxidative Stress: Novel Target for Neuroprotection
Anand Kumar Pandey, Ranjana Patnaik, Dafin F. Muresanu, Aruna Sharma and Hari Shanker Sharma

Environmental Conditions Modulate Neurotoxic Effects of Psychomotor Stimulant Drugs of Abuse
Eugene A. Kiyatkin and Hari Shanker Sharma

Central Nervous Tissue Damage after Hypoxia and Reperfusion in Conjunction with Cardiac Arrest and Cardiopulmonary Resuscitation: Mechanisms of Action and Possibilities for Mitigation
Lars Wiklund, Cecile Martijn, Adriana Miclescu, Egidijus Semenas, Sten Rubertsson and Hari Shanker Sharma

Interactions Between Opioids and Anabolic Androgenic Steroids: Implications for the Development of Addictive Behavior
Fred Nyberg and Mathias Hallberg

Neurotrophic Factors and Neurodegenerative Diseases: A Delivery Issue
Barbara Ruozi, Daniela Belletti, Lucia Bondioli, Alessandro De Vita, Flavio Forni, Maria Angela Vandelli and Giovanni Tosi

Neuroprotective Effects of Cerebrolysin, a Combination of Different Active Fragments of Neurotrophic Factors and Peptides on the Whole Body Hyperthermia-Induced Neurotoxicity: Modulatory Roles of Co-morbidity Factors and Nanoparticle Intoxication
Hari Shanker Sharma, Aruna Sharma, Herbert Mössler and Dafin Fior Muresanu

Alzheimer's Disease and Amyloid: Culprit or Coincidence?
Stephen D. Skaper

Vascular Endothelial Growth Factor and Other Angioglioneurins: Key Molecules in Brain Development and Restoration
José Vicente Lafuente, Naiara Ortuzar, Harkaitz Bengoetxea, Susana Bulnes and Enrike G. Argandoña

INDEX

Volume 103

Lost and Found in Behavioral Informatics
Melissa A. Haendel and Elissa J. Chesler

Biological Databases for Behavioral Neurobiology
Erich J. Baker

A Survey of the Neuroscience Resource Landscape: Perspectives from the Neuroscience Information Framework
Jonathan Cachat, Anita Bandrowski, Jeffery S. Grethe, Amarnath Gupta, Vadim Astakhov, Fahim Imam, Stephen D. Larson, and Maryann E. Martone

The Neurobehavior Ontology: An Ontology for Annotation and Integration of Behavior and Behavioral Phenotypes
Georgios V. Gkoutos, Paul N. Schofield, and Robert Hoehndorf

Ontologies for Human Behavior Analysis and Their Application to Clinical Data
Janna Hastings and Stefan Schulz

Text-Mining and Neuroscience
Kyle H. Ambert and Aaron M. Cohen

Applying *In Silico* Integrative Genomics to Genetic Studies of Human Disease: A Review
Scott F. Saccone

SUBJECT INDEX

Volume 104

Cross Species Integration of Functional Genomics Experiments
Jeremy J. Jay

Model Organism Databases in Behavioral Neuroscience
Mary Shimoyama, Jennifer R. Smith, G. Thomas Hayman, Victoria Petri, and Rajni Nigam

Accessing and Mining Data from Large-Scale Mouse Phenotyping Projects
Hugh Morgan, Michelle Simon, and Ann-Marie Mallon

Bioinformatics Resources for Behavior Studies in the Laboratory Mouse
Carol J. Bult

Using Genome-Wide Expression Profiling to Define Gene Networks Relevant to the Study of Complex Traits: From RNA Integrity to Network Topology
M.A. O'Brien, B.N. Costin, and M.F. Miles

Genetic and Molecular Network Analysis of Behavior
Robert W. Williams and Megan K. Mulligan

Large-Scale Neuroinformatics for *In Situ* Hybridization Data in the Mouse Brain
Lydia L. Ng, Susan M. Sunkin, David Feng, Chris Lau, Chinh Dang, and Michael J. Hawrylycz

Opportunities for Bioinformatics in the Classification of Behavior and Psychiatric Disorders
Elissa J. Chesler and Ryan W. Logan

SUBJECT INDEX

Volume 105

Optic Nerve Disease and Axon Pathophysiology
Alireza Ghaffarieh and Leonard A. Levin

Role of Electrical Activity of Neurons for Neuroprotection
Takeshi Morimoto

Molecular Control of Axon Growth: Insights from Comparative Gene Profiling and High-Throughput Screening
Murray G. Blackmore

Gatekeeper Between Quiescence and Differentiation: p53 in Axonal Outgrowth and Neurogenesis
Giorgia Quadrato and Simone Di Giovanni

Cyclin-Dependent Kinase 5 in Axon Growth and Regeneration
Tao Ye, Amy K. Y. Fu, and Nancy Y. Ip

Rho Signaling and Axon Regeneration
L. McKerracher, Gino B. Ferraro, and Alyson E. Fournier

Neuron-Intrinsic Inhibitors of Axon Regeneration: PTEN and SOCS3
Xueting Luo and Kevin K. Park

INDEX

Volume 106

Neurotrophic Factors and the Regeneration of Adult Retinal Ganglion Cell Axons
Alan R. Harvey, Jacob Wei Wei Ooi, and Jennifer Rodger

MBS: Signaling Endosomes and Growth Cone Motility in Axon Regeneration
Michael B. Steketee and Jeffrey L. Goldberg

Intrinsic Mechanisms Regulating Axon Regeneration: An Integrin Perspective
Richard Eva, Melissa R. Andrews, Elske H.P. Franssen, and James W. Fawcett

The Role of Serotonin in Axon and Dendrite Growth
Ephraim F. Trakhtenberg and Jeffrey L. Goldberg

Inflammatory Pathways in Spinal Cord Injury
Samuel David, Juan Guillermo Zarruk, and Nader Ghasemlou

Combinatorial Therapy Stimulates Long-Distance Regeneration, Target Reinnervation, and Partial Recovery of Vision After Optic Nerve Injury in Mice
Silmara de Lima, Ghaith Habboub, and Larry I. Benowitz

From Bench to Beside to Cure Spinal Cord Injury: Lost in Translation?
Andreas Hug and Norbert Weidner

SUBJECT INDEX

Volume 107

Neuromodulation: A More Comprehensive Concept Beyond Deep Brain Stimulation
Clement Hamani and Elena Moro

Computational Models of Neuromodulation
Christopher R. Butson

Neurophysiology of Deep Brain Stimulation
Manuela Rosa, Gaia Giannicola, Sara Marceglia, Manuela Fumagalli, Sergio Barbieri, and Alberto Priori

Neurophysiology of Cortical Stimulation
Jean-Pascal Lefaucheur

Neural Mechanisms of Spinal Cord Stimulation
Robert D. Foreman and Bengt Linderoth

Magnetoencephalography and Neuromodulation
Alfons Schnitzler and Jan Hirschmann

Current Challenges to the Clinical Translation of Brain Machine Interface Technology
Charles W. Lu, Parag G. Patil, and Cynthia A. Chestek

Nanotechnology in Neuromodulation
Russell J. Andrews

Optogenetic Neuromodulation
Paul S. A. Kalanithi and Jaimie M. Henderson

Diffusion Tensor Imaging and Neuromodulation: DTI as Key Technology for Deep Brain Stimulation
Volker Arnd Coenen, Thomas E. Schlaepfer, Niels Allert, and Burkhard Mädler

DBS and Electrical Neuro-Network Modulation to Treat Neurological Disorders
Amanda Thompson, Takashi Morishita, and Michael S. Okun

Neuromodulation in Psychiatric Disorders
Yasin Temel, Sarah A. Hescham, Ali Jahanshahi, Marcus L. F. Janssen, Sonny K. H. Tan, Jacobus J. van Overbeeke, Linda Ackermans, Mayke Oosterloo, Annelien Duits, Albert F. G. Leentjens, and LeeWei Lim

Ethical Aspects of Neuromodulation
Christiane Woopen

Subject Index

Volume 108

Tissue Engineering and Regenerative Medicine: Past, Present, and Future
António J. Salgado, Joaquim M. Oliveira, Albino Martins, Fábio G. Teixeira, Nuno A. Silva, Nuno M. Neves, Nuno Sousa, and Rui L. Reis

Tissue Engineering and Peripheral Nerve Reconstruction: An Overview
Stefano Geuna, S. Gnavi, I. Perroteau, Pierluigi Tos, and B. Battiston

Bone Marrow Mesenchymal Stem Cell Transplantation for Improving Nerve Regeneration
Júlia Teixeira Oliveira, Klauss Mostacada, Silmara de Lima, and Ana Maria Blanco Martinez

Perspectives of Employing Mesenchymal Stem Cells from the Wharton's Jelly of the Umbilical Cord for Peripheral Nerve Repair
Jorge Ribeiro, Andrea Gartner, Tiago Pereira, Raquel Gomes, Maria Ascensão Lopes, Carolina Gonçalves, Artur Varejão, Ana Lúcia Luís, and Ana Colette Maurício

Adipose-Derived Stem Cells and Nerve Regeneration: Promises and Pitfalls
Alessandro Faroni, Giorgio Terenghi, and Adam J. Reid

The Pros and Cons of Growth Factors and Cytokines in Peripheral Axon Regeneration
Lars Klimaschewski, Barbara Hausott, and Doychin N. Angelov

Role of Inflammation and Cytokines in Peripheral Nerve Regeneration
P. Dubový, R. Jančálek, and T. Kubek

Ghrelin: A Novel Neuromuscular Recovery Promoting Factor?
Raimondo Stefania, Ronchi Giulia, Geuna Stefano, Pascal Davide, Reano Simone, Filigheddu Nicoletta, and Graziani Andrea

Neuregulin 1 Role in Schwann Cell Regulation and Potential Applications to Promote Peripheral Nerve Regeneration
Giovanna Gambarotta, Federica Fregnan, Sara Gnavi, and Isabelle Perroteau

Extracellular Matrix Components in Peripheral Nerve Regeneration
Francisco Gonzalez-Perez, Esther Udina, and Xavier Navarro

Subject Index

Volume 109

The Use of Chitosan-Based Scaffold to Enhance Regeneration in the Nervous System
Sara Gnavi, Christina Barwig, Thomas Freier, Kirsten Haarstert-Talini, Claudia Grothe, and Stefano Geuna

Interfaces with the Peripheral Nerve for the Control of Neuroprostheses
Jaume del Valle and Xavier Navarro

The Use of Shock Waves in Peripheral Nerve Regeneration: New Perspectives?
Thomas Hausner and Antal Nógrádi

Phototherapy and Nerve Injury: Focus on Muscle Response
Shimon Rochkind, Stefano Geuna, and Asher Shainberg

Electrical Stimulation for Promoting Peripheral Nerve Regeneration
Kirsten Haastert-Talini and Claudia Grothe

Role of Physical Exercise for Improving Post-traumatic Nerve Regeneration
Paulo A.S. Armada-da-Silva, Cátia Pereira, SandraAmado, and António P. Veloso

The Role of Timing in Nerve Reconstruction
Lars B. Dahlin

Future Perspectives in Nerve Repair and Regeneration
Pierluigi Tos, Giulia Ronchi, Stefano Geuna, and Bruno Battiston

INDEX

Volume 110

The Relevance of Metals in the Pathophysiology of Neurodegeneration, Pathological Considerations
Kurt A. Jellinger

Pantothenate Kinase-Associated Neurodegeneration (PKAN) and *PLA2G6*-Associated Neurodegeneration (PLAN): Review of Two Major Neurodegeneration with Brain Iron Accumulation (NBIA) Phenotypes
Manju A. Kurian and Susan J. Hayflick

Mitochondrial Membrane Protein-Associated Neurodegeneration (MPAN)
Monika Hartig, Holger Prokisch, Thomas Meitinger, and Thomas Klopstock

BPAN: The Only X-Linked Dominant NBIA Disorder
T.B. Haack, P. Hogarth, A. Gregory, P. Prokisch, and S.J. Hayflick

Neuroferritinopathy
M.J. Keogh, C.M. Morris, and P.F. Chinnery

Aceruloplasminemia: An Update
Satoshi Kono

Therapeutic Advances in Neurodegeneration with Brain Iron Accumulation
Giovanna Zorzi and Nardo Nardocci

The Neuropathology of Neurodegeneration with Brain Iron Accumulation
Michael C. Kruer

Imaging of Iron
Petr Dusek, Monika Dezortova, and Jens Wuerfel

The Role of Iron Imaging in Huntington's Disease
S.J.A. van den Bogaard, E.M. Dumas, and R.A.C. Roos

Lysosomal Storage Disorders and Iron
Jose Miguel Bras

Manganese and the Brain
Karin Tuschl, Philippa B. Mills, and Peter T. Clayton

Update on Wilson Disease
Aggarwal Annu and Bhatt Mohit

An Update on Primary Familial Brain Calcification
R.R. Lemos, J.B.M.M. Ferreira, M.P. Keasey, and J.R.M. Oliveira

INDEX

Volume 111

History of Acupuncture Research
Yi Zhuang, Jing-jing Xing, Juan Li, Bai-Yun Zeng, and Fan-rong Liang

Effects of Acupuncture Needling with Specific Sensation on Cerebral Hemodynamics and Autonomic Nervous Activity in Humans
Kouich Takamoto, Susumu Urakawa, Kazushige Sakai, Taketoshi Ono, and Hisao Nishijo

Acupuncture Point Specificity
Jing-jing Xing, Bai-Yun Zeng, Juan Li, Yi Zhuang, and Fan-rong Liang

Acupuncture Stimulation Induces Neurogenesis in Adult Brain
Min-Ho Nam, Kwang Seok Ahn, and Seung-Hoon Choi

Acupuncture and Neurotrophin Modulation
Marzia Soligo, Stefania Lucia Nori, Virginia Protto, Fulvio Florenzano, and Luigi Manni

Acupuncture Stimulation and Neuroendocrine Regulation
Jung-Sheng Yu, Bai-Yun Zeng, and Ching-Liang Hsieh

Current Development of Acupuncture Research in Parkinson's Disease
Bai-Yun Zeng, Sarah Salvage, and Peter Jenner

Acupuncture Therapy for Stroke Patients
Xin Li and Qiang Wang

Effects of Acupuncture Therapy on Alzheimer's Disease
Bai-Yun Zeng, Sarah Salvage, and Peter Jenner

Acupuncture Therapy for Psychiatric Illness
Karen Pilkington

Acupuncture for the Treatment of Insomnia
Kaicun Zhao

Acupuncture for the Treatment of Drug Addiction
Cai-Lian Cui, Liu-Zhen Wu, and Yi-jing Li

Acupuncture Regulation of Blood Pressure: Two Decades of Research
John C. Longhurst and Stephanie Tjen-A-Looi

Effect and Mechanism of Acupuncture on Gastrointestinal Diseases
Toku Takahashi

INDEX

Volume 112

An Introduction to the Clinical Phenomenology of Tourette Syndrome
Davide Martino, Namrata Madhusudan, Panagiotis Zis, and Andrea E. Cavanna

Functional Neuroanatomy of Tics
Irene Neuner, Frank Schneider, and N. Jon Shah

Functional Imaging of Dopaminergic Neurotransmission in Tourette Syndrome
Bàrbara Segura and Antonio P. Strafella

Nondopaminergic Neurotransmission in the Pathophysiology of Tourette Syndrome
Patrick T. Udvardi, Ester Nespoli, Francesca Rizzo, Bastian Hengerer, and Andrea G. Ludolph

Reinforcement Learning and Tourette Syndrome
Stefano Palminteri and Mathias Pessiglione

Genetic Susceptibility and Neurotransmitters in Tourette Syndrome
Peristera Paschou, Thomas V. Fernandez, Frank Sharp, Gary A. Heiman, and Pieter J. Hoekstra

Pharmacological Animal Models of Tic Disorders
Kevin W. McCairn and Masaki Isoda

Animal Models Recapitulating the Multifactorial Origin of Tourette Syndrome
Simone Macrì, Martina Proietti Onori, Veit Roessner, and Giovanni Laviola

Neuroendocrine Aspects of Tourette Syndrome
Davide Martino, Antonella Macerollo, and James F. Leckman

Clinical Pharmacology of Dopamine-Modulating Agents in Tourette's Syndrome
Sabine Mogwitz, Judith Buse, Stefan Ehrlich, and Veit Roessner

Clinical Pharmacology of Nondopaminergic Drugs in Tourette Syndrome
Andreas Hartmann

Antiepileptic Drugs and Tourette Syndrome
Andrea E. Cavanna and Andrea Nani

Clinical Pharmacology of Comorbid Obsessive–Compulsive Disorder in Tourette Syndrome
Valeria Neri and Francesco Cardona

Clinical Pharmacology of Comorbid Attention Deficit Hyperactivity Disorder in Tourette Syndrome
Renata Rizzo and Mariangela Gulisano

Emerging Treatment Strategies in Tourette Syndrome: What's in the Pipeline?
C. Termine, C. Selvini, G. Rossi, and U. Balottin

Tics and Other Stereotyped Movements as Side Effects of Pharmacological Treatment
Marcos Madruga-Garrido and Pablo Mir

INDEX

Volume 113

Autism Spectrum Disorder and the Cerebellum
Esther B.E. Becker and Catherine J. Stoodley

Contribution of Long Noncoding RNAs to Autism Spectrum Disorder Risk
Brent Wilkinson and Daniel B. Campbell

Identifying Essential Cell Types and Circuits in Autism Spectrum Disorders
Susan E. Maloney, Michael A. Rieger, and Joseph D. Dougherty

Connecting Signaling Pathways Underlying Communication to ASD Vulnerability
Stephanie Lepp, Ashley Anderson, and Genevieve Konopka

MET Receptor Tyrosine Kinase as an Autism Genetic Risk Factor
Yun Peng, Matthew Huentelman, Christopher Smith, and Shenfeng Qiu

Transcriptional Dysregulation of Neocortical Circuit Assembly in ASD
Kenneth Y. Kwan

Motor Skill in Autism Spectrum Disorders: A Subcortical View
Leanne Chukoskie, Jeanne Townsend, and Marissa Westerfield

Orchestration of Neurodevelopmental Programs by RBFOX1: Implications for Autism Spectrum Disorder
Brent R. Bill, Jennifer K. Lowe, Christina T. DyBuncio, and Brent L. Fogel

Immune Dysregulation in Autism Spectrum Disorder
Elaine Y. Hsiao

Autism Susceptibility Genes and the Transcriptional Landscape of the Human Brain
Shingo Miyauchi and Irina Voineagu

INDEX

Volume 114

Modern Concepts of Focal Epileptic Networks
Premysl Jiruska, Marco de Curtis, and John G.R. Jefferys

Neocortical Focus: Experimental View
Igor Timofeev, Sylvain Chauvette, and Sara Soltani

Malformations of Cortical Development and Neocortical Focus
Heiko J. Luhmann, Werner Kilb, and Hans Clusmann

Limbic Networks and Epileptiform Synchronization: The View from the Experimental Side
Charles Behr, Margherita D'Antuono, Shabnam Hamidi, Rochelle Herrington, Maxime Lévesque, Pariya Salami, Zahra Shiri, Rüdiger Köhling, and Massimo Avoli

Limbic Networks: Clinical Perspective
Aylin Y. Reid and Richard J. Staba

Modern Concepts of Seizure Modeling
Christophe Bernard, Sebastien Naze, Timothée Proix, and Viktor K. Jirsa

Mechanisms of Ictogenesis
Thomas Blauwblomme, Premysl Jiruska, and Gilles Huberfeld

Seizure Termination
Frédéric Zubler, Andreas Steimer, Heidemarie Gast, and Kaspar A. Schindler

Epileptic Focus and Alteration of Metabolism
Jakub Otáhal, Jaroslava Folbergrová, Richard Kovacs, Wolfram S. Kunz, and Nicola Maggio

Modern Techniques of Epileptic Focus Localization
Lukas Martinkovic, Hrvoje Hecimovic, Vlastimil Sulc, Radek Marecek, and Petr Marusic

From Treatment to Cure: Stopping Seizures, Preventing Seizures, and Reducing Brain Propensity to Seize
Ivan Pavlov and Stephanie Schorge

INDEX

Volume 115

Environmental Alterations of Epigenetics Prior to the Birth
Chiao-Ling Lo and Feng C. Zhou

Transgenerational Epigenetics and Brain Disorders
Nadia Rachdaoui and Dipak K. Sarkar

The Epigenetic Landscape of Alcoholism
Harish R. Krishnan, Amul J. Sakharkar, Tara L. Teppen, Tiffani D.M. Berkel, and Subhash C. Pandey

Epigenetic Regulatory Mechanisms in Stress-Induced Behavior
Sumana Chakravarty, Salil Saurav Pathak, Swati Maitra, Nitin Khandelwal, Karisetty Bhanu Chandra, and Arvind Kumar

Epigenetics of Schizophrenia: An Open and Shut Case
David P. Gavin and Christina Floreani

Epigenetic Mechanisms in Autism Spectrum Disorder
Adrian Zhubi, Edwin H. Cook, Alessandro Guidotti, and Dennis R. Grayson

MicroRNAs and Ethanol Toxicity
Rajesh C. Miranda

INDEX

Volume 116

Introduction to Sequencing the Brain Transcriptome
Robert Hitzemann, Priscila Darakjian, Nikki Walter, Ovidu Iancu, Robert Searles, and Shannon McWeeney

Analysis Considerations for Utilizing RNA-Seq to Characterize the Brain Transcriptome
Christina Zheng, Sunita Kawane, Daniel Bottomly, and Beth Wilmot

Data Integration and Reproducibility for High-Throughput Transcriptomics
Michael Mooney and Shannon McWeeney

Coexpression and Cosplicing Network Approaches for the Study of Mammalian Brain Transcriptomes
Ovidiu Dan Iancu, Alexander Colville, Priscila Darakjian, and Robert Hitzemann

Splicing in the Human Brain
Ammar Zaghlool, Adam Ameur, Lucia Cavalier, and Lars Feuk

Understanding Complex Transcriptome Dynamics in Schizophrenia and Other Neurological Diseases Using RNA Sequencing
Xi Wang and Murray J. Cairns

The Central Role of Noncoding RNA in the Brain
Boris Guennewig and Antony A. Cooper

Genetics of Gene Expression in CNS
Robert W. Williams and Ashutosh K Pandey

Transcriptomic Changes in Brain Development
Allissa A. Dillman and Mark R. Cookson

Gene Expression in the Addicted Brain
Zhifeng Zhou, Mary-Anne Enoch, and David Goldman

RNA-Seq Reveals Novel Transcriptional Reorganization in Human Alcoholic Brain
Sean P. Farris and R. Dayne Mayfield

INDEX

Volume 117

Learning-Induced Structural Plasticity in the Cerebellum
Hiroshi Nishiyama

Cerebellar Mechanisms of Learning and Plasticity Revealed by Delay Eyelid Conditioning
Michael D. Mauk, Wenke Li, Andrei Khilkevich, and Hunter Halverson

Cerebellar Long-Term Potentiation: Cellular Mechanisms and Role in Learning
Giorgio Grasselli and Christian Hansel

The Ontogeny of Associative Cerebellar Learning
John H. Freeman

INDEX

Volume 118

Neuroimmune Mechanisms of Alcohol and Drug Addiction
Changhai Cui, David Shurtleff, and R. Adron Harris

Neuroimmune Pathways in Alcohol Consumption: Evidence from Behavioral and Genetic Studies in Rodents and Humans
Gizelle Robinson, Dana Most, Laura B. Ferguson, Jody Mayfield, R. Adron Harris, and Yuri A. Blednov

Fetal Alcohol Spectrum Disorders and Neuroimmune Changes
Paul D. Drew and Cynthia J.M. Kane

Role of Microglia in Regulation of Ethanol Neurotoxic Action
Lucy Chastain and Dipak K. Sarkar

Functions of the Chemokine Receptor CXCR4 in the Central Nervous System and Its Regulation by μ-Opioid Receptors
Bradley Nash and Olimpia Meucci

Discovery of a Novel Site of Opioid Action at the Innate Immune Pattern-Recognition Receptor TLR4
Jonathan Henry W. Jacobsen, Linda R. Watkins, and Mark R. Hutchinson

Neuroimmune Basis of Methamphetamine Toxicity
Jennifer M. Loftis and Aaron Janowsky

Marijuana Use Brain Immune Mechanisms
Guy A. Cabral and Melissa Jamerson

Interactions of HIV and Drugs of Abuse: The Importance of Glia and Host Genetic Factors
Kurt F. Hauser and Pamela E. Knapp

Neuroimmune Basis of Alcoholic Brain Damage
Fulton T. Crews and Ryan P. Vetreno

Converging Actions of Alcohol on Liver and Brain Immune Signaling
Gyongyi Szabo and Dora Lippai

Opportunities for the Development of Neuroimmune Therapies in Addiction
Lara A. Ray, Daniel Roche, Keith Heinzerling, and Steve Shoptaw

Use of Addictive Substances and NeuroHIV
Sulie L. Chang, Kaitlyn P. Connaghan, Yufeng Wei, and Ming D. Li

INDEX

Volume 119

Adenosine Receptor Neurobiology: Overview
Jiang-Fan Chen, Chien-fei Lee, and Yijuang Chern

Adenosine Receptor PET Imaging in Human Brain
Masahiro Mishina and Kiich Ishiwata

An Overview of Adenosine A_{2A} Receptor Antagonists in Parkinson's Disease
Peter Jenner

Mode of Action of Adenosine A_{2A} Receptor Antagonists as Symptomatic Treatment for Parkinson's Disease
Akihisa Mori

Adenosine Receptors and Dyskinesia in Pathophysiology
Masahiko Tomiyama

Clinical/Pharmacological Aspect of Adenosine A_{2A} Receptor Antagonist for Dyskinesia
Tomoyuki Kanda and Shin-ichi Uchida

Interaction of Adenosine Receptors with Other Receptors from Therapeutic Perspective in Parkinson's Disease
Nicolas Morin and Thérèse Di Paolo

Effects of the Adenosine A_{2A} Receptor Antagonist on Cognitive Dysfunction in Parkinson's Disease
Shin-ichi Uchida, Takako Kadowaki-Horita, and Tomoyuki Kanda

Clinical Nonmotor Aspect of A_{2A} Antagonist in PD Treatment
Masahiro Nomoto, Masahiro Nagai, and Noriko Nishikawa

Adenosine Receptors and Huntington's Disease
Chien-fei Lee and Yijuang Chern

Adenosine Receptors and Epilepsy: Current Evidence and Future Potential
Susan A. Masino, Masahito Kawamura, Jr., and David N. Ruskin

Adenosine Receptor Control of Cognition in Normal and Disease
Jiang-Fan Chen

Adenosine Receptors in Cerebral Ischemia
Alessia Melani, Anna Maria Pugliese, and Felicita Pedata

Roles of Adenosine and its Receptors in Sleep–Wake Regulation
Zhi-Li Huang, Ze Zhang, and Wei-Min Qu

Involvement of Adenosine A_{2A} Receptors in Depression and Anxiety
Koji Yamada, Minoru Kobayashi, and Tomoyuki Kanda

The Adenosine Neuromodulation System in Schizophrenia
Daniel Rial, Diogo R. Lara, and Rodrigo A. Cunha

INDEX

Volume 120

The Story of "Speed" from "Cloud Nine" to Brain Gain
Andrew Lees, Katrin Sikk, and Pille Taba

Amphetamine-Type Stimulants: The Early History of Their Medical and Non-Medical Uses
Nicolas Rasmussen

Miracle or Menace?
Mike Jay

Psychostimulants: Basic and Clinical Pharmacology
Andrew C. McCreary, Christian P. Müller, and Małgorzata Filip

Epigenetic Mechanisms of Psychostimulant-Induced Addiction
Anti Kalda and Alexander Zharkovsky

Experimental Models on Effects of Psychostimulants
Sulev Kõks

Neurologic Complications of Psychomotor Stimulant Abuse
Juan Sanchez-Ramos

Neurobehavioral Sequelae of Psychostimulant Abuse
Atbin Djamshidian

Neuropsychiatric Adverse Effects of Amphetamine and Methamphetamine
Jaanus Harro

"Addicted to Euphoria": The History, Clinical Presentation, and Management of Party Drug Misuse
Jenny Bearn and Matthew O'Brien

"Natural Amphetamine" Khat: A Cultural Tradition or a Drug of Abuse?
Nilesh B. Patel

Methcathinone "Kitchen Chemistry" and Permanent Neurological Damage
Katrin Sikk and Pille Taba

"Legal Highs" – An Emerging Epidemic of Novel Psychoactive Substances
Jolanta B. Zawilska

Psychostimulants and Artistic, Musical, and Literary Creativity
Iain Smith

Opium as a Literary Stimulant: The Case of Samuel Taylor Coleridge
Neil Vickers

INDEX

Volume 121

Alzheimer's Disease: Genomics and Beyond
Fuhai Song, Guangchun Han, Zhouxian Bai, Xing Peng, Jiajia Wang, and Hongxing Lei

The Potential of Proteomics in Understanding Neurodegeneration
Ramavati Pal, Jan Petter Larsen, and Simon Geir Moller

Proteomics Approach to Identify Biomarkers in Neurodegenerative Diseases
Annapurna Nayak, Gregory Salt, Sunil K. Verma, and Uday Kishore

Uncovering Neurodegenerative Protein Modifications via Proteomic Profiling
Xavier Gallart-Palau, Aida Serra, and Siu Kwan Sze

Comparative Proteomics for the Evaluation of Protein Expression and Modifications in Neurodegenerative Diseases
Antonio Conti and Massimo Alessio

INDEX

Volume 122

Utility of Autoantibodies as Biomarkers for Diagnosis and Staging of Neurodegenerative Diseases
Cassandra DeMarshall, Abhirup Sarkar, Eric P. Nagele, Eric Goldwaser, George Godsey, Nimish K. Acharya, and Robert G. Nagele

Metabolomics of Neurodegenerative Diseases
Alejandro Botas, Hannah Moore Campbell, Xu Han, and Mirjana Maletic-Savatic

Parkinson's Disease: *In Vivo* Brain Metabolomics by MRS
Mario Rango

Recent Advances and Applications of Metabolomics to Investigate Neurodegenerative Diseases
Clara Ibáñez, Alejandro Cifuentes, and Carolina Simó

Lipidomics of Human Brain Aging and Alzheimer's Disease Pathology
Alba Naudí, Rosanna Cabré, Mariona Jové, Victoria Ayala, Hugo Gonzalo, Manuel Portero-Otín, Isidre Ferrer, and Reinald Pamplona

INDEX

Volume 123

Unifying Mechanism of Controlling Kir3 Channel Activity by G Proteins and Phosphoinositides
Diomedes E. Logothetis, Rahul Mahajan, Scott K. Adney, Junghoon Ha, Takeharu Kawano, Xuan-Yu Meng, and Meng Cui

The Roles of Gβγ and Gα in Gating and Regulation of GIRK Channels
Nathan Dascal and Uri Kahanovitch

RGS Redundancy and Implications in GPCR–GIRK Signaling
Craig A. Doupnik

Structural Insights into GIRK Channel Function
Ian W. Glaaser and Paul A. Slesinger

Localization and Targeting of GIRK Channels in Mammalian Central Neurons
Rafael Luján and Carolina Aguado

GIRK Channel Plasticity and Implications for Drug Addiction
Ezequiel Marron Fernandez de Velasco, Nora McCall, and Kevin Wickman

GIRK Channels: A Potential Link Between Learning and Addiction
Megan E. Tipps and Kari J. Buck

Behavioral and Genetic Evidence for GIRK Channels in the CNS: Role in Physiology, Pathophysiology, and Drug Addiction
Jody Mayfield, Yuri A. Blednov, and R. Adron Harris

INDEX

Volume 124

Part I: Introductory Chapter

Neuronal Nicotinic Acetylcholine Receptor Structure and Function and Response to Nicotine
John A. Dani

Part II: Schizophrenia
The Role of Nicotine in Schizophrenia
Robert E. Featherstone and Steven J. Siegel

Neuronal α7 Nicotinic Receptors as a Target for the Treatment of Schizophrenia
Tanya L. Wallace and Daniel Bertrand

Role of the Neuregulin Signaling Pathway in Nicotine Dependence and Co-morbid Disorders
Miranda L. Fisher, Anu Loukola, Jaakko Kaprio, and Jill R. Turner

Effective Cessation Strategies for Smokers with Schizophrenia
A. Eden Evins and Corinne Cather

Part III: Mood Disorders
Role of the Brain's Reward Circuitry in Depression: Transcriptional Mechanisms
Eric J. Nestler

Nicotine Addiction and Psychiatric Disorders
Munir Gunes Kutlu, Vinay Parikh, and Thomas J. Gould

The Role of Depression in the Uptake and Maintenance of Cigarette Smoking
Janet Audrain-McGovern, Adam M. Leventhal, and David R. Strong

Part IV: Parkinson's Disease
Nicotine and Nicotinic Receptor Drugs: Potential for Parkinson's Disease and Drug-Induced Movement Disorders
Maryka Quik, Tanuja Bordia, Danhui Zhang, and Xiomara A. Perez

Part V: Alzheimer's Disease
Nicotinic Cholinergic Mechanisms in Alzheimer's Disease
Jianxin Shen and Jie Wu

INDEX

FSC
www.fsc.org
MIX
Papier | Fördert gute Waldnutzung
FSC® C083411

Kolibri 360 GmbH
Ferdinand-Jühlke-Straße 7
99095 Erfurt, Deutschland
produktsicherheit@kolibri360.de